Lecture Notes in Computer Science 6459

Commenced Publication in 1973
Founding and Former Series Editors:
Gerhard Goos, Juris Hartmanis, and Jan van Leeuwen

Editorial Board

David Hutchison
 Lancaster University, UK
Takeo Kanade
 Carnegie Mellon University, Pittsburgh, PA, USA
Josef Kittler
 University of Surrey, Guildford, UK
Jon M. Kleinberg
 Cornell University, Ithaca, NY, USA
Alfred Kobsa
 University of California, Irvine, CA, USA
Friedemann Mattern
 ETH Zurich, Switzerland
John C. Mitchell
 Stanford University, CA, USA
Moni Naor
 Weizmann Institute of Science, Rehovot, Israel
Oscar Nierstrasz
 University of Bern, Switzerland
C. Pandu Rangan
 Indian Institute of Technology, Madras, India
Bernhard Steffen
 TU Dortmund University, Germany
Madhu Sudan
 Microsoft Research, Cambridge, MA, USA
Demetri Terzopoulos
 University of California, Los Angeles, CA, USA
Doug Tygar
 University of California, Berkeley, CA, USA
Gerhard Weikum
 Max Planck Institute for Informatics, Saarbruecken, Germany

Ronan Boulic   Yiorgos Chrysanthou
Taku Komura (Eds.)

Motion
in Games

Third International Conference, MIG 2010
Utrecht, The Netherlands, November 14-16, 2010
Proceedings

Volume Editors

Ronan Boulic
Ecole Polytechnique Fédérale de Lausanne, EPFL
VRLab
1015 Lausanne, Switzerland
E-mail: Ronan.Boulic@epfl.ch

Yiorgos Chrysanthou
University of Cyprus
Dept. of Computer Science
1678 Nicosia, Cyprus
E-mail: yiorgos@cs.ucy.ac.cy

Taku Komura
University of Edinburgh
School of Informatics
Edinburgh, EH8 9YL, UK
E-mail: tkomura@ed.ac.uk

Library of Congress Control Number: 2010938325

CR Subject Classification (1998): I.2.1, I.2.9-11, I.6.8, H.5, I.3-4, H.3-4

LNCS Sublibrary: SL 6 – Image Processing, Computer Vision, Pattern Recognition, and Graphics

ISSN	0302-9743
ISBN-10	3-642-16957-0 Springer Berlin Heidelberg New York
ISBN-13	978-3-642-16957-1 Springer Berlin Heidelberg New York

This work is subject to copyright. All rights are reserved, whether the whole or part of the material is concerned, specifically the rights of translation, reprinting, re-use of illustrations, recitation, broadcasting, reproduction on microfilms or in any other way, and storage in data banks. Duplication of this publication or parts thereof is permitted only under the provisions of the German Copyright Law of September 9, 1965, in its current version, and permission for use must always be obtained from Springer. Violations are liable to prosecution under the German Copyright Law.

springer.com

© Springer-Verlag Berlin Heidelberg 2010
Printed in Germany

Typesetting: Camera-ready by author, data conversion by Scientific Publishing Services, Chennai, India
Printed on acid-free paper   06/3180

Preface

Following the very successful Motion in Games events in 2008 and 2009, we organized the Third International Conference on Motion in Games from 14–16 November 2010, in Utrecht, The Netherlands.

Games have become a very important medium for both education and entertainment. Motion plays a crucial role in computer games. Characters move around, objects are manipulated or move due to physical constraints, entities are animated, and the camera moves through the scene. Even the motion of the player nowadays is used as input to games. Motion is currently studied in many different areas of research, including graphics and animation, game technology, robotics, simulation, computer vision, and also physics, psychology, and urban studies. Cross-fertilization between these communities can considerably advance the state of the art in this area. The goal of the Motion in Games conference was to bring together researchers from these various fields to present the most recent results and to initiate collaboration. The conference was organized by the Dutch research project GATE. The conference consisted of a regular paper session, a poster session, as well as presentations by a selection of internationally renowned speakers in the field of games and simulations.

November 2010

Ronan Boulic
Yiorgos Chrysanthou
Taku Komura
Roland Geraerts
Arjan Egges
Mark Overmars

Organization

Program Chairs

Ronan Boulic — VRLab, EPFL, Lausanne, Switzerland
Yiorgos Chrysanthou — University of Cyprus, Nicosia, Cyprus
Taku Komura — Edinburgh University, UK

Local Chairs

Roland Geraerts — Games and Virtual Worlds group, Utrecht University, NL
Arjan Egges — Games and Virtual Worlds group, Utrecht University, NL
Mark Overmars — Games and Virtual Worlds group, Utrecht University, NL

Program Committee

Allbeck, Jan M. — George Mason University, USA
Amato, Nancy — Texas A&M University, USA
Badler, Norman — University of Pennsylvania, USA
Cani, Marie-Paule — INRIA, Grenoble, France
Courty, Nicolas — University of South Brittany, Vannes, France
Donikian, Stephane — IRISA, Rennes, France
Faloutsos, Petros — University of California, Los Angeles, USA
Galoppo, Nico — University of North Carolina, USA
Gross, Markus — ETH Zurich, Switzerland
Kallmann, Marcelo — University of California, Merced, USA
Kim, HyungSeok — Konkuk University, Seoul, Korea
King, Scott — Texas A&M University, USA
Lau, Manfred — JST ERATO Igarashi Design Interface Project, Japan
Laumond, Jean-Paul — LAAS, Toulouse, France
Liu, Karen — Georgia Institute of Technology, USA
Loscos, Celine — Universitat de Girona, Spain
Magnenat-Thalmann, Nadia — MIRALab, Geneva, Switzerland
Manocha, Dinesh — University of North Carolina, USA
Multon, Franck — CNRS-INRIA, France
Nijholt, Anton — Universiteit Twente, The Netherlands
O'Sullivan, Carol — Trinity College Dublin, Ireland

van de Panne, Michiel   University of British Columbia, Canada
Pelachaud, Catherine   CNRS, France
Pettrè, Julien   IRISA, Rennes, France
Shin, Sung Yong   Korea Advanced Inst. of Science and
                                     Technology, Korea
Thalmann, Daniel   EPFL, Lausanne, Switzerland
Terzopoulos, Demetri   University of California, Los Angeles, USA
Yin, KangKang   Microsoft Research Asia, China
Zhang, Jian   Bournemouth University, UK
Zordan, Victor   University of California, Riverside, USA

Sponsored by

Motion in Games 2010 was sponsored by the GATE project[1,2]

Game research for training and entertainment

---

[1] http://gate.gameresearch.nl

[2] The GATE project is funded by the Netherlands Organization for Scientific Research (NWO) and the Netherlands ICT Research and Innovation Authority (ICT Regie).

Table of Contents

Body Simulation

Simulating Humans and Lower Animals 1
  Demetri Terzopoulos

Evaluating the Physical Realism of Character Animations Using
Musculoskeletal Models 11
  Thomas Geijtenbeek, Antonie J. van den Bogert,
  Ben J.H. van Basten, and Arjan Egges

Learning Movements

Physically-Based Character Control in Low Dimensional Space 23
  Hubert P.H. Shum, Taku Komura, Takaaki Shiratori, and
  Shu Takagi

Learning Crowd Steering Behaviors from Examples (Abstract) 35
  Panayiotis Charalambous and Yiorgos Chrysanthou

Body Control

Full-Body Hybrid Motor Control for Reaching 36
  Wenjia Huang, Mubbasir Kapadia, and Demetri Terzopoulos

Pose Control in Dynamic Conditions 48
  Brian F. Allen, Michael Neff, and Petros Faloutsos

Spatial Awareness in Full-Body Immersive Interactions: Where Do We
Stand? 59
  Ronan Boulic, Damien Maupu, Manuel Peinado, and
  Daniel Raunhardt

Motion Planning

Scalable Precomputed Search Trees 70
  Manfred Lau and James Kuffner

Toward Simulating Realistic Pursuit-Evasion Using a Roadmap-Based
Approach 82
  Samuel Rodriguez, Jory Denny, Takis Zourntos, and
  Nancy M. Amato

Path Planning for Groups Using Column Generation 94
    Marjan van den Akker, Roland Geraerts, Han Hoogeveen, and
    Corien Prins

Physically-Based Character Control

Skills-in-a-Box: Towards Abstract Models of Motor Skills 106
    Michiel van de Panne

Angular Momentum Control in Coordinated Behaviors 109
    Victor Zordan

Crowds and Formation

Simulating Formations of Non-holonomic Systems with Control Limits
along Curvilinear Coordinates 121
    Athanasios Krontiris, Sushil Louis, and Kostas E. Bekris

Following a Large Unpredictable Group of Targets among Obstacles 134
    Christopher Vo and Jyh-Ming Lien

Geometry

Real-Time Space-Time Blending with Improved User Control 146
    Galina Pasko, Denis Kravtsov, and Alexander Pasko

Motion Capture for a Natural Tree in the Wind...................... 158
    Jie Long, Cory Reimschussel, Ontario Britton, Anthony Hall, and
    Michael Jones

Active Geometry for Game Characters.............................. 170
    Damien Rohmer, Stefanie Hahmann, and Marie-Paule Cani

Autonomous Characters

CAROSA: A Tool for Authoring NPCs 182
    Jan M. Allbeck

BehaveRT: A GPU-Based Library for Autonomous Characters......... 194
    Ugo Erra, Bernardino Frola, and Vittorio Scarano

Level of Detail AI for Virtual Characters in Games and Simulation 206
    Michael Wißner, Felix Kistler, and Elisabeth André

Navigation

Scalable and Robust Shepherding via Deformable Shapes 218
  Joseph F. Harrison, Christopher Vo, and Jyh-Ming Lien

Navigation Queries from Triangular Meshes 230
  Marcelo Kallmann

Motion Synthesis

Motion Parameterization with Inverse Blending 242
  Yazhou Huang and Marcelo Kallmann

Planning and Synthesizing Superhero Motions 254
  Katsu Yamane and Kwang Won Sok

Perception

Perception Based Real-Time Dynamic Adaptation of Human
Motions 266
  Ludovic Hoyet, Franck Multon, Taku Komura, and Anatole Lecuyer

Realistic Emotional Gaze and Head Behavior Generation Based on
Arousal and Dominance Factors 278
  *Cagla Cig, Zerrin Kasap, Arjan Egges, and
  Nadia Magnenat-Thalmann*

Why Is the Creation of a Virtual Signer Challenging Computer
Animation? 290
  Nicolas Courty and Sylvie Gibet

Real-Time Graphics

Realtime Rendering of Realistic Fabric with Alternation of Deformed
Anisotropy 301
  Young-Min Kang

Responsive Action Generation by Physically-Based Motion Retrieval
and Adaptation 313
  Xiubo Liang, Ludovic Hoyet, Weidong Geng, and Franck Multon

Visibility Transition Planning for Dynamic Camera Control
(Abstract) 325
  Thomas Oskam, Robert W. Sumner, Nils Thuerey, and Markus Gross

Posters

The Application of MPEG-4 Compliant Animation to a Modern Games
Engine and Animation Framework 326
   Chris Carter, Simon Cooper, Abdennour El Rhalibi, and
   Madjid Merabti

Knowledge-Based Probability Maps for Covert Pathfinding 339
   Anja Johansson and Pierangelo Dell'Acqua

Modification of Crowd Behaviour Modelling under Microscopic Level
in Panic Situation.. 351
   Hamizan Sharbini and Abdullah Bade

Expressive Gait Synthesis Using PCA and Gaussian Modeling 363
   Joëlle Tilmanne and Thierry Dutoit

Autonomous Multi-agents in Flexible Flock Formation................ 375
   Choon Sing Ho, Quang Huy Nguyen, Yew-Soon Ong, and
   Xianshun Chen

Real-Time Hair Simulation with Segment-Based Head Collision 386
   Eduardo Poyart and Petros Faloutsos

Subgraphs Generating Algorithm for Obtaining Set of Node-Disjoint
Paths in Terrain-Based Mesh Graphs 398
   Zbigniew Tarapata and Stefan Wroclawski

Path-Planning for RTS Games Based on Potential Fields.............. 410
   Renato Silveira, Leonardo Fischer, José Antônio Salini Ferreira,
   Edson Prestes, and Lucianal Nedel

Learning Human Action Sequence Style from Video for Transfer to 3D
Game Characters 422
   XiaoLong Chen, Kaustubha Mendhurwar, Sudhir Mudur,
   Thiruvengadam Radhakrishnan, and Prabir Bhattacharya

Author Index 435

Simulating Humans and Lower Animals

Demetri Terzopoulos

University of California, Los Angeles, USA

Abstract. This paper summarizes material on the simulation of humans and lower animals presented by the author in an invited talk delivered at the *Third International Conference on Motion in Games*.

1 Introduction

Artificial life, a discipline that spans the computational and biological sciences, continues to have considerable impact within computer graphics, yielding virtual worlds teeming with realistic artificial flora and fauna [1]. Artificial animals are complex synthetic organisms that have functional, biomechanical bodies, perceptual sensors, and brains with locomotion, perception, behavior, learning, and cognition centers. As self-animating graphical characters, virtual humans and lower animals are poised to dramatically advance the interactive game and motion picture industries even more so than purely physics-based simulation technologies are currently doing.

This paper reviews progress that my research group has made toward the realistic simulation of humans and lower animals. The first part of the paper describes our artificial life approach to the challenge of emulating the rich complexity of pedestrians in urban environments. This type of realistic multi-human simulation is the natural progression beyond our earlier work that addressed the simulation of lower animals, such as fish, which I will also summarize. Our artificial life approach builds upon physics, however, especially biomechanics, just as physics-based simulation builds upon geometric modeling. In the second part of the paper, I will review our efforts towards the realistic biomechanical simulation of the complex motions of lower animals and humans.

2 Artificial Life Simulation

Human simulation has become an increasingly important topic for the interactive game and motion picture industries. In this context, our artificial life paradigm combines aspects of the fields of biomechanics, perception, ethology, machine learning, cognitive science, and sociology. This first part of the paper describes our model of autonomous pedestrians, which builds upon our earlier work on the artificial life modeling of lower animals that is also reviewed.

Arcade   Main Waiting Room   Concourses and Platforms

Fig. 1. A virtual train station populated by self-animating virtual humans

2.1 Autonomous Pedestrians

We have demonstrated a reconstructed model of the original Pennsylvania (train) Station in New York City, richly populated with autonomous virtual pedestrians [2] (Fig. 1). In a departure from the literature on so-called "crowd simulation", we have developed a comprehensive human model capable of synthesizing a broad variety of activities in the large-scale indoor urban environment. Our virtual pedestrians are autonomous agents with functional bodies and brains. They perceive the virtual environment around them, analyze environmental situations, have natural reactive behaviors, and proactively plan their activities. We augment the environment with hierarchical data structures that efficiently support the perceptual queries influencing the behavioral responses of the autonomous pedestrians and also sustain their ability to plan their actions over local and global spatiotemporal scales.

Our artificial life simulation of pedestrians (Fig. 2) integrates motor, perceptual, behavioral, and, importantly, cognitive components, each of which we will review in the subsequent sections. Featuring innovations in these components, as well as in their combination, our model yields results of unprecedented fidelity and complexity for fully autonomous multi-human simulation in virtual public spaces.

Appearance and Locomotion. As an implementation of the appearance and motor levels of the character, we employ a human animation package called *DI-Guy*, which is commercially available from Boston Dynamics Inc. DI-Guy provides a variety of textured geometric human models together with a set of basic motor skills, such as strolling, walking, jogging, sitting, etc. Emulating the natural appearance and movement of human beings is a highly challenging problem and, not surprisingly, DI-Guy suffers from several limitations, mostly in its kinematic control of human motions. To ameliorate the visual defects, we have customized the motions of DI-Guy characters and have implemented a motor control interface to conceal the details of the underlying kinematic control

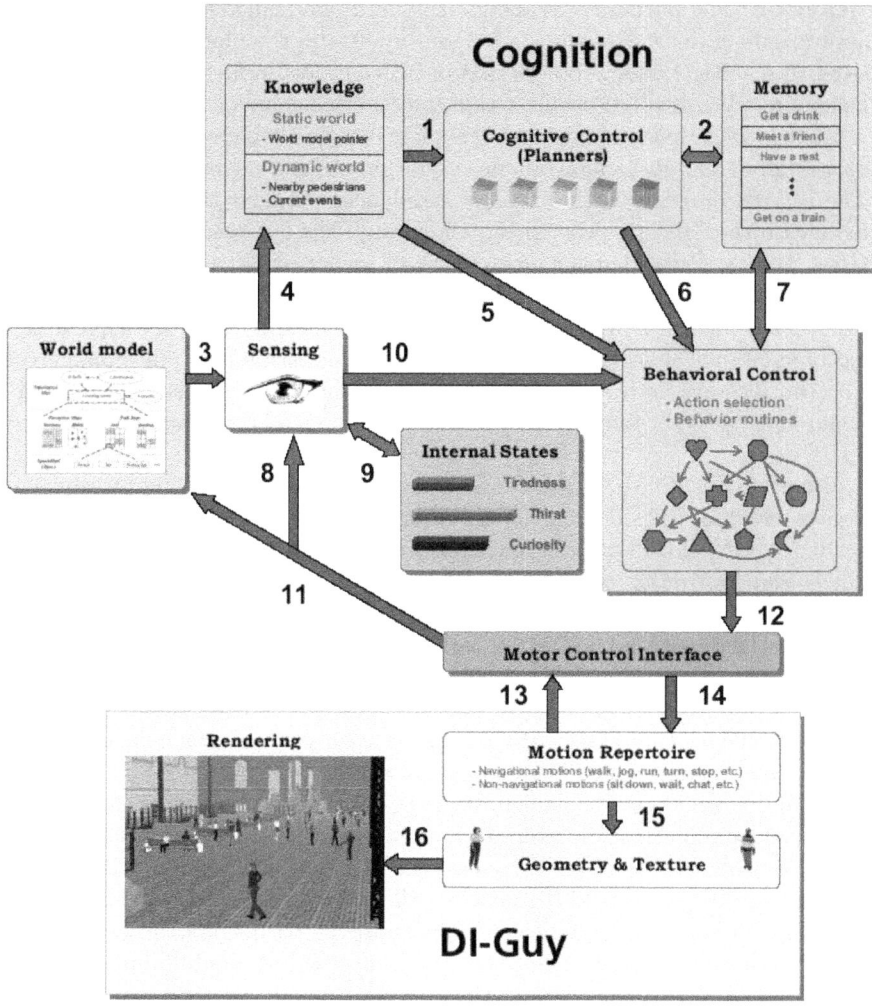

Fig. 2. Autonomous pedestrian simulation model

layer from our higher-level behavior modules, enabling the latter to be developed more or less independently.

Perception. In a highly dynamic virtual world, an autonomous intelligent character must have a keenly perceptive regard for its surroundings in order to interact with it effectively. The hierarchical world model is used extensively by pedestrians to perceive their environment, providing not only the raw sensed data (such as those obtained from perception maps), but also higher-level interpretations of perceived situations (such as those obtained from specialized objects) that are at least as important and useful to a pedestrian.

Behavior. The purpose of realistic behavioral modeling is to link perception to appropriate actions. We adopt a bottom-up strategy, which uses primitive reactive behaviors as building blocks that in turn support more complex motivational behaviors, all controlled by an action selection mechanism.

At the lowest level, we developed six key reactive behavior routines that cover almost all of the obstacle avoidance situations that a pedestrian can encounter. The first two are for static obstacle avoidance, the next three are for avoiding mobile objects (mostly other pedestrians), and the last one is for avoiding both. Given that a pedestrian possesses a set of motor skills, such as standing still, moving forward, turning in several directions, speeding up and slowing down, etc., these routines are responsible for initiating, terminating, and sequencing the motor skills on a short time scale guided by sensory stimuli and internal percepts. The routines are activated in an optimized sequential manner, giving each the opportunity to alter the currently active motor control command (speed, turning angle, etc.).

While the reactive behaviors enable pedestrians to move around freely, almost always avoiding collisions, navigational behaviors enable them to go where they desire. We developed several such routines—*passageway selection, passageway navigation, perception guided navigation, arrival-at-a-target navigation*, etc.—to address issues, such as the speed and scale of online path planning, the realism of actual paths taken, and pedestrian flow control through and around bottlenecks. Furthermore, to make our pedestrians more interesting, we have augmented their behavior repertoires with a set of non-navigational, motivational routines, such as *select an unoccupied seat and sit, approach a performance and watch, queue at ticketing areas and purchase a ticket*, etc.

An action selection mechanism triggers appropriate behaviors in response to perceived combinations of external situations and internal affective needs represented by the mental state. For example, in a pedestrian whose thirst exceeds a predetermined threshold, behaviors will be triggered, usually through online planning, to locate a vending machine, approach it, queue if necessary, and finally purchase a drink. In case more than one need awaits fulfillment, the most important need ranked by the action selection mechanism receives the highest priority. Once a need is fulfilled, the value of the associated mental state variable decreases asymptotically back to its nominal value. We instantiate different classes of pedestrians suitable for a train station environment, each class having a specialized action selection mechanism, including commuters, tourists, law enforcement officers, and buskers.

We have recently developed a decision network framework for behavioral human animation [3]. This probabilistic framework addresses complex social interactions between autonomous pedestrians in the presence of uncertainty.

Cognition. At the highest level of autonomous control, cognitive modeling [4] yields a deliberative, proactive autonomous human agent capable of applying knowledge to conceive and execute intermediate and long-term plans. A stack memory model enables a pedestrian to remember, update, and forget chains of goals. The stack couples the deliberative intelligence with the reactive behaviors, enabling a pedestrian to achieve its long-term goals. For example, a commuter can enter the station, with the goal of catching a particular train at a specific time. The cognitive model divides this complex goal into simpler intermediate goals, which may involve navigating to the ticket purchase areas to buy a ticket (which may involve waiting in line), navigating to the concourse area, possibly purchasing a drink if thirsty, sitting down to take a rest if tired, watching a street performance if interested, meeting a friend, and/or navigating to the correct stairs and descending to the proper train platform when the time comes to board a train.

2.2 Autonomous Lower Animals

As a precursor to autonomous pedestrians, we developed several lower animals, among them artificial fishes [5]. Each virtual fish is an autonomous agent with a deformable body actuated by internal muscles. The body includes eyes and a brain with motor, perception, behavior, and learning centers (Fig. 3). Through controlled muscle actions, artificial fishes swim through simulated water in accordance with hydrodynamic principles. Their articulate fins enable them to locomote and maneuver underwater. In accordance with their perceptual awareness of their virtual world, their ethological model arbitrates a repertoire of piscine behaviors, including collision avoidance, foraging, preying, schooling, and mating.

3 Biomechanical Simulation and Control

In this part of the paper, I will review our work on aspects of biomechanical simulation and control. Initially we focused on lower animals, especially the aforementioned artificial fish, before progressing to human biomechanical simulation and control.

3.1 Simulating Fish Biomechanics

To synthesize realistic fish locomotion we designed a simple, biomechanical fish model (Fig. 4) consisting of 23 nodal point masses and 91 damped springs, each of which is a Voigt uniaxial viscoelastic element comprising an ideal spring and an ideal damper connected in parallel. The damped spring arrangement maintains the structural stability of the body while allowing it to flex. Twelve of the damped springs running the length of the body are actively contractile, thus serving as muscles. Through controlled muscle actions, the artificial fishes swim through virtual water in accordance with hydrodynamic principles. Their caudal and articulate pectoral fins enable them to locomote and freely maneuver underwater. Forward caudal locomotion normally uses the posterior muscles on either side of the body, while normal turning uses the anterior muscles.

Though rudimentary compared to those in real animals, the brains of artificial fishes are nonetheless able to learn how to control the muscles of the body in order to swim freely within the virtual world and carry out perceptually guided motor tasks [6].

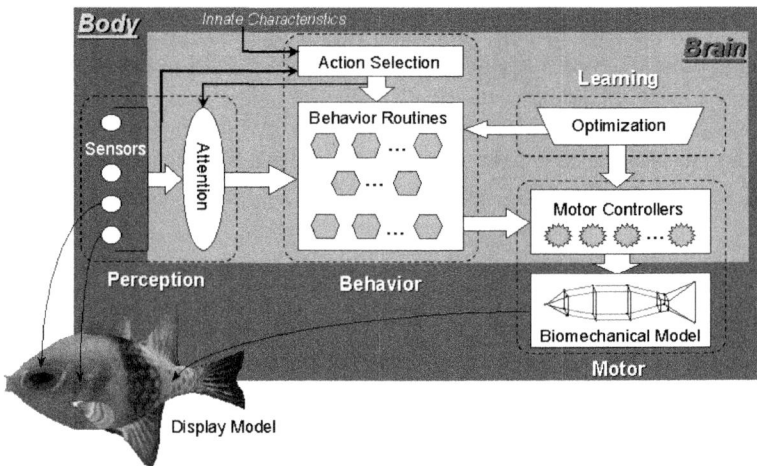

Fig. 3. Top: Artificial fishes in their physics-based virtual world; the 3 reddish fish are engaged in mating behavior, the greenish fish is a predator hunting prey, the remaining 3 fishes are feeding on plankton (white dots). Bottom: Functional diagram of the artificial fish model illustrating the body and brain submodels.

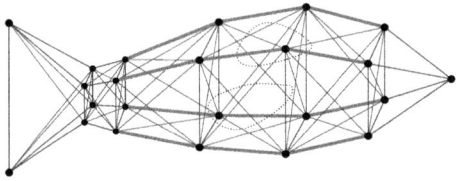

Fig. 4. Biomechanical fish model. Nodes are lumped masses. Lines are damped springs (shown at their natural lengths). Bold red lines are actively contractile damped springs, representing muscles. Dashed curves indicate caudal fins.

Fig. 5. A supine dynamic "virtual stuntman" rolls over and rises to an erect position, balancing in gravity

3.2 Dynamic Human Characters

An ambitious goal in the area of physics-based computer animation is the creation of virtual agents that autonomously synthesize realistic human motions and possess a broad repertoire of lifelike motor skills. To this end, the dynamic control of articulated, joint-torque-actuated, anthropomorphic figures subject to gravity and contact forces remains a difficult open problem. We have developed a *virtual stuntman*, a dynamic graphical character that possesses a nontrivial repertoire of lifelike motor skills [7]. The repertoire includes basic actions such as balance, protective stepping when balance is disturbed, protective arm reactions when falling, multiple ways of rising upright after a fall (Fig. 5), and several more vigorously dynamic motor skills. Our virtual stuntman is the product of a framework for integrating motor controllers, which among other ingredients includes an explicit model of pre-conditions; i.e., those regions of a dynamic figure's state space within which a given motor controller is applicable and expected to work properly. We have demonstrated controller composition with pre-conditions determined not only manually, but also automatically based on support vector machine (SVM) learning theory.

3.3 Simulating the Neck-Head-Face Complex

The neck has a complex anatomical structure and it plays an important role in supporting the head atop the cervical spine, while generating the controlled head movements that are essential to so many aspects of human behavior. We have developed a biomechanical model of the human head-neck system that emulates the relevant anatomy [8] (Fig. 6). Characterized by appropriate kinematic redundancy (7 cervical vertebrae coupled by 3-DOF joints) and muscle actuator redundancy (72 neck muscles arranged in 3 muscle layers), our model presents a challenging motor control problem, even for the relatively simple task of balancing the mass of the head atop the cervical column. We have developed a

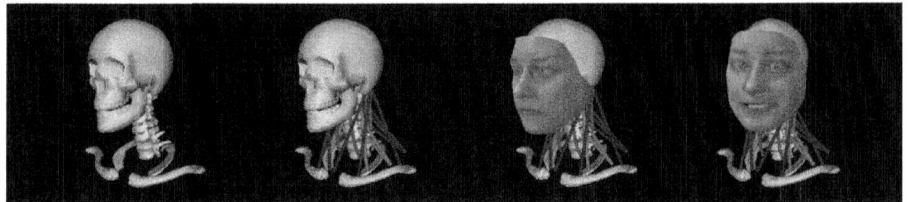

Fig. 6. Biomechanical neck-head-face model with neuromuscular controller

neuromuscular control model for the neck that emulates the relevant biological motor control mechanisms. Incorporating low-level reflex and high-level voluntary sub-controllers, our hierarchical controller provides input motor signals to the numerous muscle actuators. In addition to head pose and movement, it controls the coactivation of mutually opposed neck muscles to regulate the stiffness of the head-neck multibody system.

Taking a machine learning approach, the neural networks within our neuromuscular controller are trained offline to efficiently generate the online pose and tone control signals necessary to synthesize a variety of autonomous movements for the behavioral animation of the human head and face (the biomechanical face model described in [9]).

3.4 Comprehensive Biomechanical Simulation of the Human Body

We have been developing a comprehensive biomechanical model of the human body [10], undertaking the combined challenge of modeling and controlling more or less all of the relevant articular bones and muscles, as well as simulating the physics-based deformations of the soft tissues, including muscle bulging (Fig. 7).

In particular, we have created a jointed physics-based skeletal model that consists of *75 bones* and *165 DOFs* (degrees of freedom), with each vertebral bone and most ribs having independent DOFs. To be properly actuated and controlled, our detailed bone model requires a comparable level of detail with respect to muscle modeling. We incorporate a staggering *846 muscles*, which are modeled as piecewise line segment Hill-type force actuators. We have also developed an associated physics-based animation controller that computes accelerations to drive the musculoskeletal system toward a sequence of preset target key poses, and then computes the required activation signal for each muscle through inverse dynamics.

Our volumetric human body model incorporates detailed skin geometry, as well as the active muscle tissues, passive soft tissues, and skeletal substructure. Driven by the muscle activation inputs and resulting skeletal motion, a companion simulation of a volumetric, finite element model of the soft tissue introduces the visual richness of more detailed 3D models of the musculature. Specifically, we achieve robust and efficient simulation of soft tissue deformation within the finite element framework by decoupling the visualization geometry from the simulation geometry. A total of *354,000 Body-Centered-Cubic (BCC) tetrahedra* are

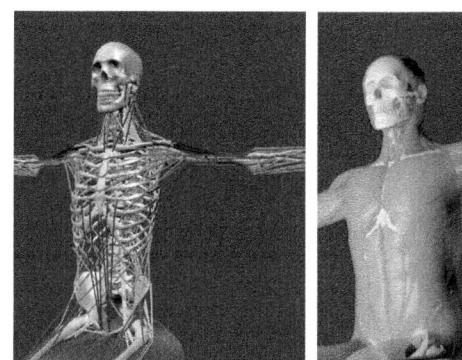
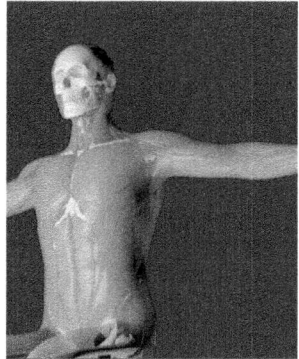
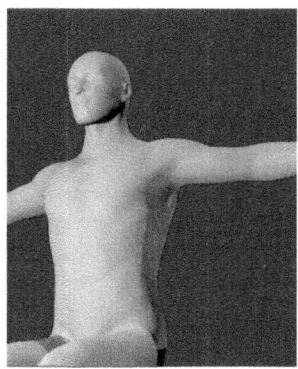

Fig. 7. Our comprehensive human body simulation is characterized by the biomechanical modeling of the relevant musculoskeletal tissues. The skeleton with 75 bones is actuated by 846 muscles (left). The motion of the skeleton and the activation level of each muscle deforms the inner soft tissue (center) and, hence, the outer skin (right).

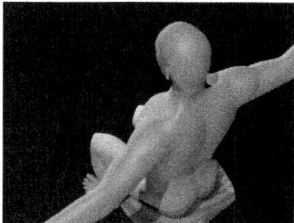
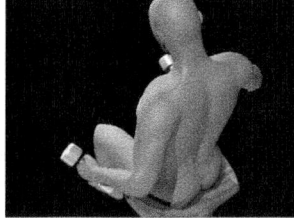
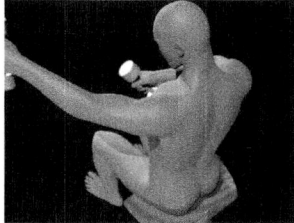

Fig. 8. An inverse dynamics motor controller drives the musculoskeletal system to track a sequence of target poses

simulated to animate the detailed deformation of the embedded high-resolution surfaces of the skin and each of the muscles.

Figure 8 demonstrates biomechanically simulated arm flexing motions with dumbbell loads.

Acknowledgements

I thank my former PhD students whose work is reviewed in this paper and whose names are cited in the following list of references.

References

1. Terzopoulos, D.: Artificial life for computer graphics. Communications of the ACM 42(8), 32–42 (1999)
2. Shao, W., Terzopoulos, D.: Autonomous pedestrians. Graphical Models 69(5-6), 246–274 (2007)

3. Yu, Q., Terzopoulos, D.: A decision network framework for the behavioral animation of virtual humans. In: Proc. ACM SIGGRAPH/EG Symposium on Computer Animation (SCA 2007), San Diego, CA, pp. 119–128 (2007)
4. Funge, J., Tu, X., Terzopoulos, D.: Cognitive modeling: Knowledge, reasoning and planning for intelligent characters. In: Proc. ACM SIGGRAPH 1999 Conf. Computer Graphics Proceedings, Annual Conference Series, Los Angeles, CA, pp. 29–38 (1999)
5. Terzopoulos, D., Tu, X., Grzeszczuk, R.: Artificial fishes: Autonomous locomotion, perception, behavior, and learning in a simulated physical world. Artificial Life 1(4), 327–351 (1994)
6. Grzeszczuk, R., Terzopoulos, D.: Automated learning of muscle-actuated locomotion through control abstraction. In: Proc. ACM SIGGRAPH 1995 Conf. Computer Graphics Proceedings, Annual Conference Series, Los Angeles, CA, pp. 63–70 (1995)
7. Faloutsos, P., van de Panne, M., Terzopoulos, D.: Composable controllers for physics-based character animation. In: Proc. ACM SIGGRAPH 2001 Conf. Computer Graphics Proceedings, Annual Conference Series, Los Angeles, CA, pp. 151–160 (2001)
8. Lee, S.-H., Terzopoulos, D.: Heads up! Biomechanical modeling and neuromuscular control of the neck. ACM Transactions on Graphics 25(3), 1188–1198 (2006); Proc. ACM SIGGRAPH 2006 Conf., Boston, MA
9. Lee, Y., Terzopoulos, D., Waters, K.: Realistic modeling for facial animation. In: Proc. ACM SIGGRAPH 1995 Conf. Computer Graphics Proceedings, Annual Conference Series, Los Angeles, CA, pp. 55–62 (1995)
10. Lee, S.-H., Sifakis, E., Terzopoulos, D.: Comprehensive Biomechanical Modeling and Simulation of the Upper Body. ACM Transactions on Graphics 28(4), 99, 1–17 (2009)

Evaluating the Physical Realism of Character Animations Using Musculoskeletal Models

Thomas Geijtenbeek[1], Antonie J. van den Bogert[2],
Ben J.H. van Basten[1], and Arjan Egges[1]

[1] Games and Virtual Worlds, Utrecht University, The Netherlands
[2] Orchard Kinetics LLC, Cleveland, OH, USA

Abstract. Physical realism plays an important role in the way character animations are being perceived. We present a method for evaluating the physical realism of character animations, by using musculoskeletal model simulation resulting from biomechanics research. We describe how such models can be used without the presence of external force measurements. We define two quality measures that describe principally different aspects of physical realism. The first quality measure reflects to what extent the animation obeys the Newton-Euler laws of motion. The second quality measure reflects the realism of the amount of muscle force a human would require to perform the animation. Both quality measures allow for highly detailed evaluation of the physical realism of character animations.

1 Introduction

Perceived realism of movement is an important component of character animation. However, its subjective nature makes it an impractical quality. Therefore, it is desirable to develop automatic evaluation methods that approximate the perceived realism of motion.

One approach in constructing such evaluation method is through analysis of results from user studies. Another approach is to develop a measure based on statistical analysis of existing realistic motions. Our approach is based on the idea that there exists a relation between the *perceived realism* and the *physical realism* of a motion. The main rationale behind this idea is that all motion we witness in real life adheres to the laws of physics.

There are several levels of accuracy at which physical realism of character animation can be measured. Some methods focus on the center of mass trajectory of a character, while others take into account the dynamics of the individual body parts. However, these approaches are all limited in their ability to describe human motion, since they do not consider the fact that the joint torques that produce motion are the direct result of muscle forces acting on these joints. The amount of torque a muscle can generate depends on both its maximum force as well as its muscle moment arm, which is pose-dependent. In addition, several muscles operate over multiple joints, creating complex dependencies between joint torques. These aspects can only be accurately described through a model of the human musculoskeletal system.

In the field of computer animation, the development of such musculoskeletal models may be considered too specialized and labor-intensive to be worth the effort. However, in biomechanics research, realistic musculoskeletal models are considered an effective tool for conducting research on human motion. As a result, significant effort has been put into the development of such models. The applicability for computer animation is limited though, because musculoskeletal simulation requires motion data to be augmented with force measurements at external contact points. We have developed a method for estimating these external force measurements, thus enabling the use of musculoskeletal models directly with purely kinematic motion data. In addition, we show how a musculoskeletal model can be employed to measure two principally different types of physical realism.

Our first measure evaluates to what degree a character motion obeys the *Newton-Euler laws of motion*, which dictate that changes in linear and angular momentum must be consistent with external forces due to gravity and contact. Examples of animations not obeying these laws are a character hanging still in mid-air, or a character standing straight at a 45 degree angle without falling. Our measure reflects these errors in a way that uniformly applies to both flight and contact stages.

Our second measure evaluates the realism of the muscle force that a human character would require to perform a motion. This measure detects situations where for example a character is moving too fast, or jumping too high. Such motions need not be in conflict with the Newton-Euler laws of motion, but are physically unrealistic because they require an excessive amount of muscle force.

Both quality measures provide objective feedback on the physical realism of an animation, with a level of detail that is unprecedented in the field of computer animation.

2 Related Work

Human character animation has received a lot of attention during the past decades. There are several classes of techniques, each having its own advantages and disadvantages. See Van Welbergen et al. [1] for an overview of different animation techniques.

The amount of research conducted on the perception of physical realism of animation is limited. O'Sullivan et al. [2] define a number of measures regarding perception of physical realism of the motions of solid objects, based on user studies. Reitsma and Pollard [3] observe through user studies that errors in horizontal velocity and added accelerations in human jumping motions are easier observed by humans than errors in vertical velocity and added decelerations. Point light experiments have shown observers can accurately estimate lifted weight [4], as well as pulled weight [5] from point light displays.

Much work has been done on physical realism in generated motions. Safonova et al. [6] show that certain basic physical properties can be conserved during motion transitions. Ikemoto et al. [7] evaluate transitions by foot skating and by evaluating the *zero-moment point*. Ren et al. [8] present a tool to evaluate the realism of animations using a set of statistical models, based on natural example data. Some work makes use of user studies to evaluate the resulting animations. For example, Van Basten and Egges [9] evaluate the relationship between posture distance metrics and perceived realism of motion transitions. Some techniques modify a existing animations to improve the physical realism. For example, Shin et al. [10] show how motions can be adjusted in such a way that they satisfy the *zero-moment point* constraint, as well as some additional physical constraints. Ko and Badler [11] attempt to achieve physical realism in generated walking motions by constraining maximum joint torque and adjusting balance in an inverse dynamics analysis.

Erdemir et al. [12] provide an overview of the different approaches and applications of musculoskeletal modeling. Veeger and Van der Helm [13] provide an exemplary insight in the complexity of musculoskeletal mobility. In recent years, a number of tools have implemented these techniques. Examples of such tools are the commercially available *AnyBody* system [14], the *Human Body Model* [15], and the open source project *OpenSim* [16]. To date, musculoskeletal models have not been used to evaluate computer animations.

3 Method

In this section we will describe in detail how a musculoskeletal model can be used to evaluate the physical realism of a character animation. We will describe how to compute measures representing the error in the *dynamics* of the animation, as well as the error in the amount of *muscle force* a human would require to perform an animation. Our method is independent of the musculoskeletal model or modeling software that is being used; we will therefore omit implementation specific aspects of musculoskeletal modeling.

3.1 Prerequisites

Musculoskeletal Model. Our evaluation method uses a model that incorporates both the skeletal and the muscular structure of the entire human body. Such a full-body musculoskeletal model can formally be described as a tuple $\mathcal{H}$:

$$\mathcal{H} = \{\mathcal{B}, \mathcal{Q}, \mathcal{U}\} \qquad (1)$$

The set $\mathcal{B}$ describes the individual body segments of the model. Each segment has specific mass, inertial properties, and a reference position and orientation. The set $\mathcal{Q}$ describes the n kinematic degrees of freedom of the model. These consist of a global position and orientation, plus those defined by the joints in the model. Each joint connects two segments in $\mathcal{B}$, with specific constraints, at a reference position and orientation. The set $\mathcal{U}$ describes the l muscles in the model. For each muscle, it describes the relation between joint angles and muscle moment arm, as well the maximum force the muscle can produce. The exact form of $\mathcal{B}$, $\mathcal{Q}$ and $\mathcal{U}$ depends on the implementation of the model and is not relevant to our evaluation method.

Animation. The input for our method is an animation **A**, consisting of a sequence of T frames:

$$\mathbf{A} = \{A_1, A_2, ..., A_T\} \quad , \quad A_t = \{q_t, \dot{q}_t, \ddot{q}_t, S_t^l, S_t^r\} \tag{2}$$

where q_t is a pose vector representing the n kinematic degrees of freedom defined in $\mathcal{Q}$; $\dot{q}_t$ and $\ddot{q}_t$ are first and second order derivatives of q_t. S_t^l and S_t^r are polygons representing the ground contact region at frame t, for both left and right foot.

3.2 Quality Measures

Our method produces the following quality measures:

1. The *dynamics error* measure, $\kappa : \mathbf{A} \rightarrow \mathbb{R}$, which is defined as the total amount of external force and moment that is required for the motion to satisfy the Newton-Euler laws of motion. In other words, it describes the amount of 'magical support' required to make a motion dynamically valid. The measure uniformly applies to both contact and flight stages. Force and moment are normalized using body mass and height, enabling comparison between character models of different weight or height.
2. The *muscle error* measure, $\lambda : \mathbf{A} \rightarrow \mathbb{R}$, which is defined as the total amount of excess muscle force required for a character to perform **A**, normalized by the total maximum force of all muscles in $\mathcal{U}$. The excess muscle force is defined as the amount of force a muscle must produce on top of its maximum capacity. The normalization enables comparison between character models of different strength.

3.3 Method Overview

Both quality measures are calculated per frame, for each $A_t \in \mathbf{A}$. This is done in four consecutive steps, as depicted in Figure 1.

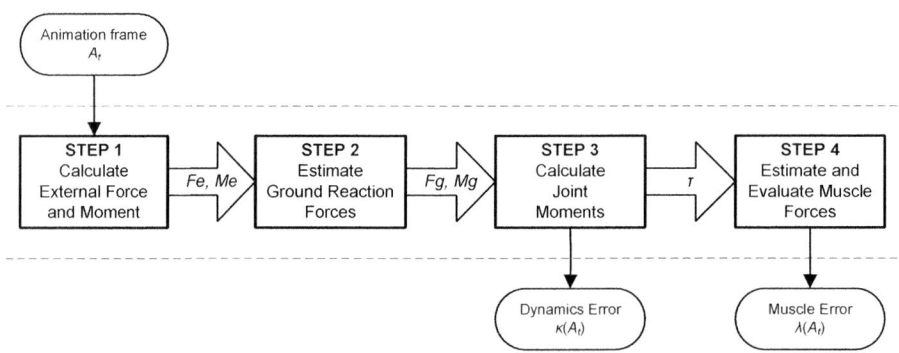

Fig. 1. Overview of the consecutive steps that are performed for each frame A_t

3.4 Step 1: Calculate External Force and Moment

In this step we compute the external force and moment F_e and M_e that act on the character at frame A_t. We do so by solving the dynamics equations of the motion defined by q_t, $\dot{q}_t$ and $\ddot{q}_t$, which can be formulated as:

$$\mathbf{M}(q_t)\ddot{q}_t + \mathbf{T}(q_t)\tau + c(q_t, \dot{q}_t) = 0 \qquad (3)$$

where $\mathbf{M}(q_t)$ is an $n \times n$ matrix that describes the pose dependent mass distribution, based on $\mathcal{B}$ and $\mathcal{Q}$. The vector τ contains the (unknown) moments and forces acting on the degrees of freedom in $\mathcal{Q}$. The vector $c(q_t, \dot{q}_t)$ are gravitational, centrifugal and Coriolis forces. $\mathbf{T}(q_t)$ is an $n \times n$ coefficient matrix that has no specific meaning and depends on the form of $\mathbf{M}(q_t)$.

The approach to constructing $\mathbf{M}(q_t)$, $\mathbf{T}(q_t)$ and $c(q_t, \dot{q}_t)$ is not relevant for our method and depends on the modeling software that is employed. The only important aspect for our method is that global position and orientation are part of the dynamics equation. After re-arranging the components in (3) we get:

$$\tau = \mathbf{T}(q)^{-1}\left[\mathbf{M}(q)\ddot{q} + c(q, \dot{q})\right] \qquad (4)$$

After solving, the external force and moment acting on the root element of the character correspond to the elements in τ related to global position and rotation, as defined in $\mathcal{Q}$. If F_r and M_r are the force and moment defined in the root coordinate frame, with the origin at r and orientation $\mathbf{R}$, then the external force and moment in the global coordinate frame, F_e and M_e, is defined as:

$$F_e = \mathbf{R}^{-1}F_r \quad , \quad M_e = \mathbf{R}^{-1}M_r + r \times (\mathbf{R}^{-1}F_r) \qquad (5)$$

3.5 Step 2: Estimate Ground Reaction Forces

In this step, we estimate the ground reaction force and moment for both left foot, F_g^l and M_g^l, and right foot, F_g^r and M_g^r. The estimate is based on external force and moment F_e and M_e, support polygons S_t^l and S_t^r, and a static friction coefficient μ. We assume that the ground plane is $y = 0$, and that the positive y-axis is pointing upwards. The ground reaction forces are bound by the following constraints:

1. At least one foot must be in contact with the ground before there can be any ground reaction force ($S_t^l \cup S_t^r \neq \emptyset$).
2. The external force F_e must point upwards, otherwise it cannot be attributed to ground contact ($F_{e,y} > 0$).
3. The horizontal component of the ground reaction force must be in agreement with the Coulomb friction model ($||F_{e,xz}|| \leq \mu ||F_{g,y}||$).
4. The origin of the ground reaction force and moment for each foot must lie within the support polygon of the corresponding foot.

In our method, we do not employ a dynamic friction model. All ground reaction forces are considered the result of static ground contact.

If constraint 1 and 2 are not met, no external forces or moments will be applied to the character during step 3 ($F_g^l = M_g^l = F_g^r = M_g^r = 0$). Otherwise, we first compute the total ground reaction force and moment, F_g and M_g, and distribute these among both feet afterwards.

We assume that the upward component of F_e can be fully attributed to ground reaction force: $F_{g,y} = F_{e,y}$. We do not limit the maximum amount of ground reaction force in the upward direction, since this is only limited by the amount of pushing force a character can produce. If such a pushing force is not realistic, we will see this reflected in the muscle error $\lambda(A_t)$.

To meet constraint 3, we limit the magnitude of the horizontal component of the ground reaction force, $F_{g,xz}$, based on friction constant μ and the vertical ground reaction force $F_{g,y}$:

$$F_{g,xz} = \begin{cases} F_{e,xz} & \text{if } ||F_{e,xz}|| \leq \mu ||F_{g,y}|| \\ \frac{\mu ||F_{g,y}||}{||F_{e,xz}||} F_{e,xz} & \text{if } ||F_{e,xz}|| > \mu ||F_{g,y}|| \end{cases} \quad (6)$$

The ground reaction moment around y, $M_{g,y}$, is also limited by contact friction between the character and the ground plane. However, since we do not expect large moments around y, we assume that any such moment is automatically countered by static friction: $M_{g,y} = M_{e,y}$.

The ground reaction force and moment applied to each feet must originate from a point on $y = 0$ that lies within the respective support polygon. This imposes a constraint on the ground reaction moment around x and z. To apply this constraint, we first define c_e as the point on $y = 0$ from which F_e and M_e would originate:

$$c_{e,x} = \frac{M_{e,z}}{F_{g,y}} \quad , \quad c_{e,z} = \frac{-M_{e,x}}{F_{g,y}} \quad (7)$$

If c_e lies outside a support polygon S, we define the origin of the ground reaction force, c_g, as the point inside S that is closest to c_e; otherwise: $c_g = c_e$. The ground reaction moments $M_{g,x}$ and $M_{g,z}$ then become:

$$M_{g,x} = -c_{g,z} F_{g,y} \quad , \quad M_{g,z} = c_{g,x} F_{g,y} \quad (8)$$

In cases where only one foot is in contact with the ground we assign F_g and M_g to that foot. In cases where both feet are in contact with the ground, the computation of (8) is performed individually for each foot. The forces and moments are then weighted according to the distance between c_e and support polygons S_t^l and S_t^r. If $d : c \times S \to \mathbb{R}$ is the distance between a point c and support polygon S, then the weighting factors ω_l and ω_r become:

$$\omega_l = \frac{d(c_e, S_t^l)}{d(c_e, S_t^l) + d(c_e, S_t^r)} \quad , \quad \omega_r = \frac{d(c_e, S_t^r)}{d(c_e, S_t^l) + d(c_e, S_t^r)} \quad (9)$$

where ω_l is the scaling factor applied to acquire F_g^l and M_g^l, and ω_r is the scaling factor applied to acquire F_g^r and M_g^r.

3.6 Step 3: Calculate Joint Moments

Now that we have estimated ground reaction force and moment for both feet, we can add them to the dynamics equation:

$$\tau = \mathbf{T}(q)^{-1}\left[\mathbf{M}(q)\ddot{q} + \mathbf{E}(q)\left[F_g^{l\,T}\ M_g^{l\,T}\ F_g^{r\,T}\ M_g^{r\,T}\right]^T + c(q,\dot{q})\right] \quad (10)$$

where $\mathbf{E}(q)$ is a $12 \times n$ coefficient matrix of which the form is dictated by $\mathbf{M}(q)$. After solving[1] for τ, F_r and M_r correspond to the elements in τ related to global position and orientation of the root element, in the coordinate frame of the root. They are the remaining external force and moment that could not be attributed to ground contact. Their magnitudes are used for the calculation of the dynamics error during frame A_t:

$$\kappa(A_t) = \frac{||F_r(A_t)||}{mg} + \frac{||M_r(A_t)||}{mgh} \quad (11)$$

where m is the subject body mass, g is the gravitational constant and h is the subject height. We employ these variables to ensure that both elements are dimensionless quantities, independent of weight and height.

3.7 Step 4: Estimate and Evaluate Muscle Forces

Each moment $\tau_i \in \tau$ is a result of the muscles acting on the degree of freedom q_i (except for those related to global position and orientation). The relation between the joint moments in τ and the vector of muscle forces, u, can be described as:

$$\tau = \mathbf{D}(q_t)u \quad (12)$$

where $\mathbf{D}(q)$ is a $l \times n$ matrix describing the muscular moment arms of the l muscles in u, derived from $\mathcal{U}$ (given pose q_t). In human characters, the number of muscles is much larger than the number of degrees of freedom ($l >> n$), making the number of solutions for u infinite. However, since our interest is to detect unrealistic muscle usage, we will seek a solution that assumes optimal load sharing between muscles. A common approach to achieve this is to minimize the sum of squared muscle stresses [17]:

$$\underset{u}{\mathrm{argmin}} \sum_{i=1}^{l}\left(\frac{u_i}{u_{max,i}}\right)^2 \quad (13)$$

This optimization is subject to both the moment arm constraints in (12) as well as to $u_i \geq 0$ (muscles cannot produce a negative force). The values for u_{max} are defined in $\mathcal{U}$ and represent the maximum strength of the animated character's individual muscles. Common values of u_{max} are well-document in biomechanics literature and are mostly derived from cadaveric data [17]. If desired, they can be adjusted to match the physique of any virtual character.

---

[1] When there are no ground reaction forces, equation (10) is equal to (4) and needs not be solved again.

Excess Muscle Force. After we have determined muscle forces u for frame A_t, we wish to evaluate if they are realistic. Our quality measure λ is a dimensionless quantity, defined as the amount of excess muscle force, u_{ex}, normalized by the sum of the elements in u_{max}:

$$\lambda(A_t) = \frac{\sum_{i=1}^{l} u_{ex,i}}{\sum_{i=1}^{l} u_{max,i}} \quad , \quad u_{ex,i} = \begin{cases} u_i - u_{max,i} & \text{if } u_i > u_{max,i} \\ 0 & \text{if } u_i \leq u_{max,i} \end{cases} \quad (14)$$

where $u_{ex,i}$ is the excess muscle force produced by muscle i, based on the maximum force defined by $u_{max,i}$.

3.8 Final Quality Measures

We have shown how to compute quality measures $\kappa(A_t)$ and $\lambda(A_t)$ for any animation frame A_t. There are numerous ways to combine these individual scores to get a score for a full animation clip $\mathbf{A}$. For our research, we have chosen to compute the average of all frames in the animation, which is sufficient when evaluating short animation clips.

4 Experimentation

To demonstrate our method, we investigate the effect of changing character body mass and animation speed on the physical realism of a set of motion captured animations. Changing the body mass of the character reflects a situation where a motion capture performer is much lighter or heavier than the virtual character. Some motions may not be physically realistic in that situation; we expect to see this reflected in muscle error λ, but not in dynamics error κ. Changing the weight of a character should not affect the degree in which its motion is subject to the Newton-Euler laws of motion. We expect that changing animation speed will affect both muscle error λ as well as dynamics error κ.

Our research will be based on the *Human Body Model*, which is a commercially available musculoskeletal model of the entire body [15]. The skeleton consists of 18 segments and 46 kinematic degrees of freedom. The muscle model includes 290 muscles. The inverse dynamics is solved using an approach described in detail by Kane et. al [18]. The squared muscle stress is minimized using the approach described by Xia and Feng [19], which is based on a recurrent neural network and has been developed with real-time performance in mind.

Animations are recorded using a 12 camera Vicon system and a custom 47 marker setup. In order to get smooth first and second order derivatives for q_t, the animation data is filtered using a second order Butterworth filter with a cut-off frequency of 4Hz. Our character ($m = 72$kg, $h = 1.70$m) has performed 16 different motions (see Figure 5). We separately vary the weight and speed multiplier in the range of 0.2 to 3.0 with steps of 0.2. We use a friction coefficient of $\mu = 1$.

5 Results

The results indicate that our measures behave as anticipated. First, there is a clear relation between body weight and muscle error (see Figure 2). This relation is stronger with more labor-intensive motions such as jumping or knee bending, and smaller for the more relaxed motions, such as waving.

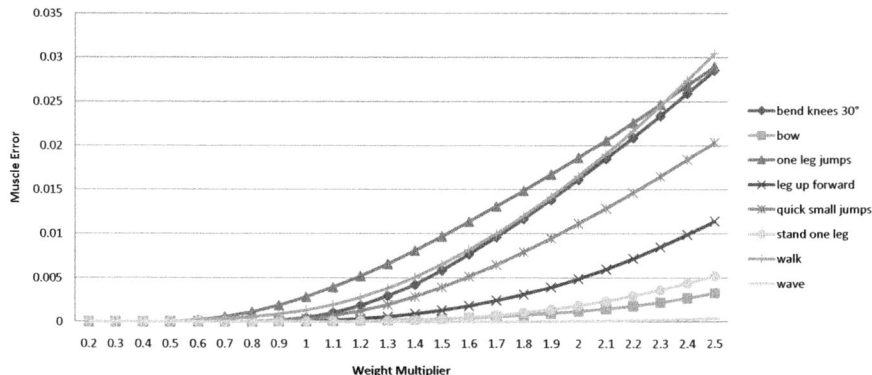

Fig. 2. Muscle error λ, with varying subject weight

We observed no significant change in the dynamics error as a result of weight change, which is in line with our expectations.

The relation between speed change and muscle error is shown in Figure 3. The relation is stronger with highly dynamic motions, such as walking. With such motions, even slowing down the animation results in a slight increase in muscle error.

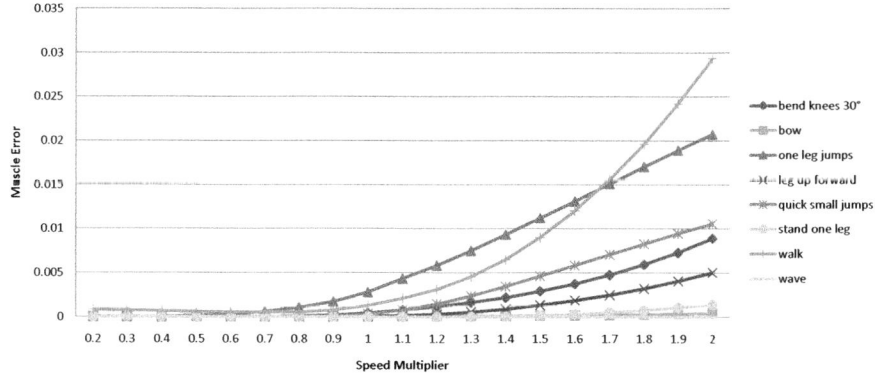

Fig. 3. Muscle error λ, with varying animation speed

The effect of speed change on dynamics error κ is shown in Figure 4. It can be seen that both slowing down and speeding up the animation leads to an increase in dynamics error, but only for dynamic motions such as walking and jumping. This shows it is not dynamically realistic to walk or jump at reduced or increased speed, without adjusting the motion. It can also be seen that increasing speed results in a larger increase in dynamics error than decreasing speed, which is in line with research by Reitsma and Pollard [3].

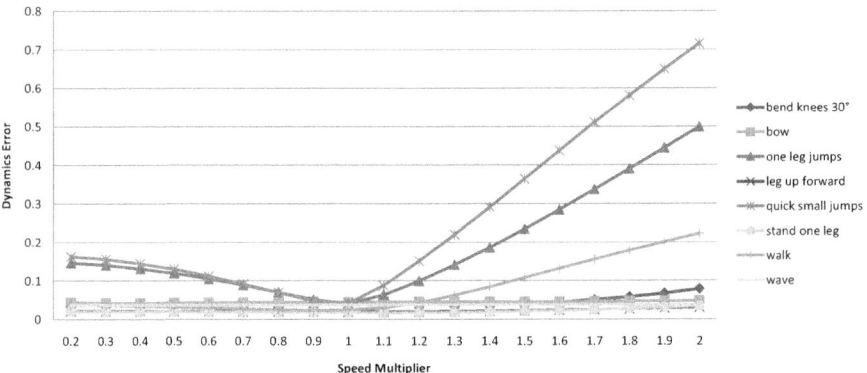

Fig. 4. Dynamics error κ, with varying animation speed

Another interesting observation is that when evaluating a sped up walking animation, there is a relatively high increase in muscle error, compared to the increase in dynamics error. However, when evaluating a sped up jumping motion, there is a high increase in dynamics error, and a relatively low increase in muscle error. One could interpreted this as follows: it is physically more realistic for a muscular animated character to walk with increased speed than to jump with increased speed.

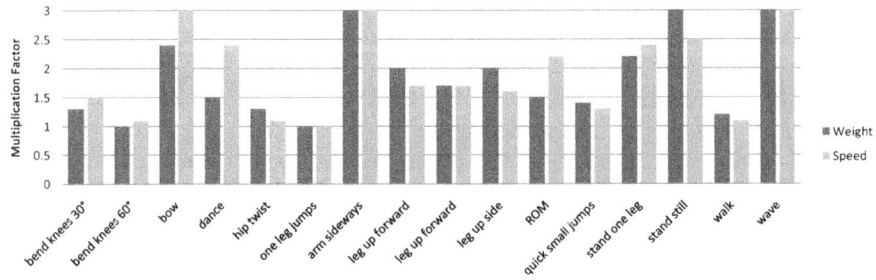

Fig. 5. Maximum weight and speed multipliers for which $\lambda \leq 0.003$. Multipliers are capped at 3; this was the highest multiplier we tested for in our experiments.

Finally, we show the maximum factor by which speed and weight can be increased before the muscle force required to perform a specific animation becomes physically unrealistic (defined by $\lambda > 0.003$). Figure 5 shows these limits for a number of animations.

6 Conclusion, Discussion and Future Work

We have demonstrated a method for evaluating the physical realism of character animations using musculoskeletal models. We have shown how ground reaction forces can be extracted from residual forces and moments. We have also shown how these residuals can be used to uniformly measure the dynamical validity of a motion, according to the Newton-Euler laws of motion. Finally, we have shown how the muscle forces estimated by a musculoskeletal model can be interpreted as a measure for realism of motion. The results of our initial experiments are promising.

We see many applications for our quality measures. For example, the measures can be used to detect good transition points for motion graphs. Also, both measures can be shown as feedback to the animator during interactive motion editing. Muscle error can even be visualized intuitively on a 3D mesh, using color animation.

Preliminary tests indicate that our estimated ground reaction forces correspond well to measured data. We plan to record additional motions with synchronized force plate data to further investigate this claim. We are aware that our method has only limited use in biomechanics research, since it cannot detect situations where both feet exert horizontal forces in different directions. However, for the purpose of measuring physical realism this is not an issue.

Our method of estimating ground reaction forces can still be improved by incorporating a more advanced friction model. At this point, we assume all ground contact is static. We are currently working on incorporating a dynamic friction model, which will make our technique suitable for detecting foot sliding, based purely on dynamics error and muscle error.

For some motions, we have found the muscle error to be above zero, even when neither weight nor speed was adjusted. We expect the reason for this is that some of the maximum muscle forces defined in our model are rather low in comparison to other models. We expect that adjusting these maximum muscle forces to more common values will resolve this issue.

Finally, an important next step in our research would be to evaluate the relationship between our quality measures and the perceived realism via user studies. It would then also make sense to compare our measures to those based on simplified dynamical models, and see if the enhanced accuracy of musculoskeletal models also leads to a more accurate measure of perceived realism.

References

1. van Welbergen, H., van Basten, B.J.H., Egges, A., Ruttkay, Z., Overmars, M.H.: Real Time Animation of Virtual Humans: A Trade-off Between Naturalness and Control. Eurographics - State of the Art Reports, 45–72 (2009)

2. O'Sullivan, C., Dingliana, J., Giang, T., Kaiser, M.K.: Evaluating the visual fidelity of physically based animations. ACM TOG 22(3), 527–536 (2003)
3. Reitsma, P.S.A., Pollard, N.S.: Perceptual metrics for character animation: sensitivity to errors in ballistic motion. ACM TOG 22(3), 537–542 (2003)
4. Runeson, S., Frykholm, G.: Visual perception of lifted weight. Journal of exp. psychology: Human Perception and Performance 7(4), 733–740 (1981)
5. Michaels, C.F., De Vries, M.M.: Higher Order and Lower Order Variables in the Visual Perception of Relative Pulling Force. Journal of Experimental Psychology: Human Perception and Performance 24(2), 20 (1998)
6. Safonova, A., Hodgins, J.K.: Analyzing the physical correctness of interpolated human motion. In: Proc. of the 2005 ACM SIGGRAPH/Eurographics Symposium on Computer Animation, pp. 171–180 (2005)
7. Ikemoto, L., Arikan, O., Forsyth, D.: Quick transitions with cached multi-way blends. In: Proc. of the 2007 Symp. on Interactive 3D Graphics and Games, pp. 145–151. ACM, New York (2007)
8. Ren, L., Patrick, A., Efros, A.A., Hodgins, J.K.: A data-driven approach to quantifying natural human motion. ACM TOG 24(3), 1090–1097 (2005)
9. van Basten, B.J.H., Egges, A.: Evaluating distance metrics for animation blending. In: Proc. of the 4th Int. Conf. on Foundations of Digital Games, pp. 199–206. ACM, New York (2009)
10. Shin, H.J., Kovar, L., Gleicher, M.: Physical touch-up of human motions. In: Proc. of the 11th Pacific Conf. on Comp. Graphics and Appl., pp. 194–203 (2003)
11. Ko, H., Badler, N.I.: Animating human locomotion with inverse dynamics. In: IEEE Computer Graphics and Applications, 50–59 (1996)
12. Erdemir, A., McLean, S., Herzog, W., van den Bogert, A.J.: Model-based estimation of muscle forces exerted during movements. Clinical Biomechanics 22(2), 131–154 (2007)
13. Veeger, H.E.J., Van Der Helm, F.C.T.: Shoulder function: the perfect compromise between mobility and stability. Journal of Biomechanics 40(10), 2119–2129 (2007)
14. Rasmussen, J., Damsgaard, M., Surma, E., Christensen, S.T., de Zee, M., Vondrak, V.: Anybody-a software system for ergonomic optimization. In: Fifth World Congress on Structural and Multidisciplinary Optimization (2003)
15. van Den Bogert, A.J., Geijtenbeek, T., Even-Zohar, O.: Real-time estimation of muscle forces from inverse dynamics. health.uottawa.ca, 5–6 (2007)
16. Delp, S.L., Anderson, F.C., Arnold, A.S., Loan, P., Habib, A., John, C.T., Guendelman, E., Thelen, D.G.: OpenSim: open-source software to create and analyze dynamic simulations of movement. IEEE Trans. on Bio-Medical Eng. 54(11), 1940–1950 (2007)
17. van Der Helm, F.C.T.: A finite element musculoskeletal model of the shoulder mechanism. Journal of Biomechanics 27(5), 551–553 (1994)
18. Kane, T.R., Levinson, D.A.: Dynamics online: theory and implementation with AUTOLEV. Online Dynamics, Inc. (1996)
19. Xia, Y., Feng, G.: An improved neural network for convex quadratic optimization with application to real-time beamforming. Neurocomputing 64, 359–374 (2005)

Physically-Based Character Control in Low Dimensional Space

Hubert P.H. Shum[1], Taku Komura[2], Takaaki Shiratori[3], and Shu Takagi[1]

[1] RIKEN, 2-1 Hirosawa, Wako, Saitama, Japan
[2] Edinburgh University, 10 Crichton Street, EH8 9AB, United Kingdom
[3] Carnegie Mellon University, 5000 Forbes Ave., Pittsburgh PA USA

Abstract. In this paper, we propose a new method to compose physically-based character controllers in low dimensional latent space. Source controllers are created by gradually updating the task parameter such as the external force applied to the body. During the optimization, instead of only saving the optimal controllers, we also keep a large number of non-optimal controllers. These controllers provide knowledge about the stable area in the controller space, and are then used as samples to construct a low dimensional manifold that represents stable controllers. During run-time, we interpolate controllers in the low dimensional space and create stable controllers to cope with the irregular external forces. Our method is best to be applied for real-time applications such as computer games.

Keywords: Character animation, dynamics simulation, dimensionality reduction, controller interpolation.

1 Introduction

Recently, physical-based character simulation is attracting attention of game producers mostly because the simulated characters can react to a changing environment. The wide variety of simulated motions makes the game characters lively and increases the attractiveness of the games. State-of-the-art techniques have shown that optimized dynamic controllers are robust against external perturbations in different environments [21], which makes them particularly suitable for interactive systems such as fighting games. However, in case the perturbations are unpredictable, adjusting or synthesizing controllers during run-time becomes very challenging.

In this paper, we propose a method to synthesize new robust controllers using pre-computed controllers. We choose a finite state machine based controller as a fundamental framework due to its robustness and easiness to design [23]. Although some researches linearly interpolate each parameter of controllers [22], the non-linear relationship among the controllers tends to cause instability in the resultant motion. Therefore, we prepare samples of stable controllers through optimization and construct a non-linear low dimensional manifold using landmark Isomap[3]. To effectively construct the control manifold, we adapt a large number of sub-optimal controllers found during the optimization process. Since these controllers give supplementary knowledge about the distribution of stable controllers, the manifold faithfully represents the stable region in the high dimensional space. During run-time, we interpolate optimal controllers in the low dimensional latent space with reference to the observed environment. Linear interpolation in the reduced space becomes a non-linear interpolation in the full control space, and hence produces robust controller interpolations.

We show that our system can produce stable controllers for characters perturbed by different external forces. Since the simulated characters can cope with arbitrary perturbations during run-time, our system is particularly suitable for situations where the external force is unpredictable, such as when several characters collaboratively push one object together.

2 Related Works

Dynamics Character Control: Designing a physically-based controller for character animation is difficult and non-trivial, as joint torques, instead of the joint angles, need to be specified. To ease the controller designing process, researchers propose to track captured motions [10, 14] or manually designed keyframes [2, 23], or use abstract models such as the inverted pendulum [7, 12, 17] to reduce the number of control variables. We apply keyframe based tracking because of its easiness of designing and implementation. Our idea to collect different controllers and composing a manifold for non-linear interpolation is also applicable for pre-designed abstract models, which may also require parameter tuning for handling unpredictable external perturbation.

Controller Optimization: The controller space is highly non-linear and non-differentiable. Traditional optimization techniques like gradient descent method can easily get caught in local minima [22]. On the contrary, as covariance matrix adaptation (CMA) does not assume a smooth space [5], it is particularly suitable for optimizing physically-based controllers [19, 21]. Since it is difficult to directly construct a controller for a drastically different environment, continuous optimization is proposed for gradually adjusting the environment during the optimization process [18, 22]. In this paper, we also apply CMA under the continuous optimization framework.

Dimensionality Reduction: Non-linear dimensionality reduction techniques have been widely applied for character control [4, 13]. We apply a method called landmark Isomap [3], which can discover the intrinsic dimensions of a given data set, and is much more computationally efficient than other methods such as Gaussian Process Latent Variable Models [9] and Isomap [16].

Run-time Synthesizing of Controllers: To deal with run-time changes in the environment and the conditions of the character, run-time controller optimization is proposed [8, 11]. Because optimization is computationally costly, it is suggested to precompute a set of controllers and apply run-time interpolation

[22]. Since interpolating controllers in the irregular controller space often introduces instability, we propose to interpolate the controllers in a reduced space constructed from samples of stable controllers.

3 The Proposed Framework

The overview of our system is shown in Figure 1. Our objective is to train a series of controllers that can handle a task (such as walking against external perturbation) with different task values (such as the amount of the perturbation). Given an initial controller, our system applies continuous optimization to train a set of controllers that can cope with an incremental task parameter (the green area). The optimal controller for the current task value is used as the initial controller in the next iteration. CMA is used to optimize a controller with respect to the next task value (the red area). Each optimization step generates a covariance matrix and a number of controllers, which are applied in the next optimization step until the optimal controller is found. During each optimization step, reasonably good controllers that can keep walking without losing balance for 20 steps are stored in a database (the blue area). After the optimization is finished, landmark Isomap is applied to analyze the controllers and create a latent space of stable controllers, which is used for interpolation during run-time.

3.1 Model and Controller

Our character model has 14 joints. The hip and shoulder joints are modelled as two dimensional universal joints, while the rest are modelled as one dimensional hinge joints. In total, our model has 18 degrees of freedom. The total body size and weight are set $160cm$ and $70kg$, respectively, and the segment size and weight are set by referring to [1].

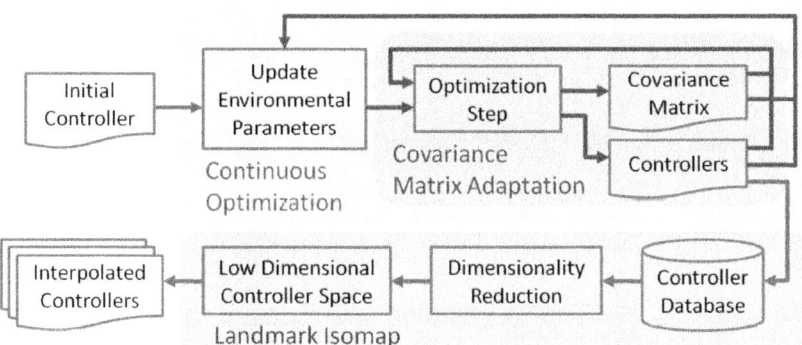

Fig. 1. We implement CMA under the continuous optimization framework. Based on the collected controller samples, we apply landmark Isomap to compute the latent space for run-time controller interpolation.

We implemented a physics-based gait controller similar to [23]. Four keyframes are used to simulate a walking motion. The target joint angles and proportional / derivative gains for each keyframe are provided by the controller. The system transits from one keyframe to another with reference to a set of predefined durations. During optimization, we optimize the joint angles and the proportional gains for all keyframes. The derivative gains are set 10% of the proportional gains. The target joint angles and gain parameters are set constant for the arms in our experiments.

3.2 Continuous Optimization

In this section, we explain how we apply continuous optimization [22] to create a series of controllers with respect to different task parameters.

The idea of continuous optimization is to iteratively optimize the controller and increase the task parameter value. The system starts with a given initial controller. In each iteration, the task parameter is incremented, making the controller unstable. The system then optimizes the controller until an optimal one is found. Such a controller will be considered as the initial controller in the next iteration with an updated task value.

Figure 2 illustrates an example of continuous optimization. The task here is to walk in the presence of an external pushing force. Let us assume the controller contains two control parameters only, which are represented by axes. Each point in the graph thus represents a controller. The grey area indicates the space that the controller is unstable and the white area indicates where is stable. The optimization starts with a manually designed controller x_0. We apply $5N$ of pushing force to the character and carry out the optimization until the optimal controller x_1 is found. Using x_1 as a starting point, we apply $10N$ of force and carry out another iteration of optimization, which returns a stable controller x_4. Further increasing the force results in a stable controller x_6.

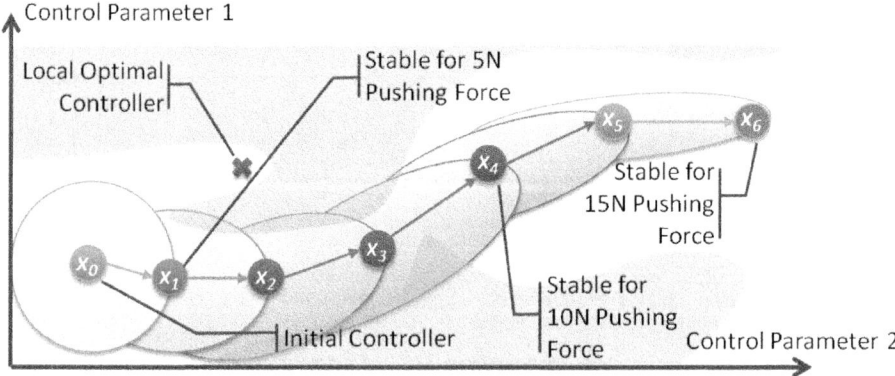

Fig. 2. Using continuous optimization, the optimization engine takes intermediate steps to create the final controller. The colored regions visualize the sampling areas calculated by covariance matrix adaptation.

Continuous optimization has two major advantages. First, since the optimal controller is used in the next iteration as the initial controller, the optimization engine explores area around the previously stable controller. This helps to maintain the similarity among the set of resultant controllers. Second, since the task parameter is updated incrementally, the optimization system is less likely to get trapped in local optimal solution. For example, in Figure 2, if we optimize a $15N$ controller directly from the initial controller x_0, there will be a high chance that the optimization gets caught at the local optimal solution indicated by the red cross. However, because the task parameter is updated incrementally, we indirectly guide the optimization process step by step and avoid the local optimal.

3.3 Covariance Matrix Adaptation

In this section, we explain the covariance matrix adaptation (CMA) method [5] that we used to optimize the controllers with respect to a given task parameter.

The main concept of CMA is to repeatedly sample and evaluate a set of controllers based on a covariance matrix and a reference controller. In contrast to stochastic search, the covariance matrix is used to reflect the intrinsic correlation between control parameters during sampling. In each optimization step, the reference controller is replaced by the best controller found in the previous step, while the covariance matrix is updated based on a set of good controllers.

The details of CMA are explained below. In each optimization step, Ω controllers are sampled with normal distribution:

$$x = x_{ref} + \sigma N(0, C) \tag{1}$$

where x_{ref} is the reference controller, σ is the standard derivation, and $N(0, C)$ is a zero mean multivariate normal distribution with the covariance matrix C. σ controls the step size of each optimization step and is set to 0.003 in our system. The samples are then evaluated using the objective function described in Section 3.4.

Based on the evaluation of the sample controllers, the system updates the covariance matrix C and the reference controller x_{ref}. Considering the best λ of the samples, the new covariance matrix is calculated by

$$C' = \frac{1}{\lambda} \sum_{i=1}^{\lambda} (x_i - x_{ref})(x_i - x_{ref})^T \tag{2}$$

where x_i represents a set containing the best λ samples, and we set $\lambda = \Omega \times 25\%$. The new reference controller is replaced by the sample with the highest score, x_{best}:

$$x'_{ref} = x_{best} \tag{3}$$

Note that in the original CMA implementation, x'_{ref} is updated as a weighted sum of the best λ samples. However, this implementation performs sub-optimally in our system because it implies interpolation in the irregular controller space.

The ellipses around the controller from x_0 to x_5 in Figure 2 represent their respective sampling areas. The initial covariance matrix at x_0 is set as uniform sampling in all dimensions. Then, the covariance matrix evolves based on the best samples. As a result, the sampling area deforms to fit in the stable area of the controller space, and hence optimization becomes more effective in the later steps.

Rank μ Update. Here, we explain how to speed up the CMA using one of it variant called rank μ update [6].

In theory, we can speed up the optimization by using fewer samples (i.e. smaller Ω). In practice, a small number of samples usually results in an unreliable covariance matrix. This is because the samples drawn are less likely to cover a large area in the controller space.

Rank μ update is an algorithm to estimate the covariance matrix under a system with small number of samples. Apart from considering the samples in the previous step (Equation 2), we calculate the covariance matrix with respect to the matrices used in all the past steps:

$$C' = \frac{1}{S}\left(\sum_{s=1}^{S-1} C_s + \frac{1}{\lambda}\sum_{i=1}^{\lambda}(x_i - x_{ref})(x_i - x_{ref})^T\right) \quad (4)$$

where C_s denotes the covariance matrix used at step s, S represent the total number of steps, and the rest of the parameters are defined as Equation 2. By this way, the matrix is updated with respect to all previous steps, thus eliminating the risk of an unreliable covariance matrix.

Noise Induction. Physical systems are usually noisy due to the randomness in contact points and control torque error, and hence the controller measurement fluctuates across simulations. For example, when the foot of the character touches the floor, a small error at the ankle orientation will result in a change in the center of pressure, and hence affect the overall balance of the character hugely.

To evaluate the controllers accurately, we simulate multiple trials and evaluate the average fitness for each of them. In each trial, we introduce a small adjustment on the initial posture and a small amount of control torque error to simulate the noise [21]. The final evaluation considering all the trials is more representative. Empirically, we found that 100 trials give a good balance between evaluation accuracy and computational cost.

3.4 Objective Function

In this section, we explain the objective function we used to evaluate the fitness of the controllers. Apart from traditional terms that consider stability, motion style and energy [20], we add an extra term to minimize the change in controlling parameters across optimization steps. The objective function is defined as a weighted sum of the following terms.

Controller Difference. We minimize the difference between the initial controller and the current controller described in Section 3.2. This ensures that the two controllers are close in the controller space, and hence results in stable control when we synthesize new controllers by interpolating them.

$$E_{difference} = -\sum_i |x^i_{initial} - x^i| \qquad (5)$$
$$i \in \{\text{angles}, \text{gains}\}$$

where $x^i_{initial}$ and x^i represent the ith controller parameter of the initial and the current controller, respectively. The control parameter consists of target angles and gains for all keyframes.

Stability. Stability is defined by the duration a character can walk without falling. We maximize the mean duration of walking from a number of simulations, and minimize their variance:

$$E_{stability\_mean} = \text{mean}(T)$$
$$E_{stability\_variance} = -\text{variance}(T) \qquad (6)$$

where T is the time the character can walk. The maximum cut-off duration is set 20 seconds, which means that we terminate the simulation there assuming the gait is sufficient stable.

Motion Style. To construct a locomotion controller that produces walking motion similar to humans, we compare the simulated cycle with the captured human data:

$$E_{style} = -\text{mean}(P - P_c) - \text{mean}(d - d_c) - \text{mean}(\Theta - \Theta_c) \qquad (7)$$

where P is the position of the joints, d is the distance the body have travelled, and Θ is the orientation of the body around the vertical axis in the simulated motion, and P_c, d_c, Θ_c are the corresponding values in the captured motion.

Gain. To avoid high gains that results in stiff movements, we also minimize the gain during optimization:

$$E_{gain} = -\sum_i x^i \qquad (8)$$
$$i \in \{\text{gains}\}$$

where x^i represent the ith gain value.

3.5 Landmark Isomap

In this section, we first give a brief explanation about the classical Isomap and the enhanced landmark Isomap. Then, we explain how we implement landmark Isomap in our system, and discuss different parameters that affect the result.

Isomap [16] is a well-known non-linear dimensionality reduction method. It uses local metric information to learn the underlying global geometry of a data set. More specifically, it creates a connectivity map for the dataset using K nearest neighbors. Each sample is represented by a node in the map, and each pair of neighbors is connected with an edge. Based on the local distance, Isomap computes a distance matrix containing the shortest distance between any two nodes using Dijkstra's algorithm. Finally, multidimensional scaling (MDS) [24] is applied to calculate the eigenvectors of the distance matrix, hence reducing its dimensions.

Landmark Isomap [3] is a speed-up version of Isomap. The two bottlenecks of the classical Isomap are the computation of the distance matrix by Dijkstra's algorithm ($O(KN^2 \log N)$) and apply MDS on it ($O(N^3)$), where N is the number of samples and K is the neighbor size. Landmark Isomap speeds up the system by designating n of the data points to be landmark points, in which $n < N$. Pairwise distances are only computed between landmark and other samples ($O(KnN \log N)$). Due to the use of a smaller distance matrix, MDS is a lot faster ($O(n^2 N)$).

In our system, we collect all the stable controller samples which were computed during the CMA optimization to compute the low dimensional latent space. Sampling only the optimal controllers will fail to create a representative low dimensional space, since the controllers are relatively sparse in the high dimensional space. Instead, we sample the controllers that have a good score in every optimization step. With the large number of good controllers, the computed low dimensional manifold will cover a large area of stable region.

For landmark Isomap, we need to designate samples to be landmarks. Ideally, the landmark should be distributed evenly among the samples, such that the error of the distance matrix is minimized. In our system, we randomly assigned 20% of the samples in each optimization step as landmarks. Since samples within each optimization step are similar to each other, this approach ensures that each sample is close to at least one landmark.

One important issue is the selection of the value K of the K nearest neighbor algorithm used for constructing the connectivity map. This heavily affects the quality of the final result. Large K can result in "short-circuit errors", which means non-similar samples getting classified as neighbors. Small K can result in too many disconnected maps. We set K to be the average number of samples selected in each optimization step, as the samples within one step are usually similar.

Run-time Controller Interpolation. During run-time, we interpolate the controllers in the low dimensional space. We first observe the task parameter during run-time, and select the two optimal controllers for the task value immediately below and above the observed value. Then, we linearly interpolate the two controllers in the low dimensional space, and inversely project the interpolated controller into the high dimensional space for character control.

Linear interpolation along the low dimensional manifold theoretically results in a more stable controller than that produced by interpolation in the full space, because the former is a constructed from samples of stable controllers.

4 Experimental Results

In this section, we describe our experiments using a pushing controller. We use Open Dynamics Engine [15] for the simulator.

We apply continuous optimization to create a set of controllers that can counteract force continuously applied at the torso from the front and the back. The optimization starts with no pushing force. Whenever an optimal controller is found, we increase the force and apply the optimization again. We generate a set of controllers that can cope with the force from $0N$ up to $130N$ with the step size of $2N$. We computed 66 optimal controllers and sampled 2822 sub-optimal controllers. The whole process takes around 20 hours using an 8 core system (Duo Intel Xeon W5590).

We computed a low dimensional manifold of stable controllers by applying landmark Isomap to the collected samples. We set K of the K nearest neighbor operation in Isomap to 10, which is the average number of samples in each CMA optimization step. We determine the dimensionality of the latent space using leave-one-out cross validation. In our pushing motion experiment, we find that the reconstruction error does not increase significantly if the number of dimension is 5 or more. Therefore, we set the reduced dimension to 5.

During run-time, we observe the force applied to the character, and interpolate the optimal controllers in the low dimensional manifold. Since the interpolation and the forward dynamics simulation are fast, our system runs in real-time even with a slow system (Intel Core 2 Duo P8600).

We experiment on the stability of the character by changing the amount of external force applied during run-time (Figure 3). The system obtains the controller to deal with the force by interpolation. We observe that when the amount of pushing force increases, the character becomes stiffer and bend the upper body against the direction of the force for counteracting it. We repeated this experiment for 200 times with different variations of external forces, and examined that the character could walk without falling for more than 98% of the simulations.

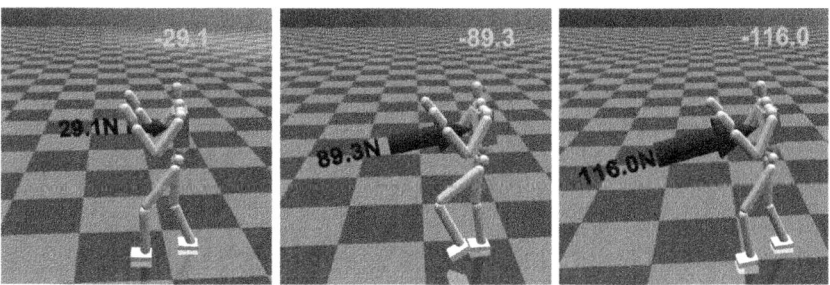

Fig. 3. The character applies the appropriate interpolated controller to counteract the pushing force. The red arrow represents the pushing force, while the green text represents the force that the controller can cope with.

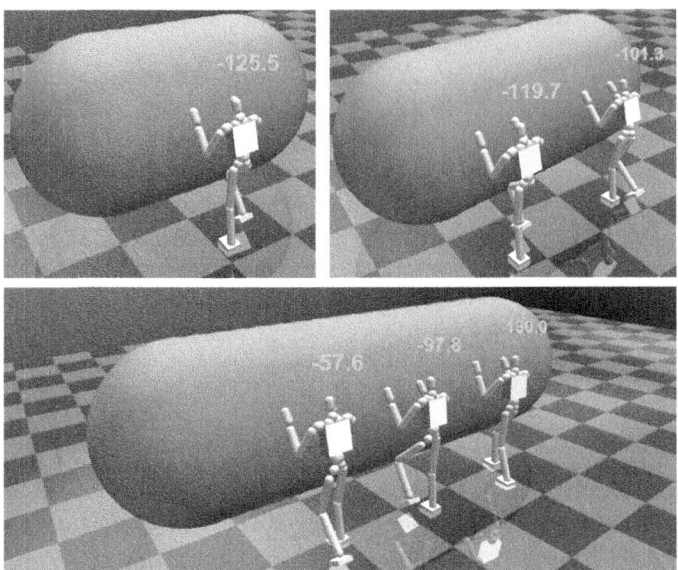

Fig. 4. Characters cooperatively push a large object. The green text represents the force that the controller can cope with.

We conduct experiments to simulate different number of characters pushing a physical object (Figure 4). The dynamic friction of the object is set to 5% of its velocity. We evaluate the force transfered from the object to the character using force sensors provided by the Open Dynamics Engine, and update the controller according to the value. When two or more characters are pushing the object cooperatively, the external force turns irregular, because the force exerted by one character indirectly affects those of the others. Still, our characters can stably walk by our controller.

5 Discussions

In this paper, we explained our method to automatically construct a low dimensional manifold for stable gait control from a rich set of samples obtained during continuous CMA optimization. The low dimensional manifold represents the stable area in the full controller space, and hence is capable of synthesizing stable controllers in the high dimensional space.

Constructing a reliable low dimensional manifold provides an important tool in dynamics control. Apart from controller interpolation shown in this paper, it is possible to apply run-time optimization in the reduced space [8, 11]. Since the dimensionality of the controller is greatly reduced, optimization can be done much more effectively.

Recent research suggests incorporating inverse pendulum model as an abstracted model of the body [17]. Our method provides a potentially better alternative for controller abstraction, since inverse pendulum model is only valid during single supporting stages, and its reliability depends on the posture of the character model during gait motion. Our low dimensional representation does not have these constraints.

Interaction among physically-controlled characters is a tough problem due to the instability of the controllers. One of our future directions is to simulate physical interactions among multiple characters.

References

[1] Armstrong, H.G.: Anthropometry and mass distribution for human analogues. Military male aviators, vol. 1 (1988)
[2] Coros, S., Beaudoin, P., van de Panne, M.: Generalized biped walking control. In: SIGGRAPH 2010: ACM SIGGRAPH 2010 Papers, pp. 1–9. ACM, New York (2010)
[3] De Silva, V., Tenenbaum, J.B.: Global versus local methods in nonlinear dimensionality reduction. In: Advances in Neural Information Processing Systems 15, vol. 15, pp. 705–712 (2003)
[4] Grochow, K., Martin, S.L., Hertzmann, A., Popović, Z.: Style-based inverse kinematics. In: SIGGRAPH 2004: ACM SIGGRAPH 2004 Papers, pp. 522–531. ACM, New York (2004)
[5] Hansen, N.: The cma evolution strategy: A comparing review. In: Towards a New Evolutionary Computation, Studies in Fuzziness and Soft Computing, vol. 192, pp. 75–102. Springer, Heidelberg (2006)
[6] Hansen, N., Müller, S.D., Koumoutsakos, P.: Reducing the time complexity of the derandomized evolution strategy with covariance matrix adaptation (cma-es). Evol. Comput. 11(1), 1–18 (2003)
[7] Kwon, T., Hodgins, J.: Control systems for human running using an inverted pendulum model and a reference motion capture sequence. In: SCA 2010: Proceedings of the 2010 ACM SIGGRAPH/Eurographics Symposium on Computer animation. Eurographics Association, Aire-la-Ville, Switzerland (2010)
[8] de Lasa, M., Mordatch, I., Hertzmann, A.: Feature-based locomotion controllers. ACM Trans. Graph. 29(4), 1–10 (2010)
[9] Lawrence, N.D.: Gaussian process latent variable models for visualisation of high dimensional data. In: NIPS, p. 2004 (2004)
[10] Lee, Y., Kim, S., Lee, J.: Data-driven biped control. ACM Trans. Graph. 29(4), 1–8 (2010)
[11] Mordatch, I., de Lasa, M., Hertzmann, A.: Robust physics-based locomotion using low-dimensional planning. ACM Trans. Graph. 29(4), 1–8 (2010)
[12] Pratt, J.E., Tedrake, R.: Velocity-based stability margins for fast bipedal walking, vol. 340, pp. 299–324 (2006)
[13] Shin, H.J., Lee, J.: Motion synthesis and editing in low-dimensional spaces: Research articles. Comput. Animat. Virtual Worlds 17(3-4), 219–227 (2006)
[14] da Silva, M., Abe, Y., Popović, J.: Interactive simulation of stylized human locomotion. In: SIGGRAPH 2008: ACM SIGGRAPH 2008 Papers, pp. 1–10. ACM, New York (2008)
[15] Smith, R.: Open dynamics engine, http://www.ode.org/
[16] Tenenbaum, J.B., Silva, V., Langford, J.C.: A global geometric framework for nonlinear dimensionality reduction. Science (5500), 2319–2323 (December)

[17] Tsai, Y.Y., Lin, W.C., Cheng, K.B., Lee, J., Lee, T.Y.: Real-time physics-based 3d biped character animation using an inverted pendulum model. IEEE Transactions on Visualization and Computer Graphics (2009)
[18] Van De Panne, M., Lamouret, A.: Guided optimization for balanced locomotion. In: Terzopoulos, D., Thalmann, D. (eds.) 6th Eurographics Workshop on Animation and Simulation, Computer Animation and Simulation, Eurographics, pp. 165–177. Springer, Wien (September 1995)
[19] Wampler, K., Popović, Z.: Optimal gait and form for animal locomotion. ACM Trans. Graph. 28(3), 1–8 (2009)
[20] Wang, J.M., Fleet, D.J., Hertzmann, A.: Optimizing walking controllers. In: SIGGRAPH Asia 2009: ACM SIGGRAPH Asia 2009 Papers, pp. 1–8. ACM, New York (2009)
[21] Wang, J.M., Fleet, D.J., Hertzmann, A.: Optimizing walking controllers for uncertain inputs and environments. ACM Trans. Graph. 29(4), 1–8 (2010)
[22] Yin, K., Coros, S., Beaudoin, P., van de Panne, M.: Continuation methods for adapting simulated skills. In: SIGGRAPH 2008: ACM SIGGRAPH 2008 Papers, pp. 1–7. ACM, New York (2008)
[23] Yin, K., Loken, K., van de Panne, M.: Simbicon: Simple biped locomotion control. ACM Trans. Graph. 26(3), Article 105 (2007)
[24] Young, F.W.: Multidimensional scaling. In: Kotz-Johnson Encyclopedia of Statistical Sciences, vol. 5, John Wiley & Sons, Inc., Chichester (1985)

Learning Crowd Steering Behaviors from Examples

Panayiotis Charalambous and Yiorgos Chrysanthou

University of Cyprus, Computer Science Department
{totis,yiorgos}@cs.ucy.ac.cy
http://graphics.cs.ucy.ac.cy

Abstract. Crowd steering algorithms generally use empirically selected stimuli to determine when and how an agent should react. These stimuli consist of information from the environment that potentially influence behavior such as an agent's visual perception, its neighboring agents state and actions, locations of nearby obstacles, etc. The various different approaches, rule-based, example-based or other, all define their responses by taking into account these particular sensory inputs.

In our latest experiments we aim to determine which of a set of sensory inputs affect an agent's behavior and at what level. Using videos of real and simulated crowds, the steering behavior (i.e. trajectories) of the people are tracked and time sampled. At each sample, the surrounding stimuli of each person and their actions are recorded. These samples are then used as the input of a regression algorithm that learns a function that can map new input states (stimuli) to new output values (speed, direction). A series of different simulations are conducted with different time varying stimuli each time in order to extract all the necessary information.

Identifying the most important factors that affect good steering behaviors can help in the design of better rule based or example based simulation systems. In addition they can help improve crowd evaluation methods.

Full-Body Hybrid Motor Control for Reaching

Wenjia Huang, Mubbasir Kapadia, and Demetri Terzopoulos

University of California, Los Angeles

Abstract. In this paper, we present a full-body motor control mechanism that generates coordinated and diverse motion during a reaching action. Our framework animates the full human body (stretching arms, flexing of the spine, as well as stepping forward) to facilitate the desired end effector behavior. We propose a hierarchical control system for controlling the arms, spine, and legs of the articulated character and present a controller-scheduling algorithm for coordinating the sub-controllers. High-level parameters can be used to produce variation in the movements for specific reaching tasks. We demonstrate a wide set of behaviors such as stepping and squatting to reach low distant targets, twisting and swinging up to reach high lateral targets, and we show variation in the synthesized motions.

1 Introduction

A fundamental research problem in motion planning for virtual characters (and robots) is to control the body in order to achieve a specified target position. It occurs frequently in game scenarios when characters must reach to grasp an object. There are three issues that may need to be addressed in generating realistic reaching motion: (1) correctness of hand placement, (2) satisfaction of the constraints on the character's body, and (3) naturalness of the whole-body movement. An ideal reaching motion requires the character's hand to follow a smooth, collision-free trajectory from its current position to the target position while satisfying the constraints on the entire body. Inspired by prior work [8, 11, 16], we consider balancing, coordinating, and variation for reaching motion that takes into account the motion of the arms, spine, and legs of an anthropomorphic character.

In particular, we develop a hierarchical control strategy for controlling a virtual character. We use analytical IK to control the hands of the character, thus providing fast and accurate control of the end effectors (hands). A stepping motion controller, coupled with a non-deterministic state machine, controls the lower limbs to move the character towards a target that is beyond its reach. The spine controller ensures the balance of the character by accounting for its center of mass, and uses rotation decomposition for simple and robust control. Finally, we develop a novel controller-scheduling algorithm that generates the coordination strategies between multiple controllers.

2 Related Work

There has been a considerable amount of research both in robotics [3, 17] and in animation [6, 10, 13–15, 19] that addresses the problem of actuator control. Techniques like Rapidly-Exploring Random Trees [4, 10] and Probabilistic Road Maps [7] focus on the correctness of the generated motion, while data-driven methods [9, 22] animate natural-looking characters. However, controllers that generate animations that are natural, varied, and work in a wide variety of scenarios (dynamic environments) have yet to be realized. The work in [8] uses hierarchical control for gesture generation. Spine rotation decomposition strategies have been presented in [16]. The work in [5] uses key posture interpolation to generate body postures and IK for limb movement.

Inverse Kinematics is often used in conjunction with motion-capture data for character animation [9]. There are three popular methods to solve IK problems. Jacobian-based numerical methods can deal with arbitrary linkages, but they are computationally intensive. Cyclic Coordinate Descent (CCD) [12] is efficient, but it tends to favor the joints near the end-effector [18]. Analytical IK methods [5, 11] for body posture control have proven to be a good solution provided the number of joints is kept small. Prior research [5, 20] models human limbs as 3 joints and 7 DOF linkages. Using this model, an analytical solution to the IK problem can be derived accurately and efficiently, and it is the chosen method for this work.

Our work differs from prior work as follows: We use a simplified character model coupled with an analytical IK solver for efficiently controlling the character's limbs. Next, we employ a coordinated control strategy for reaching motion that takes into account spine as well as limb motion. Finally, the spine rotation in our framework is decomposed into swing and twist components, which can be controlled independently.

3 Character Representation

In this section, we describe the character model that we use in our work and the controllers used to animate the character. Section 3.1 outlines the character representation (character skeleton and mesh structure). Section 3.2 describes our hierarchical control structure and Sections 3.3, 3.4, and 3.5 detail the implementation of the arm, spine (including head), and leg controllers, respectively.

3.1 Character Model

Figure 1(a) illustrates the mesh and skeleton structure of our anthropomorphic character model. The model is composed of 16 separate meshes and it is equipped with an articulated skeleton of 21 bones with 43 degrees of freedom. The meshes are bound to the skeleton through rigid skinning. The character has 4 linkages—two arms and two legs, each consisting of 3 bones and 4 joints. The extra joint is at the lowest level of the joint hierarchy (e.g., the finger tip of the arm linkage) and has 0 DOFs. The remaining three joints—the base, mid, and end joint—have a

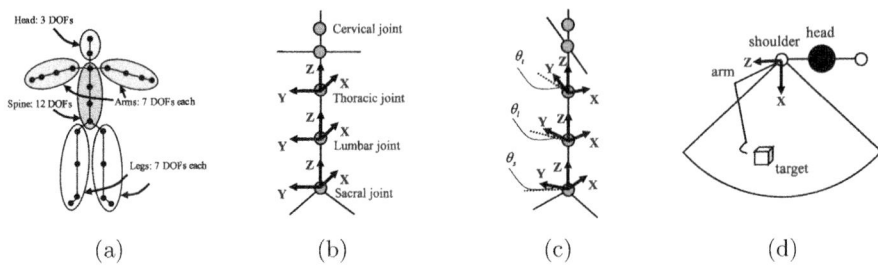

Fig. 1. (a) Character skeleton decomposition. (b) Spine modeling and local joint frames. (c) Twist rotations of all spine joints. (d) Top view of *comfortable reaching angle* (yellow). The y axis of the shoulder frame points out of the page.

total of 7 DOFs. The four linkages of the character are controlled using analytical inverse kinematics, as was described above.

The remaining joints model the spine of the character. Attempting to model the spine as a single bone would result in stiff and unnatural movement. The human spine comprises cervical, thoracic, lumbar, and sacral regions [2]. We therefore model the spine using four bones and four joints (cervical, thoracic, lumbar, and sacral) (Figure 1(b)). The spine's root joint (sacral) has 6 DOFs, with position as well as orientation control. The 2 intermediate spine joints (lumbar and thoracic) each have 3 DOFs. The neck joint (cervical) has 3 DOFs that serve to control the orientation of the character's head. Additional 0 DOF joints connect to the spine a shoulder bone, which provides attachment points for the arms, and a pelvic bone, which provides attachment points for the legs.

3.2 Hierarchical Control Structure

The character has a hierarchical skeletal structure. The spine is at the highest level of the hierarchy and it determines the position and orientation of the entire character. The head, arms, and legs are at the second level of the hierarchy. The arms and legs are controlled using analytical IK, facilitating easy control of the joint angles and specifying joint limits. The four controllers that we employ deal with different sets of hierarchical skeleton components. The arm controller enables the arm to perform a reaching motion. The spine controller controls the spine and decomposes rotations into swing and twist for simple and robust control. The head controller determines the direction in which the character's head is facing. The leg controller controls the legs and the root joint of the spine by generating stepping movements.

3.3 Arm Controller

Arm motion is a fundamental aspect of reaching, as it determines the ability and accuracy of the motion of the end-effector (hand) in reaching a desired position.

The end-effector is required to follow a smooth trajectory from its current position to the desired position, while obeying joint constraints and providing natural-looking motions (we do not deal with hand grasping motions). Modeling the trajectories of hand motion using Bezier curves [1] can yield satisfactory results, but this requires pre-captured hand motion data. We employ an ease-out strategy in which the hand trajectory is the shortest path to the target and the speed of the hand starts relatively high and gradually decreases as the hand reaches the target. This results in the hand slowing down naturally as it nears the target position and avoids "punching" the target.

The motion of the hand is governed by the starting velocity $\mathbf{v}_s$, the deceleration $\mathbf{a}$, and the total time t_t. Given a start position $\mathbf{p}_s$ and end position $\mathbf{p}_e$, the target displacement is calculated as $\mathbf{d}_h = \mathbf{p}_e - \mathbf{p}_s$. The motion of the hand has the following properties:

$$\mathbf{v}_s = v_s \frac{\mathbf{d}_h}{|\mathbf{d}_h|}, \tag{1}$$

$$\mathbf{a} = \frac{-\mathbf{v}_s}{t_t(1 - r_s)}, \tag{2}$$

$$\mathbf{d}_h = \mathbf{v}_s t_t + \frac{1}{2}\mathbf{a}(t_t(1 - r_s))^2, \tag{3}$$

where r_s, the ratio of the duration of the start phase to the total duration of the reaching motion, is used to determine the deceleration of the hand as it nears the target. Substituting equations (1) and (2) into (3), the total time t_t can be calculated as

$$t_t = \frac{2|\mathbf{d}_h|}{v_s(r_s + 1)}. \tag{4}$$

Algorithm 1 describes the arm controller logic to initialize and update the position of the hand over time. The user-defined parameters v_s and r_s can be modified in order to achieve variation in the results. We initialize $v_s = 20$ cm/s and $r_s = 0.8$.

3.4 Spine Controller

The spine of the character plays a crucial role during reaching as the character needs to bend and twist its upper body to reach targets at different positions. The spine controller determines the upper-body position and orientation of the character (six degrees of freedom). Our four joint spine model enhances the control of the character, resulting in more fluid and natural animations.

Spine Rotation Decomposition. Previous work [5] introduced swing-and-twist orientation decomposition for arm joints, enabling simple and robust control. Inspired by this work, we similarly decompose the orientations of the spine joints. The advantage of this decomposition method is that it separates the spine motion into two primitive orientations, each of which can be governed by a separate control module in order to achieve robust control.

Algorithm 1. Arm controller

Procedure *Initialize*($\mathbf{p}_s$, $\mathbf{p}_e$, v_s, r_s)
    Input: $\mathbf{p}_s$: Start hand position
    Input: $\mathbf{p}_e$: Target hand position
    Input: v_s: Start hand speed
    Input: r_s: Deceleration ratio
    Output: t_t: Total time
    Output: a: Deceleration
    Output: $\mathbf{v}_s$: Start hand velocity
    Output: $\mathbf{p}_c$: Current hand position
    Output: t_c: Current execution time
    $\mathbf{d}_h := \mathbf{p}_e - \mathbf{p}_s$
    $t_t := 2|\mathbf{d}_h|/v_s(r_s + 1)$
    $\mathbf{a} := -\mathbf{v}_s/t_t(1 - r_s)$
    $\mathbf{v}_s := v_s \mathbf{d}_h/|\mathbf{d}_h|$
    $\mathbf{p}_c := \mathbf{p}_s$
    $t_c := 0.0$

Procedure *Update*($\mathbf{v}_s$, t_t, r_s, $\mathbf{a}$, $\mathbf{p}_e$, $\mathbf{p}_c$, t_c, Δt)
    Input: $\mathbf{v}_s$: Start hand velocity
    Input: t_t: Total time
    Input: r_s: Deceleration ratio
    Input: a: Deceleration
    Input: $\mathbf{p}_e$: Target hand position
    Input: $\mathbf{p}_c$: Current hand position
    Input: t_c: Current execution time
    Input: Δt: Time step per frame
    Output: $\mathbf{v}_c$: Updated hand velocity
    Output: $\mathbf{p}_c$: Updated hand position
    Output: t_c: Updated execution time
    if $|\mathbf{p}_e - \mathbf{p}_c| < \varepsilon$ **then stop**
    if $t_c < t_t r_s$ **then** $\mathbf{v}_c := \mathbf{v}_s$
    else $\mathbf{v}_c := \mathbf{v}_s + \mathbf{a}(t_c - t_t r_s)$
    $\mathbf{p}_c := \mathbf{p}_c + \mathbf{v}_c \Delta t$
    $t_c := t_c + \Delta t$

The twist is a rotation around the z axis of the joint frame (Figure 1(c)). For the body to twist by an angle θ, the twist angle is decomposed among the three spine joints as follows:

$$\theta = \theta_s + \theta_l + \theta_t, \tag{5}$$

where θ_s is the twist angle of the sacral (root) joint, and θ_l and θ_t are the twist angles for the lumbar and thoracic joints, respectively. The twist angle θ is uniformly distributed among the three spine joints by setting $\theta_s = \theta_l = \theta_t$.

A *comfortable reaching angle* is defined for deciding the total twist angle, shown in Figure 1(d). The current reaching angle is the angle between the x axis and $\mathbf{d}_p$, the projection of the reaching vector $\mathbf{d}$ on the x-z plane of the shoulder frame, where $\mathbf{d} = \mathbf{p}_e - \mathbf{p}_{sh}$ is the vector from the shoulder to target, and

$$\mathbf{d}_p = \frac{[d_x, 0, d_z]^T}{\sqrt{d_x^2 + d_z^2}}. \tag{6}$$

A search process determines the total twist angle for which the target falls within the *comfortable reaching angle*. This angle is further decomposed into rotations of each spine joint (Figure 1(c)). For the whole spine to twist by an angle θ, each of the three lower spine joints need to twist by an angle $\theta/3$.

The swing is a rotation around an axis that lies in the x-y plane of the joint frame (Figure 2(a)). The swing axis is taken to be the one that most efficiently lowers or raises the upper body towards the target for the arm to reach the target. A plane is defined by $\mathbf{d}$ and the normal $\mathbf{y}$ of the ground plane. The swing axis $\mathbf{s}$ is defined as the normal of this plane (Figure 2(b)), which ensures the efficient movement of the upper body in the vertical direction:

$$\mathbf{s} = \mathbf{d} \times \mathbf{y}. \tag{7}$$

A *start swing distance* is defined for determining the swing angle of each joint. If the current vertical distance between the shoulder and the target is larger than

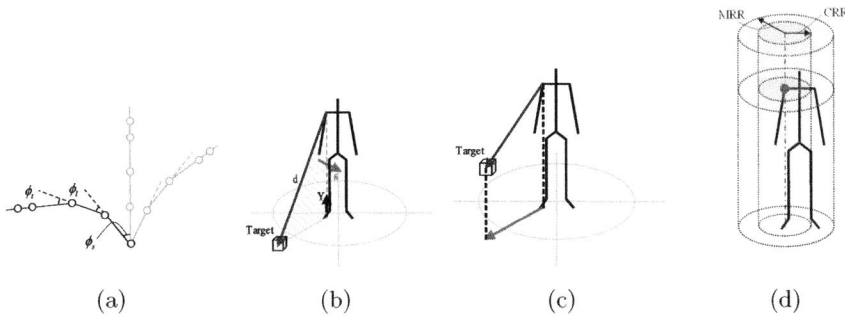

Fig. 2. (a) Swing rotation. (b) Swing axis. (c) Stepping direction (red) as the projection of arm reaching direction (blue) to ground. (d) Convenient reaching range CRR (inner cylinder) and maximum reaching range MRR (outer cylinder) of the character.

the start swing distance, the spine will swing. The swing angle ϕ is calculated by first determining the direction of the swing (clockwise or anti-clockwise) and then the three swing angles are iteratively increased until the hand is able to reach the target (Figure 2(a)).

Center of Mass Adjustment. A shift of the center of mass (COM) is observable during spine motion. Consider a reaching motion when bending the upper body down to pick up an object. In order to balance the character, the hip must move backwards in order to maintain the COM within the support polygon of the character. Similarly when reaching a high target, the hip must move forward to maintain balance.

In our framework, we adjust the position of the sacral joint, which determines the root of the spine, to generate visually appealing balancing motion during spine swing. We do not adopt kinetic methods like [11], thus achieving fast performance in spine motion synthesis. The root displacement is computed as a function of the swing angle ϕ and $\mathbf{d}_p$, projecting reach direction $\mathbf{d}$ on the x-z plane:

$$\mathbf{r} = -f(\phi)\mathbf{d}_p, \tag{8}$$

where we choose $f(\phi) = k\phi$. The constant $k = |\mathbf{r}_{\max}|/\phi_{\max}$ is the ratio of the maximal root displacement to the maximum swing angle. In our implementation, we set $\phi_{\max} = 120$ deg and $|\mathbf{r}_{\max}| = 0.1$ m.

Head Control. Intentional head motion through the control of the cervical joint of the spine can improve the liveliness of a character. The head rotates to face the target in order to help the eyes track the target. We model the head motion in two phases—a start phase and a tracking phase. During the start phase, if the head is not facing the target, the orientation of the head gradually adjusts until it aligns with the vector $\mathbf{d}_t = \mathbf{p}_e - \mathbf{p}_n$, where $\mathbf{p}_e$ is the position of the target and $\mathbf{p}_n$ is the position of the cervical joint. Given the start orientation and the target direction, the intermediate rotations are calculated by spherical linear interpolation. The speed of the head rotation is open to user specification. During the tracking phase, the head continually updates its orientation in order to track the updated target direction, which will vary due to the relative displacement between the head and the target.

3.5 Leg Controller

When synthesizing reaching motion, the focus tends to be on arm control and stepping motion is often neglected. Humans frequently step in the direction of the target in order to assist their reaching action. In doing so, the target falls within the convenient reaching range; i.e., it is neither too close nor too far from the actor. Some research has considered this topic in the context of motion planning [7]. Reference [23] considered stepping as a dynamic stability problem. We address the problem as an intentional behavior, employing a leg controller that uses a probabilistic state machine to determine the stepping motion of the virtual character.

Stepping Motions. Two kinds of stepping motions can occur during the reaching process—half stepping or full stepping. Half stepping involves only one foot step forward while the other remains at its original position. Full stepping involves both feet moving forward to assist the reaching motion. One can observe that when the step length is small, people usually perform half stepping, while a large step length often compels full stepping. However, this is also subject to individual "random" factors, which is considered in our implementation.

Two parameters govern stepping motion synthesis—step length and step direction. The direction of the step is determined by $\mathbf{d}_p$, the projection of the reaching direction $\mathbf{d}$ on the x-z plane, as shown in Figure 2(c). The step length is initialized to a empirically selected default value of 0.433 m, and it is adjusted in the stepping decision phase to produce stepping variations.

A half stepping motion comprises the movement of the spine root as well as one foot end-effector. The animation literature [21] describes the trajectory of the foot as a convex arc and the trajectory of the root as a concave arc. We simplify these two arcs as piecewise linear functions. Here, the other foot is constrained to stay in its original position.

The full stepping motion control is similar to that of half stepping, but it involves the movement of both feet. After the first foot lands on the ground, the other foot follows in the same direction. Meanwhile, the root moves slowly until the second foot reaches the spot of the first foot.

Stepping Decision State Machine. A probabilistic state machine is used to determine the stepping motion of a character that assists its reaching motion. The requirement and length of a step is determined by the *maximal reaching range (MRR)* and *convenient reaching range (CRR)* of the character, as shown in Figure 2(d). These are defined as two cylinders centered at the shoulder of the reaching arm. The stepping decision state machine works as follows:

- If the target is outside the MRR of the character, the character will perform a stepping motion.
- If the target falls between the CRR and MRR of the character, the character will step with a probability $P(\frac{d-d_c}{d_{\max}-d_c})$, where $d = \sqrt{d_x^2 + d_z^2}$ is the horizontal distance of the target from the shoulder, d_c is the convenient reaching distance, and $d_{\max}$ is the maximum reaching distance.
- If the target falls within the CRR, the character will not perform a stepping motion.

If the character is to perform a stepping motion, the choice of half-stepping or full-stepping is determined based on the principle that a character is more likely to take a full step for targets farther away and will take a half step for nearby targets. The probability P_{half} that a character will take a half step is $P_{\text{half}} = P_{\max}$ when $d = d_c$ and it decreases linearly to $P_{\min}$ as d approaches d_{max}, where $d_c < d < d_{\max}$. Meanwhile, the probability of taking a full step, $P_{\text{full}} = 1 - P_{\text{half}}$.

The step length l is computed as a function of the horizontal distance of the target from the shoulder d and the maximum step length $l_{\max}$ and minimum step length $l_{\min}$ of the character, as follows:

$$l = \begin{cases} l_{\min} & \text{if } d \leq d_c; \\ \frac{l_{\max}|d-d_c|p + l_{\min}|d-d_{\max}|(1-p)}{|d-d_c|p + |d-d_{\max}|(1-p)} & \text{if } d_c < d < d_{\max}; \\ l_{\max} & \text{if } d \geq d_{\max}, \end{cases} \quad (9)$$

where p is a normally distributed random number.

Leg Flexibility. The flexibility of the legs is an interesting motion feature that manifests itself during a reaching action. In order to pick up an object from the ground, the character needs to choose between stooping or squatting to lower its upper body. One factor in producing this motion is the flexibility of the legs. If the flexibility is low, a stoop is preferable in the reaching motion, otherwise if the flexibility is high, a squat is preferred. The squat motion involves lowering the position of the root of the spine, while the stoop motion does not. A stoop range is computed as the ratio of the length of the arm and a user-defined "flexibility parameter. If the vertical distance between the shoulder and the target exceeds this range, the root is triggered to squat down in order to reach the target.

4 Motion Controller Scheduling for Coordinated Reaching Motion

The order of movement of the arm, legs, and spine often varies with each situation and for different individuals. Hence, there exist multiple valid orderings of the individual control strategies that produce realistic animation. For example, a character may first choose to minimize the distance to the target in the horizontal space (stepping motion), then twist the body to achieve a comfortable reaching

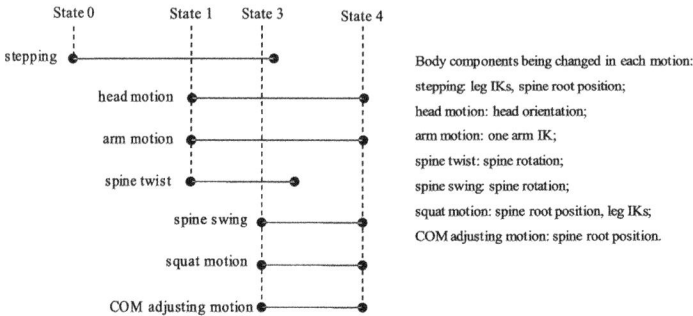

Fig. 3. Motion controller scheduling

angle (spine twist rotation), then minimize the distance in the vertical space (spine swing rotation), and finally position the hand to touch the object. This ordering of controllers may easily be changed in a different situation.

To address the non-uniqueness, we devise a motion controller scheduling strategy that selects a sequence of one or more controllers to animate a virtual character for a particular reaching task. An overview of the motion controller scheduling algorithm is presented in Figure 3. First, we identify four states in which a character may find itself during any given reaching task. In each state, one or multiple controllers are triggered to perform stepping, arm, head, or spine motion in an effort to reach the target. These states are evaluated in sequence, implicitly ordering the selection of the controllers. The state definitions and associated controller selection strategies are as follows:

- **State 0.** Check if the target is between the CRR and MRR or outside the MRR of the character. If so, trigger the stepping motion controller, which determines the step direction, step length and type of step—half step or full step; then transition to State 1.
- **State 1.** Check if the character is not performing a stepping motion; i.e., the target is within the convenient reaching range or the stepping motion has progressed to a certain time t_{st}. If so, trigger the arm and head controllers and trigger the spine twist rotation if necessary; then transition to State 2.
- **State 2.** Check if the character is not performing a spine twist motion or the twisting has progressed to a certain time t_{tw}. If so, trigger the spine swing rotation along with the COM adjusting motion. Also, trigger the squat motion if necessary; then transition to State 3.
- **State 3.** Check if the arm has reached the target or the maximum time to reach the target is up. If so, stop all the controllers. The animation is considered complete.

The values of t_{st} and t_{tw} determine the transitions between States 1 and 2 and States 2 and 3. They serve as intuitive, user-defined parameters with which to produce varied motions. The user provides values of the ratios $r_{st} = t_{st}/t_{st}^{total}$

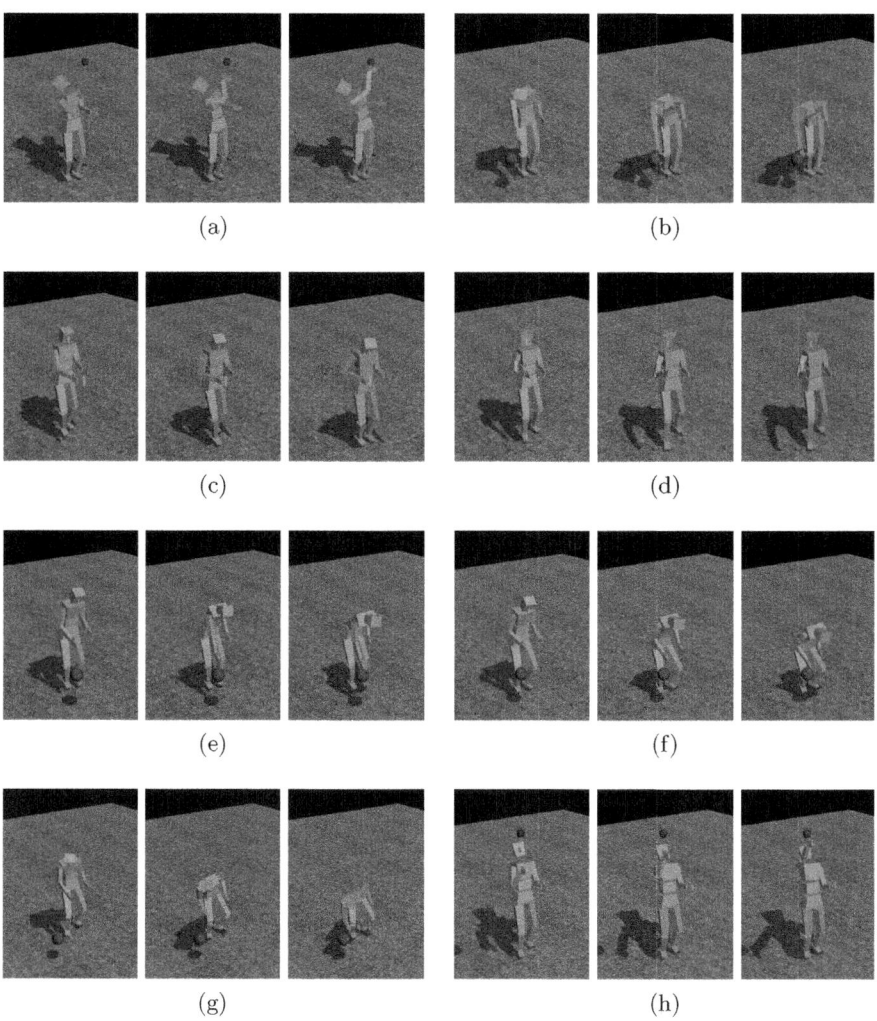

Fig. 4. A diverse set of reaching motions synthesized using our method. (a)–(b) Swing and twist spine motion: (a) The character reaches an apple to the upper-left; (b) the character reaches a basketball to the lower-right. (c)–(d) The character performs full stepping and half stepping to reach a glass. (e)–(f) The character with different leg flexibilities: (e) The character has a lower leg flexibility than in (f) as it stoops to reach a low target; (f) the character squats to reach the same low target. (g)–(h) Combined results: (g) The character steps and squats to reach a low basketball; (h) the character steps and swings up to reach a high apple.

and $r_{tw} = t_{tw}/t_{tw}^{total}$, from which t_{st} and t_{tw} are automatically computed as the points in time to transition to the next state. The scheduling algorithm can deal with most reaching tasks. For unreachable targets, the motion synthesis will stop as the hand arrives at the nearest position to the target or at time out.

5 Results

Figure 4 illustrates a diverse set of reaching motions synthesized for a full-body character using our method. For targets beyond its reach, the character chooses appropriate stepping motions and spine motions and it integrates them with head and hand motions to achieve coordinated, graceful whole-body movement. Diversity in the motion is achieved using a probabilistic model for stepping along with user defined parameters such as leg flexibility. Coordination and diversity are integrated in the motion synthesis by the modular design of the motion controllers with the systematic scheduling algorithm.

6 Conclusion and Future Work

We introduced a set of efficient motor controllers for the arms, spine, and legs of an anthropomorphic character and proposed a controller scheduling algorithm to coordinate the movements of these body parts in order to synthesize natural reaching motions. Our results demonstrate diverse reaching actions coordinated with spine and stepping motions.

More spine configurations and stepping with additional steps should be considered in future work. Like [22], motion capture data can be coupled with controllers to improve the naturalness of the synthesized motion. Moreover, we would like to extend our method to consider collision avoidance as well as to handle dynamic targets. A more intelligent control mechanism such as a planner would be required to handle space-time goals.

References

1. Faraway, J., Reed, M., Wang, J.: Modeling 3D trajectories using Bezier curves with application to hand motion. Applied Statistics 56, 571–585 (2007)
2. Gray, H.: Anatomy, Descriptive and Surgical. Gramercy, New York (1977)
3. Inoue, K., Yoshida, H., Arai, T., Mae, Y.: Mobile manipulation of humanoids: Real-time control based on manipulability and stability. In: Proc. IEEE Int. Conf. on Robotics and Automation, pp. 2217–2222 (2000)
4. Kallmann, M.: Scalable solutions for interactive virtual humans that can manipulate objects. In: Proc. 1st Conf. on Artificial Intelligence and Interactive Digital Entertainment, pp. 69–75 (2005)
5. Kallmann, M.: Analytical inverse kinematics with body posture control. Computer Animation and Virtual Worlds 19(2), 79–91 (2008)
6. Kallmann, M.: Autonomous object manipulation for virtual humans. In: SIGGRAPH 2008: ACM SIGGRAPH 2008 Courses, pp. 1–97. ACM, New York (2008)

7. Kallmann, M., Aubel, A., Abaci, T., Thalmann, D.: Planning collision-free reaching motions for interactive object manipulation and grasping. Computer Graphics Forum (Proc. Eurographics 2003) 22(3), 313–322 (2003)
8. Kallmann, M., Marsella, S.: Hierarchical motion controllers for real-time autonomous virtual humans. In: Panayiotopoulos, T., Gratch, J., Aylett, R.S., Ballin, D., Olivier, P., Rist, T. (eds.) IVA 2005. LNCS (LNAI), vol. 3661, pp. 243–265. Springer, Heidelberg (2005)
9. Kovar, L., Gleicher, M., Pighin, F.: Motion graphs. In: SIGGRAPH 2008: ACM SIGGRAPH 2008 Courses, pp. 1–10. ACM, New York (2008)
10. Kuffner Jr., J., Latombe, J.C.: Interactive manipulation planning for animated characters. In: Proc. Pacific Graphics, p. 417 (2000)
11. Kulpa, R., Multon, F.: Fast inverse kinematics and kinetics solver for human-like figures. In: 5th IEEE-RAS Int. Conf. on Humanoid Robots, pp. 38–43 (December 2005)
12. Kulpa, R., Multon, F., Arnaldi, B.: Morphology-independent representation of motions for interactive human-like animation. CG Forum 24, 343–352 (2005)
13. Lee, S.H., Sifakis, E., Terzopoulos, D.: Comprehensive biomechanical modeling and simulation of the upper body. ACM Trans. on Graphics 28(4), 99, 1–17 (2009)
14. Lee, S.H., Terzopoulos, D.: Heads up! Biomechanical modeling and neuromuscular control of the neck. ACM Transactions on Graphics 25(3), 1188–1198 (2006)
15. Liu, Y.: Interactive reach planning for animated characters using hardware acceleration. Ph.D. thesis, University of Pennsylvania, Philadelphia, PA (2003)
16. Monheit, G., Badler, N.I.: A kinematic model of the human spine and torso. IEEE Computer Graphics and Applications 11(2), 29–38 (1991)
17. Murray, R.M., Sastry, S.S., Zexiang, L.: A Mathematical Introduction to Robotic Manipulation. CRC Press, Inc., Boca Raton (1994)
18. Park, D.: Inverse kinematics,
http://sites.google.com/site/diegopark2/computergraphics
19. Shapiro, A., Kallmann, M., Faloutsos, P.: Interactive motion correction and object manipulation. In: Proc. Symp. on Int. 3D Graphics and Games, pp. 137–144 (2007)
20. Tolani, D., Goswami, A., Badler, N.I.: Real-time inverse kinematics techniques for anthropomorphic limbs. Graphical Models and Im. Proces. 62(5), 353–388 (2000)
21. Williams, R.: The Animator's Survival Kit. Faber, London (2001)
22. Yamane, K., Kuffner, J.J., Hodgins, J.K.: Synthesizing animations of human manipulation tasks. ACM Transactions on Graphics 23(3), 532–539 (2004)
23. Yoshida, E., Kanoun, O., Esteves, C., Laumond, J.P.: Task-driven support polygon humanoids. In: 6th IEEE-RAS Int. Conf. on Humanoid Robots (2006)

Pose Control in Dynamic Conditions

Brian F. Allen[1,*], Michael Neff[2], and Petros Faloutsos[1]

[1] University of California, Los Angeles
vector@cs.ucla.edu
[2] University of California, Davis

Abstract. Pose control for physically simulated characters has typically been based on proportional-derivative (PD) controllers. In this paper, we introduce a novel, analytical solution to the 2nd-order ordinary differential equation governing PD control. The analytic solution is tailored to the needs of pose control for animation, and provides significant improvement in the precision of control, particularly for simulated characters in dynamic conditions.

1 Introduction

Physical simulation holds great promise for generating realistic character animation. Over the past decade, articulated rigid body dynamics algorithms have matured from a novelty to broad adoption. However, with a few recent exceptions, physically simulated characters in games are limited to life-less rag-doll roles. One important factor limiting the adoption of physically simulated living characters is the difficulty of controlling articulated characters within an unpredictable and dynamic game environment.

For a physically animated character to engage in purposeful behaviors, appropriate torques must be applied to the character's joints to simulate the forces of activated muscle groups. Control torques must take into account the full position and desired motion of the character, along with external forces, such as gravity.

One early method proposed to simplify this control problem is pose control[9], wherein the animator specifies a set of pre-determined poses and the game animation system chooses the current pose. Animation systems can arrange the possible poses in a directed graph, with edges representing allowed transitions and runtime execution traversing this graph[7] according to user input[17] or the game's artificial intelligence (AI). Each pose specifies target positions for each controlled degree of freedom (DoF). Typically, the target velocity of joints is zero. At run-time, the character's joints apply torques to reach the target pose, with the specific amount of torque supplied computed by a low-level control law, such as the proportional-derivative (PD). Despite (or perhaps, because of) the simplicity of pose control, it is commonly used in practice and in the animation literature, where it often forms the foundation of more complex algorithms capable of impressive behaviors[7,4,12,16,11,15].

In this paper, we report a novel (to our knowledge) analytical solution for the control parameters from the second-order differential equation underlying

---
* Corresponding author.

PD control. This analytic formulation is particularly relevant to the use of PD controllers in games, where specific guarantees about character motion, despite unexpected conditions, are essential.

2 Motivation

To illustrate the difficulty arising from physical simulation in games, consider a simple, common example of a character catching a ball. We assume the need for the character to successfully catch the ball would be absolute, since it is determined by the gameplay logic. Because the character is physically simulated and pose-controlled, the game system computes an appropriate pose and the precise time at which the character must be in that pose to correctly catch the ball. But herein lies a key problem. If the character begins in a static pose (e.g., standing still) a PD-based pose controller can guarantee that the ball will be caught[1]. However, if that character is *in motion*, correct timing will be lost, the character will miss the catch, and gameplay will be unacceptably compromised.

Figure 1 shows a series of trajectories that illustrate this problem for a single joint. If the starting state is static (i.e., zero velocity) then previously proposed methods (e.g., [1]) can reach the desired pose with sufficient accuracy. However,

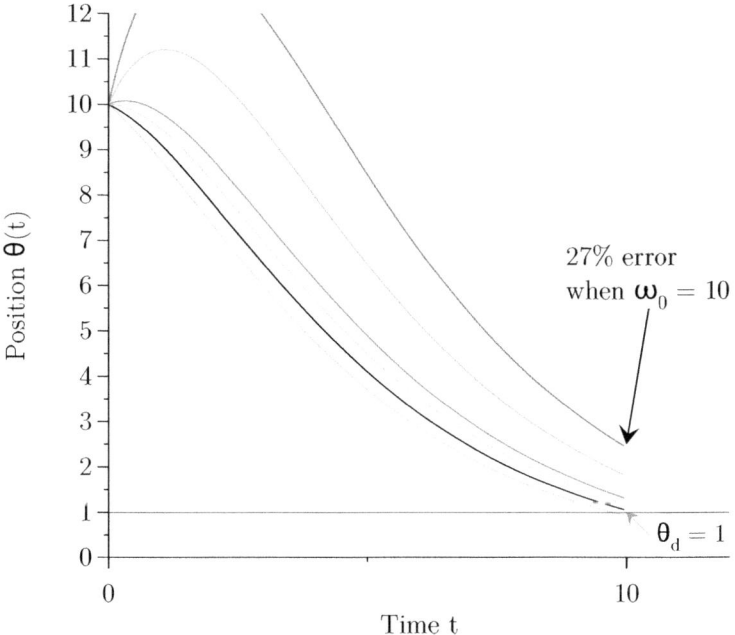

Fig. 1. Although previously proposed methods can calculate control parameters to almost exactly reach the target pose under static conditions (black trajectory, $\theta(10) = 1.045$, see [1]), they miss the desired target under dynamic conditions (i.e., non-zero initial velocities) by significant amounts

as the initial velocity diverges from zero, they can no longer reach the desired pose at the desired time– missing the catch, per the example.

One alternative approach to this example would be to compute a trajectory (such as by simple interpolation or motion planning[13]) to the catch pose and then follow that trajectory using stiffly tracking PD control. The problem here is that stiff tracking eliminates the natural response to forces and perturbations, rendering the system effectively equivalent to computationally expensive kinematic animation. Methods have been proposed to change the control parameters when a collision occurs[18], but these approaches require specific ad hoc parameters and it is unclear how such systems can deal with frequent contacts.

3 Proportional-Derivative Control Law

The PD controller for a given degree of freedom (DoF) is responsible for determining the specific amount of torque to apply at any point in time. Thus, given some current state θ, ω and a target position θ_d specified by the target pose, the PD control torque τ is computed as

$$\tau = k(\theta_d - \theta) - \gamma \omega, \qquad (1)$$

where k, γ are the parameters of the controller, stiffness and damping respectively. Different values for these parameters can result in drastically different motions, even for the same initial and target poses. The contribution of this work is to determine these parameters automatically, based only on the initial conditions and the target pose. Our approach hinges on determining an analytical expression that unique yields the necessary parameters.

In general, equation 1 has more (mathematical) degrees of freedom than the number of constraints imposed by the initial and target poses. So we begin by assuming that the desired trajectory connecting the initial and target poses neither oscillates nor takes an excessively long time. This assumption leads to a requirement that the PD controller is critically damped, which in turn specifies a quadratic relationship between the parameters, reducing the degrees of freedom and allowing the elimination of k,

$$\gamma^2 = 4\,km, \qquad (2)$$

where m is the moment of inertia about the axis of the DoF.

In general, m is a function of the mass properties of the character as well as the character's current state. As we are considering game and animation systems only, we can assume that all physical properties of the character are known. This assumption is reasonable because the character's canonical and defining representation lies in physics simulator of the game itself. Note that this is not the case for actual physical systems such as real-world robots. In those systems the physical properties must be measured or estimated, incurring unavoidable error. We take advantage of this assumption by noting that m can be computed exactly for any given state of the simulated character[5]. With this assumption, equation 1 can be written as a second-order differential equation,

$$m\frac{d^2\theta}{dt^2} + \gamma \omega + \frac{\gamma^2}{4m}\theta = k\theta_d, \qquad (3)$$

which has the homogeneous solution

$$\theta(t) = \left(\theta_0 + t\left(\omega_0 + \frac{\gamma\,\theta_0}{2m}\right)\right)e^{-\frac{\gamma t}{2m}}, \qquad (4)$$

after setting θ_d to 0 and possibly inverting θ so that $\theta_0 > 0$. Both of these transforms can be applied with no loss of generality.

In physical simulation, the current state (θ_0, ω_0) is known. The desired state (θ_d) is specified by the current target pose, and the time t_d at which the character should reach that pose is assumed to be known. Given these constraints, the final unknown parameter γ is fully constrained. That is, if a value for γ can be determined that satisfies equation 4, then that value gives a PD controller that will exactly reach the desired pose at the desired time. Unfortunately, it is known that equation 4 has no closed-form expression for γ[10], as the equation is equivalent to a linear differential-algebraic equation (DAE) with and index of one[2]. This property complicates the numerical approach, since the more efficient shooting methods (such as linear) do not apply directly.

3.1 Analytic Solution

Although no algebraic closed-form expression for γ in equation 4 exists, in this section we present an analytic expression for γ such that constraints on both the initial position and the desired position are satisfied exactly. To our knowledge, this result is novel.

The key insight allowing the solution is the rearrangement of equation 4 into the form $y = xe^x$. Expressions of this form can be manipulated using the Ω map, also called the Lambert-W function[3], defined as $x = \Omega(x)e^{\Omega(x)}$ and illustrated in figure 2. As is clear from the definition, the Ω function can be used directly to solve equations of the form $y = xe^x$ for x.

Although Ω has no closed-form expression, it can be represented analytically and is simple and efficient to compute in practice[6]. $\Omega(\cdot)$ is a multi-valued map, but the principle branch, i.e., the branch passing through the origin (shown as the solid blue line in figure 2), can be expressed as

$$\Omega_0(x) = \sum_{n=1}^{\infty} \frac{(-n)^{n-1}}{n!} x^n. \qquad (5)$$

For $x \geq -1/e, x \in \Re$, $\Omega(x)$ has at least one real solution, and for $x > 0$, $\Omega(x)$ is unique.

Using the Ω map, γ can be expressed explicitly in terms of the initial and target poses,

$$\gamma = -\frac{2m}{t_d\theta_0}\left(t_d\omega_0 + \theta_0\left(1 + \Omega\left(-\frac{\theta_d}{\theta_0}e^{-1-\frac{t_d\omega_0}{\theta_0}}\right)\right)\right). \qquad (6)$$

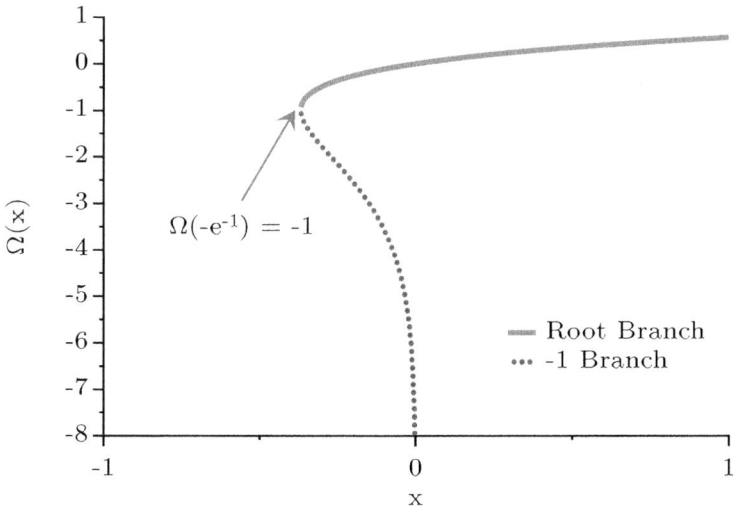

Fig. 2. The Ω map (also called the Lambert-W function) showing the values of x satisfying $x = \Omega(x)e^{\Omega(x)}$. The map is composed of two branches, the root branch Ω_0 shown in solid blue, and the Ω_{-1} branch in red dotted line.

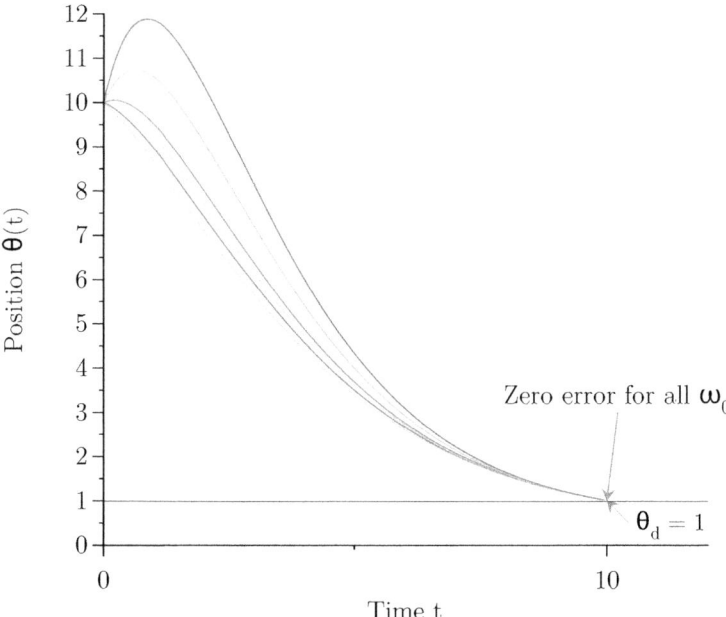

Fig. 3. The analytically computed control parameter γ provides a critically damped trajectory that exactly satisfies the target position. This graph shows the same conditions, including initial velocities, as figure 1. Note that the resulting trajectories do differ naturally under physical control according to the differences in initial state. However, in each case, the joint reaches the target position exactly at the desired time.

This expression provides the PD control parameter γ (and k by the critically damped assumption) that will yield a trajectory that exactly satisfies the initial conditions and the final position constraint $\theta(t) = \theta_d$. Figure 3 shows several example trajectories generated by varying the initial velocity but keeping the same target position and target time, $\theta(10) = 1$.

3.2 Existence of γ

Since Ω has a domain of $\leq -1/e$, γ exists (has at least one real solution) iff

$$\theta_d \leq \theta_0 e^{t_d \omega_0 / \theta_0} \tag{7}$$

with $t_d > 0$. Further, γ is unique iff

$$-(\theta_d/\theta_0)\exp(-1 - \frac{t_d \omega_0}{\theta_0}) > 0, \tag{8}$$

or equivalently when $\theta_d < 0$, since we have already assumed that $\theta_0 > 0$.

4 Results

To validate the proposed method's utility for reliable control, we simulate a simple game character with the task of blocking incoming balls. Target poses and the required timing are determined directly from the trajectory of the ball; if the target pose is reached at the required time, the ball will be successfully blocked.

The simulation is implemented using the Open Dynamics Engine[14]. The control torque is computed using equation 1 and applied directly to the shoulder joint. The PD control parameters are calculated from equation 6 and required to be critically damped, as per equation 2. The time step is 0.005 s and gravity simulated at 9.8 m/s^2.

Figure 5 shows the trajectory of motion used to block an incoming ball, corresponding to the trajectory in figure 5(e). The small circles represent the projected time and position incoming balls will cross the plain of the arm, thus defining the target pose. In comparison, a traditional PD controller with fixed control parameters is hand-tuned to correctly block the first incoming ball, but then fails to reach the later poses on-time. Figure 4 illustrates the trajectory followed using hand-tuned controllers.

To illustrate the application of the proposed method to games, dynamic disturbances are added. A series of perturbation weights drop onto the arm from a random position approximately 1.5 m above the shoulder joint. Each perturbation ball weighs 10 kg. Figure 6 illustrates the ability of the analytic solution to correctly adapt the stiffness of control as needed. The controlled arm is able to both achieve the target pose at the correct time, as illustrated by correctly blocking the incoming ball, and to allow a natural response to the perturbing weights falling from above. The arm moves with precisely the stiffness needed to reach the target pose on time.

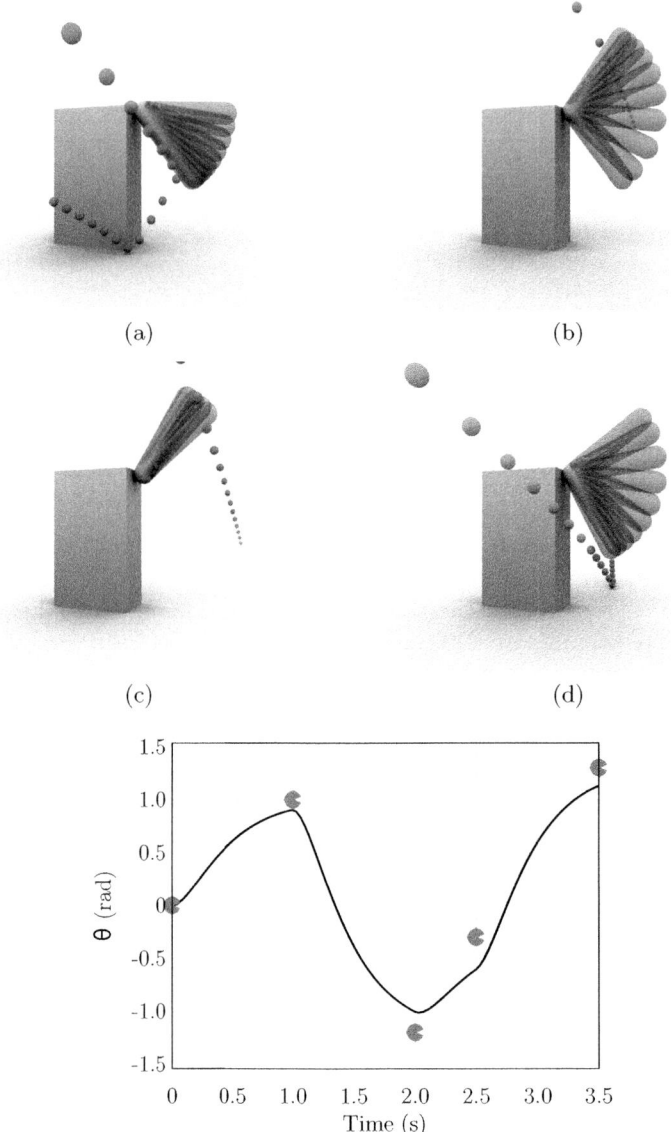

(a)　　　　　　　　　　(b)

(c)　　　　　　　　　　(d)

(e) The complete four-target pose trajectory with red circles indicating the target pose.

Fig. 4. A standard PD controller is manually tuned to block the first ball (4(a)), but misses subsequent poses (4(b), 4(c), 4(d)). Each image shows the time from the previous pose to the current time, illustrated with the stroboscopic effect.

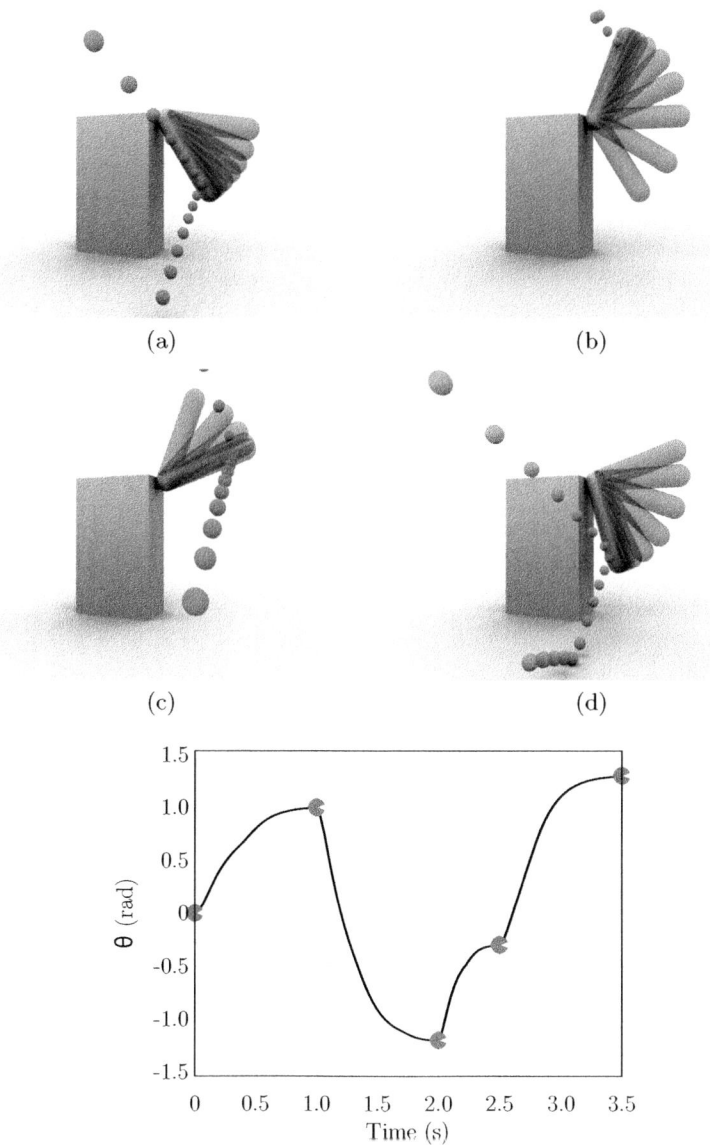

(e) The complete four-target pose trajectory with red circles indicating the target pose.

Fig. 5. The analytic solution provides the necessary PD control parameters to ensure the target poses are reached, enabling the physically controlled arm to block all four balls. Each image shows the time from the previous pose to the current time, illustrated with the stroboscopic effect.

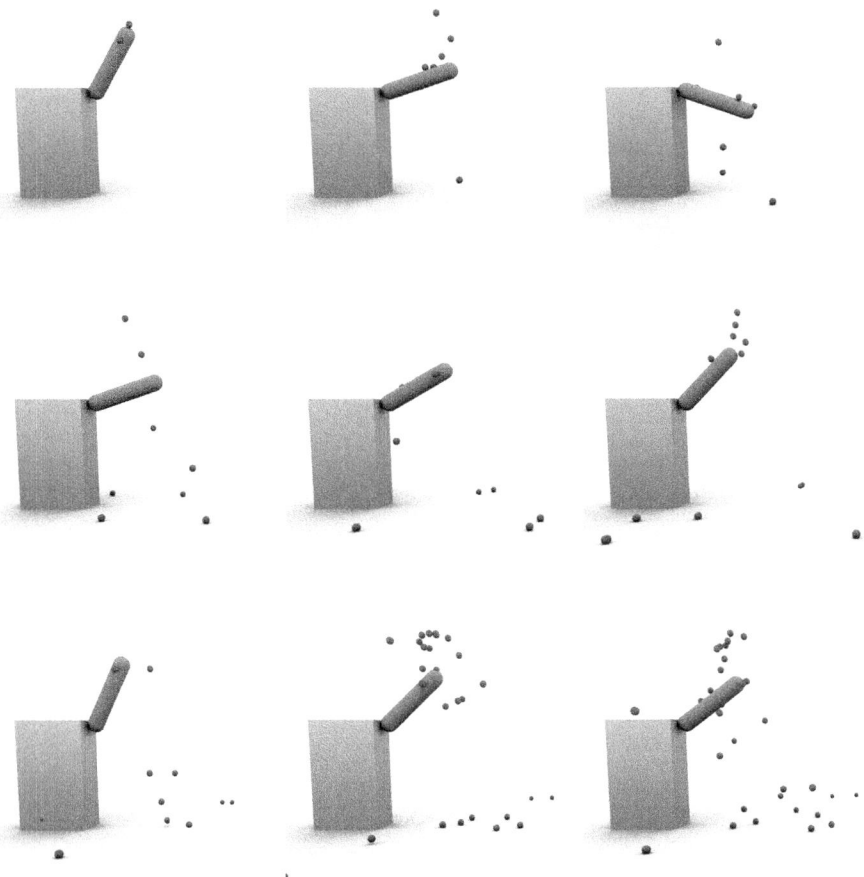

Fig. 6. A sequence of frames (read left-to-right, top-to-bottom) showing PD control under perturbation. Each falling ball is 10 kg.

5 Limitations

As one can see visually in figure 2, the domain of the Ω map does not include all reals. Therefore, some combinations of initial state and target position simply cannot be attained by a critically damped PD controller, resulting in no real solution for equation 6. This represents a fundamental limitation on all critically damped PD pose controllers, and the analytic solution at least makes these cases explicit, as in section 3.2.

Another limitation to this approach is the requirement for critically damped motion. This is a useful simplifying assumption that allows the analytical solution, but human motion is not always critically damped. Some natural human motions, such as arms swinging loosely at one's side while walking, are significantly underdamped, and thus not well suited to control by the proposed method. However, the types of motions that are likely to require precise timing within a game are the direct, intentional motions (such as reaching out to catch a ball) that are also those most nearly critically damped.

Our formulation of PD control assumes a zero target velocity. This fits naturally within the context of pose control, where motion comes to a halt as the target pose is reached. Of course, different applications would prefer control over the final velocity. Unfortunately, final velocity control is not possible within standard critically damped PD control, since equation 6 is already fully specified.

Finally, throughout we consider only a single degree of freedom. The application of precise PD control to articulated characters has been considered previously[1,8,18], and is not addressed here since the proposed method is fully compatible with any existing use of PD control where the moment of inertia can be computed.

6 Conclusion

Proportional-derivative-based pose control provides a practical means for game developers to bring motion and life to physically simulated characters in their games. This paper presents a novel, analytic solution to the differential equation governing PD control. This solution builds on the recent mathematical and numerical development of the Ω map, and as such might prove useful as a guide to solving related differential equations. Further, because of the prevalence of PD control in a variety of fields, the specific solution presented may be of value beyond games and animation.

With the wide-spread use of PD control in physically simulated games today, we believe our result to be a simple, practical improvement. Equation 6 is computationally simple and is easily added to existing pose controllers. The resulting controller provides an exact, analytic solution to pose control.

We believe such precision is important for games, where the outcome of physical simulation must foremost support the demands of gameplay. For physically simulated animation to be used for primary characters in games, precision and reproducibility must be guaranteed. In terms of the illustrative example in section 2, the ball can be caught precisely without abandoning the immersive advantages of physical simulation.

In the future, we hope to relax the requirement of critically damped control to support different damping ratios, and also we hope to address the need to reach specific target velocities, moving beyond the targeting of fixed poses.

References

1. Allen, B., Chu, D., Shapiro, A., Faloutsos, P.: On the beat!: Timing and tension for dynamic characters. In: Proceedings of the 2007 ACM SIGGRAPH/Eurographics Symposium on Computer Animation, Eurographics Association, p. 247 (2007)

2. Brenan, K., Campbell, S., Campbell, S., Petzold, L.: Numerical solution of initial-value problems in differential-algebraic equations. Society for Industrial Mathematics (1996)
3. Corless, R., Gonnet, G., Hare, D., Jeffrey, D., Knuth, D.: On the LambertW function. Advances in Computational Mathematics 5(1), 329–359 (1996)
4. Faloutsos, P., van de Panne, M., Terzopoulos, D.: Composable controllers for physics-based character animation. In: SIGGRAPH 2001: Proceedings of the 28th Annual Conference on Computer Graphics and Interactive Techniques, pp. 251–260. ACM, New York (2001)
5. Featherstone, R.: Rigid body dynamics algorithms. Springer, New York (2008)
6. Free Software Foundation (FSF): GNU scientific library
7. Hodgins, J., Wooten, W., Brogan, D., O'Brien, J.: Animating human athletics. In: Proceedings of the 22nd Annual Conference on Computer Graphics and Interactive Techniques, p. 78. ACM, New York (1995)
8. Notman, G., Carlisle, P., Jackson, R.: Context based variation of character animation by physical simulation. In: Games Computing and Creative Technologies: Conference Papers (Peer-Reviewed), p. 2 (2008)
9. Van de Panne, M., Kim, R., Fiume, E.: Virtual wind-up toys for animation. In: Graphics Interface, pp. 208–208 (1994)
10. Sanchez, D.: Ordinary Differential Equations and Stability: Theory: An Introduction. WH Freeman and Company, San Francisco (1968)
11. Shapiro, A., Chu, D., Allen, B., Faloutsos, P.: A dynamic controller toolkit. In: Sandbox 2007: Proceedings of the 2007 ACM SIGGRAPH Symposium on Video Games, pp. 15–20. ACM, New York (2007)
12. Shapiro, A., Faloutsos, P.: Interactive and reactive dynamic control. In: SIGGRAPH 2005: ACM SIGGRAPH 2005 Sketches, p. 26. ACM, New York (2005)
13. Shapiro, A., Kallmann, M., Faloutsos, P.: Interactive motion correction and object manipulation. In: ACM SIGGRAPH Symposium on Interactive 3D Graphics and Games (I3D 2007), Seattle, April 30 - May 2 (2007)
14. Smith, R.: Open dynamics engine, http://ode.org
15. Wang, J., Fleet, D., Hertzmann, A.: Optimizing Walking Controllers. ACM Transactions on Graphics, SIGGRAPH Asia (2009)
16. Yin, K., Loken, K., van de Panne, M.: Simbicon: simple biped locomotion control. In: SIGGRAPH 2007: ACM SIGGRAPH 2007 Papers, p. 105. ACM, New York (2007)
17. Zhao, P., van de Panne, M.: User interfaces for interactive control of physics-based 3d characters. In: Proceedings of the 2005 Symposium on Interactive 3D Graphics and Games, p. 94. ACM, New York (2005)
18. Zordan, V., Hodgins, J.: Motion capture-driven simulations that hit and react. In: Proceedings of the 2002 ACM SIGGRAPH/Eurographics Symposium on Computer Animation, pp. 89–96. ACM, New York (2002)

Spatial Awareness in Full-Body Immersive Interactions: Where Do We Stand?

Ronan Boulic[1], Damien Maupu[1], Manuel Peinado[2], and Daniel Raunhardt[3]

[1] VRLAB, Ecole Polytechnique Fédérale de Lausanne, station 14, 1015 Lausanne, Switzerland
[2] Universidad de Alcalá, Departamento de Automática, Spain
[3] BBV Software Service AG, Zug, Switzerland
{Ronan.Boulic,Damien.Maupu}@epfl.ch,
{Manuel.Peinado,Daniel.Raunhardt}@gmail.com

Abstract. We are interested in developing real-time applications such as games or virtual prototyping that take advantage of the user full-body input to control a wide range of entities, from a self-similar avatar to any type of animated characters, including virtual humanoids with differing size and proportions. The key issue is, as always in real-time interactions, to identify the key factors that should get computational resources for ensuring the best user interaction efficiency. For this reason we first recall the definition and scope of such essential terms as immersion and presence, while clarifying the confusion existing in the fields of Virtual Reality and Games. This is done in conjunction with a short literature survey relating our interaction efficiency goal to key inspirations and findings from the field of Action Neuroscience. We then briefly describe our full-body real-time postural control with proactive local collision avoidance. The concept of obstacle spherification is introduced both to reduce local minima and to decrease the user cognitive task while interacting in complex environments. Finally we stress the interest of the egocentric environment scaling so that the user egocentric space matches the one of a height-differing controlled avatar.

Keywords: Spatial awareness, immersion, presence, collision avoidance.

1 Introduction

Full-body interactions have a long history in Virtual Reality and start to have significant commercial successes with dedicated products. However exploiting the full-body movements is still far from achieving its full potential. Current spatial interactions are limited to the control of severely restricted gesture spaces with limited interactions with the environment (e.g. the ball of a racquet game). In most games the postural correspondence of the player with the corresponding avatar posture generally doesn't matter as long as the involvement is ensured and preserved over the whole game duration. We are interested in achieving two goals: 1) increasing the user spatial awareness in complex environment through a closer correspondence of the user and the avatar postures, and 2) impersonating potentially widely different entities ranging from a self-similar avatar to any type of animated characters, including virtual humanoids with differing size and proportions. These two goals may sound contradictory when the animated character differs markedly from the user but we are convinced it is a definitely useful long term goal to identify the boundary conditions of distorted self-avatar acceptance. For the time being, the core issue remains, as always in real-time interactions, to identify the key factors that should get computational resources for ensuring the best interaction efficiency, either as a gamer or as an engineer evaluating a virtual prototype for a large population of future users. Within this frame of mind it is important to recall the definition and scope of such essential terms as *immersion* and *presence*, while clarifying the confusion existing in the fields of Virtual Reality and Games. This is addressed in the first part of section 2 in conjunction with a literature survey relating our interaction efficiency goal to key inspirations and findings from the field of Action Neuroscience. The second part of this background section deals with the handling of collision avoidance in real-time interactions. Section 3 then briefly describe our full-body real-time postural control with proactive local collision avoidance. The concept of obstacle spherification is introduced both to reduce local minima and to decrease the user cognitive task while interacting in complex environments. Finally we stress the interest of the egocentric environment scaling so that the user egocentric space matches the one of a height-differing controlled avatar.

2 Background

Spatial awareness in our context is the ability to infer one's interaction potential in a complex environment from one's continuous sensorimotor assessment of the surrounding virtual environment. It is only one aspect of feedback and awareness considered necessary to maintain fluent collaborations in virtual environments [GMMG08]. We first address terminology issues in conjunction with a brief historical perspective. We then recall the key references about handling collision avoidance for human or humanoid robot interactions.

2.1 What Action Neuroscience Tells Us about Immersive Interactions

The dualism of mind and body from Descartes is long gone but no single alternative theory is yet able to replace it as an explanatory framework integrating both human perception and action into a coherent whole. Among the numerous contributors to alternate views to this problem, Heidegger is often acknowledged as the first to formalize the field of embodied interactions [D01] [MS99] [ZJ98] through the neologisms of *ready-to-hand* and *present-at-hand* [H27]. Both were illustrated through the use of a hammer that can be ready-to-hand when used in a standard task; in such a case it recedes from awareness as if it became part of the user's body. The hammer can be present-at-hand when it becomes unusable and has to be examined to be fixed. Most of human activities are spent according to the readiness-to-hand mode similar to a subconscious autopilot mode. On the other hand human creativity emerges through the periods in present-at-hand mode where problems have to be faced and solved.

We are inclined to see a nice generalization of these two modes in the work of the psychologist Csikszenmihalyi who studied autotelic behaviors, i.e. self-motivational activities, of people who showed to be deeply involved in a complex activity without direct rewards [NvDR08][S04]. Csikszenmihalyi asserted that autotelicity arises from a subtle balance between the exertion of available skills and addressing new challenges. He called *flow* the strong form of enjoyment resulting from the performance of an autotelic activity where one easily looses the sense of time and of oneself [S04]. This term is now in common use in the field of game design, or even teaching, together with the term of involvement. Both terms are most likely associated to the content of the interactive experience (e.g. the game design) rather than the form of the interaction (the sensory output which is generic across a wide range of game designs).

The different logical levels of content and form have not been consistently used in the literature about interactive virtual environments, hence generating some confusion about the use of the word *presence*, (e.g. in [R03][R06]). As clearly stated by Slater, presence has nothing to do with the content of an interactive experience but only with its form [S03]. It qualifies the extent to which the simulated sensory data convey the feeling *of being there* even if cognitively one knows not to be in a real life situation [S03]. As Slater puts it, a virtual environment system can induce a high presence, but one may find the designed interaction plain boring. Recently Slater has opted for using the expression *Place Illusion* (PI) in lieu of *presence* due to the existing confusion described above [RSSS09]. While PI refers to the static aspects of the virtual environment, the additional term of *Plausibility* (Psi) refers to its dynamic aspects [RSSS09]. Both constitute the core of a new evaluation methodology presented in [SSC10].

A complementary view to explaining presence/PI is also grounded on the approach from Heidegger and the more recent body of work from Gibson [FH98] [ZJ98] [ON01] both characterized by the ability *to 'do' there* [SS05]. However it is useful to recall the experimental findings from Jeannerod [J09] that questions the validity of the Gibsonian approach [ON01]. The experiment is based on the display of point-light movement sequences inspired by the original study of Gunnar Johansson on the perception of human movement [J73]. The point light sequences belong to three families: movement without/with human character, and for this latter class, without meaning (sign language) or known meaning (pantomime) for the subjects. Movements without human characters appear to be processed only in the visual cortex, whereas those with human character are treated in two distinct regions depending on whether they are known (ventral "semantic" stream) or unknown (dorsal "pragmatic" [J97] goal-directed stream). One particularly interesting case is that an accelerated human movement loses its human character and is processed only in the visual cortex. These findings have the following consequences for our field of immersive interactions when interacting with virtual humans. First it is crucial to respect the natural dynamics of human entities when animating virtual humans otherwise it may be disregarded as human altogether. Second, humans beings have internal models of human actions that are activated not only when they perform it themselves but also when they see somebody else performing it. It is hence reasonable to believe that the viewing of virtual human reproducing good quality movement activate the corresponding internal model of immersed subjects. As a final remark, viewed known and unknown human movements are treated along different neural streams, one of which might be more difficult to verbalize a posteriori in a questionnaire as it has no semantic information associated with it.

As suggested above it can be more difficult to assess presence through the usual means of questionnaires as actions performed in this mode are not performed at a conscious level. In general, alternative to questionnaires have to be devised, such as the comparison with the outcome that would occur if performed in a real world setting. Physiological measurements are particularly pertinent as Paccalin and Jeannerod report that simply viewing someone performing an action with efforts induce heart and breathe rate variations [PJ00].

Simultaneously introduced with presence, the concept of *immersion* refers to the objective level of sensory fidelity a Virtual Reality system provides [SS05]. For example, the level of visual immersion depends only on the system's rendering software and display technology [BM07]. Immersion is a necessary condition for enforcing spatial awareness. Bowman et al have chosen to study the level of visual immersion on application effectiveness by combining various immersion components such as field of view, field of regard (total size of the visual field surrounding the user), display size, display resolution, stereoscopy, frame rate, etc [BM07].

2.2 Handling Collisions for Full-Body Avatar Control

Numerous approaches have been proposed for the on-line full-body motion capture of the user, we refer the interested reader to [A] [MBRT99] [TGB00] [PMMRTB09] [UPBS08]. A method based on normalized movement allows to retarget the movement on a large variety of human-like avatars [PMLGKD08][PMLGKD09].

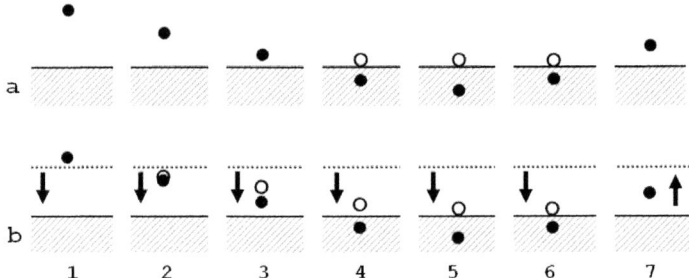

Fig. 1. (a) In the rubber band method, the avatar's body part (white dot) remains tangent to the obstacle surface while the real body part collides (black dot). (b) In the damping method [PMMRTB09], whenever a real body part enters the influence area of an obstacle (frame 2), the avatar displacement is progressively damped (2-3) to ensure that no interpenetration happens (4-6). No damping is exerted when moving away from the obstacle (7).

Spatial awareness includes the proper handling of the avatar collisions, including self-collisions [ZB94]. In fact the control of a believable avatar should avoid collision proactively rather than reactively; this reflects the behavior observed in monkeys [B00]. In case an effective collision happens between the user and the virtual environment the standard approach is to repel the avatar hence inducing a collocation error; however such an error is less disturbing than the visual sink-in that would occur otherwise [BH97] [BRWMPB06] (hand only) [PMMRTB09] (full body); both of these approaches are based on the rubberband method (Fig. 1a) except that the second one has an additional damping region surrounding selected body parts to be able to enforce the proactive damping of segments' movements towards obstacles (Fig. 1b, Fig. 2). The original posture variation damping has been proposed in [FT87] and extended in a multiple priorities IK architecture in [PMMRTB09].

Fig. 2. Selective damping of the arm movement component towards the obstacles; a line is displayed between a segment and an obstacle [PMMRTB09]

3 Smoothing Collision Avoidance

The damping scheme recalled in the previous section is detailed in Fig. 3 in the simplified context of a point (called an *observer*) moving towards a planar obstacle. Only the relative displacement component along the normal is damped. Such an approach may produce a local minima in on-line full-body interaction whenever an obstacle lies between a tracked optical marker and the tracking avatar segment (equivalent to the point from Fig. 3). In such a case the segment attraction is counter-balanced by the damping as visible on Fig. 4 top row.

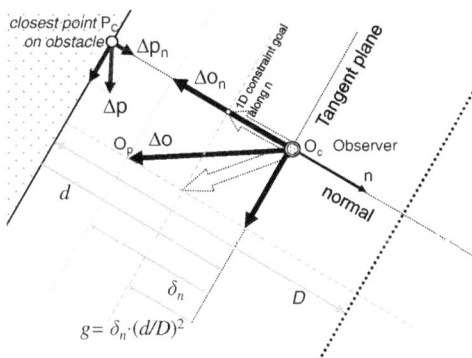

Fig. 3. Damping in the influence area of a planar obstacle for a point-shaped observer with a relative movement towards the obstacle. The relative normal displacement δ_n is damped by a factor $(d/D)^2$.

The present section proposes a simple and continuous alteration of obstacle normals so that the obstacle appears from far as if it were a sphere whereas it progressively reveals its proper normals when the controlled segment is getting closer to its surface. Obstacle shapes offering large flat surfaces are ideal candidates for the spherification as we call it (see Fig. 4 bottom row)). The concept is simple to put into practice. Instead of using the normal to the obstacle n, we replace it by n_f, a combination of n and a "spherical" normal n_s which results from taking a vector from the obstacle centroid (see Fig. 4 bottom row). We have :

$$n_f = \text{NORMALIZE}((1-k)n + k\ n_s) \quad (1)$$

$$\text{with} \quad k = \left(\frac{d}{D}\right)^{1/m}. \quad (2)$$

In Eq. 1, k is a spherification factor which ranges from 0 at the obstacle's surface, to 1 at the boundary of its influence area (d=D). The rationale for this factor is that we can "spherify" a lot when we are far away from the obstacle (k=1), but not when close to it (k<<1) because using an altered normal could lead to a collision.

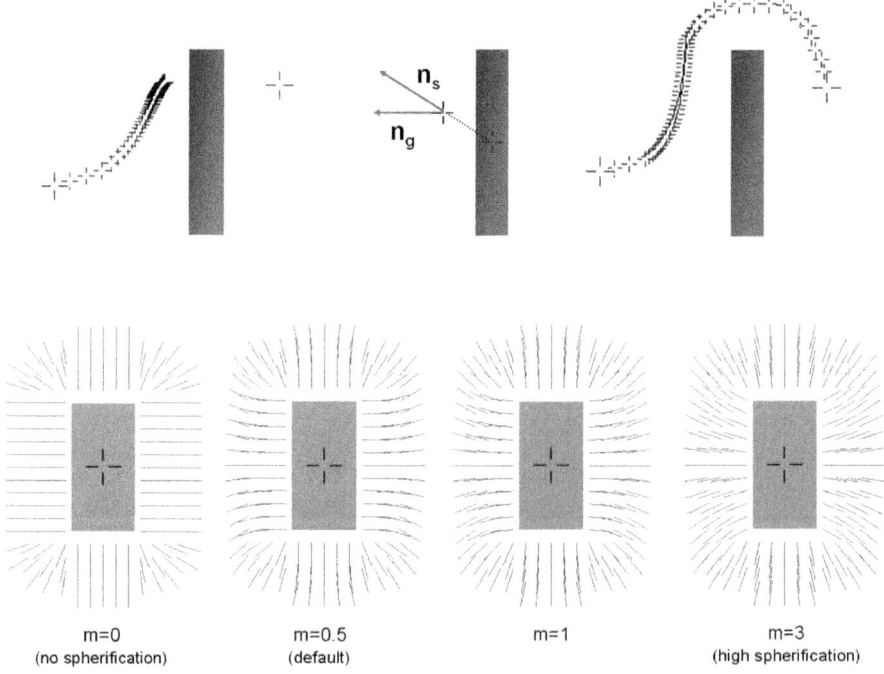

Fig. 4. (top row) local minima due to a balance of the damping with the attraction towards a goal (left), the two contributing vectors to the spherification (middle), the resulting behavior (right), (bottom row) various degrees of spherification from null (left) to hight (right)

Finally, the degree of spherification m is a user-defined parameter which can be increased to provide more aggressive spherifications that help solve difficult scenarios. Fig. 4 (bottom row) shows the normals generated by this method for a box-shaped obstacle and different values of m. Fig. 6 (top row) shows how a scenario with blockage is solved in a smooth manner thanks to the spherification of the obstacle. It is still possible to devise situations where spherification alone cannot prevent blockage but we deem the technique worthy because it offers a significant improvement in many common situations, at virtually no computation cost.

This technique has been integrated in the combined IK postural control and collision avoidance presented in [PMMRTB09]. Additional illustrations of on-line use can be seen in Fig. 6 but it is even more interesting for off-line simulation of complex reach tasks as described in [PBRM07] and briefly recalled now. The virtual human being totally autonomous in such a context, a deliberative software layer is required to avoid the local minima that are not handled by the spherification. For that purpose we introduced the *task-surface heuristic* that is used to check whether an obstacle initially lies within the triangle set built from the effector location, its reach goal and all parent joint locations (Fig. 5a,b). If such a collision check is positive, a set of candidate paths is built as described in [ABD98] and the best path is selected according to a fitness criteria dedicated to reach tasks and optimizing the shoulder joint mobility [PBRM07]. Fig. 5c highlights the selected path and successive postures when performing it; please note that the torso and the arm segments benefit from the proactive smooth collision avoidance with respect to the box.

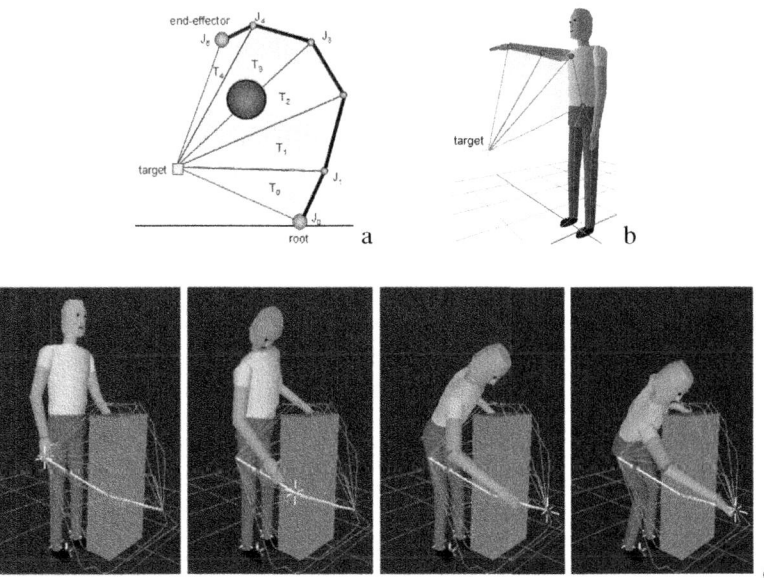

Fig. 5. (a) Triangle task heuristic for a simple chain (a), or a virtual human (b). Selected optimal path with respect to the shoulder mobility (c) [PBRM07].

Fig. 6. Illustration of the Embodiment and the Awareness issues; how should we control a height-differing avatar while automatically managing collision avoidance. (a) immersive display with active marker motion capture system, (b) control of a self-height avatar, (c) control the child avatar with the visuocentic strategy, (d) egocentric scaling strategy: the virtual environment is scaled as experienced by the child avatar [BMT09].

4 Spatial Awareness with Height-Differing Avatars

Spatial awareness is also strongly related with the relative height of the target avatar. The issue we examine in this section can be formulated as follow: how should we render the virtual environment to best convey the experience of an avatar of differing body height than the user? We consider this problem with a third person viewpoint setup as in Fig. 6a. Our motivation is to benefit from the large field of view provided by the 3m x 2,3m screen. This figure illustrates an extreme case of an adult wishing to interact in a virtual kitchen as if he were as tall as a child (Fig. 6c).

We have conducted an experiment in a simpler setup than Fig. 6 (see details in [BMT09]) for a range of reach tasks with three scaling strategies:

Reference strategy: control of a same-height avatar (display analog to Fig. 6a,b). Both the visual display and the postural sensor data (active optical markers visible in Fig. 6.a) are respectively presented and exploited at scale 1/1. Its purpose is to calibrate the reaching duration as a function of target heights.

Visuocentric scaling strategy: we could call it the *puppeteer* strategy as the user sensor data are scaled by the body heights ratio to match the body height of the avatar (display analog to Fig. 6c). The "puppeteer" subject has to rely heavily on the visual feedback to guide the avatar in the reach tasks.

Egocentric scaling strategy: This strategy does not change the sensor data but instead scales the displayed environment with the inverse ratio of body heights. This ensures an exact correspondence of the user egocentric space and the one of the controlled avatar. Hence the user *becomes* the puppet in this strategy as opposed to the puppeteer strategy.

The [BMT09] study quantified the reaching response duration as a function of a set of target heights, from very low to above the head but always reachable. The targets were either in free space or on a shelf, and the subject was controlling either a simple solid collocated with the hand (baseline) or a full body avatar.

Despite great performance variations between subjects, performances remained coherent per subject. So we performed a per-subject duration normalization, and a target height normalization by the subject body height to obtain comparable results for an analysis of variance. The egocentric strategy provided better fitted reach duration characteristics in the baseline case (a single solid was controlled). The baseline and the avatar control cases provided similar performance only when reaching in free space. The subjects were significantly slower to reach all the targets when it was displayed on a shelf although the targets are located at the same relative distance. The control of height-differing avatars with both scaling strategies led to performances consistent with the body height differences. Only for the lowest target, the visuocentric led to a significantly slower reach compared to the egocentric scaling. Overall the egocentric scaling showed less variance which suggests it is more appropriate to enforce spatial awareness.

One key finding of the study was that nearby obstacle (the shelf) can alter significantly the performance of subjects when they control an avatar instead of a simple solid. We suspect that subjects took great care of cognitively avoiding the shelf (no collision avoidance algorithm was implemented). This would support the introduction of the proactive collision avoidance algorithm to relieve the user from such management that induce delays and cognitive fatigue.

5 Conclusion

Immersive interactions stands at the intersection of various complementary research fields. On the perception-action level, the intense debates about the nature of human embodiment from the past century have highlighted that most human sensorimotor activities occur at a "not explicitly conscious" cognitive level, e.g. through motor internal models [J97, J09]. This dimension of immersive interactions is still not fully addressed in the immersive interaction. We believe it is highly desirable to reproduce the same cognitive pattern while interacting with the full-body; i.e. it is critical to free the user from any conscious cognitive load of explicitly managing the spatial relationships between the controlled avatar and the virtual environment. We have presented some of our efforts on handling collision avoidance by proactively damping body segments movement along the obstacle normal and by altering these normals to ease the obstacles bypass. Further researches will evaluate sensorimotor tasks performed by increasingly differing avatars to assess the boundaries of avatar embodiment in task-driven contexts.

Acknowledgments. This work has been partly supported by the Swiss National Foundation under the grant N° 200020-117706 and by the University of Alcalá under grant PI2005/083. Thanks to Mireille Clavien for the characters design and to Achille Peternier for the MVisio viewer and the CAVE software environment.

References

[A] Autodesk MotionBuilder, http://usa.autodesk.com
[ABD98] Amato, N.M., Bayazit, O.B., Dale, L.K.: OBPRM: An Obstacle-Based PRM for 3D Workspaces. In: WAFR 1998 (1998)
[B00] Berthoz, A.: The Brain Sense of Movement. Chapter "building coherence", Section "seeing with the skin". In: Perspective in Cognitive Neuroscience, pp. 83–86. Harward University Press, Cambridge (2000)
[BMT09] Boulic, R., Maupu, D., Thalmann, D.: On Scaling Strategies for the Full Body Interaction with Virtual Mannequins. Interacting with Computers, Special Issue on Enactive Interfaces 21(1-2), 11–25 (2009)
[BH97] Bowman, D., Hodges, L.F.: An Evaluation of Techniques for Grabbing and Manipulating Remote Objects in Immersive Virtual Environments. In: Symp. I3D, pp. 35–38 (1997)
[BM07] Bowman, D., McMahan, P.: Virtual Reality: How Much Immersion Is Enough? Computer 40(7), 36–43 (2007)
[BRWMPB06] Burns, E., Razzaque, S., Whitton, M.C., McCallus, M.R., Panter, A.T., Brooks, F.P.: The Hand is More Easily Fooled than the Eye: Users Are More Sensitive to Visual Interpenetration than to Visual-proprioceptive Discrepancy. Presence: Teleoperators and Virtual Environments 15, 1–15 (2006)
[D01] Dourish, P.: Where the action is. MIT Press, Cambridge (2001)
[FH98] Flach, J.M., Holden, J.G.: The reality of experience: Gibson's way. Presence-Teleoperators and Virtual Environments 7(1), 90–95 (1998)
[FT87] Faverjon, B., Tournassoud, P.: A Local Based Approach for Path Planning of Manipulators with a High Number of Degrees of Freedom. In: IEEE Int'l Conf. Robotics and Automation, pp. 1152–1159. IEEE Press, New York (1987)
[GMMG08] García, A.S., Molina, J.P., Martínez, D., González, P.: Enhancing collaborative manipulation through the use of feedback and awareness in CVEs. In: 7th ACM SIGGRAPH Int. Conf. VRCAI 2008. ACM, New York (2008)
[H27] Heidegger, M.: Being and Time, John Macquarrie and Edward Robinson, translated in English in 1962. Harper and Row, New York (1962)
[J97] Jeannerod, M.: The cognitive neuroscience of action. Blackwell, Malden (1997)
[J09] Jeannerod, M.: Le cerveau volontaire. Odile Jacob Sciences (2009)
[J73] Johansson, G.: Visual perception of biological motion and a model for its analysis. Perception and Psychophysics 14, 201–211 (1973)
[MBRT99] Molet, T., Boulic, R., Rezzonico, S., Thalmann, D.: An architecture for immersive evaluation of complex human tasks. IEEE TRA 15(3) (1999)
[MS99] Murray, C.D., Sixsmith, J.: The Corporeal Body in Virtual Reality. Ethos 27(3), 315–343 (1999)
[NvDR08] Nijholt, A., van Dijk, B., Reidsma, D.: Design of Experience and Flow in Movement-Based Interaction. In: Egges, A., Kamphuis, A., Overmars, M. (eds.) MIG 2008. LNCS, vol. 5277, pp. 166–175. Springer, Heidelberg (2008)
[PMMRTB09] Peinado, M., Meziat, D., Maupu, D., Raunhardt, D., Thalmann, D., Boulic, R.: Full-body Avatar Control with Environment Awareness. IEEE CGA 29(3) (May-June 2009)
[PBRM07] Peinado, M., Boulic, R., Raunhardt, D., Meziat, D.: Collision-free Reaching in Dynamic Cluttered Environments, VRLAB Technical Report (2007)

[PMLGKD08] Pronost, N., Multon, F., Li, Q., Geng, W., Kulpa, R., Dumont, G.: Interactive animation of virtual characters: application to virtual kung-fu fighting. In: International Conference on Cyberworlds 2008, Hangzhou – China (2008)

[PMLGKD09] Pronost, N., Multon, F., Li, Q., Geng, W., Kulpa, R., Dumont, G.: Morphology independent motion retrieval and control. The International Journal of Virtual Reality 8(4), 57–65 (2009)

[ON01] O'Regan, J.K., Noë, A.: A sensorimotor account of vision and visual consciousness. Behavioral and Brain Sciences 24(5), 939–1011 (2001)

[PJ00] Paccalin, C., Jeannerod, M.: Changes in breathing during observation of effortfull actions. Brain Research 862, 194–200 (2000)

[R03] Reteaux, X.: Presence in the environment: theories, methodologies and applications to video games. PsychNology Journal 1(3), 283–309 (2003)

[R06] Reteaux, X.: Immersion, Presence et Jeux Video. In: Geno, S. (ed.) Le Game Design de Jeux Video, Approches de l'Expression Videoludique, L'Harmattan, Paris (2006)

[RSSS09] Rovira, A., Swapp, D., Spanlang, B., Slater, M.: The Use of Virtual Reality in the Study of People's Responses to Violent Incidents. Front Behav. Neurosci. 3, 12 (2009)

[S03] Slater, M.: A note on presence terminology. Presence Connect 3, 3 (2003)

[S04] Steel, L.: The autotelic principle. In: Iida, F., Pfeifer, R., Steels, L., Kuniyoshi, Y. (eds.) Embodied Artificial Intelligence. LNCS (LNAI), vol. 3139, pp. 231–242. Springer, Heidelberg (2004)

[SS05] Sanchez-Vives, M.V., Slater, M.: From presence to consciousness through virtual reality. Nat. Rev. Neurosci. 6(4), 332–339 (2005)

[SSC10] Slater, M., Spanlang, B., Corominas, D.: Simulating virtual environments within virtual environments as the basis for a psychophysics of presence. In: Hoppe, H. (ed.) ACM SIGGRAPH 2010 Papers, SIGGRAPH 2010, Los Angeles, California, July 26 - 30. ACM, New York (2010)

[TGB00] Tolani, D., Goswami, A., Badler, N.I.: Real-Time Inverse Kinematics Techniques for Anthropomorphic Limbs. Graphical Models 62(5), 353–388 (2000)

[UPBS08] Unzueta, L., Peinado, M., Boulic, R., Suescun, A.: Full-Body Performance Animation with Sequential Inverse Kinematics. Graphical Models 70(5), 87–104 (2008)

[ZJ98] Zahorik, P., Jenison, R.L.: Presence as being-in-the world. Presence-Teleoper. Virtual Environ. 7, 78–89 (1998)

[ZB94] Zhao, X., Badler, N.: Interactive Body Awareness. Computer Aided Design 26(12), 861–866 (1994)

Scalable Precomputed Search Trees

Manfred Lau[1,2] and James Kuffner[1]

[1] Carnegie Mellon University, USA
[2] JST ERATO Igarashi Design Interface Project, Tokyo Japan

Abstract. The traditional A*-search method builds a search tree of potential solution paths during runtime. An alternative approach is to compute this search tree in advance, and then use it during runtime to efficiently find a solution. Recent work has shown the potential for this idea of precomputation. However, these previous methods do not scale to the memory and time needed for precomputing trees of a reasonable size. The focus of this paper is to take a given set of actions from a navigation scenario, and precompute a search tree that can scale to large planning problems. We show that this precomputation approach can be used to efficiently generate the motions for virtual human-like characters navigating in large environments such as those in games and films. We precompute a search tree incrementally and use a density metric to scatter the paths of the tree evenly among the region we want to build the tree in. We experimentally compare our algorithm with some recent methods for building trees with diversified paths. We also compare our method with traditional A*-search approaches. Our main advantage is a significantly faster runtime, and we show and describe the tradeoffs that we make to achieve this runtime speedup.

1 Introduction

Traditional A*-search methods build a search tree during runtime and perform a forward search to find a solution path. We refer to this as a *forward search* as it can solve problems with one specific start location and one goal location, and a tree is built in the forward direction from the start towards the goal. An alternative approach is to first compute this tree beforehand without considering the obstacles and start/goal locations. We then use the precomputed tree to efficiently find a solution for any configuration of obstacles and any start/goal queries. The runtime process performs a *backward search*, as it begins from the goal and attempts to backtrace a valid path towards the start.

Recent work has shown potential for this idea of precomputation [8,4,1]. However, the trees that these methods generate are so small (typically 5 or 6 depth levels) that it is either not possible or difficult to use them in real planning problems (requiring solutions of up to 50 levels in our experiments). Furthermore, they do not scale to the time and memory needed for precomputing trees of a reasonable size. Hence, generating large trees that can scale to general planning scenarios is an important problem.

This paper makes two main contributions. First, we show a *fully-developed* system of the concept of precomputation for animating virtual human-like characters, and show that our approach can be used to *efficiently* generate the motions for these characters (Figure 1). When we say *fully-developed*, we mean that: (i) we start from a method to

Fig. 1. We use our precomputation method to efficiently generate the motions for virtual human-like characters navigating in large environments

precompute large trees with diverse paths; (ii) we apply an efficient backward-tracing method presented in Lau and Kuffner [8] to use the precomputed trees to solve real planning queries (most previous methods focus on building these trees, but do not use them to show the advantage of the precomputation concept by solving actual planning problems); (iii) we show our method's advantages and tradeoffs compared with other tree building methods and traditional forward search methods; and (iv) we use the overall approach to efficiently generate the motions for virtual human-like characters navigating in large environments such as those in games and films. We call our approach Scalable Precomputed Search Trees (SPST). We focus on path planning scenarios with characters navigating in environments with obstacles.

Secondly, we show how to take a given set of actions and precompute a search tree that has *diverse paths* and that can *scale to large environments*. We also perform a set of empirical comparisons with previous methods to show the effectiveness of our approach. Although our approach is simple, it has many advantages. We can precompute a tree with a more efficient computation time than the algorithms that have been previously proposed [4,1]. Our precomputed tree can solve more planning queries than trees built with previous methods, given the same amount of memory for storing the tree. We can build a tree for any memory size available for storing it, and we can build a tree that can cover a region of arbitrary shape and size.

Our algorithm precomputes a search tree by incrementally adding a node and its corresponding edge to the tree. We use a "density" metric to essentially scatter the edges or paths of the tree evenly throughout the region that we would like to build the tree in. We show that our algorithm satisfies several desirable properties. We experimentally compare our algorithm with some recent methods for building trees with diversified paths. In addition, we compare our method with traditional A*-search approaches. Our method can solve general planning queries in large environments more than 200 times faster than A*-search methods; we show and describe the tradeoffs that we make to achieve this runtime speedup. We also test the robustness of our method by studying the effect of grid resolution on our algorithm.

2 Related Work

The precomputed trees that previous methods [8,4,1] build are too small to use for general planning problems. Lau and Kuffner [8] build trees that have a limited depth level. They showed that the concept of precomputation can lead to a faster runtime, but they showed these results either for problems with very small environments or problems requiring a two-level hierarchical approach. Using a two-level approach made it difficult to compare the advantages and disadvantages of the precomputation method against A*-search methods. They were unable to make this comparison fairly because it is difficult to build large trees effectively. While there are cases when it is beneficial to combine two-level approaches with precomputed search trees, building precomputed trees of at least a reasonable size and comparing them to traditional forward search methods are still important issues.

Green and Kelly [4] and Branicky et al. [1] describe methods to take an existing set of paths in a tree and select a smaller set from it that is as diverse as possible. These methods require at least quadratic time with respect to the number of paths; hence they can only be used for small path sets. Furthermore, they both require the existence of a set of paths from which to select from. Generating the exhaustive set of paths to select from only works for trees with small depth levels, and this is in fact the approach that we take in our experiments. For larger trees, generating an exhaustive set requires exponential time with respect to the depth level, and it is not clear how we can generate a subset of paths from which to begin selecting from. Indeed, the tree precomputation method in this paper solves this problem: how to generate such a set of paths for trees of large depths while keeping the paths diversified enough that the trees can be used to solve as many planning queries as possible.

Our method is related to sampling-based planning approaches. In our algorithm, we choose nodes from which to expand from in the same way as Rapidly-Exploring Random Trees (RRTs) [10]. We choose nodes this way for the same reason as RRTs do: so that the selected nodes will be evenly spaced and not biased towards a particular region. Our algorithm differs from RRTs as we use a metric to locally pick paths that are as evenly spread out as possible; this process does not take the obstacles into account as the tree is being precomputed. Probabilistic Roadmaps (PRMs) [5] are effective for planning in high-dimensional spaces. They first build a roadmap for a given environment and then use it to find solution paths. Our method also has a preprocessing phase, but our precomputed tree can then be used during runtime for any obstacles and any start/goal queries. An extended version of PRMs [11] first builds a tree without taking obstacles into account and later map them back into the environment. The difference in our case is that since we have a set of actions as input, we first build the tree in the action space. This is more general because each path of the tree can later fit anywhere in the environment. Finally, the key difference between our method and RRTs and PRMs is the overall precomputation concept to first *precompute a tree* and then use a *runtime backward search* to find a solution.

Our approach differs from chess-playing methods [12] that, for example, compute endgame policies in advance, since our focus is on path planning problems where characters navigate in environments with obstacles. The paths in our precomputed trees are stored and analyzed for these navigation scenarios.

3 SPST

Scalable Precomputed Search Trees (SPST) is a fully-developed system of the idea of precomputation. This section focuses on taking a given a set of actions, and efficiently precomputing trees that can scale to large environments (Algorithm 1). We also setup an environment gridmap and a goal gridmap during precomputation. During runtime, we can solve for any start/goal queries and any obstacle configuration very efficiently using a runtime backward search method. Lau and Kuffner [8] describe these gridmaps and the runtime method in more detail. At the end of this section, we briefly discuss the significance of these parts for completeness.

Notation: Let A be the set of actions, where an action represents a virtual human-like character's motion. Example motions include: walking, jogging, and turning at different angles. For the purpose of planning, we take the 2D top-down view of the character's motion, and only consider its position and orientation. There is a cost associated with each action, and it is the distance that the character travels in this 2D view. The idea is to plan for some combination of these motions, and concatenate them into a longer sequence that allows the character to get from the start to the goal. An algorithm that uses A*-search to generate motions this way was presented in [7].

We are also given whether or not each action can transition to other actions. For every i, j (i can equal j), if action a_i can transition to action a_j, $Transition(a_i, a_j)$ is true. Otherwise, it is false. Some transitions are necessary: for example, a left step must be followed by a right step. Others are for aesthetics purposes: after taking a sharp left turn, we may not want to take a sharp right turn.

Let T be the precomputed tree, n be each node of T, and N be the set of all nodes. Each node corresponds to one action, denoted by $n.action$. $n.childs$ denotes the child nodes of node n. Let e be each edge of T. Every time we add a new node n to T, we also add a corresponding edge that connects n and its parent node; we refer to this combination as a node/edge. We refer to the path that the action of a node covers as a *traced path*; more details are given below when we describe the $Trace()$ function. Let m be a 2D grid that covers the region occupied by the tree.

Precomputation of Tree: Our method can be summarized as follows: we iteratively add a node and its corresponding edge to T. At each iteration, we "randomly" select which node in the existing T to expand from, and then use a density metric to locally decide which child of the selected node to expand. This iterative strategy is greedy and leads to non-optimal solution paths. However, since it is not possible to precompute large-scale trees that can provide optimal solutions, we choose a strategy that is fast and provides near optimal solutions (as shown in Section 5). The intuition for the density metric is that since the tree is precomputed for any obstacle and any start/goal queries, a simple but effective way to increase the likelihood of finding a solution is to "scatter" the paths evenly in the region that T should be built in. Figure 2 shows a visual example of this process.

We now describe Algorithm 1 in more detail. In the PrecomputeTree() function, K is the number of nodes to be built in T, and is a parameter that can be set depending on the memory available for storing the tree. R is a 2D region that we want to build the tree in, and can be arbitrarily large and be in any shape. Note that the obstacles and goal

Algorithm 1. Precomputation of Tree

```
   function Trace(e)
 1  m_temp.Init();
 2  foreach (x,y) ∈ Path(e) do
 3      map_x = Map(x);
 4      map_y = Map(y);
 5      m_temp(map_x, map_y) = 1;
 6  return m_temp;

    function PrecomputeTree(A, K, R)
 7  T.Init(n_root);
 8  m.Init();
 9  d_overall ← 0;
10  for k = 1 to K do
11      n_near ← Nearest(N, α(k));
12      Δd_best = FLT_MAX;
13      a_best = NULL;
14      foreach a_j ∈ A do
15          if ¬Transition(n_near.action, a_j) then continue;
16          if T.AlreadyExpanded(n_near.childs, a_j) then continue;
17          e ← T.SimulateAddChild(n_near, a_j);
18          if OutsideRegion(R, Trace(e)) then continue;
19          m_current ← m ⊕ Trace(e);
20          Δd_current ← (Density(m_current) − d_overall) / Length(Trace(e));
21          if Δd_current < Δd_best then
22              Δd_best = Δd_current;
23              a_best = a_j;

24      if a_best == NULL then continue /* do not increment k */;
25      e ← T.AddChild(n_near, a_best);
26      m ← m ⊕ Trace(e);
27      d_overall ← Density(m);
28  return T;
```

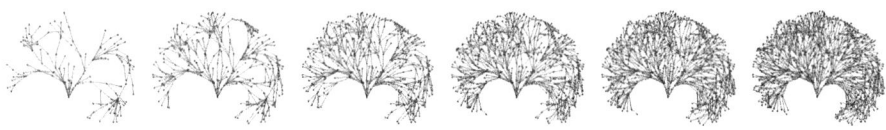

Fig. 2. This example shows the iterative addition of nodes as a tree is built. The tree on the left has 130 nodes, and each successive one has 130 additional nodes.

are not taken into account during precomputation. The function starts by initializing T with a root node (n_{root}), which is a placeholder node that contains no action and can transition to all other actions. This root node is initialized with position $(0,0)$, orientation 0 and total cost 0. It initializes the grid m by setting all its gridcells to 0. This grid provides a discretized "count" of the space that the tree covers. $d_{overall}$ is the density (described below) measure of m. We incrementally select a node/edge to add to T. $\alpha(k)$ is a randomly-sampled point in R, and $Nearest()$ selects the node in the existing set N that is nearest to $\alpha(k)$. We implement the nearest neighbor computation with a kd-tree. This randomly-sampled selection scheme is the same as in RRTs as we explained in Section 2. We then try to add a child node to n_{near}: we choose to associate this child node with an action whose traced path locally minimizes the density measure if that path is added {lines 14-23}. The $AlreadyExpanded()$ function checks to see if a_j is already a child of n_{near}. $SimulateAddChild(n_{near}, a_j)$ simulates the effect of adding a new node representing a_j as a child node of n_{near}. It does not add the new node and its corresponding edge to T here; instead it returns information about the corresponding edge (which is represented by e on line 17). The $Trace()$ function marks all the gridcells covered by $Path(e)$. $Path(e)$ {line 2} takes an edge e that connects a parent node and a child node, and generates the "traced" 2D path of motion if we start from the overall position at the parent node and take the action at the child node. $Path(e)$ then returns a set of discretized 2D points passing along this path. These points have to be chosen so that we neither generate too many points and make the algorithm inefficient, nor generate too few points and have them not cover all the gridcells that the path covers. $Map()$ {lines 3-4} maps from the coordinate system of the action/motion space to the coordinate system of the grid m_{temp}. m_{temp} has the same shape and grid structure as m. To avoid accessing all the cells of m_{temp} in each execution of $Trace()$, m_{temp} is initialized once in the algorithm, and the (map_x, map_y) points are saved for resetting m_{temp} each time. Once we have $Trace(e)$, $OutsideRegion()$ returns $true$ if at least one of the gridcells marked by $Trace(e)$ is outside of R. R has the same grid structure as m. The $\oplus$ operator performs component-wise addition to the grids. $Density(m)$ takes a grid m (which does not have to be rectangular shape) with the "count" in the gridcells labelled c_i (i from 1 to $ncells$), and computes the "density" of the paths in the tree:

$$Density(m) = \sum_j \left[\left(\frac{\sum_i c_i}{ncells} - c_j \right)^2 \right] \quad (1)$$

Intuitively, a smaller density value means that the paths are more evenly spread out. $Length()$ is the distance that the "traced" 2D path travels. Since we have discretized this path, we compute the number of gridcells that the discretized set of 2D points cover. The reason for dividing by the length is to normalize for the length of the traced path when considering the density value. $AddChild(n_{near}, a_{best})$ adds a new node representing a_{best} as a child node of n_{near}, and also adds the corresponding edge. Each node maintains the overall 2D position and orientation after taking the action it represents. Each node also maintains the total cost of taking all the actions from the tree's root node to that node.

Precomputation of Gridmaps: We setup two gridmaps that are important for the speed of the runtime search. An *environment gridmap*, a 2D grid, is placed over the region of the precomputed tree and we initialize its gridcells to 0. The intuition for this gridmap is that we will map the obstacles to this grid and thereby the tree during runtime. The discretized grid is then used for efficient runtime collision checks.

We place another grid, a *goal gridmap*, over the tree. For each gridcell, we consider the nodes in that cell and their corresponding paths. Each node corresponds to a unique path if we trace the path from the tree's root node to that node (by following the path traced by applying the actions at each node). For each gridcell, we sort all the nodes/paths in that cell by the total cost of each path. The intuition for the goal gridmap is that we want to start searching with the lowest cost path first during runtime.

Runtime Backward Search: At runtime, we first map the obstacles to the environment gridmap, and mark the cells with obstacles. We then map the goal position to the goal gridmap, and find the cell that the goal belongs to.

Let the sorted set of nodes in that cell of the goal gridmap be N_{goal}. For each node in N_{goal}, we try to follow its path towards the start or the tree's root node by continuously following every node's parent node. We mark each node that we have visited (colored red in the figure on the right) as this process occurs (all the nodes are originally unmarked). In the figure, the dotted square is the cell (of the goal gridmap) that the goal is in. The backward-tracing process for each node in N_{goal} will stop in one of three cases: (1) it arrives at a node (the one colored black) whose corresponding action collides with an obstacle, in which case we stop the backward-tracing and try the next lowest cost node in N_{goal}; (2) it arrives at a node that was previously marked as visited, in which case we also stop the backward-tracing and try the next lowest cost node in N_{goal}; or (3) it arrives at the tree's root node, in which case the path we have just traced (the nodes colored green) is the solution. The runtime process returns the lowest cost collision-free path that is available in the precomputed tree.

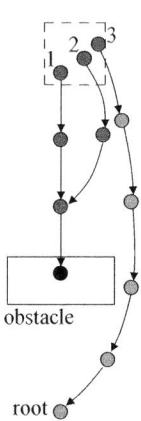

4 Properties of SPST

Precomputation of Tree: The execution time of the tree precomputation is $O(K(\log K + \|A\| * F))$. The $\log K$ term comes from the kd-tree nearest neighbor computation. F is due to a faster way to compute the $Density(m)$ value: instead of iterating through all the gridcells of m, we keep a frequency count of the values in each gridcell and compute $Density(m)$ using this information. F is the largest value with at least a count of one; it starts at 0 and increases as k increases. F is a function of K, R, the cell sizes of m, and A (the space that each action covers). In practice, the $K * \|A\| * F$ term is more significant than the $K \log K$ term. In our experiments, the largest K we used is about $2e6$, A is about 10-20, and the largest F we have is about 500.

Given a specific precomputed tree, our approach is not complete. However, we have a weaker notion of "complete"-ness *with respect to the given tree*: if there is a solution *in the precomputed tree*, the algorithm will find it in finite time; if there is no solution *in the precomputed tree*, it will stop and report failure in finite time. We now show that given enough time (and memory), all the nodes in the exhaustive tree will eventually be expanded. We define $Exh(d)$ to be the exhaustive tree with finite depth d and finite average branching factor b. We define a notion called **Probabilistic Expansion**:

$$\lim_{k \to \infty} P(\, n_i \text{ will be expanded} \mid n_i \in Exh(d) \,) = 1 \qquad (2)$$

where d can be arbitrarily large. Algorithm 1 follows this notion of Probabilistic Expansion.

Proof: We prove by contradiction: there is at least one node, n, that is not expanded. If n's parent node is not expanded (but exists in the exhaustive tree), we instead set n to be its parent node. We continue this until n's parent node is expanded. We now have an unexpanded node n whose parent node p is expanded. We must always at least have such a case because the tree's root node must be expanded at the $k = 1$ iteration. Let

$\mu()$ be the measure of volume in a metric space [9] and $V(p)$ be the Voronoi region of p. We must have $\mu(V(p)) > 0$ regardless of the number of nodes in the current tree and k, since the tree has a finite size. Let the branching factor of p be b, which is finite. As $k \to \infty$, we must eventually sample $V(p)$ b times (recall that we only sample from finite region R), and n must be expanded. ∎

Runtime Backward Search: The runtime backward search method was presented in Lau and Kuffner [8]; we provide further analysis of it here. The execution time of this method is $O(N_{goal_{largest}} * d_{largest})$. $N_{goal_{largest}}$ is the largest number of nodes/paths in one gridcell among all the cells of the goal gridmap, given a precomputed tree. It is typically in the thousands, and up to a few tens of thousands. $d_{largest}$ is the largest depth in the given precomputed tree. It typically lies between ten and fifty. This execution time explains the efficiency of the runtime search.

It is interesting to note that the runtime method searches through the smallest number of nodes, for a given precomputed tree and planning query. The runtime backward-tracing is a "lazy" way to discover nodes that cannot be reached for a given goal and obstacle configuration.

5 Experimental Evaluation

We have developed a system that uses our precomputation approach to generate motions for virtual human-like characters navigating in large environments (Figure 1). One main result is that our approach has a runtime that is very efficient. We now present experimental evaluation to show the effectiveness and robustness of our method.

Comparison of Tree Precomputation Methods: First, we compare our algorithm (Algorithm 1) with four other recent methods. The key here is to compare the trees that are built. We only build trees that fit in a relatively small environment in this part, since it is not clear how we can use some of the other methods to build large trees.

Table 1. Comparison of Tree Precomputation Methods

Method	% success						precomputation time (sec.)						density value				path cost	
	250	100	50	25	10	5 KB	250	100	50	25	10	5 KB	50	25	10	5 KB	mean	std
SPST	73.44	71.75	69.48	67.88	54.55	39.38	1.2	0.8	0.6	0.5	0.4	0.4	28,752	11,403	4,162	1,650	110.09	8.80
original PST	60.37	40.30	39.12	21.92	21.16	6.32	0.043	0.021	0.014	0.011	0.009	0.008	109,039	47,976	11,098	5,542	103.79	10.40
Bramicky et al. [1] I-P	65.09	54.38	46.04	40.22	23.02	10.37	720	126	38	14	7	5	66,270	23,059	6,960	3,298	102.19	2.28
Bramicky et al. [1] I-E	35.33	26.22	21.84	10.79	7.76	4.72	780	138	49	21	11	9	120,155	42,329	13,347	5,309	121.87	20.13
Green and Kelly [4]	68.89	66.27	62.48	57.93	48.82	31.87	11460	1800	480	80	19	6	55,019	16,526	4,285	1,841	112.02	8.22

We generate a large number of random planning queries and try to use the trees precomputed with the different methods to solve them. We select a fixed starting position and orientation, and generate random goal positions. This is equivalent to generating random start/goal queries. Since we build an exhaustive tree with 5 depth levels with which to select paths from for the purposes of some of the other tree building methods, these methods can only solve queries within the region covered by the exhaustive 5-level tree (we let R be this region for SPST, which explains the tree's shape for SPST in Figure 3). Hence we select random goal positions within R so we can perform a fair comparison. We generate obstacles randomly by randomly generating the number of

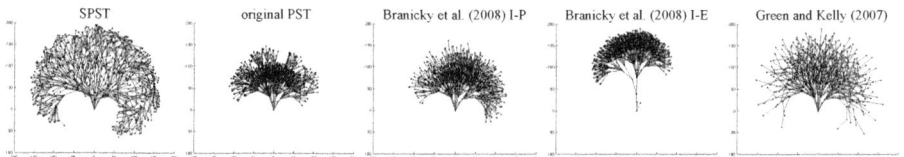

Fig. 3. Examples of precomputed trees used in our comparison. All trees have the same number (826) of nodes. Each tree's root is at (0,0), and the paths move in a forward (or up in the figure) direction because the input actions/motions allow the character to move forward and/or slightly turn left/right. Note that many paths overlap because of the tree's structure.

obstacles, the positions and orientation of each one, and the sizes of each one given that they have a rectangular shape. Each obstacle must at least overlap with R. We use the same set of random queries for all of the methods; we did not include queries where the start and/or goal collide with obstacles.

In Table 1, all the methods use the (same) runtime backward search technique described in this paper, since the last three methods only provide algorithms to build the tree; the difference is in the tree precomputation technique. "original PST" is the technique in Lau and Kuffner [8]. "I-P" stands for Inner-Product and "I-E" stands for Inclusion-Exclusion. For the last three methods in the table, we first built the exhaustive tree with 5 depth levels and then selected a subset of paths using each method. Note that SPST can have paths with depth levels larger than 5; for the last three methods, the exhaustive tree with larger depth levels cannot be built because of its size, and it is not clear how to pick a subset of potential longer (than depth 5) paths to choose from. For all methods, we tried to choose parameters that give the best results. We build trees with varying sizes: the numbers in the top row are the memory in KB that we use to store the tree. We use the same amount of memory to store each node of the tree for all methods, so the trees for each column has the same number of nodes. "% success" is the % of the 1186 total planning queries that can be solved. We also tried to solve this set of queries with the exhaustive tree of 5 depth levels. It took about 2 MB to store this tree and the % success rate was 71.16. The percentages for SPST can be higher than 71.16 since the precomputed trees for SPST can have paths longer than 5 depth levels. The "precomputation time" is the time for building the trees only. We used a 2.4 GHz machine with 1 GB of RAM. The "density value" is from the $Density()$ formula. The "path cost" columns are for the 50 KB case; we have similar results for the other cases. We took the queries (248 of them) where all methods found a solution and compared the costs of these solutions. We normalize the costs for the exhaustive tree case (the optimal case) to be 100, and normalize the other costs correspondingly. We then computed the mean and standard deviation of all the normalized costs for each method (so 100 % is optimal).

The results show that, based on the % success rates, the ranking of the methods starting with the best is: SPST, Green and Kelly [4], Branicky et al. [1] I-P, original PST, and Branicky et al. [1] I-E. This is true for all memory sizes. The precomputation time for SPST is longer than that for original PST. However, the precomputation can be done beforehand, and the time for SPST is still reasonable. In contrast, the other three methods' times are significantly slower; their times increase at such a rate that it is difficult to use them in practice for large trees, and we chose to build trees with depth levels of 5 (very small) for this set of experiments just so we can compare the methods. The density values justify our use of the density metric. A smaller density value tends to correspond to a higher % success rate, which matches our intuition that scattering the paths of the tree evenly is more likely to lead to a precomputed tree that can solve more planning queries. The tradeoff of SPST here is that it provides non-optimal, but near-optimal solutions.

Figure 3 explains some of the results in Table 1. "original PST" and "Branicky I-P" tend to keep shorter and thereby smaller-cost paths. On the other hand, "Branicky I-E" seem to prefer longer paths, and hence their solutions are likely to be further away from optimal. "Green and Kelly" build trees that has more diverse paths. However, its precomputation time is the longest, and is not practical for trees of large sizes. SPST builds the most diverse trees in the sense that their paths are spread out over the region R, in this case the region covered by the 5-level exhaustive tree. Our results show that: *our simple and randomized-based method is efficient and achieves the diversity that we need.* This suggests that *the effectiveness of sampling-based methods also applies to our paradigm of motion planning with precomputation.*

Comparison between Precomputation Concept and A*-search Methods: Secondly, we explore the benefits and tradeoffs of the overall precomputation approach along with the runtime backward search, as compared to traditional A*-search methods. For these experiments, we use relatively larger environments and build trees of a much larger scale.

We generate random planning queries as before, except that we use a much larger R region (5-10x larger) and generate a larger number of obstacles. We created one additional test environment with a C-shaped obstacle (similar to the "deep local minima" example in [2]). The random queries contain a mix of simple and complex cases, and this C-shape obstacle case is a complex example with local minima. Since A*-search and SPST search in different directions, we place the start/goal positions differently in the two cases so that the direction is always moving "into" the C-shape, which makes the problem more difficult.

In Table 2, SPST took 199 seconds to precompute the 25 MB tree and 477 seconds to precompute the 50 MB tree. The "runtime" of SPST is only for the runtime backward search. "collision checks" is the number of collision checks performed. "% success" is the % of 1774 total queries that each method found a solution for. The top set of results are all percentages. We took the queries (1661 of them) where all methods found

Table 2. Comparison between Precomputation and A*-search Methods. Top set of results: from random planning queries. Bottom set: from C-shaped obstacle case.

Method	runtime	collision checks	% success	path cost
A*-search	100.00	100.00	97.91	100.00
wA* (w=2)	79.42	12.31	97.91	105.12
SPST (50 MB)	0.47	7.27	94.76	113.80
SPST (25 MB)	0.44	4.23	93.63	115.48
A*-search	2,411,505	2,885,740	N/A	786
wA* (w=2)	1,276,468	1,559,632	N/A	846
SPST (25 MB)	461	210	N/A	884

a solution and compared the runtime, collision checks, and path cost of these solutions. We normalize these values (runtime, collision checks, path cost each separately) for the A*-search case (the optimal case) to be 100, and normalize the other values correspondingly. We then computed the mean of all the normalized values for each method, and reported these means in the table (top set). The bottom set of results are actual values. The runtime in that case is in μs.

The main benefit of SPST over A*-search methods is the significantly faster runtime (>200 times for the random planning queries). SPST has fewer collision checks than A*-search, although a more greedy version (weighted A*) can also lead to fewer collision checks. The main tradeoffs of SPST are that it gives up completeness and optimality of A*-search. Completeness can be seen in the "% success" column: SPST's rates are a few % smaller. The "% success" of SPST must be smaller than that of A*-search, because SPST is only able to find solution paths that are in the precomputed tree. Hence it is still encouraging that SPST is only slightly worse here. Optimality can be seen in the "path cost" column: SPST's path costs are near-optimal, and usually about 10-15 % higher than the optimal costs. In general, as we increase the memory size of the tree, the % success rate increases and the path cost % decreases to the "optimal" percentages. The user can adjust the tree's memory size to explore this tradeoff. The purpose of the C-shaped obstacle case is to make sure that the better results do not just come from simple queries in the random set. This is true as SPST achieves an even faster runtime and fewer number of collision checks for this case.

Table 3. Effect of grid resolution on runtime cost and success rate

Method	runtime			% success		
	270x270	540x540	1080x1080	270x270	540x540	1080x1080
A*-search	100.00 104880	100.00 151770	100.00 340195	97.91	97.91	97.91
SPST (25 MB)	0.41 333	0.52 690	0.80 2652	93.63	94.31	94.76

Effect of Grid Resolution: Thirdly, the obstacle avoidance between the characters and the objects in the environment depend on the grids that we use. We empirically study the effect of different grid resolutions on the runtime cost and success rate of finding a solution, for A*-search and SPST. We used the same experimental setup as for the comparison between A*-search and SPST above. We changed only the grid resolution and kept the other variables the same. The grid resolution here refers to the one for the *environment gridmap*; we adjust the resolution for the other gridmaps accordingly.

Table 3 shows the results from our experiments. The success rate is the percent of 1774 total queries that each method found a solution for. For the runtime results, there are two values in each entry. The first value is a percentage, and the second one is the average time for the success cases in μs. To compute the percentages, we took the queries where both methods found a solution and compared the runtime of these solutions. We normalize these values for the A*-search case (the optimal case) to be 100, and normalize the other values correspondingly. We then computed the mean of all the normalized values for each method, and reported these means in the table. In general, we found that a finer grid resolution leads to an increase in runtime. This makes sense intuitively as the time for mapping the obstacles to the grid takes longer. We also found that a finer grid resolution leads to an increase in the success rate. Intuitively, as the obstacle representation gets finer, there is more space that is represented as collision free, and there is a higher chance that more paths become collision free.

6 Discussion

We have presented SPST, a fully-developed system that uses the concept of precomputing a search tree, instead of building it during runtime as in the traditional A*-search approaches. We show that this concept can be used to *efficiently* generate the motions for virtual human-like characters navigating in large environments such as those in games and films. We demonstrate that our tree precomputation algorithm can scale to the memory and time needed for precomputing trees of large sizes. Our approach has a significantly faster runtime than A*-style forward search methods. We view SPST to be one approach among many motion/path planning approaches; the user should understand its benefits and tradeoffs before deciding whether or not to use it.

There has recently been a growing interest in the issue of path diversity [6,3]. It would be one possibility of future work to compare these methods with ours. As our method and previous methods all take a greedy approach, another possible direction for future work is to develop a more formal justification of the metrics that the different methods use.

References

1. Branicky, M.S., Knepper, R.A., Kuffner, J.: Path and trajectory diversity: Theory and algorithms. In: Int'l Conf. on Robotics and Automation (May 2008)
2. Chestnutt, J., Kuffner, J., Nishiwaki, K., Kagami, S.: Planning biped navigation strategies in complex environments. In: Proceedings of the 2003 Intl. Conference on Humanoid Robots (October 2003)
3. Erickson, L., LaValle, S.: Survivability: Measuring and ensuring path diversity. In: IEEE International Conference on Robotics and Automation (2009)
4. Green, C., Kelly, A.: Toward optimal sampling in the space of paths. In: 13th Intl. Symposium of Robotics Research (November 2007)
5. Kavraki, L.E., Svestka, P., Claude Latombe, J., Overmars, M.H.: Probabilistic roadmaps for path planning in high-dimensional configuration space. Int'l Transactions on Robotics and Automation, 566–580 (1996)
6. Knepper, R.A., Mason, M.: Empirical sampling of path sets for local area motion planning. In: International Symposium on Experimental Robotics. IFRR (July 2008)
7. Lau, M., Kuffner, J.J.: Behavior planning for character animation. In: 2005 ACM SIGGRAPH / Eurographics Symposium on Computer Animation, pp. 271–280 (August 2005)
8. Lau, M., Kuffner, J.J.: Precomputed search trees: Planning for interactive goal-driven animation. In: 2006 ACM SIGGRAPH / Eurographics Symposium on Computer Animation, pp 299–308 (September 2006)
9. LaValle, S.M.: Planning Algorithms. Cambridge University Press, Cambridge (2006), http://planning.cs.uiuc.edu/
10. Lavalle, S.M., Kuffner, J.J.: Rapidly-exploring random trees: Progress and prospects. Algorithmic and Computational Robotics: New Directions, 293–308 (2001)
11. Leven, P., Hutchinson, S.: A framework for real-time path planning in changing environments. Intl. J. Robotics Research (2002)
12. Russell, S., Norvig, P.: Artificial Intelligence: A Modern Approach. Prentice Hall, Englewood Cliffs (2002)

Toward Simulating Realistic Pursuit-Evasion Using a Roadmap-Based Approach[*]

Samuel Rodriguez, Jory Denny, Takis Zourntos, and Nancy M. Amato

Parasol Lab, Dept. Computer Science and Engineering, Texas A&M University
{sor8786,jorydenny,takis,amato}@tamu.edu

Abstract. In this work, we describe an approach for modeling and simulating group behaviors for pursuit-evasion that uses a graph-based representation of the environment and integrates multi-agent simulation with roadmap-based path planning. We demonstrate the utility of this approach for a variety of scenarios including pursuit-evasion on terrains, in multi-level buildings, and in crowds.

Keywords: Pursuit-Evasion, Multi-Agent Simulation, Roadmap-based Motion Planning.

1 Introduction

The chase between a pursuer and an evader is an exciting display of dynamic interaction and competition between two groups, each trying to outmaneuver the other in a time-sensitive manner. When this is further extended to include teams of pursuers, this problem combines cooperative problem solving with fast-paced competition in scenarios that are often seen in the real world. Both of these aspects have a wide range of applications, and are one reason for the prevalence of the pursuit/evasion problem in such a wide range of disciplines.

In this work, we describe an approach for agent-based pursuit-evasion games in which agents are equipped with a roadmap for navigation that encodes representative feasible paths in the environment. Agents are also equipped with a set of behaviors which define the actions they will take given their knowledge of the environment. The agents and behaviors are versatile and tunable so that the kind of agent and the behaviors being applied can be adjusted with a set of parameters that define them. This very general representation of the problem allows us use the same approach for pursuit-evasion in a range of scenarios, including pursuit-evasion games in crowds, terrains and multi-level buildings.

---

[*] This research supported in part by NSF awards CRI-0551685, CCF-0833199, CCF-0830753, IIS-096053, IIS-0917266, NSF/DNDO award 2008-DN-077-ARI018-02, by the DOE NNSA under the Predictive Science Academic Alliances Program by grantDE-FC52-08NA28616, by THECB NHARP award 000512-0097-2009, by Chevron, IBM, Intel, HP, Oracle/Sun and by King Abdullah University of Science and Technology (KAUST) Award KUS-C1-016-04.

2 Related Work

In this section we describe work that is relevant to the pursuit evasion problem that we consider. A classical, purely graph-based approach [17] looks at the problem of pursuit-evasion on a graph in which agents move on edges between nodes in the graph and where the pursuer has the goal of occupying the same node as the evader. The problem has been looked at in polygonal environments [8], with much interest in finding exact bounds on the number of pursuers needed [16]. An approach based on exploring undiscovered portions of the frontier is described in [24].

Roadmap-based pursuit-evasion, where the pursuer and evader share a roadmap and play different versions of the pursuit-evasion game is described in [10]. These games include determining: 1) if the pursuer will eventually collide with the evader, 2) if the pursuer can collide with the evader before the evader reaches a goal location and 3) if the pursuer can collide with the evader before the evader collides with the pursuer in a dog fight scenario. In this work, a wider range of capture conditions are supported than in most previous approaches.

In [2,3,4], the benefits of integrating roadmap-based path planning techniques with flocking techniques were explored. A variety of group behaviors, including exploring and covering, were simulated utilizing an underlying roadmap. This paper builds on that work with a focus on roadmap-based pursuit and evasion behaviors.

A number of other forms of the problem have been considered. A graph-based approach to the pursuit-evasion problem is given in [15] where agents use either blocking or sweeping actions by teams of robots to detect all intruders in the environment. Large teams of robots have been considered to detect intruders with communication allowed between other agents within some range and with no map of the environment [13]. Agents deal with limited sensing information by building gap navigation trees with gap sensors in [20]. The idea of using probabilistic shadow information spaces for targets which move out of a pursuing agent's field-of-view has also been considered [25]. Other forms of collaboration between searching agents has been considered where agents utilize frontier information [5] with communication to switch between roles [6].

The level of visibility between agents plays a role in what the agents detect in the environment and what kinds of environments can be handled. The problem of restricting or limiting the amount of visibility is considered in [9] with the field of view being variable in [7]. Teams of robots with different kinds of vehicles and sensing information is looked at in [12]. The problem of pursuit-evasion on height maps, which restrict agent visibility is described in [14].

Crowds of agents moving in an environment [18,21,22] present many interesting challenges for pursuit-evasion, including large numbers of moving obstacles and when allowing for the fact that the crowd can block visibility. Sophisticated, path planning techniques have been created for planning the motion of agents in crowded areas [22], some with specific applications of pursuit-evasion scenarios [19].

3 Framework Overview

Many of the previously proposed approaches make restrictive assumptions. Examples include focusing on polygonal environments, strictly on a graph, or planar environments. Our roadmap-based approach allows us to handle many classical pursuit-evasion problems along with many that have not been considered in the literature. This includes agents moving in 3D environments such as on a terrain, in multi-level environments or in areas with crowds. Here we describe our overall system and approach to this problem.

3.1 Problem Definition

The traditional pursuit/evasion problem that we consider consists of one set of agents, the pursuers H, attempting to capture another set of agents, the evaders E. An evader $e \subset E$ is considered captured if a non zero set of pursuers $h \subset H$ fulfills some predefined distance requirement. The pursuers and evaders may have different levels of environmental knowledge, sensing abilities and capabilities, which affect the overall time to capture. The pursuit we study is a complete chase which consists of searching for a target agent and maintaining visibility until a capture can occur. The evading agents take adversarial actions which involve fleeing from the pursuing agents that have been detected and improving on hiding locations when undetected.

3.2 Approach

In an attempt to explore a range of the entire pursuit/evasion spectrum we have developed an approach that allows for quick and easy development of strategies for different versions of the problem. Our approach gives the user full control over a number of parameters including agent and behavioral properties. This ability allows us to explore a wide spectrum and provides the basis for a framework that provides a great deal of flexibility and control to the user over interesting pursuit/evasion motions and simulations.

Our approach uses a real-time simulation that allows for the movement of agents using the roadmap, complex pursuit/evasion strategies, and interesting agent interactions with both the environment and other agents. We have designed and implemented a simulation infrastructure for storing and manipulating the following information: agents, behaviors, the environment, groupings, and relationships between groups.

The outline of our simulation loop is given in Algorithm 1. It is important to note that we use a general concept of individual agent, calling them a group (with no subgroups). In this way our framework can be more general, especially when creating behaviors since a behavior created for individual agents can easily be applied to a grouping of agents, with some additional logic used to ensure grouping restrictions are maintained.

The behavior that the agent is equipped with will determine how the agent reacts throughout the simulation. These behaviors determine the actions that

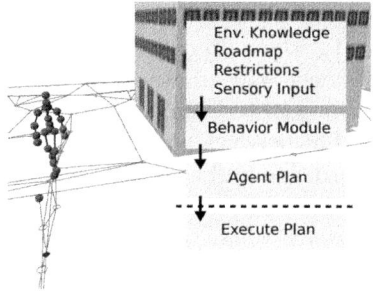

Algorithm 1. General Simulation Loop

Input: simulator *sim*, environment *env*
1: $groups_{all}$ = sim.getAllGroups()
2: **for** $g \in groups_{all}$ **do**
3:     $g \rightarrow$ applyBehaviorRule(*env*)
4: **end for**
5: **for** $g \in groups_{all}$ **do**
6:     updateState(g)
7: **end for**
8: ResolveStateWithAgents_Environment($groups_{all}$, *env*)
9: Evaluation

the agent or groups of agents take. At each time step, all groups update their state based on the last plan that was either generated or updated by the behavior rule. The state is then resolved which includes with other agents and the environment (i.e., preventing collision) as well handling any interactions that may have occurred between groups. This includes sending any communication messages that may have been generated throughout the behavior process or roadmap re-weighting. Since all interactions are delayed and only processed at the end of each time step, the environment remains constant in relation to all the agents during the individual time step. By processing the interactions in this manner, we ensure that the order in which the agents are processed does not affect their behavior.

The environment types that we have traditionally focused on are 2D environments consisting of polygonal obstacles. In this work, we extend our roadmap-based approach to 3D environments where agents can move on surfaces which alter the agent's height component along with affecting visibility. This extension allows us to handle 3D environments consisting of terrain and multi-level environments such as buildings. A terrain is a surface consisting of polygons in 3-dimensions that may represent hills and valleys whereas a building may be represented as a connected set of these surfaces which the agent may move on.

3.3 Roadmaps

Roadmaps are graph representations of an environment that encode feasible, collision-free paths through the environment. Our approach allows these graphs to be generated in variety of ways. For traditional 2D environments, we used probabilistic roadmaps [11], which are generated by first sampling nodes in the free space of the environment and creating paths between nodes that satisfy certain constraints (e.g. collision-free connections). A simple local planner is used to connect nearby nodes. The network of nodes and paths form a graph that the agent is able to query in order to find a path from its current position to an end goal. Depending on the agents being simulated, we allow enhancements to the PRM approach which improve where and how sampling is done – for example near obstacle boundaries [1] or near the medial axis of the environment [23].

Agents navigating in environments consisting of multi-level surfaces such as terrains or buildings need the ability to map these spaces as well. We allow the agents to move on the surfaces by sampling nodes on each surface. Connections are allowed between nodes on the same surface in which the straight line along that surface, projected to a 2-dimensional plane, remains completely within the projected polygon. Connections are then made between surfaces that are connected based on an input configuration which allows us to map multi-level surfaces and terrains along with connections between surfaces and the default 2D surface.

Roadmaps can also encode global information which may be useful to groups of agents, such as by storing information about an area to avoid, or by associating certain zones in the environment with certain groups of agents. Additionally, agents can then perform searches on the roadmap based on actions and observations that other agents they are cooperating with have made. These graphs/roadmaps are an essential component for our pursuit/evasion scenarios. Other benefits of the roadmap include the low computational cost of querying them once they are created and repairing them when the environment has changed. Rather than having to check for collisions during every timestep (which would be very computation-intensive), agents can query a valid roadmap quickly for paths guaranteed to be both available and collision-free.

The roadmaps are used to generate paths through the free areas of the environment that will guide an agent from some start to goal location. These paths can be adhered to strictly or act as a guide that is optimized based on agent specific criteria. In many of the examples we study, we allow the agents to use the paths from the roadmap as merely a guide through the environment. Agent path optimization parameters allow the agent to, e.g., improve the path to more quickly reach a goal location or to modify it to allow for greater clearance. As the agent is navigating through the environment following a path, another parameter determines the distance required for a subgoal to be considered reached. These parameters play an important role in determining how closely an agent adheres to the roadmap and the time it takes an agent to reach a goal in the environment.

3.4 Visibility between Agents

The agent's ability to detect other agents in the environment can be affected by a number of factors. In simple 2D problems, the agent's visibility to other agents may only be restricted by the obstacles in the environment. Agent capabilities can also determine the amount of the environment that can be sensed by setting a maximum view radius and angle. Visibility can be further restricted by considering other agents that may block the view of other agents in the environment, i.e., the 3D representation of all agents used as potential obstacles in visibility checks. When surfaces are included in environments, these surfaces may also block visibility from one agent to another. We have included each of these aspects in our visibility checks which allows us to handle complex surfaces and crowd examples.

4 Pursuit Behavior

Our general pursuit strategy has four stages: location of the target, creation of a pursuit plan, acting upon the plan, and then evaluating pursuit status. A wide range of pursuit strategies can be employed by enabling certain customizations of the general pursuing strategy. The most basic pursuit strategy will chase a target, once located, in a direct path until it either captures the target or the target is no longer visible.

4.1 Search Behaviors

The searching behaviors that agents use to locate a target can greatly influence the pursuing agents effectiveness. We have implemented a number of roadmap-based searching behaviors which allow groups of agents to utilize the roadmap to effectively cover an environment [3]. The goal of the searching behaviors is for the agents in the group to visit as much of the environment as possible. For these behaviors, agents use the roadmap to obtain pathways to free areas of the environment. The effectiveness of the search behavior has an underlying dependence on the quality of the roadmap used by the agents. Agents can either locally try to find nodes that have been least visited or search for some random area in the environment.

4.2 Enhancements

The basic pursuit strategy consists of an agent chasing a target toward the target's current location. Many improvements to this basic pursuit strategy can be enabled which potentially improve the success rate of a group of pursuing agents. Pursuing agents can "head off" an evading agent by planning their path to intercept it; we refer to this as a heading behavior. Pursuing agents can also avoid over re-planning by only creating a new pursuit plan when the evading agent has moved far enough away from the previously planned pursuit route. This can prevent erratic motion of the pursuing agent.

Enabling communication between pursuing agents can improve a pursuing agent's effectiveness. For example, communication can alert a pursuing agent to the location of an evading agent. Communication between agents can also allow for one agent, the leader, to request other agents to follow that agent when searching the environment. This type of communication allows for potentially more coordination between pursuing agents.

5 Evading Behavior

Our evasion strategy attempts to reduce the likelihood of being or staying visible to opposing agents. In this behavior, an agent generates hiding locations within a certain range in the environment. These positions are then evaluated, or scored. Different scoring functions are used depending on whether or not pursuers are present. A new hiding location can then be selected if one is found such that the score of the new location achieves some predefined percentage increase over the score of the current hiding location being used. The new goal and path are updated if a point is found that sufficiently improves the score.

5.1 Scoring Hiding Locations

In the case of an evading agent being undetected by pursuing agents, the goal of the evading agent is to decrease its likelihood of being detected. The score for each hiding location evaluated is determined by the visibility to all other potential hiding locations. A hiding location with low visibility is preferred over one with high visibility, i.e., a location that is blocked by objects in the environment is preferred.

When an evading agent is detected, five criteria are used for scoring hiding locations:

- **Distance to Pursuers:** This value will determine how much an agent prefers locations that are further away from the visible agents. The distance a hiding location is from all visible pursuing agents can be accumulated and then scaled by the evading agent's view radius.
- **Direction to Pursuers:** The direction to a hiding location is computed and scored such that directions away from pursuing agents are preferred.
- **Distance To Boundary:** The distance to the boundary of a hiding location can be factored in to the final score by weighting the factor when the potential hiding location is within some predefined distance to the boundary. This factor will allow agents to prefer hiding locations away from the boundary and reduce the likelihood of being trapped by the boundary.
- **Visibility Restricted Surfaces:** The weight of this factor can be determined by how much a hiding location is obscured by surfaces in the environment. In this way, agents would prefer hiding locations that are not visible to pursuing agents.
- **Visibility Restricted by Other Agents:** Other agents can be used when scoring hiding locations. A hiding location can be weighted such that an agent prefers a location that has other agents in between pursuing agent locations and the hiding location.

Table 1. Pursuit/Evasion on a terrain with hills and valleys obscuring visibility in the environment

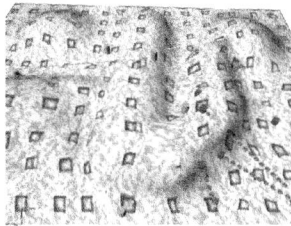

Scenario	Num Capt.	Time To Capture	Min/Max Time Chasing	Min/Max Time Hidden	Prop. Time Hidden
basic	1	3686.7	(1902,3189)	(1538,3461)	0.5261
heading	3.25	1363.5	(1056,2792)	(409,2556)	0.3044
send target	4	1028	(1133,2520)	(523,3980)	0.4169

Each of these factors can have associated weights so that evading agents can be tuned to prefer certain kinds of evasion tactics. These tunable factors allow for our adversarial evaders to exhibit potentially very different behaviors and are different from many approaches that assume the evading agents have a set evasion strategy.

6 Experiments

We present results for pursuit/evasion scenarios that have not received as much attention as the traditional problems. These involve more complex environments where the visibility of the agents is restricted by the environment. We also show examples where some interesting interaction takes place between the groups of agents being simulated. In the following, we show results in terrains, crowds, and a multi-level environment. We also show an example of interacting behaviors where agents play a game of tag with varying pursuit and evasion behaviors. Results show the average number of captures, the average time to capture, minimum and maximum time spent chasing among the pursuing agents, minimum and maximum time hidden for the evading agents and the proportion of time hidden. These values give insight into how involved each agent was in the pursuit process both in the chasing aspect and the searching of the environment. The maximum number of time steps used throughout any simulation is 5000.

7 Terrain

The terrain environment, shown in Table 1, consists of multiple hills and valleys with dimensions of 195 by 150 units and a maximum height 18.5 units. This provides numerous hiding locations for the evading agents. In this experiment 5 pursuing and 5 evading agents react to one another in the pursuit evasion game until the pursuing agents capture the evading agents. The pursuit behaviors tested in this environment are: basic pursuit, heading off the evading agents, and sharing targets between pursing agents which are also heading off. Pursuing agents have a maximum velocity of 4, view radius of 40 and height of 6 units.

Table 2. Pursuit/Evasion in a crowd of other agents (5 pursuing, 5 evading, 30 in crowd). Pursuing agents shown in Top Left corner of image, evading agents shorter, blue cylinders at center and crowd are tall, green cylinders.

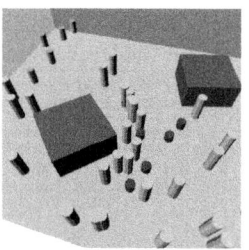

Scenario	Num Capt.	Time To Capture	Min/Max Time Chasing	Min/Max Time Hidden	Prop. Time Hidden
basic	3	1013.2	(19,376)	(1152,4727)	0.5889
heading, share target	5	448.4	(126,543)	(418,1744)	0.4104

Evading agents have a maximum velocity of 6, view radius of 50 and height of 3. In this scenario, the advantage lies with the evading agents where they are faster and can sense more of the environment than the pursuing agents.

In Table 1, it can be seen that using only basic pursuit results in fairly poor performance since the evading agents can easily outmaneuver the pursuers. When pursuing agents attempt to head off the evading agents, the performance improves overall resulting in more captures and lower time to capture. Also, sharing a target among the pursuing agents greatly improves the pursuing agents' effectiveness. Utilizing the most effective enhancements (heading off and sharing targets) allows agents to restrict the number of evasive actions.

8 Crowd

In an environment with a crowd, the evading agents can hide among crowd members. While the test environment is very basic, including the crowd in the problem adds another level of complexity. The environment dimensions are 110 by 100 units with the members of the crowd tall enough to obscure visibility. The pursuit behaviors tested are: a) basic pursuit, and b) using a heading off and shared target behavior. The basic pursuit again doesn't perform as well as using a heading off behavior along with a shared target. The most enhanced behavior results in the best performance with the highest number of captures and lowest average time to capture. This behavior among pursuing agents also alerts other agents to a target that may end up hiding in the crowd, but the subsequent searching in the crowd often resulted in a capture.

9 Multi-level Environment

The multi-level environment is a very complex environment consisting of two main floors connected to each other in two ways. One connection is a long ramp, in the background in the environment shown in Table 3. The other connection consists of multiple surfaces ramping to the second floor. Two basic pursuit

Table 3. Pursuit/Evasion in a multi-level building example

Scenario	Num Capt.	Time To Capture	Min/Max Time Chasing	Min/Max Time Hidden	Prop. Time Hidden
basic	5	784.8	(1103,2453)	(728,1349)	0.2428
send target	3	1236	(284,1267)	(1457,3921)	0.5082

Table 4. Pursuit/Evasion applied in a game of tag. Two pursuing agents start in each type of behavior (basic, heading) and attempt to capture an evading agent. An evading agent that is captured becomes part of the pursuing agent's team, executes the same behavior and can be used to capture. The final number of agents in each pursuing agent behavior is shown.

	Pursuit 1	Pursuit 2	Num. Left Evading	Time Complete
basic/basic	12.2	12.2	0.6	4708
heading/basic	17.8	6.8	0.4	4780
heading/heading	12.4	12.4	0.2	3606

behaviors are tested a basic pursuit and pursuit with a shared target. It is interesting to note that in this example, sharing a target among pursuing agents does not result in better performance. In this environment, agents independently searching the environment increase the chance that the pursuing agents will trap agents on the upper floor. This reduces the potential for the evading agents to find valid evasive actions. Agents that are independently searching, also reduce the chance of an evading agent to stay in a hiding location for very long.

10 Tag Game

The game of tag played in this scenario is a very simple extension of the standard pursuit/evasion game. In this scenario, two sets of pursuing agents with the same capabilities start out with predefined pursuit behaviors. The evading agents in the environment have the goal of avoiding both sets of pursuing agents. Once an evading agent is captured, it becomes part of the capturing agent's team and begins executing the same pursuit behavior. In Table 4, pursuit behavior comparisons are shown for basic vs. basic pursuit, heading vs. basic pursuit, and heading vs. heading pursuit behaviors. We initialize each set of pursuing agents with two pursuers. The numbers shown in Table 4 reflect the final number of agents performing the behaviors, average number of agents left evading and the average time to complete the game.

When each type of pursuit behavior is tested against the other, on average the same number of agents end up in each pursuit group. When comparing heading against basic pursuit behaviors, it can be seen that many more agents end up in the heading behavior group which shows that using a heading behavior is much more effective in this scenario. Another interesting aspect is that when both pursuit behaviors use a heading behavior, the average time until the game is complete (all agents are caught) is much lower, also showing the effectiveness of this pursuit behavior.

11 Conclusion

This paper describes a agent- and roadmap-based approach to pursuit-evasion which facilitates the study of heuristic pursuit-evasion in interesting scenarios including crowds, terrains, and multi-level environments. The presented results demonstrate this this approach is quite effective in these environments, which have more complex visibility and reachability constraints that are typically studied in pursuit-evasion games.

References

1. Amato, N.M., Bayazit, O.B., Dale, L.K., Jones, C.V., Vallejo, D.: OBPRM: An obstacle-based PRM for 3D workspaces. In: Robotics: The Algorithmic Perspective, Natick, MA, pp. 155–168. A.K. Peters, Wellesley (1998); Proc. Third Workshop on Algorithmic Foundations of Robotics (WAFR), Houston, TX (1998)
2. Bayazit, O.B., Lien, J.-M., Amato, N.M.: Better flocking behaviors using rule-based roadmaps. In: Proc. Int. Workshop on Algorithmic Foundations of Robotics (WAFR), pp. 95–111 (December 2002)
3. Bayazit, O.B., Lien, J.-M., Amato, N.M.: Better group behaviors in complex environments using global roadmaps. Artif. Life, 362–370 (December 2002)
4. Bayazit, O.B., Lien, J.-M., Amato, N.M.: Roadmap-based flocking for complex environments. In: Proc. Pacific Graphics, pp. 104–113 (October 2002)
5. Burgard, W., Moors, M., Fox, D., Simmons, R., Thrun, S.: Collaborative multi-robot exploration. In: Proc. IEEE Int. Conf. Robot. Autom. (ICRA), pp. 476–481 (2000)
6. Durham, J.W., Franchi, A., Bullo, F.: Distributed pursuit-evasion with limited-visibility sensors via frontier-based exploration. In: Proc. IEEE Int. Conf. Robot. Autom. (ICRA), pp. 3562–3568 (2010)
7. Gerkey, B.P., Thrun, S., Gordon, G.: Visibility-based pursuit-evasion with limited field of view. Int. J. Robot. Res. 25(4), 299–315 (2006)
8. Guibas, L.J., Latombe, J.c., Lavalle, S.M., Lin, D., Motwani, R.: Visibility-based pursuit-evasion in a polygonal environment. International Journal of Computational Geometry and Applications, 17–30 (1997)
9. Isler, V., Kannan, S., Khanna, S.: Randomized pursuit-evasion with limited visibility. In: Proc. ACM-SIAM Symposium on Discrete Algorithms, pp. 1060–1069 (2004)
10. Isler, V., Sun, D., Sastry, S.: Roadmap based pursuit-evasion and collision avoidance. In: Proc. Robotics: Sci. Sys., RSS (2005)

11. Kavraki, L.E., Švestka, P., Latombe, J.C., Overmars, M.H.: Probabilistic roadmaps for path planning in high-dimensional configuration spaces. IEEE Trans. Robot. Automat. 12(4), 566–580 (1996)
12. Kim, H.J., Vidal, R., Shim, D.H., Shakernia, O., Sastry, S.: A hierarchical approach to probabilistic pursuit-evasion games with unmanned ground and aerial vehicles. In: Proc. IEEE Conf. on Decision and Control, pp. 634–639 (2001)
13. Kolling, A., Carpin, S.: Multi-robot pursuit-evasion without maps. In: Proc. IEEE Int. Conf. Robot. Autom. (ICRA), pp. 3045–3051 (2010)
14. Kolling, A., Kleiner, A., Lewis, M., Sycara, K.: Solving pursuit-evasion problems on height maps. In: IEEE International Conference on Robotics and Automation (ICRA 2010) Workshop: Search and Pursuit/Evasion in the Physical World: Efficiency, Scalability, and Guarantees (2010)
15. Kolling, A., Carpin, S.: Pursuit-evasion on trees by robot teams. Trans. Rob. 26(1), 32–47 (2010)
16. Lavalle, S.M., Lin, D., Guibas, L.J., Latombe, J.c., Motwani, R.: Finding an unpredictable target in a workspace with obstacles. In: Proc. IEEE Int. Conf. Robot. Autom. (ICRA), pp. 737–742 (1997)
17. Parsons, T.D.: Pursuit-evasion in a graph. In: Theory and Applications of Graphs. Lecture Notes in Mathematics, vol. 642, pages 426–441. Springer, Heidelberg (1978)
18. Reynolds, C.W.: Flocks, herds, and schools: A distributed behavioral model. In: Computer Graphics, pp. 25–34 (1987)
19. Sud, A., Andersen, E., Curtis, S., Ming, L., Manocha, D.: Real-time path planning for virtual agents in dynamic environments. In: Virtual Reality Conference (March 2007)
20. Tovar, B., Murrieta-Cid, R., LaValle, S.M.: Distance-optimal navigation in an unknown environment without sensing distances. IEEE Transactions on Robotics 23(3), 506–518 (2007)
21. Treuille, A., Cooper, S., Popovi, Z.: Continuum crowds. ACM Trans. Graph. 25(3), 1160–1168 (2006)
22. van den Berg, J., Patil, S., Sewall, J., Manocha, D., Lin, M.: Interactive navigation of multiple agents in crowded environments. In: Proc. Symposium on Interactive 3D Graphics and Games - I3D 2008 (2008)
23. Wilmarth, S.A., Amato, N.M., Stiller, P.F.: MAPRM: A probabilistic roadmap planner with sampling on the medial axis of the free space. In: Proc. IEEE Int. Conf. Robot. Autom. (ICRA), vol. 2, pp. 1024–1031 (1999)
24. Yamauchi, B.: Frontier-based exploration using multiple robots. In: International Conference on Autonomous Agents (Agents 1998), pp. 47–53 (1998)
25. Yu, J., LaValle, S.M.: Probabilistic shadow information spaces. In: Proc. IEEE Int. Conf. Robot. Autom. (ICRA), pp. 3543–3549 (2010)

Path Planning for Groups
Using Column Generation*

Marjan van den Akker, Roland Geraerts, Han Hoogeveen, and Corien Prins

Institute of Information and Computing Sciences, Utrecht University
3508 TA Utrecht, The Netherlands
{marjan,roland,slam}@cs.uu.nl,
C.R.Prins@students.uu.nl

Abstract. In computer games, one or more groups of units need to move from one location to another as quickly as possible. If there is only one group, then it can be solved efficiently as a dynamic flow problem. If there are several groups with different origins and destinations, then the problem becomes $\mathcal{NP}$-hard. In current games, these problems are solved by using greedy *ad hoc* rules, leading to long traversal times or congestions and deadlocks near narrow passages. We present a centralized optimization approach based on Integer Linear Programming. Our solution provides an efficient heuristic to minimize the average and latest arrival time of the units.

1 Introduction

Path planning is one of the fundamental artificial intelligence-related problems in games. The path planning problem can be defined as finding a collision-free path, traversed by a unit, between a start and goal position in an environment with obstacles. Traditionally, this problem and its variants were studied in the field of robotics. We refer the reader to the books of Choset *et al.* [3], Latombe [11], and LaValle [12] for an extensive overview.

The variant we study is the problem of finding paths for one or more *groups* of units, such as soldiers or tanks in a real-time strategy game, all traversing in the same (static) environment. Each group has its own start and goal position (or area), and each unit will traverse its own path. We assume that the units in a group are equal with respect to size and speed. The objective is to find the paths that minimize the average arrival times of all units.

Current solutions from the robotics field can be powerful but are in general too slow for handling the massive number of units traversing in the ever growing environments in real-time, leading to stalls of the game. Solutions from the games field are usually fast but greedy and *ad hoc*, leading to long traversal times or congestions and deadlocks near narrow passages, in particular when two groups meet while moving in opposite directions. Obviously, such solutions have a negative impact on the gameplay.

---

* This work was partially supported by the ITEA2 Metaverse1 (www.metaverse1.org) Project.

One of the first solutions for simulating (single) group behavior was introduced by Reynolds in 1987 [17]. His influential boids model, comprising simple local behaviors such as separation, cohesion and alignment, yielded flocking behavior of the units. While this model resulted in natural behavior for a flock of birds or school of fish moving in an open environment, they could get stuck in cluttered areas. Bayazit *et al.* [2] improved this model by adding global navigation in the form of a roadmap representing the environment's free space. While the units did not get stuck anymore, they could break up, losing their coherence. By following a point that moves along a backbone path centered in a two-dimensional corridor, coherence was guaranteed by the method proposed by Kamphuis and Overmars [9]. In their method, the level of coherence was controlled by two parameters, namely the corridor width and the group area.

When multiple units are involved, possible interference between them complicates the problem, and, hence, some form of coordination may be required to solve the global problem. From the robotics field, two classes of methods have been proposed. Centralized methods such as references [18,19] compute the paths for all units simultaneously. These methods can find optimal solutions at the cost of being computationally demanding, usually making them unsuitable for satisfying the real-time constraints in games. Decoupled methods compute a path for each unit independently and try to coordinate the resulting motions [15,21]. These methods are often much quicker than centralized methods but the resulting paths can be far from optimal. Also hybrid methods such as references [7,13] have been proposed. A variant to solving the problem is called prioritized motion planning [14,23]. According to some prioritization scheme, paths are planned sequentially which reduces the problem to planning the motions for a single unit. It is however not clear how good these schemes are.

Our main contribution is that we propose a centralized as well as computationally efficient solution for the path planning planning problem with groups. This solution translates the problem into a dynamic multi-commodity flow problem on a graph that represents the environment and uses column generation to identify promising paths in this graph. More concretely, it provides the division of characters at each node (and each time step). A series of divisions can be considered as a *global* path. The assignment of such a path to a specific unit and performing *local* collision avoidance between units is handled by an external local method such as the Predictive model of Karamouzas *et al.* [10] or the Reciprocal Velocity Obstacles of van den Berg *et al.* [22]. When characters follow the same global path, Kamphuis' method [9] can be used to introduce coherence if desired. Our solution can be used to handle difficult situations which typically occur near bottlenecks (e.g. narrow passages) in the environment. It is efficient because it provides a global distribution of the paths. As far as we know, it is the first method that combines the power of a centralized method with the speed and flexibility of a local method.

Our paper is organized as follows. In Section 2, we show that path planning for one group can be solved to optimality as a dynamic flow problem. Computing the distribution of paths for multiple groups is more difficult (i.e. $\mathcal{NP}$-hard). We propose a new heuristic solution that solves the corresponding dynamic multi-commodity flow problem in Section 3. We conduct experiments on some hard problems in Section 4 and show that they can be solved efficiently. In Section 5, we discuss the applicability of this technique for path planning in games, and we conclude our paper with Section 6.

2 Path Planning for One Group

We are given one group of units, who all need to move from their origin p to their destination s. We assume that all units have equal width and speed. Our goal is to maximize the number of units that have reached q for each time t; using this approach, we automatically minimize both the average arrival time and the time by which all units have reached the destination. We further assume that the environment in which the units move is static.

To solve the problem, we first need a directed graph that resembles the free space in the environment. There are several ways to create such a graph. One possibility is to use tiles but this may lead to unnatural paths. A better alternative is to use a waypoint graph [16] in combination with a navigation mesh [6]. No matter how the graph has been constructed, we determine for each arc (i, j) in the graph its traversal time $l(i, j)$ as the time it takes to traverse the arc, and we determine its capacity c_{ij} as the number of units that can traverse the arc while walking next to each other. For instance in [6], the traversal time can be computed as the edge length divided by the maximum velocity and the capacity by the minimum clearance along the arc divided by the character's width. We choose the time unit as the time a unit has to wait until it can leave after the previous one. The path planning problem can then be modeled as a *dynamic flow problem* for which we have to determine a so-called *earliest arrival flow* from the origin to the destination. This problem can be solved by a classic algorithm due to Ford and Fulkerson [4], with a small adaptation due to Wilkinson [24]. The algorithm by Ford and Fulkerson computes a dynamic flow in an iterative version: given an optimal dynamic flow for the problem with $T - 1$ periods, an optimal dynamic flow for the T-period problem is constructed. Even though we do not have a deadline T but a number of units that have to go to the destination, we can use this algorithm by increasing the deadline each time by one time unit until all units have arrived.

When an optimal solution to the T-period problem has been constructed, Ford and Fulkerson's algorithm [4] splits it up in a set of *chain-flows*, which can be interpreted as a set of compatible paths in the graph. The flow (units in our case) are then sent through the graph following the chain-flows, where the last unit leaves the origin such that it arrives at the destination exactly at time T. Although the decomposition in chain flows maximizes the number of units that have arrived at the destination at time T, this solution does not need to be optimal when it is cut off at time t, even though their algorithm did find it as an intermediate product. Wilkinson [24] described a way to store the intermediate information of the algorithm to find an earliest arrival flow.

3 Path Planning for Multiple Groups

In this section, we consider the path planning problem for multiple groups of units. For each group, we are given the origin, the destination, and the size of the group. The goal is to minimize the average arrival time of all units. We assume that the graph that we use to model the problem is directed, that the capacities are constant over time, and that all units are available at time zero. At the end of this section, we describe what to do if these assumptions do not hold.

Since there are different groups with different origins and/or destinations, we do not have a dynamic flow problem anymore, but a *dynamic multi-commodity flow problem*, which is known to be $\mathcal{NP}$-hard in the strong sense. We present a new heuristic for the problem that is based on techniques from (integer) linear programming. We refer the reader to reference [25] for a description of this theory. The basic idea is that we formulate the problem as an integer linear program (ILP), but we restrict the set of variables by eliminating variables that are unlikely to get a positive value anyway. In this way, we make the problem tractable, without loosing too much on quality.

Instead of using variables that indicate for each arc at each time the number of units of group k that traverse this arc (an arc formulation), we use a formulation that is based on *paths* for each origin-destination pair. A path is described by the arcs that it uses and the times at which it enters these arcs. Here we require that the difference in the entering times of two consecutive arcs (i,j) and (j,k) on the path is no less than the traversal time $l(i,j)$ of the arc (i,j); if this difference is larger than $l(i,j)$, then this implies that there is a waiting time at j. Initially, we assume that there is infinite waiting capacity at all vertices. The advantage of using a formulation based on path-usage instead of arc-usage is twofold. First of all, we do not have to model the 'inflow = outflow' constraints anymore for each arc, time, and group. Second, we can easily reduce the number of variables by ignoring paths that are unlikely to be used in a good solution.

Suppose that we know all 'possibly useful' paths for each origin-destination pair. We can now model our path planning problem as an integer linear programming problem as follows. First, we introduce two sets of binary parameters to characterize each path $s \in S$, where S is the set containing all paths. The first one, which we denote by d_{ks}, indicates whether path s does connect origin/destination pair k (then d_{ks} gets value 1), or does not (in which case d_{ks} has value 0). The second set, which we denote by b_{ats}, keeps track of the time t at which arc a is entered by s: it gets value 1 if path s enters arc a at time t, and it gets value 0, otherwise. Note that these are parameters, which are fixed in advance, when the path s gets constructed. More formally, we have

$$d_{ks} = \begin{cases} 1 \text{ if path } s \text{ connects origin/destination pair } k \\ 0 \text{ otherwise} \end{cases}$$
$$b_{ats} = \begin{cases} 1 \text{ if path } s \text{ enters arc } a \in A \text{ at time } t \\ 0 \text{ otherwise.} \end{cases}$$

As decision variables we use x_s for each path $s \in S$, which will denote the number of units that follow path s. We use c_s to denote the cost of path s, which is equal to the arrival time of path s at its destination. We formulate constraints to enforce that the desired number y_k of units arrive at their destination for each origin/destination pair k and to enforce that the capacity constraints are obeyed. We define K as the number of origin/destination pairs, and we denote the capacity of arc $a \in A$ by u_a. We use T to denote the time-horizon; if this has not been defined, then we simply choose a time that is large enough to be sure that all units will have arrived by time T. This leads to the following integer linear program (ILP):

$$\min \sum_{s \in S} c_s x_s \quad \text{subject to}$$

$$\sum_{s \in S} d_{ks} x_s = y_k \quad \forall k = 1, \ldots, K$$

$$\sum_{s \in S} b_{ats} x_s \leq u_a \quad \forall a \in A; t = 0, \ldots, T$$

$$x_s \geq 0 \text{ and integral} \quad \forall s \in S.$$

Obviously, we do not know the entire set of paths S, and enumerating it would be impracticable. We will make a selection of the paths that we consider 'possibly useful', and we will solve the ILP for this small subset. We determine these paths by considering the LP-relaxation of the problem, which is obtained by removing the integrality constraints: the last constraint simply becomes $x_s \geq 0$ for all $s \in S$. The intuition behind taking the relaxation is that we use it as a guide toward useful paths, since the problems are so close together that a path which will be 'possibly useful' for the one will also be 'possibly useful' for the other. The LP-relaxation can be solved quickly because there is a clear way to add paths that improve the solution. We solve the LP-relaxation through the technique of column generation, which was first described by Ford and Fulkerson [5] for the multi-commodity flow problem.

Column Generation

The basic idea of column generation is to solve the linear programming problem for a restricted set of variables and then add variables that may improve the solution value until these cannot be found anymore. We can start with any initial set of variables, as long as it constitutes a feasible solution.

Given the solution of the LP for a restricted set of variables, we check if the current solution can be improved, and, if this the case, which paths we should add. It is well-known from the theory of column generation, in case of a minimization problem, that the addition of a variable will only improve the solution if its *reduced cost* is negative; if all variables have non-negative reduced cost, then we have found an optimal solution for the entire problem. In our case, the reduced cost of a path s, characterized by the parameters d_{ks} $(k = 1, \ldots, K)$ and b_{ats} $(a \in A; t = 0, \ldots, T)$ has reduced cost equal to $c_s - \sum_{k=1}^{K} \lambda_k d_{ks} - \sum_{a \in A} \sum_{t=0}^{T} \pi_{at} b_{ats}$, where λ_k $(k = 1, \ldots, K)$ and π_{at} $(a \in A; t = 0, \ldots, T)$ are the shadow prices for the corresponding constraints; these values follow from the solution to the current LP. The reduced cost takes the 'combinability' of the path s into account with respect to the current solution.

Since we are testing whether there exists a feasible path with negative reduced cost, we compute the path with minimum reduced cost. If this results in a nonnegative reduced cost, then we have solved the LP-relaxation to optimality; if the outcome value is negative, then we can add the corresponding variable to the LP and iterate. The problem of minimizing the reduced cost is called the *pricing problem*.

We break up the pricing problem into K sub-problems: we determine the path with minimum reduced cost for each origin/destination pair separately. Suppose that we consider the problem for the lth origin/destination pair; we denote the origin and destination by p and q, respectively. Since we have $d_{ls} = 1$ and $d_{ks} = 0$ for all $k \neq l$, the term $\sum_{k=1}^{K} \lambda_k d_{ks}$ reduces to λ_l, and we ignore this constant from now on. The resulting objective is then to minimize the adjusted path length $c_s - \sum_{a \in A} \sum_{t=0}^{T} \pi_{at} b_{ats}$. We will solve this as a *shortest path* problem in a directed acyclic graph.

We construct the following graph, which is called the *time expanded* graph. The basis is the original graph, but we add a time index to each vertex: hence, vertex i in the original graph corresponds to the vertices $i(t)$, with $t = 0, \ldots, T$. Similarly the arc (i, j) with traversal time $l(i, j)$ results in a series of arcs connecting $i(t)$ to $j(t + l(i, j))$. We further add waiting arcs $(i(t), i(t + 1))$ for each i and t. The length of the arc is chosen such that it corresponds to its contribution to the reduced cost. As c_s is equal to the arrival time of the path s in the destination, this term contributes a cost $l(i, j)$ to each arc $(i(t), j(t + l(i, j)))$ and cost 1 to each waiting arc. With respect to the term $-\sum_{a \in A} \sum_{t=0}^{T} \pi_{at} b_{ats}$, suppose that arc a corresponds to the arc (i, j). Then b_{ats}, with $a = (i, j)$, is equal to 1 if the path uses the arc $(i(t), j(t + l(i, j)))$ and zero otherwise; and, therefore, this term contributes $-\pi_{at}$ to the length of the arc $(i(t), j(t+l(i,j)))$, given that $a = (i, j)$.

Summarizing, we put the length of the arc $(i(t), j(t+l(i,j)))$ equal to $l(i, j) - \pi_{at}$, where $a = (i, j)$; the waiting arcs $(i(t), i(t + 1))$ simply get length 1. The path that we are looking for is the shortest one from $p(0)$ to one of the vertices $q(t)$ with $t \in \{0, \ldots, T\}$. We use the A^* algorithm [8] to solve this problem. We compute the reduced cost by subtracting λ_l. If this path has negative reduced cost, then we add it to the LP. Since we know the shortest paths from $p(0)$ to each vertex $q(t)$, we do not have to restrict ourselves to adding only the path with minimum reduced cost, if there are more paths with negative reduced cost. If in all K sub-problems the shortest paths have non-negative reduced cost, then the LP has been solved to optimality.

Obtaining an Integral Solution

Most likely, some of the variables in our optimal LP solution will have a fractional value, and, hence, we cannot follow this solution, as we cannot send fractions of units along a path. We use a heuristic to find a good integral solution. Since we have generated 'useful' paths when we solved the LP, it is a safe bet that these are 'useful' paths for the integral problem as well. Hence, we include all these paths in the ILP. Since there is no guarantee that these paths will enable a feasible solution to the ILP, we add some paths, which we construct as follows. First, we round down all decision variables, which leads to an integral solution that satisfies the capacity constraints, but in which too few units will arrive at their destination. For the remaining units, we construct additional paths using Cooperative A* by Silver [20]. These paths are added to the ILP, which is then solved to optimality by the ILP-solver CPLEX [1].

Extensions

Time constraints on the departure and arrival. It is possible to specify an earliest departure and/or latest possible arrival time for each group of units. These can be incorporated efficiently in the paths by restricting the time expanded graph. The only possible drawback is, if we put these limits too tight, that we may make the problem infeasible. Since the problem of deciding whether there is a feasible solution is $\mathcal{NP}$-complete, we apply a computational trick. We replace the y_k in the constraint that y_k units have to move from the origin to the destination by $y_k + Q_k$, where Q_k is an artificial variable measuring the number of units of group k that did not reach their destination. We now add a term $\sum_{k=1}^{K} w_k Q_k$ to the objective function, where w_k is a large penalty weight, which makes it unattractive for the units not to reach their target.

Changes in the environment. A change in the environment may lead to a change in the capacity of an arc (for example that it drops to zero if the arc gets closed) or to a change in the traversal time in a certain period. If we know the changes beforehand, then these are incorporated efficiently in our model. A change in the capacity can be modeled by making the capacity of arc a time dependent; the right-hand side of the capacity constraint then becomes u_{at} instead of u_a. A change in the traversal times can be modeled by changing the arcs in the graph that we use to solve the pricing problem.

Undirected edges in the graph. An undirected edge can be traversed both ways, which makes it much harder to model the capacity constraint. If, for example, the traversal time is l and we want to send x units through the edge at time t, then this is possible only if the number of units that start(ed) to traverse the edge from the other side at times $t - l + 1, t - l + 2, \ldots, t + l - 1$ does not exceed the remaining capacity. To avoid having to add this enormous number of constraints, we split such an edge e in two arcs, e_1 and e_2, which have a constant capacity over time. We do not fix the capacity distribution beforehand, but we make it time-independent by putting the capacities equal to u_{e_1} and u_{e_2}, which are two non-negative decision variables satisfying that $u_{e_1} + u_{e_2}$ is equal to the capacity of the edge. We can modify this time-independent capacity distribution a little by making u_{e_1} or u_{e_2} equal to zero until the first time it can be reached from any origin that is part of an origin/destination pair which is likely to use this arc. Similarly, in case of edges with capacity 1, we can fix $u_{e_1} = 1$ or $u_{e_2} = 1$

during given periods in time. Splitting an undirected edge results in the addition of two variables and one constraint.

4 Experiments

In this section, we will describe the experiments we have conducted. In particular, we investigated the efficiency of our solution on three difficult problems. The solution from Section 3 was implemented in C++ using the ILOG CPLEX Concert Technology library version 11.100 for solving the LPs and ILPs [1]. All the experiments were run on a PC (CentOS linux 5.5 with kernel 2.6.18) with an Intel Core 2 Duo CPU (3 GHz) with 2 GB memory. Only one core was used.

Each experiment was deterministic and was run a small number of times to obtain an accurate measurement of the average integral running times (in ms). These times include the initialization of the algorithm (such as the data structures, CPLEX, heuristics, building the initial LP), the column generation, the solving of the LP and ILP, path finding, and making an integer solution.

One Group

Fig. 1(a) shows the problem where one group moved from node 0 to node 6. The experiment was carried out for a single group with 100 through 500 units. Because it may be inefficient to let all units use the shortest path (e.g. when the capacity of the shortest path is low), it may be better to let some units take an alternative path. Indeed, as is shown in Fig. 1(b), the group (with 100 units) was split to minimize the average arrival times. The algorithm took $10ms$ for 100 units, $40ms$ for 200 units, $100ms$ for 300 units and $250ms$ for 400 units. Even with 500 units the algorithm took less than half a second. Note that these times can be distributed during the actual traversal of the paths, yielding real-time performance. In a game situation, the units should already start moving when the algorithm is executed to avoid stalls.

Two Groups Moving in Opposite Directions

In the following case, as is displayed in Fig. 2(a), two groups moved in opposite directions while switching their positions. One group started at node 0 and the other one at node 3. Since the arcs had limited capacities, the units had to share some arcs. There were two different homotopic paths between these two nodes, and both paths could be used by only 5 units per timestep. On the left side we placed 10 units and on the right side we had 50 units. Computing the solution took only $10ms$. Also other combinations were tested, e.g. 20 versus 50 units ($20ms$), 20 versus 100 units ($40ms$), 40 versus 100 units ($70ms$), and 40 versus 200 units ($230ms$). The latter case is visualized in Fig. 2(b). Here, most units from the right side used the lower path, while some used the upper path. All the units from the left side used the upper path. Again, these running times were sufficiently low for real-time usage.

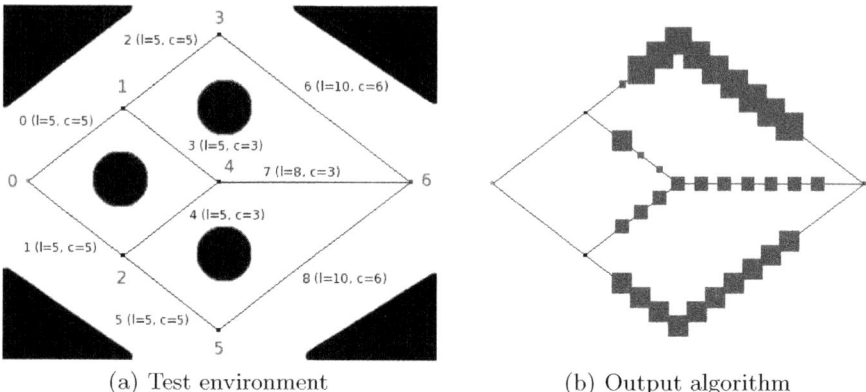

(a) Test environment   (b) Output algorithm

Fig. 1. (a) The environment used for testing the division of units among the arcs. The (large) red numbers show the node numbering and the black numbers show the arc numbering. For every arc we give the length l and capacity c. (b) The output of the algorithm for 100 units at timestep 16. The pink squares symbolically represent the units, and the width of a square is proportional to the number of units.

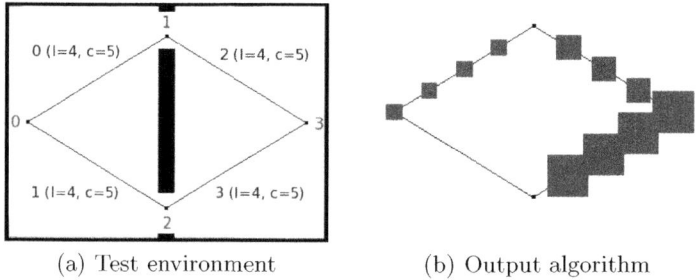

(a) Test environment   (b) Output algorithm

Fig. 2. (a) The environment and graph used for testing two groups moving in opposite directions, i.e. one group starts at node 0 and the other one starts at node 3. (b) The output of the algorithm for 40 versus 200 units at timestep 3.

Four Groups with Many Units Moving in a Big Graph

We created a large graph whose structure was a raster with arcs between the raster points. The length of these arcs was set to 3 with capacity 20. We refer the reader to Fig. 3 for an illustration of this graph and the output of the algorithm. In every corner we placed 100 units that needed to move to their diagonally opposite corners. Computing their paths took $510ms$. We also tested the algorithm with 1000 units placed at each corner (where the capacities were scaled with the same proportion), which took $480ms$. The results clearly illustrated the scaling power of the algorithm as it did not slow down when both the number of units and the capacities were scaled proportionally.

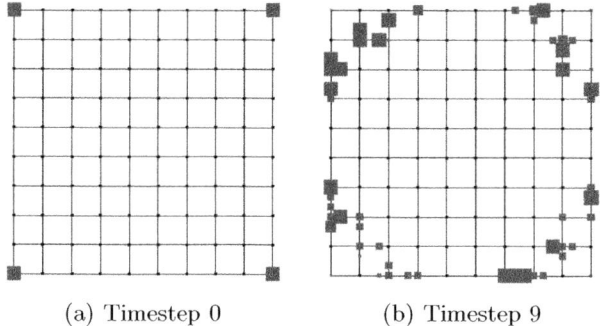

(a) Timestep 0   (b) Timestep 9

Fig. 3. (a) The test environment is empty and the graph's nodes lie on a raster. (b) The output of the algorithm is displayed at timestep 0 and 9.

5 Applicability in a Game

There are three main challenges that we face when we want to apply our technique in a game. First, all motions should be computed in real-time. Second, the environment may change while the units are moving, thereby making our solution infeasible. Third, how to satisfy game-specific constraints? For example, in combat games, it may be undesirable that a single unit (or a small troop of units) follows a path that separates them from the main group. Moreover, there can be other units with different sizes and velocities (for example soldiers and tanks). Below we will describe how we can tackle these issues, at the expense of a small loss in quality of the solution.

There are several ways to reduce the running time. The first one is to find a feasible solution by rounding instead of by solving the ILP. Since rounding down will reduce the number of units for which we find paths, we artificially increase the number of units that need to move from each origin to destination. An additional advantage of this increase is, if we have more routes for the units than we need after rounding, that we can select the best paths in a post-processing step. If necessary, we can reduce the running time even further by quitting the column generation before we have solved the LP-relaxation to optimality. In any case, the remainder of the computations can be done while the first units start moving to avoid stalls.

As mentioned in Section 3, changes in the environment are incorporated easily in the algorithm, because we can usually reuse the large part of the solution that is not affected by the change. Hence, in such a case, we initialize with the former set of paths, adjust the capacity constraints, and add additional paths in the column generation phase. We let the units depart from their current position as much as possible, but a group that is spread over an edge is artificially grouped in one or both of the edge's endpoints to reduce the number of origins; this is not a major violation of the truth since most units cannot leave from this origin immediately due to a limited capacity, and during this waiting time they can move to the origin from their current position. If there are obvious targets for sabotage, like bridges, then we can already incorporate these possibilities by computing additional paths circumventing these edges while the current solution is being executed in the game.

Avoiding isolated units can be achieved by post-processing a solution in which too many units move. If this is not satisfactory, then we can remove these isolated paths from the set of available paths and resolve the LP. If there are other types of units like tanks in the game, then we could first find paths for the tanks, and, given this solution, find paths for the units subject to the remaining capacity.

6 Conclusion

We have presented a centralized method based on techniques from ILP for path planning problems involving multiple groups. The crux is that the LP-relaxation can be solved quickly by using column generation. The solution to the LP-relaxation can be used as a basis to construct a heuristic solution. We have described a method to find a good approximation by solving a restricted ILP. If the instance is so big that solving the ILP would require too much time, then we can still use the solution to the LP-relaxation to find a solution by clever rounding. The units can then already start moving according to this solution while a good solution for the remaining units is determined in the meantime. We further have described ways to address the special constraints that are posed upon us by a game.

In future work, we will integrate two efficient local collision-avoidance models [10,22] to test whether our solution leads to visually pleasing motions. We think that our solution enhances the gameplay in difficult path planning situations involving one or multiple groups.

References

1. CPLEX 11.0. User's manual. Technical report, ILOG SA, Gentilly, France (2008)
2. Bayazit, O., Lien, J.-M., Amato, N.: Better group behaviors in complex environments using global roadmaps. Artificial Life, 362–370 (2002)
3. Choset, H., Lynch, K., Hutchinson, S., Kantor, G., Burgard, W., Kavraki, L., Thrun, S.: Principles of Robot Motion: Theory, Algorithms, and Implementations, 1st edn. MIT Press, Cambridge (2005)
4. Ford Jr., L., Fulkerson, D.: Constructing maximal dynamic flows from static flows. Operations Research 6, 419–433 (1958)
5. Ford Jr., L., Fulkerson, D.: A suggested computation for maximal multi-commodity network flows. Management Science 5, 97–101 (1958)
6. Geraerts, R.: Planning short paths with clearance using explicit corridors. In: IEEE International Conference on Robotics and Automation, pp. 1997–2004 (2010)
7. Ghrist, R., O'Kane, J., LaValle, S.: Pareto optimal coordination on roadmaps. In: International Workshop on the Algorithmic Foundations of Robotics, pp. 171–186 (2004)

8. Hart, P., Nilsson, N., Raphael, B.: A formal basis for the heuristic determination of minimum cost paths. IEEE Transactions on Systems Science and Cybernetics 4, 100–107 (1968)
9. Kamphuis, A., Overmars, M.: Finding paths for coherent groups using clearance. In: Eurographics/ACM SIGGRAPH Symposium on Computer Animation, pp. 19–28 (2004)
10. Karamouzas, I., Heil, P., van Beek, P., Overmars, M.: A predictive collision avoidance model for pedestrian simulation. In: Egges, A. (ed.) MIG 2009. LNCS, vol. 5884, pp. 41–52. Springer, Heidelberg (2009)
11. Latombe, J.-C.: Robot Motion Planning. Kluwer, Dordrecht (1991)
12. LaValle, S.: Planning Algorithms. Cambridge University Press, Cambridge (2006)
13. LaValle, S., Hutchinson, S.: Optimal motion planning for multiple robots having independent goals. Transaction on Robotics and Automation 14, 912–925 (1998)
14. Li, Y.: Real-time motion planning of multiple agents and formations in virtual environments. PhD thesis, Simon Fraser University (2008)
15. Peng, J., Akella, S.: Coordinating multiple robots with kinodynamic constraints along specified paths. International Journal of Robotics Research 24, 295–310 (2005)
16. Rabin, S.: AI Game Programming Wisdom 2. Charles River Media Inc., Hingham (2004)
17. Reynolds, C.: Flocks, herds, and schools: A distributed behavioral model. Computer Graphics 21, 25–34 (1987)
18. Sánchez, G., Latombe, J.-C.: Using a PRM planner to compare centralized and decoupled planning for multi-robot systems. In: IEEE International Conference on Robotics and Automation, pp. 2112–2119 (2002)
19. Schwartz, J., Sharir, M.: On the piano movers' problem: III. Coordinating the motion of several independent bodies: The special case of circular bodies moving amidst polygonal obstacles. International Journal of Robotics Research 2, 46–75 (1983)
20. Silver, D.: Cooperative pathfinding. In: Artificial Intelligence for Interactive Digital Entertainment, pp. 117–122 (2005)
21. Siméon, T., Leroy, S., Laumond, J.-P.: Path coordination for multiple mobile robots: A resolution complete algorithm. IEEE Transactions on Robotics and Automation 18, 42–49 (2002)
22. van den Berg, J., Lin, M., Manocha, D.: Reciprocal velocity obstacles for real-time multi-agent navigation. In: IEEE International Conference on Robotics and Automation, pp. 1928–1935 (2008)
23. van den Berg, J., Overmars, M.: Prioritized motion planning for multiple robots. In: IEEE/RSJ International Conference on Intelligent Robots and Systems, pp. 2217–2222 (2005)
24. Wilkinson, W.: An algorithm for universal maximal dynamic flows in a network. Operations Research 19, 1602–1612 (1971)
25. Wolsey, L.: Integer Programming. Wiley, New York (1998)

Skills-in-a-Box: Towards Abstract Models of Motor Skills

Michiel van de Panne

Department of Computer Science,
University of British Columbia
van@cs.ubc.ca

Abstract. When creating animated characters, we wish to develop models of motion that can generalize beyond a specific set of recorded motions. Significant progress has been made towards this end, including the development of controllers of growing sophistication for physically-simulated characters. In this talk, we take stock of some of the recent advances in the control of locomotion, with an eye towards identifying some of the common elements that might help define a future era of 'downloadable skills'. We argue in favor of learned memory-based models that allow for skills to be developed and integrated in an incremental fashion.

1 Introduction

Abstract models and manipulations lie at the core of computer graphics today. They allow us to work with reflectance functions, deformation handles, and elasticities instead of pixels, triangles, and individual animation frames. A long-standing goal in character animation has been the creation of abstract models capable of describing rich families of human and animal motions. One approach is to attempt to model motor skills, such as running or playing tennis, which can then be used together with a physics-based simulation as a highly general basis for creating motion. What then is a good 'file-format' or language that allows such skills to be efficiently expressed and shared? How can these skills be authored or captured from data? We attempt to provide an integrated view of a number of the insights gained by ourselves[11,10,3,2,1] and others[4,9,7,8,6,5] into these problems over the past several years. Our focus is on biped locomotion, given its important role as a starting point for nearly any set of skills.

2 Locomotion Skills

We first review two techniques that provide simple and robust control for physics-based walking, SIMBICON[11] and GBWC[1] (Generalized Biped Walking Control). With these techniques and related work by others, we argue for a 'motor memory' abstraction that is augmented by simple feedback loops. We show that full knowledge of the dynamics is not necessary when computing the control, although it can be implicitly incorporated with the use of learned feed-forward control when modeling the skilled execution of dynamic motions. Basic state abstraction, i.e., knowledge of center of mass and point of support, PD controllers, and knowledge of the kinematics are sufficient to achieve flexible and highly robust control of walking.

Using the previously-described techniques as a starting point, we then show it is possible to learn a wide range of related locomotion tasks using continuation methods [10]. A key lesson is that a highly effective way to develop a skill is to begin with a 'sweet spot' with a limited-but-robust repertoire of motion, and to gradually expand this to then yield rich, integrated repertoires of motion. This yields parameterized skills that can climb tall steps, step over obstacles, push heavy furniture, and walk on ice. Using a set of offline stepping-stone exercises, we can further learn how to walk in highly constrained environments[3]. Lastly, we can learn how to optimally integrate sequences of walking, running, and turning steps to quickly reach a target in the environment, all the while remaining highly robust to perturbations [2].

3 Conclusions

In summary, we argue for motor-memory-plus-simple-feedback as being a flexible and effective model for simulated locomotion skills. Incremental learning-and-integration of motions then offers a pathway towards a much wider set of integrated motions. We argue for learning-based approaches as being necessary to achieve scalable development of ever-broader skill-sets for agile simulated characters and robots.

References

1. Coros, S., Beaudoin, P., van de Panne, M.: Generlized Biped Walking Control. ACM Trans. on Graphics (Proc. SIGGRAPH) 29(4), Article 130 (2010)
2. Coros, S., Beaudoin, P., van de Panne, M.: Robust task-based control policies for physics-based characters. ACM Trans. on Graphics (Proc. SIGGRAPH ASIA) 28(5), Article 170 (2009)
3. Coros, S., Beaudoin, P., Yin, K., van de Panne, M.: Synthesis of constrained walking skills. ACM Trans. on Graphics (Proc. SIGGRAPH ASIA) 27(5), Article 113 (2008)
4. Hodgins, J., Wooten, W., Brogan, D., O'Brien, J.: Animating human athletics. In: Proc. ACM SIGGRAPH, pp. 71–78 (1995)
5. de Lasa, M., Mordatch, I., Hertzmann, A.: Feature-based locomotion controllers. ACM Trans. on Graphics (Proc. SIGGRAPH) 29(4), Article 131 (2010)
6. Lee, Y., Kim, S., Lee, J.: Data-driven biped control. ACM Trans. on Graphics (Proc. SIGGRAPH) 29(4), Article 129 (2010)
7. Macchietto, A., Zordan, V., Shelton, C.R.: Momentum control for balance. ACM Trans. on Graphics (Proc. SIGGRAPH) 28(3) (2009)
8. Muico, U., Lee, Y., Popovic', J., Popovic', Z.: Contact-aware nonlinear control of dynamic characters. ACM Trans. on Graphics (Proc. SIGGRAPH) 28(3), Article 28(3), 81 (2009)

9. Wang, J., Fleet, D.J., Hertzmann, A.: Optimizing walking controllers. ACM Trans. on Graphics (Proc. SIGGRAPH Asia) (2009)
10. Yin, K., Coros, S., Beaudoin, P., van de Panne, M.: Continuation methods for adapting simulated skills. ACM Transactions Graph. (Proc. SIGGRAPH) 27(3) (2008)
11. Yin, K., Loken, K., van de Panne, M.: SIMBICON: Simple biped locomotion control. ACM Trans. on Graphics (Proc. SIGGRAPH) 26(3), Article 105 (2007)

Angular Momentum Control in Coordinated Behaviors

Victor Zordan

University of California, Riverside

Abstract. This paper explores the many uses of angular momentum regulation and its role in the synthesis of coordinated motion generated with physically based characters. Previous investigations in biomechanics, robotics, and animation are discussed and a straightforward organization is described for distinguishing the needs and control approaches of various behaviors including stepping, walking and standing balance. Emphasis is placed on creating robust response to large disturbances as well as on the types of characteristic movements that can be generated through the control of angular momentum.

Keywords: Character animation; Physics models; Behavior control.

1 Introduction

Natural movement of humans reveals a high degree of coordination in which the entire body moves in concert to perform a given task. Even seemingly simple behaviors such as basic walking and balancing show that the arms, trunk, and legs work together to produce the signature motions we identify as humanlike. Making characters (and robots) that move with the sophistication of humans has been a long standing goal of many researchers over several decades. And recent advances in computer animation, biomechanics, and robotics have begun to explore the ability to produce whole-body coordinated motions in the control of humanoids. Beyond ad-hoc heuristics that function to perform a given task, the drive to uncover the core principles of human coordination is an increasingly interesting motivation for research in the control of animation for characters.

One exciting trend in the investigation of human motion control is the exploration of the contribution of angular momentum in the production of behaviors. Clear examples in humans, such as windmilling (to aid in balance), point to the obvious presence of angular momentum in movement with gross rotational components. Less obvious is that precise regulation appears in human motions with seemingly little rotational component. For example, recent studies in biomechanics have shown that careful angular momentum regulation appears in normal walking [1]. This finding is particularly interesting because it is not obvious that walking should include very precise control over whole-body rotation. With such compelling evidence and findings in biomechanics, there is surprisingly little exploration of angular momentum as it applies to control in physics based human animation.

In this paper, we revisit the effects of angular movement in various tasks, both observed in humans and controlled in humanoid robots and simulations. We summarize the related findings from biomechanics, robotics, and animation and draw correlations between them in order to propose a division of angular momentum controllers into two categories based on behavior. We highlight testbeds with example implementations for each category as well as describe a supervisor which uses angular momentum to perform planning. We conclude with a discussion of the open questions and possibilities related to the use of angular momentum in the control of coordinated behavior.

2 Related Work

2.1 Angular Momentum Control in Biomechanics

Popovic, Herr and several colleagues [1,2,3,4,5] have collectively performed the most thorough study of the role of angular momentum in human movement to date. They postulate that *whole-body (spin) angular momentum may be regulated directly by the central nervous system.* In their investigations, they have studied walking motion in depth and report observing surprisingly small angular momentum values in straightline walking which lead to very small (2 degree) angular excursions over entire cycles of normal subjects' walking. Based on such observations, they propose that walking is regulated to have *Zero Spin* (ZS) angular momentum about the center of mass (CM). For walking and other "ZS" behaviors, their hypothesis is that both angular momentum and its time derivative are regulated to remain close to zero. Several of their various findings support this hypothesis. Along with data analysis and models of human subjects, these researchers have also spelled out the value of regulating angular momentum in control for humanoid robots and they have implemented and described a handful of simulations.

Along with ZS behaviors such as walking, this research team has also suggested that there are behaviors which have *Non-Zero Spin* (NZS) angular momentum [1]. Within this group they include "large and rapid turning motions" as well as motions in response to "sufficiently large disturbances". Interestingly, they also report that, upon entering the conditions of these behaviors, human subjects exhibit an observable switch in control strategy away from the ZS control described for walking [5]. One of the few related investigations thus far is in the strategies used for turning in walking behaviors which indicates that a significant non-zero spin angular momentum is induced (largely by the swing leg) during normal turning. Also, in a recent paper [3], they propose that humans modulate spin momentum "to enhance CM control." Through simulation they show that a simplified, seven-link humanoid can induce momentum to bring its CM within its support after starting from a statically instable balance state (i.e. where the CM starts from rest outside of the support.) Their finding is supported by human examples of the same phenomena and *indicates a distinct need to be able to induce angular momentum* in order to regain stable balance.

We postulate that humans may be regulating angular momentum constantly to carry out a wide class of rotationally rich "NZS" behaviors. Herr and Popovic support this hypothesis in their analysis of a human performing a "Hula" action - a dynamic but sustained and stable motion which shows an order of magnitude more angular momentum than walking [4]. Beyond this sparse set of examples, such NZS behaviors have not been studied (with respect to angular momentum and control) by the biomechanics research community to date.

2.2 Angular Momentum Control in Robotics

Robotics researchers have postulated ways in which control of angular momentum can increase controller robustness while lead to coordinated motions for humanoid robots. Much of the recent effort in this area follows from the work of Kajita et al. [6]. Their "resolved momentum control" strategy appears to be the first in which angular momentum is controlled simultaneously with linear momentum (i.e. control over the CM), added deliberately to make behavior control easier. In this work, they tout the benefit of combined momenta to "describe the macroscopic behavior of the entire robot" independent of its structure. In this work, and similar follow-on work [7], they show results applied to humanoid robots where the angular momentum about the vertical axis is driven to near zero for activities such as a kick and walking.

Goswami and Kallem [8] support angular momentum guidance as a robust method for controlling biped robots. They suggest that a controller might directly guide the time derivative of angular momentum, $\dot{H}$, and state that this term is "physically central to rotational instability and intuitively more transparent to the phenomena of tipping and tumbling" than derived quantities such as ZMP and centroid moment pivot (defined below.) They also describe theoretical strategies for regaining stability, each reducing to a unified control approach: return to a condition where $\dot{H} = 0$, that is leading from an NZS to ZS state.

Several papers appear in the robotics literature that employ some variant of angular momentum control. While the proposed control laws vary, all propose simple heuristic-based control laws crafted for specific effects. Kajita and colleagues set the angular momentum to be zero for control of their humanoid robot [6]. Abdallah and Goswami use a momentum controller to absorb disturbance effects [9]. They suggest that during large external perturbations humans absorb impact by preserving momentum for a specified period of time. After the impact has been absorbed, the character recovers its posture. Stephens employs a bang-bang control to use the body like a flywheel, applying maximum torque as necessary [10]. One common theme in all of these papers is that each treats the control of angular momentum as a damper (that is, to dissipate a disturbance). Collectively, they show that simple control laws can be very effective. Notably however, no exploration of *sustained* NZS behavior control appears in robotics, to our knowledge.

2.3 Angular Momentum Control in Animation

Most of the full-body control work that has been proposed in computer animation to date has employed some mechanism for controlling the CM and/or its derivatives often by generating a reference position and, possibly, a reference velocity. This has lead to a host of controllers for activities including walking, leaping, running, and standing (balance). While heuristic, tuned controllers have been shown to be very robust, they often appear robotic. In recent years, such manual approaches have given way to more automatic controllers which compute joint torques for the full-body based on a relatively small set of control inputs - usually some combination of reference joint trajectories, CM trajectories, and constraints (e.g. to keep contact forces within friction limits). While it can be shown that controlling the CM while maintaining the ground reaction forces (GRF) to remain within a friction cone can yield indirect changes in angular momentum, this phenomena is generally discouraged in an optimization framework in lieu of energy efficiency unless it is ultimately "necessary". That is, only under extreme conditions (e.g. when the CM is close to or on the boundary of the support) will a character exhibit noticeable changes associated with angular momentum.

Only a handful of papers have investigated control which drives angular momentum in an explicit manner. Kudoh and colleagues propose a controller that employs angular momentum to constrain the (predicted) zero moment point (ZMP) to remain in the support [11]. Their animations exhibit a reaction similar to those seen in CM control with GRF constraints because their system induces momentum only when the ZMP is at the edge of the support. Macchietto et al. [12] uses a sustained NZS controller that produces response to disturbances that resist fast changes to the ZMP while directing it to a safe desired location. More detail on this research is outlined in Section 4. de Lasa et al. [13] follow ZS behavior characteristic described in biomechanics ($H = 0$) in their control for locomotion but also induce momentum about the vertical axis ($H_z > 0$) for turning jumps, consistent with NZS control. Ye and Liu [14] apply a ZS strategy for all behaviors and succeed in producing a large number of motions, all with visible coordinated rotation across the body. [15] employs a hybrid approach similar to Kajita et al. [6] for generating steps ($H_z = 0$). Using this technique, control over vertical spin momentum is shown to yield characteristic arm movement for walking without the need for special treatment or a specific joint reference trajectory.

3 Mechanics of Momentum

In the absence of external force, the linear and angular momenta of a system, denoted L and H, are conserved. Momentum change comes from the ground reaction forces (GRF) and force due to gravity as well as additional external forces if they are present. In the case of flat ground and no additional external forces, we can summarize the momenta change simply. Assuming the aggregate GRF force, f, is applied at position, p, then $\dot{L} = mg + f$ and $\dot{H} = s \times f$ where g is

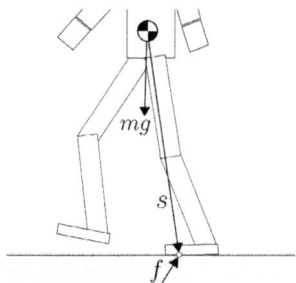

Fig. 1. Static force analysis for a standing character

the gravitational constant, m is the total mass, and c is the CM of the character. $s = p - c$ as seen in Figure 1.

By simple inspection, we can see that the derivative of the linear momentum is the same as the mass-scaled CM acceleration. Note this coupling implies that control over the CM (and its time derivatives) as has been seen in most previous control approaches for locomotion across disciplines is equivalent to control of linear momentum (and its derivatives), assuming mass is constant. We can also see that the angular momentum change completely defines the relationship between the position of the center of mass, c, based on the aggregate GRF applied at point p. Finally, together with the GRF (applied at p) these two momenta rates can be integrated to yield the *complete rigid motion of the character about its CM*.

As a quick aside, what is the point p? In biomechanics, this point is commonly called the center of pressure (CP) or the point where aggregate continuous forces is applied equivalently (i.e. without adding additional torque). In robotics, this point is often considered synonymous with the ZMP, and it is, in the case of flat horizontal ground. However, the ZMP can be shown to differ from the CP when the ground is irregular [8,16]. Formally, the ZMP is the point where the net moment about the GRF has zero-value in the horizontal components, which is the CP - on flat ground. In literature in robotics and animation, the ZMP is often "predicted" to be outside of the support - which is physically inconsistent with its definition. Because there is less confusion and misuse of the meaning of the term, we opt to use the term "CP" in lieu of "ZMP" whenever possible. See [16] for a lengthier discussion on the topic.

In the related biomechanics and robotics literature, researchers point out that for a given (non-zero) GRF only one possible position for p will yield zero angular momentum change (assuming flat ground.) They call this point the centroid moment pivot (CMP) [16,8,2]. Unlike the ZMP or CP, this point is a derived value which is defined in a manner in which it may or may not reside within the support. The beauty of this term is that, just as the location of the CM projected on the ground plane can indicate stability in a standing character, the CMP can be used to indicate *rotational stability*. If the CMP is within the support, then the CP could theoretically be aligned with the CMP and a zero angular momentum could be realized. However, if the CMP is computed to be outside of the support then the CP cannot meet the CMP. To avoid tipping, angular momentum must be added to the system and the character will invariably experience some amount of whole-body rotation (for non-zero GRF.)

This characteristic leads us to a mechanism for cleanly dividing the two classes of control identified, namely ZS behaviors and NZS behaviors. As the CMP crosses the edge of the support, we infer that the control changes from one to the other depending on the direction. We organize the remainder of this paper based on this distinction. First, we highlight an approach for controlling NZS behaviors through an analysis of response to large disturbances in regaining stability in balance. Next, we explore aspects of ZS behaviors, specifically as they apply in the production of steps and walking. Finally, we describe a mechanism developed for planning which exploits knowledge about the angular momentum independent of the behavior control approach used.

4 Balance Control

In our recent investigation in standing balance, we explore a method for controlling the CP and the CM simultaneously in order to respond in the presence of large disturbances [12]. To this extent, we propose a multi-objective technique which synthesizes animation according to three objectives, namely an objective for tracking a reference motion (taken from human motion capture) and two objectives that control changes in linear and angular momenta. By converting each objective into a set of desired accelerations, we combine them into a single quadratic optimization problem which can be solved in a straightforward and efficient manner. The resulting accelerations are applied to the character based on torques computed from a floating-base hybrid inverse dynamics calculation. In our results, we show a suite of animations which reveal full-body coordinated response to a host of different conditions and inputs.

One of the contributions of this work is in the method used to control the CP and CM simultaneously. Exploiting the described relationship between momenta and stability, at each run of the optimization we convert target CP and CM values to a pair of desired changes in momenta. By accounting for the dependencies created through the requirement of a unified GRF, the two desired momenta terms are computed to create a consistent desired change in the character's center of mass and full-body rotation. In contrast, computing these terms separately would lead to the objectives competing and both could not be fully realized. Notably, de Lasa and colleagues get around this issue in their work through a prioritization approach, placing the linear momentum objective ahead of the angular momentum objective in priority [13].

From the perspective of ZS/NZS angular momentum classification, our balance control approach falls succinctly in the NZS category because the technique is inherently asking for non-zero momenta changes which lead to non-zero angular momentum behavior. In contrast to other controllers, our approach is unique in that we purposefully drive angular momentum changes that will control the

CP to move in a smooth, deliberate manner throughout the entire behavior. Kudoh et al [11] instead put barrier constraints on the ZMP which is quite similar to constraining the CM with friction limits. As Kudoh's controller constrains the ZMP to stay within the support, it only "turns on" angular momentum control when the ZMP hits the support boundary. In this condition, the situation is already at an extreme, and bang-bang like control response is needed to quickly move the ZMP away from the support edge. However, human motion doesn't appear to support this approach [3]. Instead a smooth, less extreme ZMP trajectory seems desirable based on the data. A more common strategy for responding to disturbances is to damp the movement of the ZMP or the angular momentum present but without explicitly guiding the ZMP/CP it can still head to the edge and lead to rotational instability. These approaches treat the NZS case as if it is simply a transition back to ZS control.

In our control approach, we bring the (ZMP/)CP toward the center of support, but do so gradually and smoothly. By damping its motion, we avoid abrupt changes that lead quickly to rotational instability (e.g. tipping) and by persistently guiding the CP back to the center of support the controller directs the motion of the character to a safer stance. When we turn off the angular momentum objective, our system exhibits balance which is both less robust based on the size of impact and contains fewer of the rotational artifacts we observe in humans related to balance, for example raising the arms in response to a push backward. The extent of our system's ability to control the angular momentum is highlighted in Figure 2.

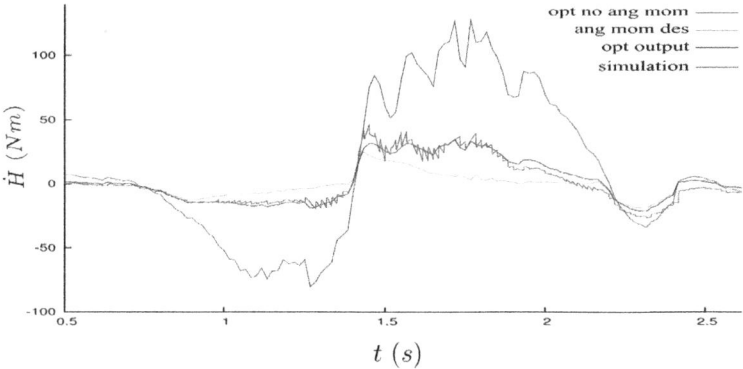

Fig. 2. Response to a large impulse. Angular momentum change observed under the no-control condition (red) is highly reduced when control is applied. The discrepancy between the desired (green) and the optimization solution (blue) accounts for the other objectives, tracking and moving the CM. The simulation (purple) closely follows the optimization request. The jagged simulation response is due to the fact that the optimization is updated at a lower frequency, revealing that assumptions about GRF and so on quickly become stale - this could be remedied if we were willing to run the optimization in lockstep with the simulation.

5 Locomotion

In a more recent publication [15], we explore a momentum-based strategy that is aligned with the ZS category of control. In that paper, we expand the described framework for balance control to produce stepping behaviors produced by specifying trajectories for the CM and the swing foot. The desired motion of the swing foot is translated into a joint-angle reference trajectory and in doing so we can use the same solver implemented for the balance paper. However, instead of guiding the CP as we did previously, we specify angular momentum directly similar to the scheme described by Kajita et al. in their resolved momentum control paper [6]. Specifically, we enforce $\dot{H}_z = 0$ while the other spin momentum terms are uncontrolled. The effect of this controller is to restrict twist (yaw) rotation, thereby maintaining a consistent facing direction while allowing rotation in roll and pitch. Since one goal was to stop in presence of disturbances, the

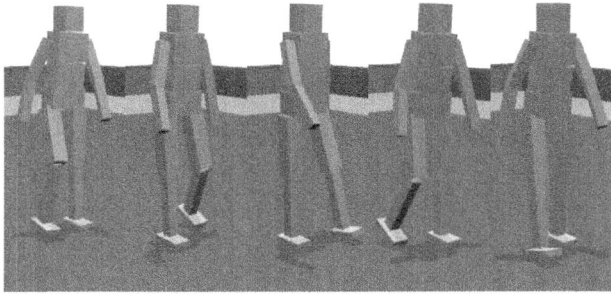

Fig. 3. Arm swing is produced which reasonably imitates what is present in humans even though the tracked trajectory for this animation is a single arms-at-side pose

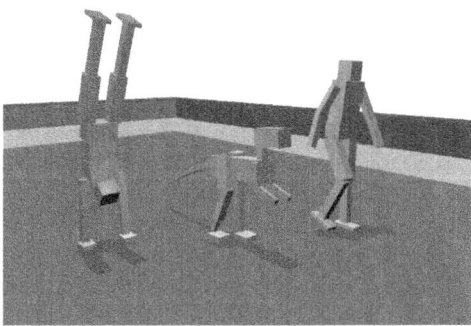

Fig. 4. Various walking gaits. In these animations, additional leg swing (left) and tail motion (center) act functionally the same as the arm swing movements seen in the basic walker (right). These "secondary" motions are generated automatically from angular momentum control. (Parameters for the dinosaur model from [17].)

latter is a means of allowing the character to yield to any momentum changes following an impact which is a behavior identified as humanlike by Abdallah and Goswami [9]

By specifying a series of equal length steps, this controller can also produce a walking gait. While the method of angular momentum regulation is more relaxed than other control techniques (such as [13] and others from robotics), the controller successfully produces walking behavior with distinct humanlike characteristics (see Figure 3). Further, the approach does not require parameters that are specific to the character and so we can use the same system to generate walking for characters with different morphologies (see Figure 4).

6 Planning

As a final example, we showcase a method for employing angular momentum at the planning stage of control. In our recent work in stepping, we developed a supervisor which guides the step behavior controller. By combining and expanding the momenta change equations outlined in Section 3, we can compute the position p on the ground plane, assuming we're given the remaining terms. Using p, we can determine when the support is about to rotate and, therefore, when a step is necessary to prevent tipping. However, knowing that a *will* occur, it is potentially too late to react reasonably, especially following a large disturbance. Thus to better provide anticipatory control, we add dampers which dissipate momenta, as such $\dot{L}_{des} = -d_l \cdot L$ and $\dot{H}_{des} = -d_h \cdot H$. And the result is a momentum driven, easily tunable predictor for when and where to take a step. Namely, we compute p_{des} as:

$$p_{x_{des}} = c_x + \frac{d_l \cdot L_x}{f_z} c_z + \frac{d_h \cdot H_y}{f_z} \quad (1)$$

$$p_{y_{des}} = c_y + \frac{d_l \cdot L_y}{f_z} c_z - \frac{d_h \cdot H_x}{f_z} . \quad (2)$$

where $f_z = \dot{L}_z + mg$. If this value is outside of the support, we take a step. With a scaled set of gain values, we can also use these equations to choose where to step.

The strength and uniqueness of this supervisor is that it takes into account the angular momentum present which can be substantial following a large disturbance. Another perk of this supervisory controller is that it is easily tunable to create desired effects since it only depends on a small number of gain values [15]. In contrast, other step supervisors including inverted pendulum and capture point models [18,19] ignore angular momentum in choosing where to step. As we show in the paper, if we simply drop the angular momentum term from the calculation, the supervisor reduces to a very similar structure as the capture point and we see a significant drop in the capability of the resulting character in handling disturbances [15]. We conclude from our findings that consideration of angular momentum is quite valuable in high level planning as well as low level behavior control.

7 Discussion

In our momentum explorations thus far, we have shown how it is possible to use momentum to produce coordinated motion such as balance, stepping, and walking through a multi-objective framework. Several key observations were made during the course of these efforts. Among the most important, we found that balance and locomotion control is a problem concerned with the aggregate dynamics of the articulated-body system; what happens with each individual body within this black box is less important than the combined effect of all. In this way, control can be delegated to other objectives (for example, the style of the animation take from motion capture). Control inputs for momentum are inherently low dimensional independent of the dimensionality (or morphology) of the character which seems to pair nicely with this finding. Next, while rotational movement adds a great deal of visual finesse to a motion, we have seen that it is relatively easy to solicit with guidance over the angular momentum as evidenced by the many successful yet simple control laws proposed thus far. With such a strong potential for producing rich, realistic movement, it is our goal to thoroughly explore the maintenance of angular momentum as an explicit and deliberate feature of control.

The concepts described in this paper highlight the importance of angular momentum control over whole-body coordinated behaviors in all human activities. The underlying hypothesis is that control over angular momentum is a critical mechanism which is being employed in nature. This hypothesis is supported by compelling evidence published mostly within the last five years in biomechanics and robotics. To date, all but a small handful of animation researchers have ignored or greatly simplified the whole-body rotational aspects of behavior control. Recent findings in regards to control over angular momentum and its ability to create coordinated and more humanlike movement make it difficult to continue to justify arguments for this choice.

To the animation community, perhaps the greatest value of momentum-based is simplification: full body, coordinated control is reduced to the specification of two basic, well-behaved momenta signals. Practical techniques for controlling these signals have been proposed and successfully tested. And now, a new set of questions arise. How are humans using momentum control to produce movements, if indeed they are? How can animators best harness benefits of this phenomena? What are the driving signals and pertinent features? What are the limits of this control paradigm? Thus far, we have seen very little in terms of speculations but yet there is evidence of tremendous untapped potential, suggesting great opportunity.

8 Conclusion

In this paper, we highlight findings from the literature and our explorations with respect to the relationship of angular momentum and coordinated behavior in physics-based character animation control. We present a basic ZS/NZS classification for distinguishing behaviors that exhibit distinct strategies in angular momentum regulation and discuss recently proposed techniques for using angular momentum in control in the context of this breakdown as well as summarize our experimental results and those of others in animation. Notably, sustained NZS control is virtually non-existent across discipline thus far.

We conclude that angular momentum control is a general tool for guiding coordinated action through the purposeful, deliberate changes to whole-body angular momentum. Biomechanists have shown that humans carefully regulate angular momentum in common activities and preliminary investigations from several disciplines support that angular momentum control, even in simple forms, can achieve a high degree of coordination and robustness in responses to a variety of conditions. Further, resulting secondary effects caused by such controllers, such as arm swing in humanoids, produce motions that bear strong likenesses to those observed in natural, real-world movement. We predict that with such mechanisms for control, physically based character motion can become more flexible and believable in addition to new behavior controllers becoming easier to construct.

Acknowledgments. The author wishes to acknowledge the co-authors of related research, C. C. (James) Wu, Adriano Macchietto, and Christian Shelton for their input and help with the production of figures. Also, thanks to Paul Kry and Ambarish Goswami for helpful discussions that lead to a more integrated understanding of this material.

References

1. Popovic, M., Hofmann, A., Herr, H.: Angular momentum regulation during human walking: biomechanics and control. In: Proceedings of the IEEE International Conference on Robotics and Automation, Citeseer, pp. 2405–2411 (2004)
2. Popovic, M., Hofmann, A., Herr, H.: Zero spin angular momentum control: definition and applicability. In: 2004 4th IEEE/RAS International Conference on Humanoid Robots, pp. 478–493 (2004)
3. Hofmann, A., Popovic, M., Herr, H.: Exploiting angular momentum to enhance bipedal center-of-mass control. IEEE Trans. Rob. Autom. (2007)
4. Herr, H., Popovic, M.: Angular momentum in human walking. Journal of Experimental Biology 211 (2008)
5. Farrell, M., Herr, H.: Angular Momentum Primitives for Human Turning: Control Implications for Biped Robots. In: IEEE Humanoids Conference (2008)
6. Kajita, S., Kanehiro, F., Kaneko, K., Fujiwara, K., Harada, K., Yokoi, K., Hirukawa, H.: Resolved momentum control: humanoid motion planning based on the linear and angular momentum. In: Intelligent Robots and Systems, IROS (2003)
7. Ahn, K., Oh, Y.: Walking Control of a Humanoid Robot via Explicit and Stable CoM Manipulation with the Angular Momentum Resolution. In: 2006 IEEE/RSJ International Conference on Intelligent Robots and Systems, pp. 2478–2483 (2006)
8. Goswami, A., Kallem, V.: Rate of change of angular momentum and balance maintenance of biped robots. In: IEEE Int. Conf. Robotics and Automation, ICRA (2004)

9. Abdallah, M., Goswami, A.: A biomechanically motivated two-phase strategy for biped upright balance control. In: IEEE Int. Conf. Robotics and Automation (2005)
10. Stephens, B.: Humanoid Push Recovery. In: IEEE-RAS International Conference on Humanoid Robots (2007)
11. Kudoh, S., Komura, T., Ikeuchi, K.: Stepping motion for a humanlike character to maintain balance against large perturbations. In: IEEE Int. Conf. Robotics and Automation (2006)
12. Macchietto, A., Zordan, V., Shelton, C.R.: Momentum control for balance. ACM Transactions on Graphics 28, 80:1–80:8 (2009)
13. de Lasa, M., Mordatch, I., Hertzmann, A.: Feature-based locomotion controllers. ACM Transactions on Graphics 29, 131:1–131:10 (2010)
14. Ye, Y., Liu, C.K.: Optimal feedback control for character animation using an abstract model. ACM Transactions on Graphics 29, 74:1–74:9 (2010)
15. Wu, C.C., Zordan, V.: Goal-directed stepping with momentum control. In: ACM SIGGRAPH / Eurographics Symposium on Computer Animation, pp. 113–118 (2010)
16. Popovic, M., Goswami, A., Herr, H.: Ground Reference Points in Legged Locomotion: Definitions, Biological Trajectories and Control Implications. The International Journal of Robotics Research 24 (2005)
17. Coros, S., Beaudoin, P., van de Panne, M.: Robust task-based control policies for physics-based characters. ACM Transactions on Graphics 28 (2009)
18. Pratt, J., Carff, J., Drakunov, S., Goswami, A.: Capture Point: A Step toward Humanoid Push Recovery. In: Proceedings of the IEEE-RAS/RSJ International Conference on Humanoid Robots (2006)
19. Rebula, J., Canas, F., Pratt, J., Goswami, A.: Learning Capture Points for Bipedal Push Recovery. In: IEEE International Conference on Robotics and Automation, ICRA 2008, pp. 1774–1774 (2008)

Simulating Formations of Non-holonomic Systems with Control Limits along Curvilinear Coordinates

Athanasios Krontiris, Sushil Louis, and Kostas E. Bekris

Computer Science and Engineering Department
University of Nevada, Reno - Reno, NV, 89557, USA
bekris@cse.unr.edu

Abstract. Many games require a method for simulating formations of systems with non-trivial motion constraints, such as aircraft and boats. This paper describes a computationally efficient method for this objective, inspired by solutions in robotics, and describes how to guarantee the satisfaction of the physical constraints. The approach allows a human player to select almost an arbitrary geometric configuration for the formation and to control the aircraft as a single entity. The formation is fixed along curvilinear coordinates, defined by the curvature of the reference trajectory, resulting in naturally looking paths. Moreover, the approach supports dynamic formations and transitions from one desired shape to another. Experiments with a game engine confirm that the proposed method achieves the desired objectives.

Keywords: Path Planning, Flocking and Steering Behavior, Physics-based Motion, Navigation and Way-finding.

1 Introduction

Many computer games and simulations involve the modeling of agents, such as airplanes and boats, which have to move in formation to reflect the behavior of their real world counterparts. For example, military games can utilize formations to move teams of airplanes before attacking a user specified target. This paper is especially motivated by training simulations for military officers, which must model realistically the maneuvers of opposing and friendly aircraft and boats. The need to model formations arises as a primary requirement during the development of such simulations.

An important characteristic regarding the motion of airplanes and boats is the presence of non-holonomic constraints as well as limits in velocity and steering angle. For instance, aircraft have to maintain a minimum velocity and cannot turn beyond a maximum curvature. These constraints complicate the motion planning process for these systems, especially when a formation has to be maintained. Traditionally, games sacrifice either the accurate modeling of the motion constraints or the accurate maintenance of a formation so as to achieve computationally inexpensive and simple solutions. This paper is motivated by the objective of simulating an aircraft formation while respecting the physical constraints of the aircraft, modeled as non-holonomic systems with limits in velocity and curvature. A low computational cost solution is proposed that builds upon motion planning solutions for formations of mobile robots [3].

In the proposed approach, the user specifies a destination for a team of aircraft. A reference point for the formation moves independently so as to achieve this goal and the aircraft move in response so as to maintain the formation. The parameters of the formation are expressed as the difference between aircraft and the reference point in curvilinear coordinates rather than the usual rectilinear coordinates, as shown in Figs. 1 and 2. The curvature of the curvilinear coordinates is the instantaneous curvature of

Fig. 1. Blue lines correspond to the trajectory of airplanes. Triplets of white markers point to the positions of the aircraft at the same points in time.

the reference point. This approach allows the exact, geometric computation of the velocity and curvature for each follower aircraft based on the reference point's control parameters. The computation, however, may violate the velocity or curvature limits for the airplanes. This paper shows how to impose constraints on the motion of the reference agent so that the limits are always satisfied for the followers. It also provides details for the correct implementation of the scheme. Simulated results illustrate that the proposed scheme allows a human player to guide on-the-fly aircraft formations defined over curvilinear coordinates.

Background. Many research challenges in computer games deal with the motion of multiple agents, such as crowd simulation [4, 12, 15, 19, 20], pedestrian formations [8, 22], shepherding and flocking behaviors [17, 26]. It has been argued that formations can help to model crowd motions such as tactical arrangements of players in team sports [24]. Kinematic formations can be defined using a general framework for motion coordination that employs generalized social potential fields [11]. Nevertheless, most work on popular game agents, such as people, employs simple motion models. It is important to explore algorithms for agents who obey complex motion constraints (e.g., non-holonomic ones), and dynamics [6].

Formations for aircraft are typically studied in aerospace engineering and robotics. Motion planning and control techniques, developed originally in these areas, are today mature enough to be used in computer games [5]. The various approaches to formation control can roughly be divided into three categories: (i) Behavior based approaches [1, 2, 10] design simple motion primitives for each agent and combine them to generate complex patterns through the interaction of several agents. The approach typically does not lend itself to a rigorous analysis but some schemes have been proven to converge [13]. (ii) Leader-follower approaches [3, 7, 9, 25] designate one or more agents as leaders and guide the formation. The remaining agents follow the leader with a predefined offset. (iii) Rigid-body type formations [16] maintain a constant distance between the agents' configurations during all motions.

Control theory in particular has been used to study aircraft formations with tools such as input-output feedback linearization for leader-follower formations [9], where the stability of the controller is typically the issue [25]. Control-theory is often integrated with graph theory to define controllers that allow the transition between different formations [7]. Graphical tools have also been used to study whether there exist non-trivial trajectories for specified formations given the agent's kinematic constraints [23]. Moreover, a formation space abstraction has been defined for permutation-invariant formations of robots that translate in the plane [14]. More recently, there have been methods for distributed task assignment so that mobile robots can achieve a desired formation [18].

This paper builds upon a motion planning method for formations of mobile robots [3], which appears appropriate for gaming applications. It allows a human player to control the formation as a single entity by defining a desired goal. In contrast to the optimization and linearization tools typically employed for formations, the technique is geometric and exact. It provides equations so that each follower tracks the reference trajectory but instead of maintaining a fixed distance to the leader, it adapts the shape during turns so as to satisfy the non-holonomic constraints. When a formation is turning, the method automatically plans for followers on the "outside" to speed up and robots on the "inside" to slow down, resulting in natural and pleasant formations to the human eye. The approach is also very fast, allowing for good scalability and real-time operation. Moreover, it supports almost any arbitrary geometric formation and is homogeneous, since each agent uses the exact same algorithm given certain parameters. Furthermore, it supports dynamic formations, which allow transitions from one shape to another.

Contribution. Beyond proposing this method as a way for achieving formations for systems with non-holonomic constraints in games , this paper contributes a series of algorithmic improvements: **1)** The controls of the leader are computed on the fly based on user-specified targets. In the previous work, the knowledge of a precomputed path is utilized to maintain the formation. **2)** The approach models non-holonomic systems with a minimum velocity, which cannot move backwards, such as airplanes. Then it defines constraints in the leader's motion that allow the followers to respect such physical limits as well as guarantee formation maintenance. Note that the original controller would violate such constraints, since it can return paths for mobile robots that required them to move backwards or turn with a very high curvature. **3)** The approach extends the application of the limits for the formation's leader to the case of dynamic formations, which can be used so as to switch between user-specified static formations. **4)** The paper describes the implementation of the overall technique within a typical graphics rendering loop and provides the related implementation details.

2 Problem Setup

The model employed is that of a simple car, that is a planar, first-order, non-holonomic system, with state parameters $[x_i, y_i, \theta_i]$ (Cartesian coordinates and orientation). The controls correspond to the velocity u_i and steering angle ϕ_i, which relates the path's curvature K_i ($K_i = \tan(\phi_i)$). For system i:

$$\dot{x}_i = u_i \cos \theta_i, \qquad \dot{y}_i = u_i \sin \theta_i, \qquad \dot{\theta}_i = u_i \tan \phi_i, \qquad (1)$$

where the velocity and curvature are limited similar to an aircraft:

$$|K_i| \leq K_{max}, \qquad 0 < u_{min} \leq u_i \leq u_{max} \qquad (2)$$

It is also possible to consider different limits for each aircraft.

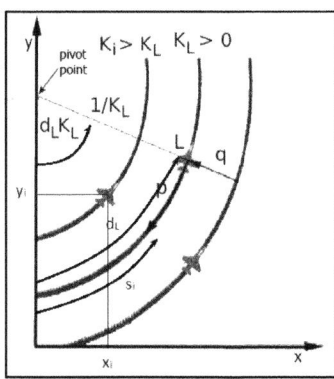

Fig. 2. An illustration of a formation along curvilinear coordinates

The objective is to maintain a static formation defined over curvilinear coordinates, which are a coordinate system in which the coordinate lines may be curved. The formation is defined relative to a reference agent, which is directly controlled by the human player and follows controls u_L and K_L. The reference agent can correspond to a leader aircraft or to a virtual reference point. Then, the curvature of the curvilinear coordinates becomes the instantaneous curvature of the reference agent. For example, if airplane L in Fig. 2 is the leader, then the middle blue line corresponds to the "x" axis of the curvilinear system, and the line perpendicular to the airplane is the "y" axis. To maintain a static formation, each follower aircraft must maintain p_i distance along the leader's curved trajectory and q_i distance along the perpendicular direction, as in Fig.2. Finally, the variable d_L denotes the distance the reference agent has covered and $s_i = d_L + p_i$ denotes the distance that the projection of follower i has covered along the leader's trajectory.

A human player must be able to select n aircraft, which are flying independently in the game, specify the type of formation and a goal that the team must achieve. Then the aircraft must move so as to reach the vicinity of the goal while maintaining the formation along curvilinear coordinates and respecting Eq. 1 and 2. The human player should be able to change the goal on the fly and the aircraft should adapt their trajectory.

3 Aircraft Formations along Curvilinear Coordinates

The approach automatically selects a reference point, which in the resulting formation will end up not having any aircraft ahead of it. This point moves independently to reach the goal specified by the user. For an obstacle-free environment, this can be achieved with a PID controller. The steering angle ϕ_L of the leader is changed to minimize the difference between θ_L and the direction to the goal. The controller makes the leader to gradually turn towards the goal and then execute a straight-line path. As the leader approaches the goal, it reduces velocity and initiates a circular trajectory around it defined by the maximum curvature and the original position from which the controller was initiated. The leader respects the control limits of Eq. 2. Different types of planners and controllers can be substituted for the leader, especially for the case of environments with obstacles, but the current solution achieves low computational overhead for planar aircraft formation, which typically operate in obstacle-free environments. Furthermore, the approach described here can be directly applied when the player "flies" one of the airplanes or when the target for the reference point is a moving goal, such as another airplane or a target on the ground.

3.1 Equations for Maintaining a Static Formation

Consider a follower aircraft, which must retain a constant distance p_i and q_i from the leader along curvilinear coordinates as described in Section 2. Then, given Figure 2 and basic trigonometry on angle $s_i K_L(s_i)$ at the pivot point, the coordinates of the follower aircraft can be expressed as:

$$x_i = (\frac{1}{K_L(s_i)} - q_i) \cdot sin(s_i K_L(s_i)) \qquad y_i = \frac{1}{K_L(s_i)} - (\frac{1}{K_L(s_i)} - q_i) \cdot cos(s_i K_L(s_i)).$$

If q_i, p_i are constant and $K_L(s_i)$ is instantaneously constant at s_i, the first derivatives of the Cartesian coordinates are:

$$\dot{x}_i = \dot{s}_i(1 - q_i K_L(s_i)) \cdot cos(s_i K_L(s_i)) \qquad \dot{y}_i = \dot{s}_i(1 - q_i K_L(s_i)) \cdot sin(s_i K_L(s_i))$$

where $\dot{s}_i = \frac{d(d_L+p_i)}{dt} = \dot{d}_L = u_L(d_L)$. Thus:

$$u_i(s_i) = \sqrt{\dot{x}^2 + \dot{y}^2} = u_L(d_L)(1 - q_i K_L(s_i)) \qquad (3)$$

Note that $u_L(d_L)$ is the current velocity of the leader at d_L but $K_L(s_i)$ is defined as the projection of the follower's position on the leader's trajectory (i.e., the curvature of the leader at distance p_i along its trajectory). To compute the follower's curvature, it is necessary to compute the second order derivatives of the Cartesian coordinates, since $K = \frac{\dot{x}\ddot{y} - \ddot{x}\dot{y}}{(\dot{x}^2 + \dot{y}^2)^{3/2}}$, which results in:

$$K_i(s_i) = \frac{K_L(s_i)}{1 - q_i K_L(s_i)} \qquad (4)$$

The advantage of considering curvilinear coordinates is that it is possible to exactly compute Eqs. 3, 4 that provide the controls of the followers as functions of the reference agent's controls. The follower requires access only to the past leader curvature $K_L(s_i)$ and current velocity $u_L(d_L)$. For $K_L(s_i)$ to be known, it has to be that $p_i < 0$, which is the reason why the reference agent is selected to be ahead of every other aircraft. Otherwise, a follower requires the curvature of the reference agent into the future. When the leader's trajectory is computed in real-time, this is not available.

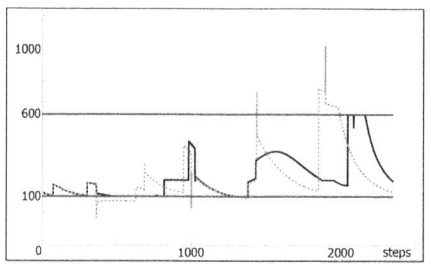
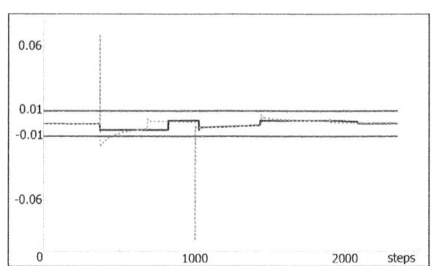

Fig. 3. (left) Velocity of the left follower in Fig. 4, (right) steering angle of same aircraft, (both) green line: the control value when the limits are ignored for the follower, blue: the control value given the proposed fix, red: the limits respected by the leader.

Using Eqs. 3 and 4 it is possible to maintain a perfect formation, such as the triangular one implied by Fig. 2, when the leader moves along a straight line. When the leader turns, the Cartesian distances between the aircraft change and the formation becomes flexible. The aircraft "outside" of the leader's turn have lower curvature and higher velocity. Aircraft "inside" of the leader's turn will have higher curvature and lower velocity.

3.2 Respecting the Control Limits

There is an important issue that arises with the above approach when the leader's trajectory is computed in real time. It is possible that the controls for the follower computed using Eqs. 3 and 4 will violate the control limits specified by Eq. 2.

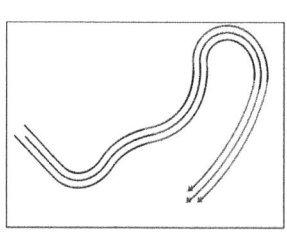

Fig. 4. A benchmark formation trajectory

For instance, consider the trajectory displayed in Figure 4 for a triangle formation, where the two followers have parameters $p_i = -70$ and $|q_i| = 90$. This trajectory can be achieved by the proposed approach only if the control limits in Eq. 2 are not respected. Fig. 3 shows that for certain control limits respected by the leader, Eqs. 3 and 4 will result in control values for the followers that violate the same limits. The proposed idea in this paper is to compute online limits for the control parameters of the leader so as to guarantee that the result of Eqs. 3 and 4 will never violate the limits of Eq. 2. This is not as straightforward as it appears, since the followers' controls depend on the current velocity of the leader $u_L(d_L)$ and a past curvature value $K_L(s_i)$. Thus, the leader's controls become correlated over time.

Constraints for the leader's curvature. Given Eqs. 4 and 2, it has to be true for a follower that:

$$\left| \frac{K_L(s_i)}{1 - q_i K_L(s_i)} \right| \leq K_{max}$$

Consider tentatively only right turns, for which $K > 0$. If the leader turns right using maximum curvature then the right side followers (for which $q_i > 0$) will have to use even greater curvature, which will violate the above constrain. Consequently, for $K > 0$ it has to be that:

$$\max\left(\frac{K_L(s_i)}{1 - q_i K_L(s_i)}\right) \leq K_{max} \quad \Rightarrow \quad \frac{K_L(s_i)}{1 - \max(q_i) K_L(s_i)} \leq K_{max}$$

Let's denote the maximum positive q_i value as: $q_{max}^+ = \max(q_i)$, then:
- if $(1 - q_{max}^+ K_L(s_i)) > 0$ then: $K_L(s_i) \leq \frac{K_{max}}{1 + q_{max}^+ K_{max}}$
- if $(1 - q_{max}^+ K_L(s_i)) < 0$ then: $K_L(s_i) \geq K_{max} \cdot (1 - q_{max}^+ K_L(s_i))$

But for the second case, it has been assumed that $K_L(s_i) > 0$, $K_{max} > 0$ and $(1 - q_{max}^+ K_L(s_i)) < 0$. Consequently, the second equation is always true since $K_L(s_i)$ is greater than something negative. Thus, only the first case imposes a constraint on the leader's curvature. For the case that the leader turns left, the computations are symmetrical. If $q_{max}^- = \min(q_i)$ denotes the maximum negative value for q_i, the overall constraint for the leader's curvature is:

$$\frac{-K_{max}}{1 - q_{max}^- K_{max}} \leq K_L(s_i) \leq \frac{K_{max}}{1 + q_{max}^+ K_{max}} \quad (5)$$

Constraints for the leader's velocity given past curvature values. Given Eqs. 3 and 2, it has to be true for a follower that:

$$0 < u_{min} \leq u_L(d_L)(1 - q_i K_L(s_i)) \leq u_{max}$$

For the lower bound, and given that $u_L(d_L) > 0$, it has to be true that:

$$\min_i(u_L(d_L)(1 - q_i K_L(s_i))) \geq u_{min} \quad \Rightarrow \quad u_L(d_L)(1 - \max_i(q_i K_L(s_i))) \geq u_{min}$$

Then there are two cases:
- If $(1 - \max_i(q_i K_L(s_i))) > 0$ then: $u_L(d_L) \geq \frac{u_{min}}{1 - \max_i(q_i K_L(s_i))}$
- If $(1 - \max_i(q_i K_l(s_i))) < 0$ then: $u_L(d_L) \leq \frac{u_{min}}{1 - \max_i(q_i K_L(s_i))}$

The second case implies that $u_L(d_L)$ must be less than or equal to something negative. This should never be true, however, for the aircraft. For that reason, and in order to guarantee that $(1 - \max_i(q_i K_L(s_i))) > 0$, it has to be that:

$$|K_L(s_i)| < \frac{1}{\max_i(|q_i|)} \quad (6)$$

This means that a follower cannot be further away from the leader along the perpendicular direction than the pivot point defined by the maximum curvature value. Otherwise, a follower has to move backwards.

For the upper bound, and given that $u_L(d_L) > 0$, it has to be true that:

$$\max_i(u_L(d_L)(1 - q_i K_L(s_i))) \leq u_{max} \quad \Rightarrow \quad u_L(d_L)(1 - \min_i(q_i K_L(s_i))) \leq u_{max}$$

Due to Eq. 6, it is true that $1 - \min_i(||q_i K_l(s_i)||) > 0$. Then this implies: $u_L(d_L) \leq \frac{u_{max}}{1 - \min_i(q_i K_L(s_i))}$

By combining the lower bound and the upper bound, it is possible to define the constraint for the leader's velocity:

$$\frac{u_{min}}{1 - \max_i(q_i K_L(s_i))} \leq u_L(d_L) \leq \frac{u_{max}}{1 - \min_i(q_i K_L(s_i))} \qquad (7)$$

Constraints for the leader's curvature to allow future velocities. Notice that it is possible the lower bound in Eq. 7 to become greater than the upper bound. This means that given past curvatures choices at different points in time, which depend on the p_i parameters of the followers, it is possible at some point the leader not to have any valid velocity to choose from, given Eq. 7. To guarantee that this will never happen it has to be always true that:

$$\frac{u_{min}}{1 - \max_i(q_i K_L(s_i))} < \frac{u_{max}}{1 - \min_i(q_i K_L(s_i))}$$

The minimum upper bound is achieved for the maximum value of $1 - \min_i(q_i K_L(s_i))$. But given Eq. 6: $max\{1 - \min_i(q_i K_L(s_i))\} = 2$, thus the upper bound in Eq. 7 is at least $\frac{u_{max}}{2}$. Thus it is sufficient to guarantee that:

$$\frac{u_{min}}{1 - \max_i(|q_i K_L(s_i)|)} < \frac{u_{max}}{2} \Rightarrow \max_i(|q_i K_L(s_i)|) \leq \frac{u_{max} - 2u_{min}}{u_{max}} \Rightarrow$$

$$|K_L(s_i)| \leq \frac{u_{max} - 2u_{min}}{u_{max} \max_i(|q_i|)} \qquad (8)$$

Given the above equation, it has to be that $u_{max} \geq 2u_{min}$ to guarantee the existence of a valid curvature. Overall, if the leader satisfies Eqs. 5, 6, 7 and 8, then followers that execute Eqs. 3 and 4 will always satisfy Eq. 2.

3.3 Dynamic Formations

Allowing the shape of the formation to change over time might be desirable for multiple reasons. For instance, the user might want to change the formation or the team might have to fit through a small opening. Dynamic formations can also be useful for assembling a static formation from a random configuration. To achieve dynamic formations, the approach allows the coordinates $[p_i, q_i]$ to be functions of time or distance along the leader's trajectory. Following the derivation in the original work on formations along curvilinear coordinates [3] the controls for the followers in the case of dynamic formations are:

$$u_i = Q u_L(d_L) \qquad (9)$$

$$K_i = \frac{1}{Q}(K_L(s_i) + \frac{(1 - q_i(s_i) K_L(s_i))(dq_i^2/ds_i^2) + K_L(s_i)(dq_i/ds_i)^2}{Q^2}) \qquad (10)$$

where $Q = \sqrt{(\frac{dq_i}{ds_i})^2 + (1 - q_i(s_i) K_L(s_i))^2}$. Once the functions $q_i(s_i)$, $\frac{dq_i}{ds_i}(s_i)$ and $\frac{dq_i^2}{ds_i^2}(s_i)$ are defined for the above controllers, it is possible to use them together with any reference trajectory. For example, these functions can be defined so as to change the formation from a square to a line, magnify the current shape or even permute positions within the same formation [3].

The control limits must still be satisfied during the execution of dynamic formations. Note that in the case of static formations, the limits on the leader's curvature (Eqs. 5, 6 and 8) could be computed once given the value $max_i|q_i|$ of the static formation. For the dynamic formation, q_i can be dynamic and the approach has to update the constraint for the leader's curvature online by using the $q_i(s_i)$ function for the followers at that point in time. A conservative approximation for the dynamic formation is to consider the maximum q_i achieved by any of the followers during the execution of the entire dynamic formation ($max_i|q_i(s_i)|$, $\forall s_i$). Then the constraint for the leader's velocity (Eq. 7) is recomputed every iteration as in the case of a static formation but for the latest value of the $q_i(s_i)$.

4 Implementation

In order to evaluate the proposed approach, a game engine developed at the University of Nevada, Reno for educational purposes was utilized. The engine is built over the Python-Ogre platform using Python as the programming language. This engine provides the opportunity to select aircraft and direct them to a desired goal. As is often the case in game engines, the engine repetitively calls a function that executes the necessary operations for physics, AI and rendering that models a constant amount of simulation time dt. Each aircraft is modeled as a different entity and has a specific amount of time to complete its computations.

Recall that equations 3 and 4 require the follower to utilize the leader's current velocity $u_L(d_L)$ together with the leader's past curvature $K_L(s_i)$. For static formations, during simulation step dt, the follower's projection along the leader's trajectory must move the same amount as the leader does, so as to retain a constant curvilinear distance p_i. This means that the duration for which a specific curvature value followed by the leader in the past is not going to be the same as the duration that the follower will use it to compute its own curvature. This is because the two agents are moving with different velocities during the corresponding time windows. The correct duration for which a curvature is going to be followed can be computed by finding the distance for which the leader executed the same trajectory and dividing with the current velocity of the follower. It might happen that during a single simulation step dt, the follower will have to switch multiple controls if the leader is moving faster.

Fig. 5 illustrates this issue. Assume the leader and the follower start at the positions shown in Fig.5(top). Each box represents the time window that the leader was using a specific set of controls. The numbers are the distances that the leader covered within each time window. For instance, the leader is

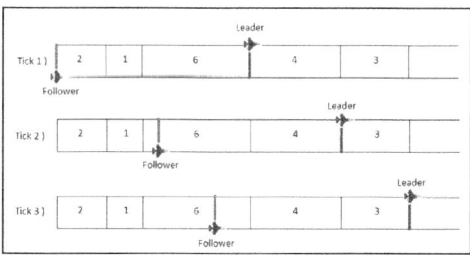

Fig. 5. A follower may be forced to switch controls during a single simulation step

about to cover a distance of 4 units. In a static formation, the same distance must be followed by the follower's projection along the leader's trajectory during the same window. But in the past and for the same distance of 4 units, the leader had switched three different curvature values (for distances 2, 1 and 6). The duration for which the follower will execute the n_{th} control is $t_n = box\_distance_n/u_L$, where $u_L(t)$ is the leader's current velocity. For the consecutive step, it happens that the follower will execute a single control, since the leader has currently slowed down compared to its past velocity.

5 Results

Variety of Formations: The proposed approach is able to accommodate a large variety of geometric configurations for formations. The only limitation is due to Eq. 6, which means that in order to allow non-zero values for the leader's curvature, the $\max(|q_i|)$ value has to be limited. For reasonable values of curvature constraints and size of aircraft, the limitation on $max(|q_i|)$ is such that allows followers in a considerable distance from the leader relative to the aircraft size. Fig. 6 provides examples of formations achieved.

Errors in Formation Maintenance: The formation in Figure 4 for $p = -60$ and $|q| = 50$ was also used to evaluate the amount of error in the achieved formation. The following results considered three versions for testing formations.

- In version 1 the aircraft do not respect control limits, so they can follow any formation. The implementation employs a constant control for the follower during each simulation step using the leader's current velocity and the leader's past curvature at the beginning of the step. Thus, this step ignores the discussion in section 4.
- Version 2 properly implements the time window approach described in Section 4 but the aircraft respect control limits. The leader respects only the original limits of Eq. 2 but not the limits proposed in Section 3.
- Version 3 corresponds to the proposed approach.

Figure 7 provides the error between the desired formation and the achieved formation at each simulation step for the trajectory in Fig. 4. Similar results have been acquired for hundreds of trajectories. If the steps in Section 4 are not followed or the limits on the leader's controls are not respected, the formation

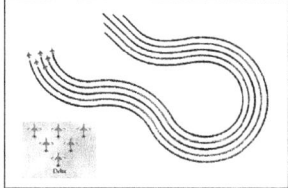 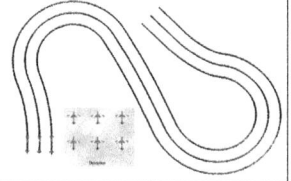 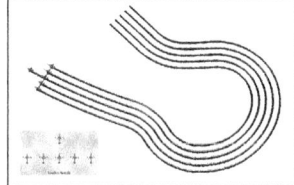

Fig. 6. Check: http://www.raaf.gov.au/roulettes/images/formations.jpg for formations tested. (left) Delta type. (middle) Domino type. (right) Leader Benefit type.

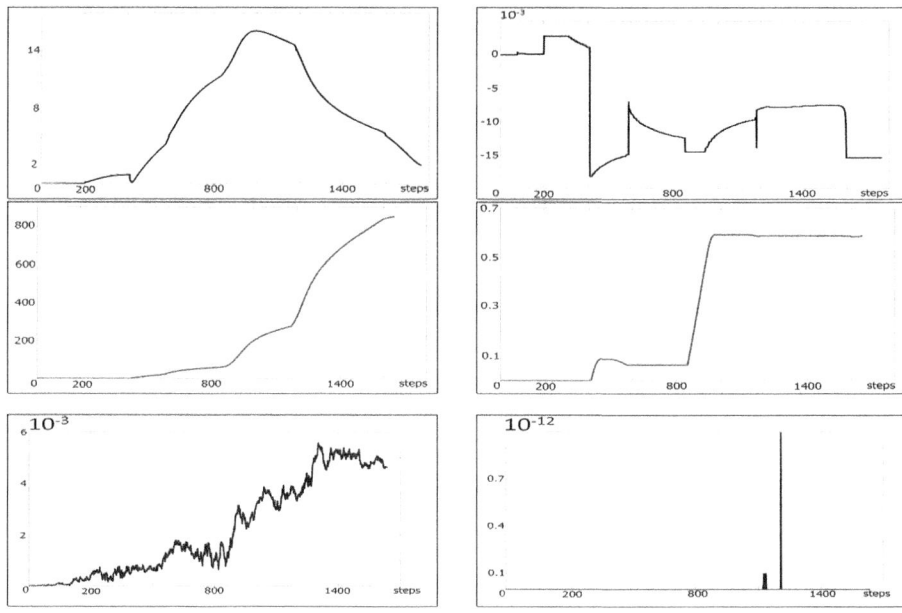

Fig. 7. These figures evaluate the errors for: (top) version 1, (middle) version 2, (bottom) version 3. (left) the error in Cartesian coordinates (unit is pixels) between the followed path and the correct path, (right) Error in orientation (unit is radians).

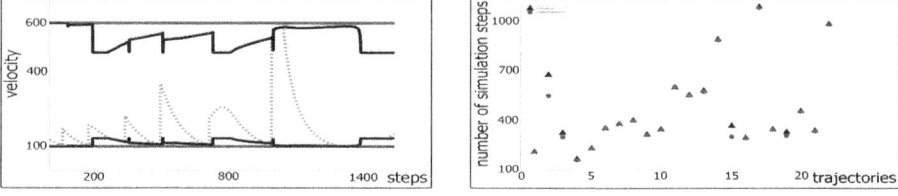

Fig. 8. (left) Red lines: general velocity limits for the aircraft, Blue: velocity limits for the leader that guarantee formation maintenance, Green: actual velocity of the leader (right) # of steps to reach a destination for aircraft formations that don't respect control limits (red dots) and aircraft that do (blue triangles)

cannot be followed and the error increases considerably. For the proposed approach, the error remains well below 10^{-2} pixels for many minutes of simulation.

Effects of Control Limits: Figure 8 studies how constrained a team of aircraft becomes when it has to respect the control limits specified in this paper. The left image shows the velocity limits for the leader for the trajectory in Fig. 4 and shows that still a large portion of the velocity parameters is valid but care has to be taken for the extreme values. The right image shows how much time it takes for the triangular formation to reach various random targets if it is not constrained or constrained. Sometimes, the more physically realistic, constrained

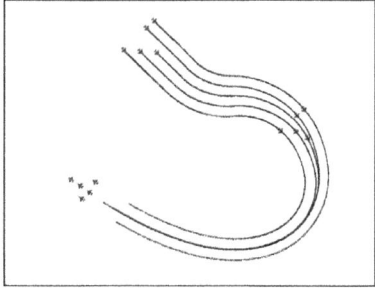 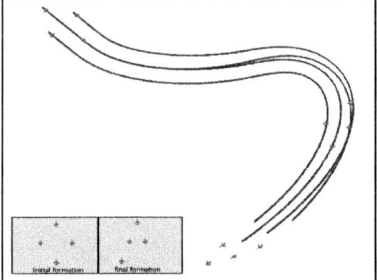

Fig. 9. Dynamic formations along curved trajectories for five and four airplanes

formation has to follow a longer trajectory to reach a target due to the introduction of the constraints for the leader. But the difference is not significant and does not justify not respecting the control limits during the simulation process.

Dynamic Formations: Figure 9 shows the extension of the approach to the case of dynamic formations.

6 Discussion

This paper has outlined a framework for simulating planar formations along curvilinear coordinates for systems with non-holonomic constraints. A previous method for mobile robots [3] has been extended to handle systems with curvature limits and minimum velocity. The implementation employs a PID controller for computing the controls of the reference agent but allows for the user to directly control one of the aircraft or to assign a mobile target to the formation. Future work includes integration of the formation algorithm with path planners and reactive methods (e.g., [21]) to get formations that avoid static (e.g., mountains) or dynamic obstacles (e.g., missiles or other aircraft). A 3D version is of significant interest for aircraft formations, as well as an extension to higher-order aircraft to allow for continuous or smooth velocity.

References

1. Balch, T., Arkin, R.: Behavior-based formation control for multi-robot teams. IEEE Transactions on Robotics and Automation 14(6), 926–939 (1998)
2. Balch, T., Hybinette, M.: Social potentials for scalable multi-robot formations. In: IEEE International Conference on Robotics and Automation, pp. 73–80 (April 2000)
3. Barfoot, T.D., Clark, C.M.: Motion planning for formations of mobile robots. Robotics and Autonomous Systems 46(2), 65–78 (2004)
4. van den Berg, J., Guy, S.J., Lin, M.C., Manocha, D.: Reciprocal n-body collision avoidance. In: International Symposium on Robotics Research, ISRR (September 2009)

5. Bottesi, G., Laumond, J.P., Fleury, S.: A motion planning based video game. Tech. rep., LAAS CNRS (October 2004)
6. Brogan, D.C., Hodgins, J.K.: Group behaviors for systems with significant dynamics. Autonomous Robots 4, 137–153 (1997)
7. Desai, J., Ostrowski, J., Kumar, V.J.: Modeling and control of formations of non-holonomic mobile robots. Transactions on Robotics and Aut. 17(6), 905–908 (2001)
8. Ennis, C., Peters, C., O'Sullivan, C.: Perceptual evaluation of position and orientation context rules for pedestrian formations. In: Symposium on Applied Perception in Graphics and Visualization (APGV 2008), pp. 75–82 (2008)
9. Fierro, R., Belta, C., Desai, K., Kumar, V.J.: On controlling aircraft formations. In: IEEE Conf. on Decision and Control, Orlando, FL, pp. 1065–1070 (Decmber 2001)
10. Fredslund, J., Mataric, M.J.: A general, local algorithm for robot formations. IEEE Transactions on Robotics and Automation 18(5), 846–873 (2002)
11. Gayle, R., Moss, W., Lin, M.C., Manocha, D.: Multi-robot coordination using generalized social potential fields. In: ICRA, Kobe, Japan, pp. 3695–3702 (2009)
12. Geraerts, R., Kamphuis, A., Karamouzas, I., Overmars, M.: Using the corridor map method for path planning for a large number of characters. In: Egges, A., Kamphuis, A., Overmars, M. (eds.) MIG 2008. LNCS, vol. 5277, pp. 11–22. Springer, Heidelberg (2008)
13. Jadbabaie, A., Lin, J., Morse, A.S.: Coordination of groups of mobile agents using nearest neighbor rules. IEEE Trans. on Automatic Control 8(6), 988–1001 (2003)
14. Kloder, S., Hutchinson, S.: Path planning for permutation-invariant multirobot formations. IEEE Transactions on Robotics 22(4), 650–665 (2006)
15. Lau, M., Kuffner, J.: Behavior planning for character animation. In: ACM SIGGRAPH / Eurographics Symposium on Computer Animation, SCA (2005)
16. Lewis, M., Tan, K.H.: High-precision formation control of mobile robotis using virtual structures. Autonomous Robots 4(4), 387–403 (1997)
17. Lien, J.M., Rodriguez, S., Malric, J.P., Amato, N.: Shepherding behaviors with multiple shepherds. In: International Conf. on Robotics and Automation (2005)
18. Michael, N., Zavlanos, M.M., Kumar, V., Pappas, G.J.: Distributed multi-robot task assignment and formation control. In: ICRA, pp. 128–133 (May 2008)
19. Patil, S., van den Berg, J., Curtis, S., Lin, M., Manocha, D.: Directing crowd simulations using navigation fields. IEEE Transactions on Visualization and Computer Graphics, TVCG (2010)
20. Pelechano, N., Allbeck, J., Badler, N.: Controlling individual agents in high-density crowd simulation. In: ACM SIGGRAPH / Eurographics Symposium on Computer Animation (SCA), San Diego, CA, vol. 3, pp. 99–108 (August 3-4, 2007)
21. Snape, J., Manocha, D.: Navigating multiple simple-airplanes in 3d workspace. In: IEEE International Conference on Robotics and Automation, ICRA (2010)
22. Stylianou, S., Chrysanthou, Y.: Crowd self organization, streaming and short path smoothing. In: Computer Graphics, Visualization and Computer Vision (2006)
23. Tabuada, P., Pappas, G.J., Lima, P.: Motion feasibility of multi-agent formations. IEEE Transactions on Robotics 21(3), 387–392 (2005)
24. Takahashi, S., Yoshida, K., Kwon, T., Lee, K.H., Lee, J., Shin, S.Y.: Spectral-based group formation control. Comp. Graph. Forum: Eurographics 28, 639–648 (2009)
25. Tanner, H.G., Pappas, G.J., Kumar, V.J.: Leader-to-formation stability. IEEE Transactions on Robotics and Automation (2004)
26. Vo, C., Harrison, J.F., Lien, J.M.: Behavior-based motion planning for group control. In: Intern. Conf. on Intelligent Robots and Systems, St. Louis, MO (2009)

Following a Large Unpredictable Group of Targets among Obstacles

Christopher Vo and Jyh-Ming Lien

Department of Computer Science, George Mason University,
4400 University Drive MSN 4A5, Fairfax, VA 22030 USA
{cvo1,jmlien}@gmu.edu
http://masc.cs.gmu.edu

Abstract. Camera control is essential in both virtual and real-world environments. Our work focuses on an instance of camera control called target following, and offers an algorithm, based on the ideas of *monotonic tracking regions* and *ghost targets*, for following a large coherent group of targets with unknown trajectories, among known obstacles. In multiple-target following, the camera's primary objective is to follow and maximize visibility of multiple moving targets. For example, in video games, a third-person view camera may be controlled to follow a group of characters through complicated virtual environments. In robotics, a camera attached to robotic manipulators could also be controlled to observe live performers in a concert, monitor assembly of a mechanical system, or maintain task visibility during teleoperated surgical procedures. To the best of our knowledge, this work is the first attempting to address this particular instance of camera control.

Keywords: Motion planning, camera planning, target following, group motion monitoring, monotonic tracking regions, ghost targets.

1 Introduction

In *multiple-target following*, the camera's primary objective is to follow and maximize visibility of multiple moving targets. Multiple-target following is essential in both virtual and real-world environments. For example, in video games, a third-person view camera may be controlled to follow a group of characters through complicated virtual environments. In robotics, a camera attached to robotic manipulators could be controlled to observe a swarm of mobile robots, or live performers in a concert; monitor assembly of a mechanical system; or maintain task visibility during teleoperated surgical procedures.

In general, it is difficult for a user to manually control the camera while also concentrating on other critical tasks. Therefore, it is desirable to have an *autonomous camera system* that handles the camera movement. This paper focuses on an instance of camera control called *multiple-target following*, and offers an algorithm for *autonomous* following a large coherent group of 10 ~ 100 targets (such as a crowd or a flock) with unknown trajectories, among known obstacles.

The camera control problem has been studied extensively in both robotics and computer graphics because of its broad applications, such as dynamic data visualization [21], robotic and unmanned vehicle teleoperation [13], and video games [16]. Unfortunately, many of these methods are not applicable to follow a large group of targets in real-time *among obstacles*. For example, there is a large body of work in robotics, where researchers have studied similar problems such as pursuit and evasion, visual servoing [6] and cooperative multi-robot observation of multiple moving targets (CMOMMT) [19]. However, these strategies usually apply to environments with sparse or no obstacles. While there exist methods that do consider occlusion, e.g. [15], they still only consider situations where the target's trajectory is known *a priori*, or are only applicable to follow a few (2 or 3) targets [14,17]. Many researchers have also considered the case where both trajectory and environment are unknown [11,1]. The main idea of these motion planners is to greedily minimize the *escaping risk* or maximize the *shortest escaping distance* of the target. In all of these methods, the camera trajectory can be computed efficiently, but is usually sub-optimal because only local information about the environment is considered, and the time horizon for planning is very short. Moreover, to the best of our knowledge, these motion strategies are all designed to track a single target.

In computer graphics, much work on camera planning is script-based [5], purely reactive [12,6], or mostly focuses on problems with predefined target trajectories [2,9]. Many of these methods are based on constraint solving, objective satisfaction or both. They are mostly designed for offline use and take a long time (usually seconds) to find a single camera placement.

The main goal of this paper is to provide an initial investigation into this important problem. Due to the nature of the aforementioned applications (video games, mobile robot swarms, virtual prototyping and group control), our investigation will focus on following a coherent group of targets by a single camera. Maintaining the visibility of a coherent group in some aspects is easier than tracking a single target because there is more than one target that the camera can follow. However, tracking a group is more difficult if the camera needs to maximize the number of visible targets over time. Difficulty also stems from the fact that a group can assume different shapes (e.g., forming a long line in a narrow corridor and a blob in an open area), clutter around the obstacles, or even split into multiple sub-groups (for a short period of time).

Our Work and Main Contributions. In this paper, we present a motion strategy that allows a single camera to robustly follow, at interactive rates, a large group (e.g., with 100 targets) among obstacles. In the aforementioned applications, some information about the environment is usually available. We believe that, given this information about the environment, the planner should be able to perform deeper *lookahead* in the search space and therefore provide better real-time camera following strategies even when the motion of the targets is unknown. The main idea is to preprocess the given environment offline to generate a data structure called *monotonic tracking regions* (MTRs) (defined later in Section 4.1) that can be used to assist real-time planning. This representation allows the camera to plan more efficiently by reducing the possible target movements to a smaller, discrete space. This significant increase in efficiency allows us to generate and evaluate multiple alternative plans in real-time. Our method also uses target coherency to better predict target positions.

In addition to the MTR-based motion strategy, we also present three new strategies (in Section 5) extended from the existing single-target techniques. We present reactive, sampling-based [3], and escaping-risk-based [18] methods. Our experimental results show that MTR-based method performs significantly better than the other strategies, especially in the environments with small obstacles and sinuous tunnels. A preliminary version of this work on both searching and tracking large group can be found at [22]. To the best of our knowledge, this work is the first attempting to address this particular instance of camera control.

2 Related Work

There exist many methods for following a target with known trajectory, such as work done by LaValle et al. [15] using dynamic programming, and by Goemans and Overmars [10] using probabilistic motion planner. In this section, we will review strategies for tracking a target with unknown trajectory. There are also extensive studies on multiple object tracking, e.g., [20], in which the goal is to maintain a belief of where the target(s) are using Kalman or particle filters. On the contrary, our goal is to maintain the visibility of targets. From our review, no method has focused on the problem of following a large group.

There exists some work considering tracking targets with unknown trajectories in a known environment [17,4]. The general idea is to partition the space into non-critical regions in which the camera can follow the target without complex *compliant motion*, that is, rotating the line of visibility between the camera and the target around a vertex of an obstacle. The main benefit of this line of work is the ability to determine the *decidability* of the camera tracking problem [4].

Recently, Li and Cheng [16] have proposed a real-time planner that tracks a target with unknown trajectory. Their main ideas include a *budgeted roadmap method* with lazy evaluation. Geraerts [8] proposed the idea of using the *corridor map* to track a single target with the corridors.

In computer graphics, camera planning is often viewed as a constraint satisfaction problem, and so there have been several attempts to represent the problem so that it can be solved efficiently with constraint satisfaction techniques. For example, several works use the idea of *screen space* or *image space* constraints, e.g., Gleicher and Witkin [9]. There are a number of works which involve the use of metaheuristics to compute optimal positions or trajectories for the camera. For example, Drucker and Zeltzer [7] used an A* planner to compute a camera path. Along the path, the orientation of the camera is then solved frame-by-frame to satisfy given constraints.

Some studies have focused on developing reactive behaviors for real-time camera motion. For example, Halper et al. [12] introduced a camera planner that predicts state based on the past trajectory and acceleration. They also proposed the idea of PVR (potential visibility region) for visibility computation, and a pipelined constraint solver.

3 Preliminaries

In this section, we formally define the problem that we attempt to solve and notations used throughout the paper. We assume that the workspace is populated with known obstacles represented by polygons. These polygons are the projection of 3D objects that can potentially block the camera's view. This projection essential reduces our problem to a 2D workspace. We further assume that, initially, at least one of the targets is visible by the camera C, and, during the entire simulation, the targets T exhibit certain degree of coherence in their motion, and it is also possible that T can split into multiple subgroups for a short period of time (similar to a flock of birds). The targets are either controlled by the user or by another program, so the trajectories of the targets are not known in advance. However we assume that the size of T and T's maximum (linear) velocity v_T^{max} are known. The position $x_\tau(t)$ and the velocity $v_\tau(t)$ of a target $\tau \in T$ at time t are known only if τ is visible by the camera.

The camera C also has a bounded linear velocity v_C^{max}. The exact configuration of this view range at time t, denoted as $\mathcal{V}_C(t)$, is defined by the camera's view direction $\theta_C(t)$ and location $x_C(t)$. The position x_C of the camera is simply governed by the following equation: $x_C(t+\triangle t) = x_C(t) + \triangle t \cdot v_C(t)$, where $v_C(t)$ is the camera's velocity at time t.

Given the positions of the targets and the position of the camera, one can compute the camera's view direction so that the number of targets inside the view range is maximized. Therefore, the problem of target following then is reduced to find a sequence of velocities $v_C(t)$:

$$\arg\max_{v_C(t)} \left(\sum_t \text{card}(\{T' \subset T \mid X_{T'}(t) \subset \mathcal{V}_C(t)\}) \right), \quad (1)$$

subject to the constraints that, $v_C(t) \leq v_C^{max}, \forall t$, and $x_C(t)$ is collision free.

The main ideas of the proposed method are to (1) identify regions with simple (monotonic) topological feature so that the planner can repetitively use the same data structure and strategy to follow the target group, and (2) utilize the fact the targets form a coherent group.

4 Monotonic Tracking Regions

The first step of the proposed method decomposes the environment into a set of *monotonic tracking regions* (MTRs). These regions usually look like tunnels and may overlap with each other. Intuitively, in these tunnel-like regions, the camera can *monotonically* maintain the visibility by moving forward or backward along a trajectory that *supports* the tunnel. More specifically, the main property MTR is that each MTR is topologically a *linear subdivision* so that the problem of camera following in an MTR can be represented as a *linear programming problem*. Note that such an MTR needs not to be convex or star shaped, and, in fact, it can have an arbitrary number of turns (like a sinuous tunnel). Moreover, MTR decomposition usually creates much fewer components than convex or star-shaped decompositions but, as we will see later, still provide similar functionality in target tracking. More precise definition of MTR and the process of computing these regions will be given in Section 4.1. In Sections 4.2 and 4.3, we will discuss how to track the target in MTR and ghost targets, respectively.

4.1 Build Monotonic Tracking Regions

Definition. We let a region M_π be a 2D generalized cylinder defined with respect to a *supporting path* π. We say π is a supporting path of M_π if every point $x \in M_\pi$ can see a subset of π. Because of this property, M_π can essentially be viewed as a linear object and the camera can see every point in M_π by moving along π. Specifically, we define M_π as:

$$M_\pi = \{x \mid \exists y \in \pi \text{ s.t. } \overline{xy} \subset C_{free}\}, \qquad (2)$$

where $\overline{xy}$ is an open line segment between x and y, and C_{free} is the free space (i.e., the area without obstacles). Furthermore, we define the subset of π visible by x as: $V_\pi(x) = \{y \in \pi \mid \overline{xy} \subset C_{free}\}$. Note that $V_\pi(x)$ can have one or multiple connected components. Finally, we say a region $M_\pi \in C_{free}$ is an MTR supported by π if

$$|V_\pi(x)| = 1, \forall x \in M_\pi, \qquad (3)$$

where $|X|$ is the number of connect components in a set X. Because each $x \in M_\pi$ can see only an interval of π, we can compactly represent the visible region (called visibility interval) of x as a tuple $V_\pi(x) = (s, t), 0 \leq s \leq t \leq 1$, if we parameterize π from 0 to 1. Fig. 1 shows an example of MTR and its supporting path π.

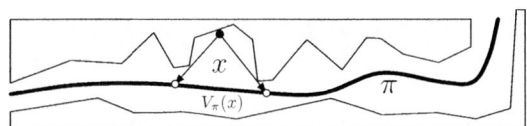

Fig. 1. An example of monotonic tracking regions defined by π

With the definition in hand, our task here is to first find a set of supporting paths whose MTRs that will cover C_{free}, and, next, from a given path π, we compute the associated MTR and the visibility interval $V_\pi(x)$ for every point x in the MTR. We will describe these two steps in detail next.

Constructing supporting paths. Our strategy here is to find the homotopy groups G of the C_{free}. We propose to use the medial axis (MA) of the C_{free} to capture G because of its several interesting relationships with MTRs. First of all, we can show that the *retraction region* of every edge π on the MA forms an MTR (supported by π).

Lemma 1. *The retraction region $R \subset C_{free}$ of an MA edge forms an MTR.*

Proof. Let π be an edge on the medial axis MA of C_{free}. The retraction region $R \subset C_{free}$ of π is simply a set of points that can be continuously retracted to π by a retraction function $r : R \to \pi$ [23]. By definition, given an arbitrary point $x \in R$, the largest circle c centered at the point $r(x)$ with x on c's boundary must be empty. Therefore, it follows naturally that each point in R must be able to see at least a point on π.

Now we briefly show that each point x can only see a consecutive region of π. We prove this by contradiction. Assuming that x can see multiple intervals of π. This means that there must be an obstacle between x and π. However, this contradicts the fact that x can be retracted to π in a straight line. □

Therefore, the supporting paths are simply constructed by extracting the edges from the MA of a given environment.

Constructing mtrs. Given an edge π of MA, its retraction region R forms an MTR supported by π. However, simply using R as π's MTR can be overly conservative. The set of points that satisfy Eq. 2 and 3 is usually larger than R. To address this issue, we iteratively expand R by considering the points adjacent to R until no points can be included without violating the definition of MTR.

Next, we compute the visibility interval for every point in an MTR. The brute force approach that computes the visibility interval for each point from scratch is no doubt time consuming. To speed up the computation, our approach is based on the following observation.

Observation 41. *If x and x' are (topological) neighbors, and x is further away from π than x' is, then $V_\pi(x) \subset V_\pi(x')$.*

For example, in Fig. 1, imagining a point x' below x, x' can see a larger interval of π than x does. That is if we can compute the visibility intervals $V_\pi(x')$ for all the points x' on π, then we should be able to obtain the visibility intervals for x that are $\triangle d$ away from x' by searching inside $V_\pi(x')$.

Dominated mtrs. The exact MA of a given environment can contain small (and in many cases unnecessary) features and result in many small MTRs. In many cases, these MTRs are unnecessary and should be removed. This is when an MTR is dominated by another MTR. We say MTR M' is dominated by another MTR M if $M' \subset M$. In our implementation, we use an approximate MA [23] to avoid small features, and then identify and remove dominated MTRs.

4.2 Follow the Targets in an MTR

The motivating idea behind decomposing the environment into a set of MTRs is that the target following problem in MTR can be solved much easily than that in the original environment. In fact, as we will see in this section, the camera can solve a long time horizon plan in MTR using linear programming.

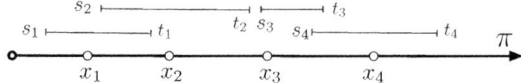

Fig. 2. Make predictions for the next $h = 4$ future steps

Follow a single target. To simplify our discussion, we will first describe how a single target can be tracked in MTR. Let $x_\tau(t)$ be the current position of the target τ. Since we know the cureent speed of the target, we can estimate the positions $x_\tau(t+\triangle t)$ in the next time step. In order to keep the target in the view, the camera's next position $x_C(t + \triangle t)$ must be: $x_C(t + \triangle t) \in V_\pi(x_\tau(t + \triangle t))$. Note that this estimation can be applied to an arbitrary value of $\triangle t$. However, when $\triangle t$ is bigger, the position of the target becomes less accurate.

Let $I_i = V_\pi(x_\tau(t+i\cdot\triangle t)) = (s_i, t_i)$. Here i is an integer from 1 to h, where h is the user-defined time horizon. Recall that both s_i and t_i are parameters on the parameterized path π. In order to follow the target for h steps, the planner needs find a sequence of camera locations x_i from a sequence of parameterized intervals such that every point x_i is in its corresponding interval I_i without violating the constraint on the camera's max speed (see Fig. 2). This can be done by solving a h dimensional linear programming problem:

$$\begin{aligned} &\min t_h - x_h \\ &\text{s.t. } s_i \leq x_i \leq t_i \\ &\quad 0 \leq (x_{i+1} - x_i) \leq \frac{v_C^{max}}{|\pi|}, \forall x_i \end{aligned} \quad (4)$$

where $v_C^{max}/|\pi|$ is the maximum *normalized* distance that the camera can travel on π. Finally, the camera's future locations are simply $x_C(t + \triangle t \cdot i) = \pi(x_i)$.

Note that the rationale behind the minimization of $(t_h - x_h)$ is that when the target moves further away beyond h steps in the future, the camera will have better chance of keeping the target in the view when it is located closer to t_h along the path π. We call the above linear programming problem the *canonical following problem*. Solving a canonical following problem can be done efficiently since h is usually not large ($h = 20$ is used in our experiments) given that modern linear programming solvers can handle thousands of variables efficiently.

It is possible that the linear programming problem has no feasible solution. We reduce the plan horizon iteratively until a solution is found.

Follow multiple targets. Now, we extend this canonical following problem to handle multiple targets T. Let $x_T(t)$ be the current positions of T. Similar to the case of a single target, we estimate the positions $x_T(t+\triangle t)$ in the next time step. In order to see a least one target, the camera must move so that

$$x_C(t + \triangle t) \in I(\triangle t) = \bigcup_{x \in x_T(t+\triangle t)} V_\pi(x) .$$

To simplify our notation, let $I_i = I_i(i \cdot \triangle t) = (s_i, t_i)$. By placing the camera in I_i, we can guarantee that at least one target is visible. However, our goal is to maximize the number of visible targets, at least over the planning horizon. To do so, we segment I_i into j sub-intervals I_i^j, each of which can see n_i^j targets. Then our goal is to pick a sub-interval from each I_i so that the total number of visible targets is maximized while still maintaining the constraint that the minimum distance between I_i^j and I_{i+1}^k is smaller than $v_T^{max}\triangle t$. Fortunately, this optimization problem can be solved greedily by iteratively adding the sub-interval with the largest n_i^j without violating the constraint. Once the sub-interval from each I_i is identified, the problem of finding the camera positions is formulated as a linear programming problem in the same way as Eq. 4.

When the targets and the camera moves (but still in the same MTR), we may need to repetitively solve the canonical following problem. Instead, we use *lazy update*. That is, we only update the old solution if it cannot satisfy all the constraints in Eq. 4.

4.3 Ghost Targets

Our planner takes advantage of the fact that the camera is following a group of somewhat coherent targets. When the number of the targets visible by the camera is smaller than the total number of the targets, the planner will generate a set of *ghost targets* in the invisible regions. The positions of the ghost targets are estimated based on the following assumptions: (1) targets tend to stay together as a group, and (2) invisible targets are in C-free outside the visibility region of the camera. Therefore, even if targets are invisible, they must be in some occluded regions nearby. Note that the planner does not distinguish if a target is visible or is a ghost. Therefore the planning strategy described in the previous sections remains the same. Our experiments show that the idea of ghost targets significantly increases the visibility.

5 Discussion and Experimental Results

5.1 Three Additional Following Strategies

We also developed three additional group following strategies. These strategies are extensions of the existing methods which are originally designed to track a single target. Since there is no prior work on group following, we will also use these strategies to compare against the proposed MTR camera.

Reactive Camera. This reactive camera determines its next configuration by placing the visible targets as center in the view as possible based only on the targets' currently positions. The motivation is that by placing the visible targets at the center of its view, the camera will have better chance to make invisible targets visible.

IO Camera. IO camera is a sampling-based method extended from [3]. At each time step, given the visible targets T, the planner first creates k point sets P_T, where k is a parameter. Each point set contains $|T|$ predicted target positions.

The predicted position of each target τ is sampled from the region visible from τ, and is at most $(v_T^{max} \cdot \triangle t)$ away from τ. The planner then creates a set P_C of camera configurations that are at most $(v_C^{max} \cdot \triangle t)$ away from C. To decide the next camera configuration, we simply determine $\arg\max_{x \in P_C} (\sum_{X \in P_T} \mathrm{vis}(x, X))$, where $\mathrm{vis}(x, X)$ is the number of points in X visible by x. To simplify our discussion, we will use the notation IO-k to denote an IO camera that samples k point sets for target prediction (e.g., in Fig. 3).

VAR Camera. VAR camera is based on [18]. Here, we first obtain a coarse representation of the environment by sampling a grid of discs in C-free. A roadmap is formed over the intersections of the discs. Finally, visibility is computed between each pair of discs with Monte-Carlo raytracing. Our VAR method is a hybrid approach that uses the constructed visibility information in a reactive behavior when the camera has good visibility of the targets (more than 50% of the targets are currently visible by the camera), and uses visibility-aware path planning from [18] to plan short, alternative paths to reach predicted locations of targets when the camera is far away.

The reactive behavior in VAR computes a waypoint for the camera on each frame. First, we find a disc D_c that is closest to the camera, and a disc D_r (from the pre-computed roadmap) that represents an imminent occlusion risk. That is, the disc D_r is one that is in the direction of the visible targets' velocity, closest to the centroid of the visible targets, and whose visibility from the camera's disc is less than 100%. A waypoint is selected along the ray extending from D_r passing through D_c. The visibility-aware path planner is the same as described in [18]. It uses an A* algorithm on the pre-computed roadmap with a heuristic function to compute a path to the D_r which simultaneously minimizes distance traveled and maximizes visibility of the target.

5.2 Experiments and Results

In our experiments, the target group is constantly moving toward a random goal, which is not known by the camera. If all targets are invisible, the camera will stay stationary. Throughout the experiments, we measure the performance of the cameras by computing the *normalized visibility* which is the ratio of the visible targets during the entire simulation. Every data point presented in the plots in this section is an average over 32 runs, each of which is a 10000 time-step simulation. We set the planning horizon $h = 20$ for all MTR cameras.

We perform our experiments in four workspaces (Fig. 3). These workspaces are designed to test the cameras in various conditions, such as large open space (disc), open space with narrow gaps (bars), small irregular obstacles with many narrow gaps (islands) and long sinuous narrow passages (tunnel). Both islands and tunnel environments are considered difficult as the targets tend to separate around the obstacles or hide behind a bend in the passage.

mtr outperforms other strategies. Our first experiment in Fig. 3 shows strong evidence that MTR camera consistently performs better than the other

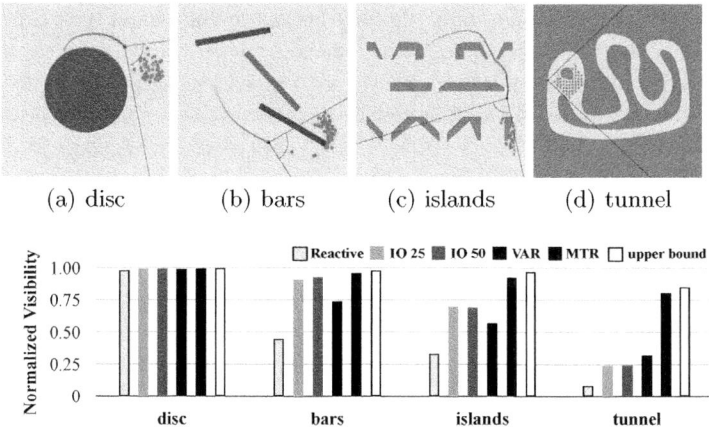

Fig. 3. Following 50 targets in four environments. Normalized visibility is the ratio of the visible targets during the entire simulation. Each bar is an average over 30 runs, each of which is a 9000 time-step simulation.

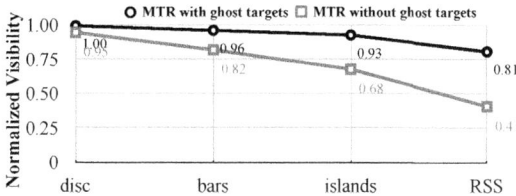

Fig. 4. Comparing MTR cameras with and without ghost target

cameras when following 50 targets in all environments. Note that we also include data called *upper bound* obtained from an MTR camera that has no speed limitation, i.e., it can move to the best configuration instantly. The strong performance of MTR is further supported by the small difference between the MTR camera and the upper bound in all four environments.

It is clear that the reactive camera performs worst, except in the disc environment. The VAR camera is the second worst, except in the tunnel environment. However, as we will see, the VAR camera seems to handle large and fast moving targets better because of its ability in estimating risks. Although IO cameras perform well in some situations, IO-25 in the bar environment is more than 200 times slower than VAR ($\approx$2600 fps) and 12 times slower than MTR ($\approx$157 fps), thus cannot be used in many applications, such as real-time task monitoring and video game. There is no significant difference between IO-25 and IO-50.

Ghost targets boost performance. In this experiment shown in Fig. 4, we attempt to estimate quantitively the performance gain due to the idea of *ghost targets* (GS). Our result shows that the performance gain is more significant in more difficult environments. If compare to the (very time consuming) IO cameras in Fig. 3, MTR without GS is only slightly better in the island and tunnel environments, but there are significant differences between the IO cameras and MTR with GS. In the future, it will be interesting to measure the performance gain by applying GS to reactive and VAR cameras.

6 Conclusion

In this paper, we contribute an online camera planning method, called MTR, which is suitable for autonomous following of multiple targets with unknown trajectories among known obstacles. The idea is to preprocess the known environment offline to obtain *monotonic tracking regions* (MTRs) which can then be used to increase the efficiency of the online planner. In addition to MTR, we present three new methods for camera planning that are extensions of existing single-target work, and we compare the performance of MTR with each of these methods. In the tested scenarios, MTR performs significantly better than all other methods, particularly in situations where environments are cluttered with many obstacles or long sinuous tunnels. We also show that adding *ghost targets* to MTR increases the performance.

Acknowledgment

This work is partially supported by NSF IIS-096053 and Autodesk. We also thank the anonymous reviewers for the useful comments.

References

1. Bandyopadhyay, T., Li, Y., Ang Jr., M., Hsu, D.: A greedy strategy for tracking a locally predictable target among obstacles. In: Proc. IEEE Int. Conf. on Robotics & Automation, pp. 2342–2347 (2006)
2. Bares, W.H., Grégoire, J.P., Lester, J.C.: Realtime constraint-based cinematography for complex interactive 3d worlds. In: AAAI 1998/IAAI 1998, pp. 1101–1106. AAAI, Menlo Park (1998)
3. Becker, C., González-Banos, H., Latombe, J.C., Tomasi, C.: An intelligent observer. In: The 4th International Symposium on Experimental Robotics IV, pp. 153–160. Springer, London (1997)
4. Bhattacharya, S., Hutchinson, S.: On the existence of nash equilibrium for a two-player pursuit-evasion game with visibility constraints. Int. J. of Rob. Res. 57, 251–265 (2009)
5. Butz, A.: Anymation with cathi. In: Proceedings of the 14th Annual National Conference on Artificial Intelligence (AAAI/IAAI), pp. 957–962 (1997)
6. Courty, N., Marchand, E.: Computer animation: a new application for image-based visual servoing. In: Proceedings 2001 ICRA, IEEE International Conference on Robotics and Automation, vol. 1, pp. 223–228 (2001)
7. Drucker, S.M., Zeltzer, D.: Intelligent camera control in a virtual environment. In: Proceedings of Graphics Interface 1994, pp. 190–199 (1994)

8. Geraerts, R.: Camera planning in virtual environments using the corridor map method. In: Egges, A., Geraerts, R., Overmars, M. (eds.) MIG 2009. LNCS, vol. 5884, pp. 194–209. Springer, Heidelberg (2009)
9. Gleicher, M., Witkin, A.: Through-the-lens camera control. In: SIGGRAPH 1992: Proceedings of the 19th Annual Conference on Computer Graphics and Interactive Techniques, pp. 331–340. ACM, New York (1992)
10. Goemans, O., Overmars, M.: Automatic generation of camera motion to track a moving guide. In: International Workshop on the Algorithmic Foundations of Robotics, pp. 187–202 (2004)
11. Gonzalez-Banos, H., Lee, C.Y., Latombe, J.C.: Real-time combinatorial tracking of a target moving unpredictably among obstacles. In: Proceedings IEEE International Conference on Robotics and Automation, vol. 2, pp. 1683–1690 (2002)
12. Halper, N., Helbing, R., Strothotte, T.: A camera engine for computer games: Managing the trade-off between constraint satisfaction and frame coherence. Computer Graphics Forum 20(3), 174–183 (2002)
13. Hughes, S., Lewis, M.: Robotic camera control for remote exploration. In: CHI 2004: Proceedings of the SIGCHI Conference on Human Factors in Computing Systems, pp. 511–517. ACM, New York (2004)
14. Jung, B., Sukhatme, G.: A region-based approach for cooperative multi-target tracking in a structured environment. In: IEEE/RSJ International Conference on Intelligent Robots and Systems, vol. 3, pp. 2764–2769 (2002)
15. LaValle, S., Gonzalez-Banos, H., Becker, C., Latombe, J.C.: Motion strategies for maintaining visibility of a moving target. In: Proceedings IEEE International Conference on Robotics and Automation, vol. 1, pp. 731–736 (1997)
16. Li, T.Y., Cheng, C.C.: Real-time camera planning for navigation in virtual environments. In: Butz, A., Fisher, B., Krüger, A., Olivier, P., Christie, M. (eds.) SG 2008. LNCS, vol. 5166, pp. 118–129. Springer, Heidelberg (2008)
17. Murrieta-Cid, R., Tovar, B., Hutchinson, S.: A sampling-based motion planning approach to maintain visibility of unpredictable targets. Auton. Robots 19(3), 285–300 (2005)
18. Oskam, T., Sumner, R.W., Thuerey, N., Gross, M.: Visibility transition planning for dynamic camera control. In: SCA 2009: Proceedings of the 2009 ACM SIGGRAPH/Eurographics Symposium on Computer Animation, pp. 55–65 (2009)
19. Parker, L.E.: Distributed algorithms for multi-robot observation of multiple moving targets. Auton. Robots 12(3), 231–255 (2002)
20. Schulz, D., Burgard, W., Fox, D., Cremers, A.B.: People Tracking with Mobile Robots Using Sample-Based Joint Probabilistic Data Association Filters. The International Journal of Robotics Research 22(2), 99–116 (2003)
21. Seligmann, D.D., Feiner, S.: Automated generation of intent-based 3d illustrations. SIGGRAPH Comput. Graph. 25(4), 123–132 (1991)
22. Vo, C., Lien, J.M.: Visibility-based strategies for searching and tracking unpredictable coherent targets among known obstacles. In: ICRA 2010 Workshop: Search and Pursuit/Evasion in the Physical World (2010)
23. Wilmarth, S.A., Amato, N.M., Stiller, P.F.: MAPRM: A probabilistic roadmap planner with sampling on the medial axis of the free space. In: Proc. of IEEE Int. Conf. on Robotics and Automation, vol. 2, pp. 1024–1031 (1999)

Real-Time Space-Time Blending with Improved User Control

Galina Pasko[1], Denis Kravtsov[2,⋆], and Alexander Pasko[2]

[1] British Institute of Technology and E-commerce, UK
[2] NCCA, Bournemouth University, UK
dkravtsov@bmth.ac.uk

Abstract. In contrast to existing methods of metamorphosis based on interpolation schemes, space-time blending is a geometric operation of bounded blending performed in the higher-dimensional space. It provides transformations between shapes of different topology without necessarily establishing their alignment or correspondence. The original formulation of space-time blending has several problems: fast uncontrolled transition between shapes within the given time interval, generation of disconnected components, and lack of intuitive user control over the transformation process. We propose several techniques for more intuitive user control for space-time blending. The problem of the fast transition between the shapes is solved by the introduction of additional controllable affine transformations applied to initial objects in space-time. This gives more control to the user. The approach is further extended with the introduction of an additional non-linear deformation operation to the pure space-time blending. The proposed techniques have been implemented and tested within an industrial computer animation system. Moreover, this method can now be employed in real-time applications taking advantage of modern GPUs.

1 Introduction

The shape transformations in computer animation and games include linear transformations, non-linear deformations, and metamorphosis or morphing (transformation between two given shapes). We consider the following aspects of the general type metamorphosis problem: two given initial shapes can have arbitrary topology; the alignment or overlapping of the shapes is not required; one-to-one correspondence is not established between the boundary points or other shape features.

A brief survey of existing approaches to shape metamorphosis is given in the next section. Implicit surfaces [2] and FRep solids [8] seem to be most suitable for the given task. The solution to the above stated metamorphosis problem has been proposed in [11], where a space-time blending operation was introduced, which is a geometric bounded blending [9] operation in a higher-dimensional space. Bounded blending is a special operation performing blending union set

⋆ Corresponding author.

theoretic operation only within a specific area specified by a bounding solid (i.e. the blended shape exists only within some solid). In contrast to other existing metamorphosis operations based on the interpolation, space-time blending is based on increasing the dimensionality of an object, function-based bounded blending [9], and consequent cross-sectioning for animation.

The original formulation of space-time blending in [11] has several problems: fast uncontrolled transition between shapes within the given time interval, generation of disconnected components during the metamorphosis and lack of intuitive user control over the transformation process. The problem of fast transition between shapes was addressed in [10] using smoothing of the generalized half-cylinders which undergo bounded blending. Here, we describe several additional techniques aimed to user-controllable space-time blending. First, additional controllable affine transformations are applied to initial objects in space-time. Then, the approach is extended by additional deformation operations in space-time applied to the initial half-cylinders. Both affine transformations and deformations provide more control to the user and help avoid the generation of disconnected components. Further extensions such as automatic generation of control points for deformations, dynamic texturing and non-linear time sampling are briefly discussed. Finally, the described approach to controllable space-time blending is implemented on GPU to allow interactive rendering rates.

2 Other Works

The existing approaches to 2D metamorphosis include physically-based methods [15,16], star-skeleton representation [17], warping and distance field interpolation [3], wavelet-based [21] and surface reconstruction methods [18]. A detailed survey on 3D metamorphosis can be found in [7]. Practically all existing approaches are based on one or several of the following assumptions: equivalent topology, polygonal shape representation, shapes alignment and possibility of establishing vertex-to-vertex correspondence. The desired type of behavior can be obtained using skeletal implicit surfaces [20] with manual or automatic establishment of correspondence between scalar field source points. Metamorphosis of more general shapes based on skeletal implicit surfaces can be handled using hierarchical tree structures similar to those used in FRep. Galin et al. [5] proposed to automatically establish correspondence between such tree structures. Trees of the source and the target objects can be matched with additional manual control from the animator. Metamorphosis of arbitrary FRep objects can be described using the linear (for two initial solids) or bilinear (for four initial solids) function interpolation [4], but it can produce poor results for misaligned objects with different topology. Turk and O'Brien [19] proposed a more sophisticated approach based on interpolation of surface points (with assigned time coordinates) using radial basis functions in 4D space. This method is applicable to misaligned surfaces with arbitrary topology thus satisfying some of the requirements to the general type metamorphosis mentioned earlier. However, for the initially given implicit surfaces this requires time consuming surface sampling and interpolation steps. Aubert and Bechmann [1] proposed a method allowing them to produce metamorphosis sequences involving topological changes of the intermediate shapes. This is achieved through the application of a free-form deformation applied to a higher-dimensional space-time object. On the other hand, the topological complexity of the resulting shapes is quite limited, because they have to be produced by the user-controlled FFD of the initial shapes.

3 Metamorphosis Using Space-Time Blending

A blending operation in shape modeling is used to generate smooth transition between two given shapes (curves or surfaces). In a number of works the term "blending" is used to designate metamorphosis of 2D shapes, but we use it here in the way traditional to geometric and solid modeling. Blending set-theoretic operations (intersection, union, and difference) provide an approximation of the exact results of the pure set-theoretic operations by smoothing sharp edges and vertices. In the case of blending union of two disjoint solids with additional material, a single resulting solid with a smooth surface can be obtained. This property of the blending union operation is the basis of the approach to shape metamorphosis called space-time blending [11]. Space-time blending is technically a bounded blending operation [9] performed in space-time on objects defined by continuous real functions. Let us illustrate the approach proposed in [11] step by step for the given 2D shapes (Figs. 1-4):

1. two initial shapes are defined on the xy-plane (union of two disks and a cross in Fig. 1);
2. each shape is considered as a cross-sections of a generalized half-cylinder in 3D space (a generalized half-cylinder bounded by a plane from one side) as it is shown in Fig. 2;
3. the axes of both generalized half-cylinders are parallel to some common straight line in 3D space, for example, to the coordinate z-axis, and the bounding planes of two half-cylinders are placed at some distance to give space for making the blend;
4. apply the added material bounded blending union operation to the half-cylinders (Fig. 3);

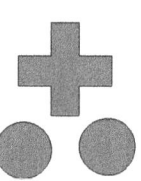

Fig. 1. Initial 2D shape (union of two disks) and final 2D shape (cross) for metamorphosis

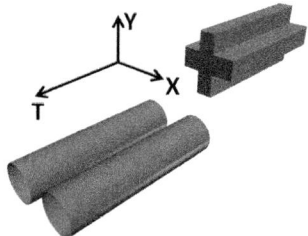

Fig. 2. Two generalized half-cylinders with the given 2D shapes as cross-sections

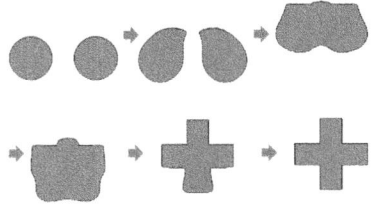
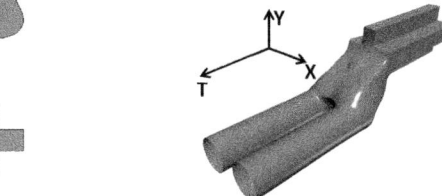

Fig. 3. Cross-sections of the bounded blending union steps of metamorphosis

Fig. 4. Bounded blending union of two generalized half-cylinders. Note the sharp upper and lower edge in the areas of the boundaries of the half-cylinders.

5. adjust parameters of the blend in order to obtain a satisfactory intermediate 2D shape in one or several 2D cross sections by planes orthogonal to z-axis (Fig. 4);
6. considering additional z-coordinate as time, make consequent orthogonal cross-sections along z-axis (Fig. 4) and combine them into 2D animation.

As it was mentioned, a bounded blending operation is applied in order to achieve the desired result. The following definition of the bounded blending set-theoretic operation is used:

$$F_{bb}(f_1, f_2, f_3) = R(f_1, f_2) + disp_{bb}(f_1, f_2, f_3), \qquad (1)$$

where $R(f_1, f_2)$ is an R-function corresponding to the type of the set-theoretic operation [13,8], the arguments of the operation $f_1(X)$ and $f_2(X)$ are defining functions of two initial solids, and $disp_{bb}(f_1, f_2, f_3)$ is a displacement function depending on the defining function of the bounding solid $f_3(X)$. The formulation for blending operations with the blend bounded by an additional bounding solid [9] was used with the following displacement function:

$$disp_{bb}(r) = \begin{cases} \frac{(1-r^2)^3}{1+r^2}, r < 1 \\ 0, r \geq 1 \end{cases}, r^2 = \begin{cases} \frac{r_1^2}{r_1^2 + r_2^2}, r_2 > 0 \\ 1, r_2 = 0 \end{cases}, \qquad (2)$$

where $r_1^2(f_1, f_2) = \left(\frac{f_1}{a_1}\right)^2 + \left(\frac{f_2}{a_2}\right)^2$, $r_2^2(f_3) = \begin{cases} \left(\frac{f_3}{a_3}\right)^2, f_3 > 0 \\ 0, f_3 \leq 0 \end{cases}$,

where the numerical parameters a_1 and a_2 control the blend symmetry, and a_3 allows us to control the influence of the function f_3 on the overall shape of the blend.

The application of the bounded blending union to the generalized half-cylinders is illustrated in figs. 3 and 4. The half-cylinders are bounded by the planes $z = -1$ and $z = 1$ to make the gap $[-1, 1]$ along z-axis between them. The bounding solid for the blend in this case is an infinite slab orthogonal to z-axis and defined by the function f_3 as an intersection of two half-spaces with

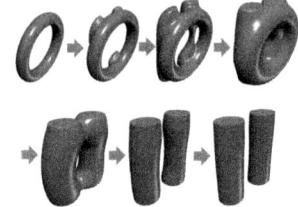

Fig. 5. Linear metamorphosis of a torus into the union of two cylinders (shown at the left)

Fig. 6. Space-time blending of a torus and a union of two cylinders

the definitions $z \geq -10$ and $z \leq 10$. The blending displacement from the exact union of two generalized half-cylinders takes zero value at the boundaries of the bounding solid (planes $z = -10$ and $z = 10$). This operation results in the exact initial 2D shapes obtained at the cross-sections outside the bounding solid. The parameters a_0 - a_3 of the bounded blend influence the blend shape and respectively the shape of the intermediate cross-sections. The main idea of space-time blending between 3D shapes is the same: each shape is considered a 3D cross-section of a 4D half-cylinder defined in space-time. Fig. 5 demonstrates intermediate steps of 3D metamorphosis with the simple linear interpolation between the defining functions and fig. 6 illustrates space-time blending applied to the same objects.

4 Control Techniques for Space-Time Blending

The main problem with the applied bounded blending (fig. 4) and the resulting animation (fig. 3) is that the most significant part of the shape transformation happens in the [-1,1] time interval with the intermediate 2D shape changing very fast from the initial to the final cross-section, which appears as a shape "jump" in the produced animation. The main reason for this "jump" is that the bounded blending is applied to a generalized half-cylinder bounded by a plane orthogonal to the axis. This set-theoretic subtraction applied to a cylinder results in the sharp edge of the half-cylinder boundary (as seen in fig. 2). This sharp edge remains a significant feature of the blended half-cylinders (see edges at the top and bottom parts of the shape in fig. 4). To avoid the fast transition between the shapes in the given interval, it was proposed in [10] to use smoothing of half-cylinders which undergo bounded blending.

Other problems of the original space-time blending are the generation of disconnected components during the metamorphosis and lack of intuitive user control over the process, where changing the parameters of bounded blending is not enough to achieve the desirable effect. To address these problems, we propose to add two more types of time-dependent transformations with their own control parameters: affine transformations (section 4.2) and deformations with control points (section 4.3).

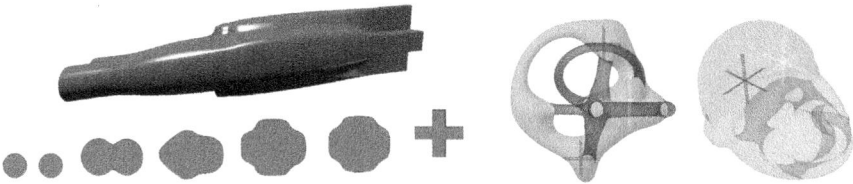

Fig. 7. Frames of animation based on the bounded blending (note the movement of the intermediate shape in the direction of the final shape and overall smoother transition)

Fig. 8. Examples of extracted "thick features"

4.1 Applying Affine Transformations

From the previous figures it can be observed that space-time cylinders are aligned with the time axis. In fact, we can extrude initial 2D shapes along an arbitrary axis in space-time. One way is the interpolation along the axis going through the centers of both objects (fig. 7). This is similar to the first order interpolation of the offsets used to shift both objects. This can be especially useful when both objects are placed at a significant distance from each other. In certain circumstances it is desirable to manipulate the transition from one shape to another in a specific way. For instance, the user may want to align particular features of the shapes or have more control over the intermediate transformation process when sizes of the shapes vary significantly. For these situations, we introduce rotation and scale affine transformations defined along the time axis (figs. 9, 10). This also helps reducing the influence of fast transitions of the defining function values of both objects and thus results in smoother transition. Without these transformations the volume of the intermediate shape needs to be significantly increased to avoid having disjoint components. But increase of the volume leads to even faster transitions between the shapes. Thus affine transformations provide more control over the interim modifications of the shape as well as help reducing the rate of the transition. All required time-dependent affine transformations can be generated automatically based on the estimated bounding shapes of the objects. Alternatively, the user has an opportunity to modify these parameters in order to achieve the desired effect.

We can apply the same technique to 3D objects. Fig. 11 demonstrates two initial 3D objects and a number of steps of the transition between them. We have applied affine transformations to adjust the size and location of the intermediate object.

4.2 Applying Additional Deformations

For certain models and values of the parameters used for the definition of space-time blending, it is possible to observe disjoint components of source and destination objects appearing during the transformation process (Fig. 12). One way of resolving this issue is the addition of user controlled deformations. The

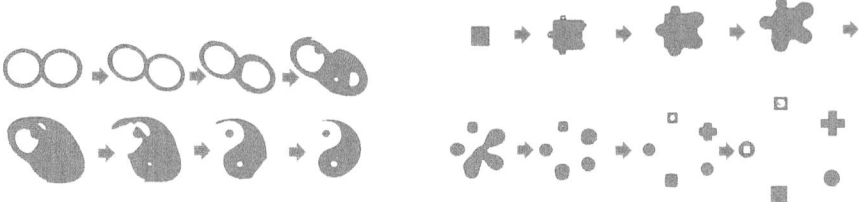

Fig. 9. User guided rotation around time axis to align object features

Fig. 10. User guided scale along time axis

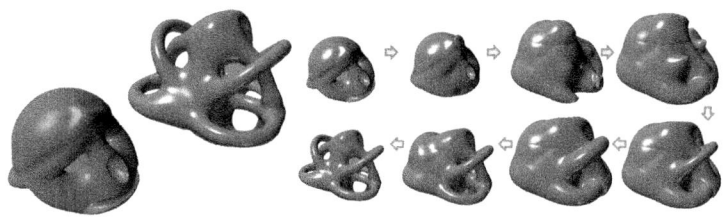

Fig. 11. Space-time blending between 3D shapes (on the left) extended by affine transformations (translation and scaling)

appearance of the disconnected component in fig. 12 can be explained by the significant difference in the distances between the initial torus and the final union of two cylinders. The transition can be improved via the introduction of time-dependent deformations in addition to the pure space-time blending. We can apply time-dependent deformations in the direction from the source object to the destination object. For the example of fig. 12, this can be done with the help

Fig. 12. Disjoint components seen during the metamorphosis using space-time blending

Fig. 13. Metamorphosis of a torus into the union of two cylinders using two additional deformations

of a non-linear space mapping ("warping") intuitively controlled by two points [14] as illustrated by fig. 13. Although the user can define the control points for the deformation interactively, these points can be also generated automatically based on the properties of the objects being blended. To do so we need to find a set of internal points with extreme values of the defining function. These points are located inside "thick features" of the model, i.e. the areas situated at the the extreme distances from the object's boundary:

$$\mathbf{D}_{src}^{p} = \bigcup_{i=1}^{N_{src}} p_{srci} : F_{src}(p_{srci}) > 0,$$
$$F_{src}(p_{srci}) > F_{src}(p_{srci} + \partial p); \|\partial p\| > 0 \qquad (3)$$

$$\mathbf{D}_{dst}^{p} = \bigcup_{j=1}^{N_{dst}} p_{dstj} : F_{dst}(p_{dstj}) > 0,$$
$$F_{dst}(p_{dstj}) > F_{dst}(p_{dstj} + \partial p); \|\partial p\| > 0$$

where $\mathbf{D}$ is a set of N points ($N_{src} = N_{dst}$) used to define non-linear space-mapping, F_{src} and F_{dst} are defining functions of the source and destination objects respectively. We find the locations of the aforementioned points performing distance transform of the functional object using Euclidean metrics, i.e. performing voxelization of the data set to retrieve a data set containing Euclidean distances to the surface of the resulting object. For every voxel in this dataset we analyze its 26 neighbours. If any of the neighbours is located further from the surface than the current one, we remove the voxel. The resulting voxels satisfy the requirement defined in equation 3. After this initial filtering similar to "erosion" operation we consider internal voxels with the highest distance value and remove all the voxels in their neighbourhood bounded by the distance to the surface. The desired number of points is then extracted from a subset of points located in the proximity of these "thick features" of the object. The user can choose the number of points retrieved in this fashion and give a hint of how close to each other he wants the retrieved points to be (fig. 8). The retrieved points are located on the medial surface of the object.

Once extreme internal points are retrieved, we need to establish correspondence between the points of the source and destination object. This is be done using the distance metrics and mutual locations of these points. The pairs of points are then used for the definition of the time-dependent deformation. Location of every destination point used in the deformation is defined over time as follows:

$$\widehat{p}_{dst_i}(t) = p_{src_i} \cdot (1 - f(t)) + p_{dst_i} \cdot f(t); f(t) \in [0; 1]$$

Thus, areas around fixed number of selected local centers of the source object will be shifted over time in the direction of the areas of interest of the destination object. This can help avoiding disjoint components, as we perform further space-time blend between the objects already modified around local centers. The main advantage of this approach is the opportunity for the user to have more control over the intermediate phases of the metamorphosis. The user can choose which parts of the objects should be modified first and the trajectory of their motion.

Transitions between particular features can be achieved with a set of non-linear deformations applied in arbitrary order. This results in another powerful expressive tool given to the animator. Besides, the balance between the deformation and the pure space-time blending can be found by selecting appropriate time schedules for both processes. The example shown in fig. 13 demonstrates two simple deformations applied to the initial torus. Each deformation is directed towards the center of one of the cylinders and fills intermediate space between the objects, thus resulting in the smoother and more natural transition.

4.3 Dynamic Texturing and Non-linear Time Sampling

Information used for affine transformations can also be employed for the improvement of time-dependent texture coordinates generation (Fig. 14). This approach is based on the method for static texture coordinates calculation described in [12]. Color at point p on the surface is defined as

$$\begin{aligned} C(p,t) = & f(n(p),t) \cdot T(S_{xy}(p,t)) + \\ & g(n(p),t) \cdot T(S_{xz}(p,t)) + \\ & h(n(p),t) \cdot T(S_{yz}(p,t)) \end{aligned}$$

where $n(p)$ is the normal evaluated at point p, $T(u,v)$ is the texture color at point (u, v), f, g, h are scalar blending functions based on normal direction and current time value. 2D coordinates used for texture sampling are defined in the following way:

$$S_{ij}(p,t) = (|\alpha_i(t) \cdot p_i + dp_i(t)|, |\alpha_j(t) \cdot p_j + dp_j(t)|)$$

where α is the scaling coefficient, $i, j \in \{x, y, z\}$ and dp is the offset retrieved from affine transformations. Employing this information can help to reduce the sliding of the projected textures on the surface and align it with the dimensions of the objects.

5 Implementation and Discussion

It is also worth mentioning that any of the aforementioned parameters used to define space-time blending can be made time-dependent. The user can dramatically change the transition varying blending parameters over time, thus adding another degree of control. This, for instance, could be used to increase the influence of source or destination object at any given moment of time or dramatically modify the intermediate object.

Due to non-linear nature of defining functions of the objects and the properties of the bounded blending operation, transition between the objects can not be expected to be a linear process. But we can adjust the visual rate of this transition performing non-uniform sampling over time. In the simplest case the time step can be adjusted depending on the estimated change of the area or volume of the shape as follows:

Table 1. Average frames per second values for different models on an NVIDIA GeForce 8800 Ultra, 768 MB of RAM

Grid resolution for polygonization	Ape to torii	Pumpkin to coach
64x64x64	200	110
128x128x128	60	30

1. perform discretization
2. estimate area or volume
3. estimate dV = area difference or volume difference
4. update time based on dV
5. repeat step 1

Discretization can be done using a simple voxelization or polygonization procedure in 2D or 3D space. It allows us to get an approximate measure of space occupied by the intermediate object. This information can be used to estimate the rate of change of the occupied area/volume at different time instants. The described process helps linearizing the resulting transition adding or removing fixed amount of material at every step.

We have further developed the approach to general type shape metamorphosis on the basis of the bounded space-time blending. In this paper, we extended the proposed approach in three directions: additional control of the metamorphosis is introduced using affine transformations for more precise control over the transition between the objects and an additional deformation operation. We have implemented our method and integrated it into a popular animation modeling system Autodesk®Maya®(Fig. 15). It allows us to comfortably work with the model continually receiving visual feedback. We are able to produce 2D and 3D animation sequences in near-real time. This is rather important, as it allows

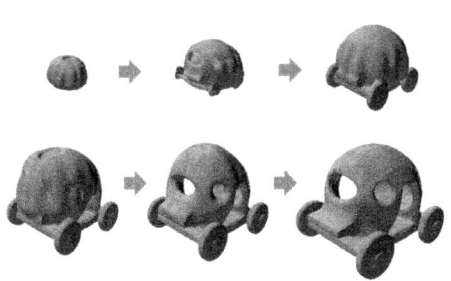

Fig. 14. Pumpkin to coach space-time blending with applied textures

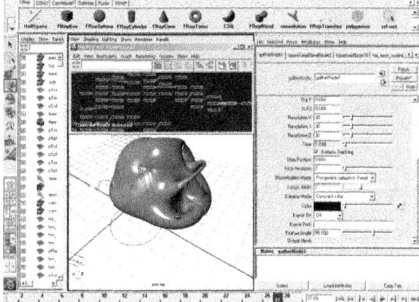

Fig. 15. Controlled space-time blending plug-in for Maya system

performing a large number of iterative modifications of the model to achieve visually pleasing results. Finally, parameters of the model are exported as a kernel for NVIDIA®CUDA®SDK. All the evaluations and rendering of the model are performed on the GPU in real-time using the approach described in [6]. The frame rates of dynamic model evaluation are shown in Table 1. These frame rates include the time required for function evaluation in the volume, polygonization and rendering. Model "Ape-to-torii" (fig. 11) consists of 53 FRep entities, while model "Pumpkin-to-coach" (fig. 14) is assembled using 65 FRep entities.

We believe that this approach can appear to be useful for computer animation and games, as it allows us to produce interesting transition effects between arbitrary 2D/3D objects. The proposed technique can be used to visualise the metamorphosis between complex entities created by the artist or between user-generated objects in an easy way in real-time.

There are a number of issues that require additional consideration. First of all, space-time blending does not by itself guarantee the elimination of unwanted disjoint components. It only provides a method allowing the user to overcome this problem and have more control over the intermediate process of shape transformation. It is thus desirable to extract the topological critical points in space-time. This information can then be employed for the generation of more detailed animation sequences in the interval of the fast shape transitions. This information could also prove to be useful for the creation of a tool simplifying the exploration of 4D and higher dimensional objects. We also plan to research the problem of volumetric attributes transfer in order to apply extended space-time blending to hypervolume objects. Another direction of further research is the application of space-time blending for advanced shape modeling operations such as lofting.

References

1. Aubert, F., Bechmann, D.: Animation by deformation of space-time objects. Computer Graphics Forum 16, 57–66 (1997)
2. Bloomenthal, J.: Introduction to Implicit Surfaces, 1st edn. The Morgan Kaufmann Series in Computer Graphics. Morgan Kaufmann, San Francisco (August 1997)
3. Daniel, C., Levin, D., Solomovici, A.: Contour blending using warp-guided distance field interpolation. In: VIS 1996: Proceedings of the 7th Conference on Visualization 1996, p. 165. IEEE Computer Society Press, Los Alamitos (1996)
4. Fausett, E., Pasko, A., Adzhiev, V.: Space-time and higher dimensional modeling for animation. In: CA 2000: Proceedings of the Computer Animation, Washington, DC, USA, p. 140. IEEE Computer Society, Los Alamitos (2000)
5. Galin, E., Leclercq, A., Akkouche, S.: Morphing the blobtree. Comput. Graph. Forum 19(4), 257–270 (2000)
6. Kravtsov, D., Fryazinov, O., Adzhiev, V., Pasko, A., Comninos, P.: Polygonal-Functional Hybrids for Computer Animation and Games. In: GPU Pro: Advanced Rendering Techniques, pp. 87–114. AK Peters Ltd., Wellesley (2010)
7. Lazarus, F., Verroust, A.: Three-dimensional metamorphosis: a survey. The Visual Computer 14(8/9), 373–389 (1998)
8. Pasko, A., Adzhiev, V., Sourin, A., Savchenko, V.: Function representation in geometric modeling: Concepts, implementation and applications. The Visual Computer (11), 429–446 (1995)

9. Pasko, G., Pasko, A., Kunii, T.: Bounded blending for function-based shape modeling. IEEE Computer Graphics and Applications 25(2), 36–45 (2005)
10. Pasko, G., Pasko, A., Ikeda, M., Kunii, T.: Advanced Metamorphosis Based on Bounded Space-time Blending. In: MMM 2004: Proceedings of the 10th International Multimedia Modelling Conference, pp. 211–217. IEEE Computer Society, Los Alamitos (2004)
11. Pasko, G., Pasko, A., Kunii, T.: Space-time blending. Computer Animation and Virtual Worlds 15(2), 109–121 (2004)
12. Peytavie, A., Galin, E., Merillou, S., Grosjean, J.: Arches: a Framework for Modeling Complex Terrains. Computer Graphics Forum (Proceedings of Eurographics) 28(2), 457–467 (2009)
13. Rvachev, V.: Methods of Logic Algebra in Mathematical Physics. Naukova Dumka, Kiev (in Russian)
14. Schmitt, B., Pasko, A., Schlick, C.: Shape-driven deformations of functionally defined heterogeneous volumetric objects. In: GRAPHITE 2003: Proceedings of the 1st International Conference on Computer Graphics and Interactive Techniques in Australasia and South East Asia, p. 127. ACM, New York (2003)
15. Sederberg, T., Greenwood, E.: A physically based approach to 2-d shape blending. In: SIGGRAPH 1992: Proceedings of the 19th Annual Conference on Computer Graphics and Interactive Techniques, pp. 25–34. ACM, New York (1992)
16. Sederberg, T., Gao, P., Wang, G., Mu, H.: 2-d shape blending: an intrinsic solution to the vertex path problem. In: SIGGRAPH 1993: Proceedings of the 20th Annual Conference on Computer Graphics and Interactive Techniques, pp. 15–18. ACM, New York (1993)
17. Shapira, M., Rappoport, A.: Shape blending using the star-skeleton representation. IEEE Comput. Graph. Appl. 15(2), 44–50 (1995)
18. Surazhsky, T., Surazhsky, V., Barequet, G., Tal, A.: Blending polygonal shapes with different topologies. Computers and Graphics 25(1), 29–39 (2001)
19. Turk, G., O'Brien, J.: Shape transformation using variational implicit functions. In: SIGGRAPH 1999: Proceedings of the 26th Annual Conference on Computer Graphics and Interactive Techniques, pp. 335–342 (1999)
20. Wyvill, B., McPheeters, C., Wyvill, G.: Animating Soft Objects. The Visual Computer 2(4), 235–242 (1986)
21. Zhang, Y., Huang, Y.: Wavelet shape blending. The Visual Computer 16(2), 106–115 (2000)

Motion Capture for a Natural Tree in the Wind

Jie Long, Cory Reimschussel, Ontario Britton,
Anthony Hall, and Michael Jones*

Department of Computer Science, Brigham Young University
michael.jones@byu.edu

Abstract. Simulating the motion of a tree in the wind is a difficult problem because of the complexity of the tree's geometry and its associated wind dynamics. Physically-based animation of trees in the wind is computationally expensive, while noise-based approaches ignore important global effects, such as sheltering. Motion capture may help solve these problems. In this paper, we present new approaches to inferring a skeleton from tree motion data and repairing motion data using a rigid body model. While the rigid body model can be used to extract data, the data contains many gaps and errors for branches that bend. Motion data repair is critical because trees are not rigid bodies. These ideas allow the reconstruction of tree motion including global effects but without a complex physical model.

1 Introduction

We address the problem of animating natural trees in games with greater accuracy but without additional computational overhead compared to techniques based on velocity or force textures–such as [5]. We believe that motion capture is one way to accomplish this goal. Motion capture of tree motion in the wind is difficult because the tree branching structure is both important and difficult to model and because branches are non rigid bodies at large deflections.

Accurate animation of trees is important to both CG animators and forestry ecologists. CG animators can use plausible and directable models of tree motion in digital storytelling. In a game, trees moving in the wind can be used to emphasize weather or create a sense of foreboding. Forestry ecologists can use models of tree motion to design pruning methodologies that maximize yield while minimizing windthrow potential.

Many approaches have been taken to modeling tree structure and geometry. Recent photo-based approaches to tree modeling [8,14,19] are particularly relevant to this work. Photo-based methods plausibly recreate 3D natural tree models by approximating the branching structure of photographed trees. However, these models are created without considering tree motion. This means that the branching structure may not match the motion of the tree.

Prior work in animating 3D tree models focuses on recreating branch motion due to wind turbulence. Wind turbulence has been simulated and has been

---
* Corresponding author.

synthesized from the frequency spectrum of turbulence created by tree crowns. Simulation-based models create tree motion based on the tree's biomechanical characteristics and wind dynamics [1,5]. Spectral approximation describes tree swaying and wind velocity field using some computer generated noise. These systems include techniques based on photographs [20] or videos [3], and some parameter-based spectral models [5,17].

Each approach is insufficient. While simulation models capture visually important wind/tree effects, such as crown sheltering, they require expensive computations that are not currently feasible in interactive applications such as games. Spectral approximation ignores sheltering effects and requires significant user intervention–but is computationally efficient. . Rather than being based on actual captured tree motion, simulation models and spectral approximations are both theoretically based.

In this paper, we present novel approaches to tree structure estimation from motion capture data and tree motion repair using interpolation. We approximate the tree structure using a minimal spanning tree over position and movement data collected during motion capture. We detect and repair the collected data using interpolation techniques based on curve fitting and machine learning. The resulting tree model and animations are realistic recreations of a tree moving in the wind and include sheltering effects while supporting fast playback. We avoid modelling wind fields explicitly, since their end effect is measured directly in the motion of the leaves and branches. Since this represents only the initial stages of applying motion capture to the problem of tree animation, we focus simply on the motion of one specific tree subject. We leave for future work questions such as how the results might scale to other trees or subject models. The animations resulting from this work can be seen in the video which accompanies this paper.

2 Related Work

In this section we discuss closely related work in tree modeling, tree animation and motion capture.

2.1 Static Tree Modeling

Static tree models describe tree shapes including topology, texture and geometry for the trunk, branches and leaves. Tree models for motion capture data need to capture the branching structure of a specific tree such that the captured motion looks plausible when animating the model. This is a unique challenge in tree modeling that has not been addressed by prior work.

Position-aware L-systems [13] have been used with some success to create models of specific plants but these results are difficult to reproduce. The processes of controlling the branching structure using the silhouette and setting the rule parameters is difficult.

Photo-based approaches [8,14,19] can produce plausible tree shapes which match a given tree but estimate the internal branching structure using methods such as particle flow [8]. Estimates of the internal branching structure are not sensitive to the motion of the original tree. We use similar methods based on photographs to create a bounding volume for the tree shape. In addition to images, we also use motion capture data to recreate a plausible internal branching structure in which points contained in one branch have similar movement.

Diener et al approximate shrub structure based on single-camera video data of a shrub in the wind [3]. Diener uses a clustering method to identify clumps of the shrub with similar motion then builds a skeleton which corresponds to the clustering. Our approach is similar but we skip the clustering step and build a skeleton directly from the marker positions and motion data in 3D rather than 2D video data.

2.2 Animation of Trees

Prior work in tree animations rely primarily on simulation-based methods and spectral approximations. Both approaches produce plausible tree movements in the wind while ignoring some effects to remain tractable. Most of these methods simplify the complex dynamics of leaf-wind interaction, which is the primary cause of branch motion. One study [16] found that much of the motion of a branch could be accounted for by the presence or absence of leaves. Motion capture obviates the dynamic model but introduces several additional problems.

Simulation-based methods use computational fluid dynamics to simulate the effect of wind on trees. Akagi and Kitajima [1] do allow trees to influence the wind using a two-way coupled model based on the Navier-Stokes equations, with an additional term for external forces. The simulation is based on a stable approach [18] to the marker and cell method. Akagi and Kitajima use virtual resistive bodies to account for tree structures smaller than the grid resolution and add adaptive resolution and a boundary conditions map to improve performance by allocating grid resolution only where needed. Simulation-based methods are currently too computationally expensive for use in games.

Spectral approximations of trees in wind use approximations to the recorded spectra of wind passing through trees to generate motion. This method was first used by Shinya and Fournier [17] and later by Chuang [2], Habel [5] and Zhang [21]. Other work also relies on approximations in the frequency domain but uses different techniques to approximate turbulence [9,10,18]. Spectral methods have also been combined with physical simulation [11,21]. Spectral approximations result in plausible motion and are efficient enough for games but ignore the bidirectional wind/tree interactions, such as sheltering effects. These effects are important for visual realism and are captured using motion capture. Our objective is to create animation data which can be used as efficiently as textures but which are more accurate.

More recent work [4,5] animates tree motion in a computationally economical way. Diener [4] simplifies the wind model using a pre-computed wind projection basis taken from vibration modes rather than a harmonic oscillator model. As with Habel [5], the wind is assumed to be spatially uniform for a single tree. At run time, the wind load is estimated for all nodes on a tree relative to the wind projection basis, and this can be combined with a level-of-detail model to render a forest of thousands of trees in real time. Each of these methods ignores the effect in return of trees on the wind and therefore omits all forms of sheltering. Another less significant problem is that the turbulence used in these models matches actual turbulence only in the frequency domain and not necessarily in the time domain. While many turbulence patterns share frequency spectrums with those created by tree crowns, only one pattern matches the spatial properties of the actual turbulence created by a specific tree in a specific wind. Our work captures the motion of a tree as it moves in the turbulence created by that tree.

Our work is similar to video-based approaches in that we capture and analyze tree motion. Unlike video [3] or image-based [20] approaches, we obtain a motion path for a cloud of points in 3D rather than applying 2D motion to 3D skeletons. Our methods may also yield new insights into how to use video data in the animation of trees.

2.3 Motion Capture

Motion capture for trees is more difficult than performance capture of human subjects because trees are both rigid and non-rigid (depending on the applied force among other factors) and have more complex and less predictable topologies.

Motion capture systems have been widely used for human or animal performance capture [15,22]. We use a method similar to Kirk [7] to automatically generate rigid skeletons from optical motion capture data. Since tree branches are both rigid and non-rigid, the data do not contain a constant distance between markers. We use a rigid body algorithm to solve the marker indexing problem. Because some of the data is collected from non-rigid motion, this introduces additional noise and gaps in the data. A central contribution of this paper is a way to repair this data for tree motion. Another approach to this problem would be to investigate marker indexing algorithms for non-rigid bodies. Doing so may reduce the amount of noise and gaps in the motion data.

3 Motion Capture

We use an optical motion capture system to collect position and motion data from which we reconstruct tree structure and movement. For this paper, the system consists of 12 OptiTrack FLEX:V100 cameras arranged in a circle in a 4m by 4m room indoors. A cherry tree sapling with height 2m was used as the test subject. The tree was placed in the center of these cameras and a fan is used to create wind at different speeds near the tree. The system has not been deployed for trees larger than 3m and we believe it would be impractical for large trees. We believe it would be more practical to explore methods for extrapolating small tree motion to create large tree motion than it would be to capture large tree motion.

Reflective markers are placed on each branch and some leaves. The arrangement of markers on a single branch segment depends on the flexibility of the

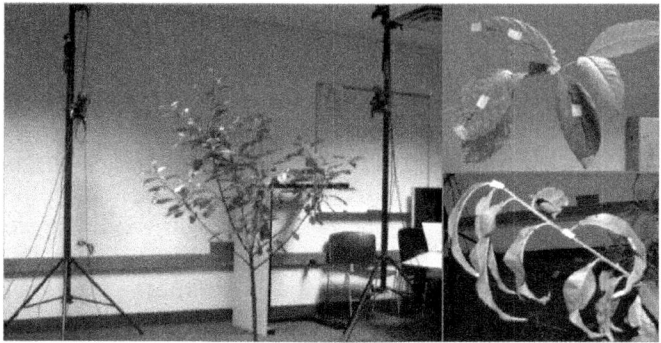

Fig. 1. Motion capture set up for a natural tree using an optical motion capture system

branch. If the branch is thin and flexible, the distance between markers is about 8cm; for a rigid branch, such as the trunk, the distance between markers is about 15cm. Placing markers more closely together allows us to approximate a flexible branch as a series of rigid linkages. This results in cleaner motion data with fewer gaps, since the motion capture system depends upon a user-defined set of fixed-length rigid links in order to track and label the markers as they move. The benefits of this approach are especially evident under higher wind speeds, when branches begin to flex and bow. This placement strategy assumes that the tree crown is sparse. Trees with dense crowns will require a different strategy.

We placed approximately 100 markers on the tree to collect branch motion. The 3D positions of all reflective markers are recorded at 100 frames/sec. Leaf motion is recorded separately using three markers on each leaf in a smaller representative sample. Fig. 1 shows the arrangement of markers for both branches and leaves motion capture.

4 Build Static Tree Geometries

One significant problem with motion capture of trees, compared to human performance capture, is that the branching structure, or topology, of a tree is less predictable and more complex than that of a human. A minimal spanning tree algorithm is used with a cost function derived from motion data to create a plausible branching structure. The branching structure is plausible when animating it with the captured motion looks plausible. The cost function is one of the contributions of this paper.

4.1 Skeleton and Topology Estimation

Fig. 2 summarizes the process of estimating the skeleton and topology. This process has three steps. First, hierarchical clustering eliminates replicated recorded markers in each frame. Next, we use position and motion data for each marker

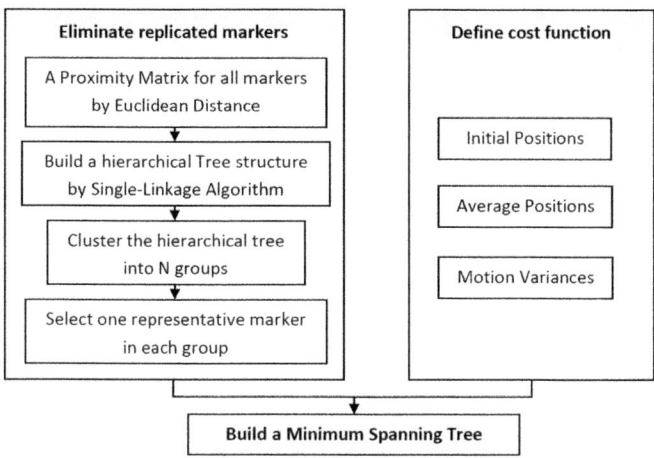

Fig. 2. Steps in building tree skeleton.

to define a cost function. The cost function is used to estimate the plausibility of merging two different markers. A minimum spanning tree algorithm uses a different cost function to connect the markers into a tree-like skeleton which will have plausible motion when animated using the captured motion data.

The first step, shown on the left side of Fig. 2, is to eliminate duplicate, yet slightly different, recorded positions of a single marker. We use Euclidian distance as the clustering metric. The single linkage algorithm groups markers into a hierarchy of n clusters, where n is the number of markers originally placed on the tree. Within each frame, a cluster is reduced to a single representative marker. When all frames have the correct number of markers, we further refine the representative position of each marker, either by choosing its position in the tree's rest pose, or by averaging its position over time. The tree skeleton will be built from the n representative marker positions. This skeleton will be used for the entire capture sequence.

The second step, shown on the right side of Fig. 2, computes costs for creating connections in the control skeleton between different pairs of markers based on the recorded position and motion data. Connection costs are computed for pairs of representative markers with one marker from each cluster. The cost function consists of three elements: initial position, average position over time, and variance of position over time. We assume that the branch motion is periodic. So the average position is similar to the position, also variance reflects the amplitude of the movement. The initial positions are recorded when there is no wind and the tree is stationary. The average positions are calculated as shown in the next equation in which m is the number of recorded frames, and p_i is the 3D position for a marker at the ith frame:

$$\bar{d} = \frac{1}{m}\Sigma_{i=1}^{m} p_i.$$

The variance in position is similarly defined as:

$$\sigma^2 = \frac{1}{m}\Sigma_{i=1}^{m}(p_i - \bar{d})^2.$$

Let α, β, and γ be constant weighting parameters, then the cost to connect markers M_a and M_b is given by:

$$\omega = \alpha||p_{M_a} - p_{M_b}|| + \beta||\bar{d}_{M_a} - \bar{d}_{M_b}|| + \gamma||\sigma_{M_a} - \sigma_{M_b}||.$$

The cost to connect markers M_a and M_b is low when M_a and M_b are close in both position and movement.

In the third step, we use Prim's MST algorithm with the node at the bottom of the trunk as a starting point to build the tree skeleton. Pairs of markers with similar position information and movements have low connection cost and are connected in the skeleton. This skeleton is directly taken as the input tree structure for rendering in the next step.

Fig. 3 shows the importance of each part of the cost function. The right side of Fig. 3 shows a tree created using just the change in variance as the cost function. This cost function results in branch tips connected to branch tips because variance increases as one moves along a tree from trunk to branch tips. The middle tree was created using only positional information. While the structure is accurate for much of the tree, several points are connected incorrectly across the middle of the tree. This will result in un-plausible motion when animated using the motion capture data. Using both metrics, along with average position, results in a more accurate model shown on the right.

By combining these parameters, we connect markers with similar position and movement. For a tree with 98 clusters of markers, 66.26% of the resulting connections are correct when compared with the actual tree. Most errors are from connecting markers in the correct branch segments but at the wrong junction points within the branch segment. This cost function occasionally connects markers from different branches but which share close positions and movements. In these cases, the motion of physically adjacent branch tips is similar and the resulting animation is still plausible.

4.2 Geometry of Branches and Leaves

After the tree skeleton is created, the next step is to generate the geometric mesh. The marker points in a single branch are used as control points to create a curve. A second curve is placed at the first marker point in the branch and oriented to the first two points in the branch. A closed circular shape is swept along the profile curve to create a NURBS surface. The profile and shape curves are discarded, leaving just the branch geometry.

Then we bind the mesh to the skeleton. This step is separated from the previous steps so that the artist has more flexibility to modify the automatic mesh before it is bound to the skeleton. After any needed updates, the mesh is bound to the geometry. Once the geometry is bound, the artist again has the flexibility to manually tweak the binding.

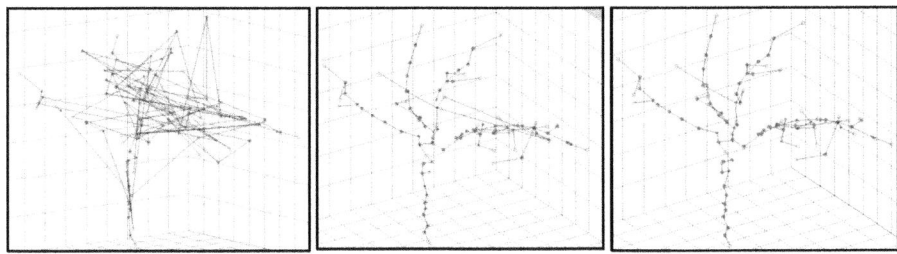

Fig. 3. Reconstructed tree topologies using variance, initial position and a combination of variance, initial position and average position

Finally, a 3D point cloud inferred from photographs guides the manual placement of leaves. The leaves are placed to fill the volume occupied by the original tree. The tree volume is created using inverse volumetric rendering [14] applied to 37 photographs taken from a known camera position. The resulting 3D point cloud is exported to a 3D modeling package and, after manually matching the tree skeleton with this point cloud, we manually place leaves on branches while remaining in the recorded crown volume.

5 Build Tree Motion

In this section, we will describe branch motion repair and leaf motion synthesis. Branch motion repair is the process of identifying and eliminating errors, gaps and noise from the motion capture data. The resulting motion will be used to animate the 3D tree model created in the prior section.

We used the rigid body algorithm which was shipped as part of the NaturalPoint Arena software to convert unindexed point clouds into an animated skeleton. Because tree branches are non rigid at large deflections, the resulting motion contains more gaps and errors than one might expect to find for rigid body motion capture. We use linear interpolation, a filter and a machine learning algorithm to repair the resulting motion. The NaturalPoint Arena softare provides some interpolation processes to fix motion gaps, but requires the user to manually identify gap regions and select a correction method. We automate gap detection and correction with different methods, depending on the gap size. A machine learning based method for addressing large gaps is one of the contributions of this paper. We use a standard curve fitting technique for small gaps.

5.1 Filter-Based Noise Detection and Removal

For some non-rigid motion, the rigid body motion capture system introduces anomalous artifacts to the motion signal, resulting in sporadic popping motions of certain leaves and twigs. These artifacts are detected using convolution-based filtering techniques, and are replaced by fitting Bezier curves over the corresponding sections of the motion signal.

5.2 Small Gaps in Data

Small gaps in data are short sequences of 100 frames or less in which no position data is recorded for a marker. Small gaps occur when a marker becomes occluded or is otherwise lost. Linear interpolation is used to repair small gaps because linear interpolation can be done quickly and is good enough for these gaps.

Linear interpolation predicts missing marker positions based on the positions of neighbors. For a marker with missing position data, we find the two nearest neighbors with available position data. Then we compute Euclidean distances among the positions of these three markers and a velocity for each marker. Different distance metrics can be used. By doing linear interpolation according to the positions and velocities, we estimate the position for the missing marker.

Linear interpolation works well if all three markers have similar movement. However, if the motions of two different, but adjacent, missing markers have their positions interpolated from the same set of nearby neighbors, the resulting interpolated motion may not preserve each marker's unique periodic motion. This may happen even though we aim to make the interpolated motion fit smoothly with the existing motion for each marker. However, losing periodic movements for a short period of time when repairing a small gap still results in visually plausible motion.

5.3 Large Gaps in Data

Repairing large gaps in data is done using a more sophisticated interpolation scheme so that the synthesized motion has good periodic properties. Large gaps in data refer to gaps which comprise more than 40% of the entire motion trace collected for a single marker. A machine learning algorithm builds a function which is used to infer motion which is used to fill large gaps.

Given the connection between two adjacent markers, motion data for both markers at low wind speed, and motion data for one marker at high speed, the machine learning algorithm trains a support vector machine(SVM) and defines a correlation function. This approach is based on the observation that good data is captured for all markers at low wind speed but that large gaps appear in the data for some markers at high wind speed. The SVM learns a correlation between data collected at low wind speeds. This relationship is used to estimate missing motion at high wind speeds–under the assumption that the relationship is not sensitive to wind velocity.

The tree skeleton structure is used to find the nearest topological neighbor with motion data for both high and low speeds. In most cases, markers at branch tips have missing data while markers at the branch base have the required data. This is because markers at the branch base move more rigidly than markers on tips. In these cases, the marker at a tip has large gaps in motion data and its nearest neighbor in the direction of the branch base often lies on the same branch.

Sequential minimal optimization(SMO) [6,12] trains a SVM which defines the correlation function between the two markers' position in low speed. In order to

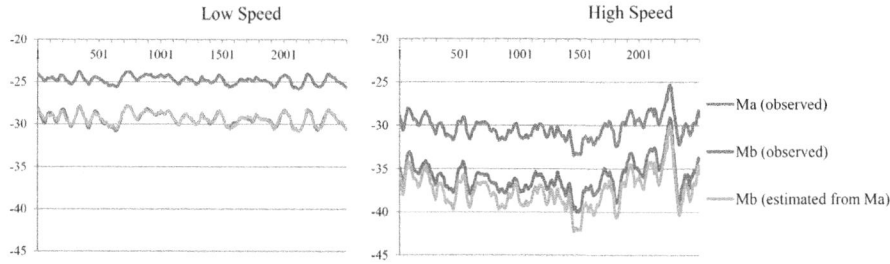

Fig. 4. Predicted versus actual displacement for a marker and low and high wind speeds

improve the precision of the correlation function and to avoid phase differences, the motion data from each series is sorted in ascending order of displacement. Let M_a be the nearest neighbor to M_b, which is a marker with missing motion at high speed. M_a and M_b both have motion data in low speed. A learned function F estimates the position of M_b given M_a. Position data from M_a recorded at high wind speeds is given to F which then estimates M_b's position at high wind speeds.

Fig. 4 shows the estimated and actual position for one marker at low and high wind speeds. The vertical axis is the displacement and the horizontal axis is the frame number. In this figure, the motion of marker M_a at low wind speed, which is the topmost trace on the left, is used with the recorded motion of marker M_b, which is the lower trace on the left, to learn a function which predicts the position of M_b given the position of M_a. For comparison, we placed the predicted position of M_b on the graph as well. The predicted position of M_b closely matches the actual position at low wind speeds. At high wind speeds, shown on the right, we held back the recorded position of M_b and predicted the position of M_b in each frame given only the position of M_a. The predicted position of M_b at high speeds closely matches the actual position of M_b but tends to overestimate the amount of displacement in M_b.

5.4 Leaf Motions

Motion data applied to leaves is based on motion captured from only a few leaves. This motion is scaled and offset to simulate a greater variety and randomness of leaf motion. The leaf geometry deforms along motion curves applied at the end, at the middle and near the stem.

The complexity of leaves moving in the wind precludes any attempt to correlate leaf movement with the movement of the branch it is on. Leaves can be quite turbulent or almost still on a branch that is either very still or sweeping any position through its arcs of movement. However, motion of the leaves is scaled with the branch motion to suggest that they are driven by the same wind. These two motion sets can also be decoupled for an artist to achieve a particular effect.

6 Results

The final animation is shown in the video which accompanies this paper. In that video, most motion capture artifacts have been removed and the motion looks reasonable. Results in skeleton estimation and motion repair, which are the main contributions of this paper, were given in the preceding sections.

7 Conclusion and Future Work

A plausible tree skeleton can be reconstructed using a minimal spanning tree algorithm over a cost function defined using position and motion data. The skeleton is plausible in the sense that replaying the capture motion on the skeleton looks realistic. Gaps and errors in motion capture data for trees can be replaced with data interpolated from neighboring branch motion. These are important steps toward realizing motion capture of trees for tree animation in games. Motion capture of tree motion is a good match for motion in games because the resulting motion is realistic but requires only replaying, rather than simulating, actual motion.

We had hoped to get better results with the repaired data and the rigid body algorithm we used. Based on these results, we believe that investigating other approaches to processing the point cloud are more promising than repairing the errors caused by using the rigid body algorithm we used.

Future work could take several interesting directions. One of these is to avoid defining rigid bodies for each branch while capturing motion, by defining the tree as a non-rigid body, which is a truer representation of its natural form. More work needs to be done to be sure the algorithm scales well for capturing the motion of other tree subjects as well as for transferring captured motion from one tree to another. By capturing data from multiple trees at once, the interactions among them could be studied and applied to simulate groups of trees or even forests.

References

1. Akagi, Y., Kitajima, K.: Computer animation of swaying trees based on physical simulation. Computers and Graphics 30(4), 529–539 (2006)
2. Chuang, Y.Y., Goldman, D.B., Zheng, K.C., Curless, B., Salesin, D.H., Szeliski, R.: Animating pictures with stochastic motion textures, pp. 853–860. ACM, New York (2005)
3. Diener, J., Reveret, L., Fiume, E.: Hierarchical retargetting of 2d motion fields to the animation of 3d plant models. In: ACM-SIGGRAPH/EG Symposium on Computer Animation (SCA), pp. 187–195. ACM-SIGGRAPH/EUROGRAPHICS (2006)
4. Diener, J., Rodriguez, M., Baboud, L., Reveret, L.: Wind projection basis for real-time animation of trees. Computer Graphics Forum (Proceedings of EUROGRAPHICS 2009), 28(2) (March 2009)

5. Habel, R., Kusternig, A., Wimmer, M.: Physically guided animation of trees. Computer Graphics Forum (Proceedings EUROGRAPHICS 2009) 28(2), 523–532 (2009)
6. Keerthi, S., Shevade, S., Bhattacharyya, C., Murthy, K.: Improvements to platt's smo algorithm for svm classifier design. Neural Comput. 13(3), 637–649 (2001)
7. Kirk, A., O'Brien, J., Forsyth, D.: Skeletal parameter estimation from optical motion capture data. In: CVPR 2005, pp. 782–788 (June 2005)
8. Neubert, B., Franken, T., Deussen, O.: Approximate image-based tree-modeling using particle flows. ACM Transactions on Graphics (Proc. of SIGGRAPH 2007) 26(3), 88 (2007)
9. Ono, H.: Practical experience in the physical animation and destruction of trees, pp. 149–159. Springer, Budapest (1997)
10. Ota, S., Fujimoto, T., Tamura, M., Muraoka, K., Fujita, K., Chiba, N.: 1/fb noise-based real-time animation of trees swaying in wind fields. In: Proceedings of the Computer Graphics International (CGI 2003), pp. 52–59 (2003)
11. Ota, S., Tamura, M., Fujimoto, T., Muraoka, K., Chiba, N.: A hybrid method for real-time animation of trees swaying in wind fields. The Visual Computer 20(10), 613–623 (2004)
12. Platt, J.: Fast training of support vector machines using sequential minimal optimization, pp. 185–208 (1999)
13. Prusinkiewicz, P., Mundermann, L., Karwowski, R., Lane, B.: The use of positional information in the modeling of plants. In: Proceedings of SIGGRAPH 2001, pp. 289–300 (August 2001)
14. Reche-Martinez, A., Martin, I., Drettakis, G.: Volumetric reconstruction and interactive rendering of trees from photographs. ACM Trans. Graph. 23(3), 720–727 (2004)
15. Rosenhahn, B., Brox, T., Kersting, U., Smith, D., Gurney, J., Klette, R.: A system for marker-less human motion estimation. In: Kropatsch, W.G., Sablatnig, R., Hanbury, A. (eds.) DAGM 2005. LNCS, vol. 3663, pp. 230–237. Springer, Heidelberg (2005)
16. Rudnicki, M., Burns, D.: Branch sway period of 4 tree species using 3-d motion tracking. In: Fifth Plant Biomechanics Conference, STFI-Packforsk, Stockholm, Sweden (2006)
17. Shinya, M., Fournier, A.: Stochastic motion under the influence of wind. In: EUROGRAPHICS 1992, vol. 11, pp. 119–128. Blackwell Publishers, Malden (1992)
18. Stam, J.: Stochastic dynamics: Simulating the effects of turbulence on flexible structures. In: EUROGRAPHICS 1997, vol. 16, pp. 119–128. Blackwell Publishers, Malden (1997)
19. Tan, P., Zeng, G., Wang, J., Kang, S.B., Quan, L.: Image-based tree modeling. In: SIGGRAPH 2007: ACM SIGGRAPH 2007 Papers, p. 87. ACM, New York (2007)
20. Wu, E., Chen, Y., Yan, T., Zhang, X.: Reconstruction and physically-based animation of trees from static images. In: Computer Animation and Simulation 1999, pp. 157–166. Springer, Heidelberg (September 1999)
21. Zhang, L., Song, C., Tan, Q., Chen, W., Peng, Q.: Quasi-physical simulation of large-scale dynamic forest scenes. In: Nishita, T., Peng, Q., Seidel, H.-P. (eds.) CGI 2006. LNCS, vol. 4035, pp. 735–742. Springer, Heidelberg (2006)
22. Zordan, V., van der Horst, N.: Mapping optical motion capture data to skeletal motion using a physical model. In: Proceedings of the 2003 ACM SIGGRAPH/EUROGRAPHICS Symposium on Computer Animation, pp. 245–250 (May 2003)

Active Geometry for Game Characters

Damien Rohmer[1,2], Stefanie Hahmann[1], and Marie-Paule Cani[1,2]

[1] Grenoble Universités, Laboratoire Jean Kuntzmann
[2] INRIA

Abstract. Animating the geometry of a real-time character is typically done using fast methods such as smooth skinning or coarse physically-based animation. These methods are not able to capture realistic behaviors such as flesh and muscles bulging with constant volume or fine wrinkling of animated garments. This paper advocates the use of active geometric models, applied on top of the current geometric layer, to mimic these behaviors without requiring any expensive computation. Our models fit into the standard animation pipe-line and can be tuned in an intuitive way thanks to their geometric nature.

Keywords: Geometry, Procedural model, Constant volume, Wrinkles.

1 Introduction

Animating characters with proper skin and clothing deformations is an open challenge in the field of computer games, where visual realism is desirable to improve the users immersion while all computations are to be done in real time. In particular, skin deformations should not produce any unwanted loss of volume when the character articulates and cloth surface should not elongate or shrink during deformations. Although a physically-based simulation could be used to maintain these constraints, it would lead to stiff equations and would hardly be applicable in real-time. The deformation methods most often used in real-time applications are therefore: smooth skinning (SSD), despite of its well know artifacts; and very coarse, low-rigidity simulations when animating the dynamics of floating garments is required. These methods are not sufficient to get the flesh and muscle bulging behavior and the dynamic wrinkles one would expect.

This paper advocates the use of *active geometric models*, placed on top of the pre-existing layers, to add visual realism to animated characters at little extra cost. Active geometric models are procedural layers aimed at maintaining a given geometric constraint over time. They act by deforming and possibly refining the current geometry on the fly, just before rendering. We illustrate this concept by presenting two examples of such models, first introduced in [1, 2]. The first one, based on volume control, is used to generate appropriate skin bulges and skin wrinkles when a character articulates. The second one, based on a measure of isometry with a rest shape, dynamically adds cloth-looking wrinkles automatically. Using these geometric models also eases user control: one just needs to specify deformation profiles for the constant-volume skin, and a single thickness parameter for the wrinkling cloth.

Fig. 1. A sitting elephant with bulging belly and a jumping character with a wrinkling dress modeled using active geometric layers to reduce skin volume variations and cloth shrinkage

The remainder of this paper develops as follows: Section 2 is a brief state of the art on skin and cloth animation. Section 3 describes our method for enhancing SSD with volume constraints. Section 4 discusses the addition of plausible cloth wrinkles on top of a coarse simulated mesh used to animate floating garments. We conclude by discussing the results in Section 5.

2 Previous Work

2.1 Animating a Character's Skin

Smooth skinning deformation (SSD) is the standard method for skinning an articulated character in real-time: scalar weights relative to each bone of the articulated structure are associated with the mesh. Each mesh vertex is repositioned during animation using the weighted combination of the positions relatif to bone's local frame. This fast method unfortunately suffers from a number of artifacts, such as the loss of volume when a joint rotates or twists (see [3]).

Numerous attempts were made to improve SSD. A first family of approaches uses a pre-computation to best fit extended SSD parameters to some example pauses. The fitting can be performed on vector corrections [4], triangle deformations [5], matrices of influences [6], or on the position and weights of extra joints within the skeleton [7]. While such approaches enable to model complex local behavior of the mesh during animation, they require good skin models for designing a set of rest pauses to start with, and are limited to the range of motion they span. Therefore unexpected motion in a video game may lead to unrealistic results.

A second family of approaches uses improved interpolation within the SSD framework at little extra cost. It includes rigidity constraints on the medial axis [8], non-linear matrix blending [9] or the use of the dual quaternions [10]. However no bulging effect (such as in fig. 1 left) due to the constant volume behavior of flesh and muscles can be modeled.

Closer to our goals, basic SSD was enhanced to get some dynamic behavior [11] and to generate wrinkling effects for skin or fitting garments [12], however without enforcing any constant volume or surface isometry constraint. Lastly, a constant volume extension of SSD was proposed [13]. It requires streamline integration of vertex positions and is thus time consuming and difficult to use in a standard animation pipeline. Contrary to this approach, we rely on standard SSD enhanced with constant volume skinning as independent post-processing corrections at each frame.

2.2 Animating Character's Clothing

Despite of a huge amount of research, realistic cloth simulation [14–16] still needs too much computing resources. Using coarse meshes or soft springs in mass-spring simulations reduces computation time but fails capturing typical cloth wrinkling behavior, since the surface tends to elongate or shrink rather than fold.

A number of geometric approaches were recently developed to model cloth material. The most recent ones [17, 18] give impressive results but require training examples and can only model the deformations herein captured. The range of possible deformation is limited as well for pre-set procedural models such as the one based on cloth buckling around cylinders [19]. A subdivision process followed by vertex correction based on local minimization was used in [20] but requires a minimization step on highly refined meshes. Procedural wrinkle maps were investigated in [21, 22] where wrinkle magnitude automatically adapts to triangle compression. However interpolating wrinkle directions on a texture map leads to unrealistic fading in and out of wrinkles during animation. Closer to our work, cloth wrinkles were defined as curves on the input cloth surface and used as deformers to generate animated wrinkles [23]. However curve shapes and influence radii were manually defined for a specific number of frames, requiring expert design skills and significant user time. Similarly to this approach, we encode wrinkles as surface curves, but use automatic algorithms to control their strength and their dynamic merging behavior over time.

3 Character Skinning with Volume Control

This section describes the active geometric layer we add in top of standard SSD to get constant volume muscle and flesh deformation while providing some intuitive shape control. This work was first published in [1], which we refer to for more details.

3.1 Goals

Volume preservation for animated characters raises specific issues:

First, volume preservation should be local, as the human body does not inflate far away (as a balloon would do) when locally compressed. The loss of volume due to SSD typically takes place near articulations and should be restored there.

Second, the volume correction provided by our new active geometric layer should allow some local shape control, since muscles and fatty tissue can require quite different bulging profiles, from a smooth inflation to some wrinkling profile.

Our solution is a weighted volume correction step where the weights specify how much restoring a given local volume variation should affect the positions of the different mesh vertices. These weights can be pre-computed from the skinning weights as in [24] or be more precisely specified by distributing the amount of correction along the axes of bone-based local frames, thanks to 1-dimensional shape profiles.

3.2 Setting the Volume to a Fixed Value

The volume enclosed in a closed triangular mesh $\overline{S}$ can be expressed as a trilinear function of its vertices $\overline{\mathbf{p}}_i = (x_i, y_i, z_i) \in \mathbb{R}^3$, $i = 1, \ldots, N$, as:

$$V = \sum_{\text{face}(l,m,n) \in S} \frac{z_l + z_m + z_n}{6} \begin{vmatrix} (x_m - x_l) & (y_m - y_l) \\ (x_n - x_l) & (y_n - y_l) \end{vmatrix}, \quad (1)$$

where $\text{face}(l, m, n) = \Delta(\overline{\mathbf{p}}_l, \overline{\mathbf{p}}_m, \overline{\mathbf{p}}_n)$ is a triangle of $\overline{S}$. Let S be the pointwise deformed surface mesh with vertex vector $\mathbf{p} = (\mathbf{p}_x, \mathbf{p}_y, \mathbf{p}_z) \in \mathbb{R}^{3N}$. The volume correction step consists of computing a correction vector $\mathbf{u} = (\mathbf{u}_x, \mathbf{u}_y, \mathbf{u}_z) \in \mathbb{R}^{3N}$ applied to S such that the volume $V(\overline{\mathbf{p}}) = V(\mathbf{p} + \mathbf{u})$ of the corrected mesh equals the specified target value V_{target}. Note that any change of volume could be animated as well by prescribing a varying target volume over time.

We achieve volume correction by solving the constrained weighted least-squares problem:

$$\begin{cases} \min \ \|\mathbf{u}\|_\gamma^2 \\ \text{subject to} \ \ V(\mathbf{p} + \mathbf{u}) = V_{target}. \end{cases} \quad (2)$$

where $\|u\|_\gamma^2 := \sum_{k=1}^{N} u^2/\gamma_k$, γ_k. $\gamma = (\gamma_1, \ldots, \gamma_N)$ is a set of weights which specify the distribution of the correction over the mesh vertices.

To efficiently solve equation (2), the volume constraint is decomposed and processed separately on the 3 axes x, y and z, which gives a linear expression to each constraint [25], as follows:

First, u_x is computed as a solution of (2), but where u_y, u_z are set to fixed values. The constraint thus becomes linear. We derive an analytical expression of the solution using Lagrangien multipliers: $\mathbf{u}_x = \mu_0 \Delta V \frac{\nabla_x V}{\|\nabla_x V\|_{1/\gamma}^2}$, where $\Delta V := V(\mathbf{p}) - V_{\text{target}}$. The scalar value μ_0 determines the percent of volume correction in x-direction. Iterating this correction step in $y-$ and $z-$direction, with μ_1 and μ_2 percent respectively of the volume correction (where $\mu_0 + \mu_1 + \mu_2 = 1$)

insures that the final surface restores the target volume. The weighted closed form solution for the correction vector $\mathbf{u}_k$ at vertex k is finally given by

$$\mathbf{u}_k = (u_x^k, u_y^k, u_z^k) = \gamma_k \Delta V \left(\mu_0 \frac{\nabla_\mathbf{x} V(\mathbf{p}_k)}{\|\nabla_\mathbf{x} V\|_{1/\gamma}^2}, \mu_1 \frac{\nabla_\mathbf{y} V^*(\mathbf{p}_k)}{\|\nabla_\mathbf{y} V^*\|_{1/\gamma}^2}, \mu_2 \frac{\nabla_\mathbf{z} V^{**}(\mathbf{p}_k)}{\|\nabla_\mathbf{z} V^{**}\|_{1/\gamma}^2} \right), \quad (3)$$

where $\nabla_\mathbf{x} V(\mathbf{p}_k)$ designates the kth entry of the gradient vector $\nabla_\mathbf{x} V$ and analogously for y and z. $\nabla_\mathbf{x} V$ is obtained by differentiating (1). V^* and V^{**} denote the volumes of the intermediate surfaces obtained after correction in $x-$ and $y-$direction respectively.

3.3 Getting Organic-Looking Volume Corrections

The volume correction method we just presented can be easily tuned to generate a variety of effects from muscle-like local bulges when a bone articulates to fatty-tissue looking bulges with folds and creases when the volume needs to be restored in a large compressed region such as the elephant belly.

While the mesh vertices weights γ are used to distribute the correction of a volume loss due to the action of a given bone, we also use the scalars (μ_0, μ_1, μ_2) introduced above to improve control over the direction of the volume correcting deformation.

This is done by successively applying the method above in local frames $(\mathbf{e}_0^b, \mathbf{e}_1^b, \mathbf{e}_2^b)$ associated with each bone b of the skeleton as illustrated in fig. 2 Left. This allows volumes variations due to the action of each bone to be processed one after the other, and the amount of correction $(\mu_0^b, \mu_1^b, \mu_2^b)$ (in percent) along the three axes to be controlled individually for each bone.

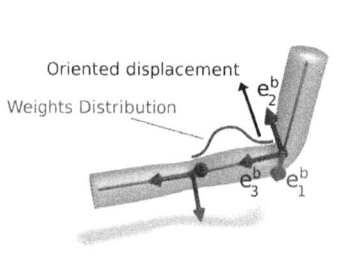
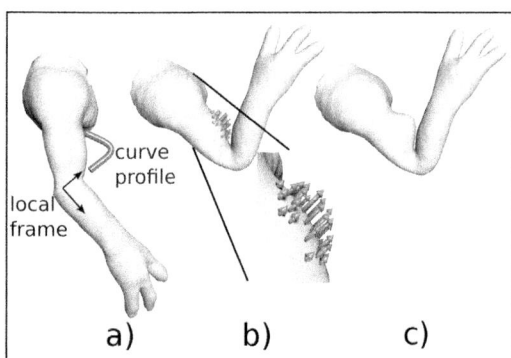

Fig. 2. Left: Local frames defined for an articulated cylinder deformed by smooth skinning using three segments. Right: Muscle effect using unidirectional inflation: a) A curve profile is designed in the local frame associated to the bending joint. b) Application of classical SSD. c) Volume correction vectors are computed using eq. (3).

In practice, the set of mesh weights in each local frame is controlled through user-specified 1D profile curves. They can be defined independently for each of the three axes. An intuitive control of the final deformation induced by volume correction is thus provided.

Selecting $\mu_2 \in [0.5, 1]$ to act only in the positive direction of a given axis of the local frame enables to mimic muscle effect as illustrated in fig. 2a-c.

More complex shapes can be modeled by using more complex profile curves or by combining them. For instance, fold effects can be achieved using oscillating curves. This can be obtained by setting $\gamma_k(\mathbf{p}_k^L) = \sin^2(\omega \pi x_k^L) \exp(-(\|\mathbf{p}_k^L\|/\sigma)^2)$, where ω modulates the frequency of the oscillations. In the cases of belly bulges of the elephant, in fig. 1 Left, we choose $\omega = 2$.

4 Active Geometry for Wrinkling Cloth

This section explains how a low quality cloth animation can be quickly enhanced by plausible, animated cloth-like wrinkles. This work was first presented in [2], which we refer to for details.

In the following the appearance of the generated wrinkles is controlled through a single, intuitive parameter: the smallest radius of curvature $R_{\min}$ that the cloth can exhibit, which is linked to the thickness and stiffness of the fabric the user wants to mimic.

4.1 Computing Shrinkage over an Animated Cloth Mesh

Suppose we have some mesh animation for cloth, either generated using a coarse physically based simulation or any other deformation method such as SSD. Let $\overline{S}$ be the cloth mesh at rest, before motion starts. Ideally, the cloth deformation should preserve isometry with respect to this initial shape. Therefore, almost no stretch (compression or elongation) should occur between $\overline{S}$ and any deformed versions S of this mesh during the animation sequence. The idea is to compute the main shrinkage direction as a continuous vector field, from the information of triangle stretch from $\overline{S}$ to S. Then, wrinkles will be generated orthogonally to the local directions of compression. The framework of this method is illustrated in fig. 3.

To measure triangle stretch, we express the 2D affine transformation from a triangle in $\overline{S}$ to its counterpart in S by the 2×2 matrix T. Let $(\mathbf{u}_1, \mathbf{u}_2)$ and $(\overline{\mathbf{u}_1}, \overline{\mathbf{u}_2})$ be the 2D edge vectors of the triangle in the local frames of S and $\overline{S}$ respectively. Then the transformation matrix T is given by $T = [\mathbf{u}_1, \mathbf{u}_2][\overline{\mathbf{u}_1}, \overline{\mathbf{u}_2}]^{-1}$. The 2×2 positive definite matrix $U = \sqrt{T^T T}$ called *stretch tensor* measures the amount of compression and elongation conveyed by the transformation T. If λ, the smallest eigenvalue of U, is such that $\lambda < 1$, the triangle is compressed along the corresponding eigenvector $\mathbf{e}$ (fig. 3b).

As cloth should wrinkle orthogonally to local directions or compression, we use interpolation to define a continuous *wrinkle vector field* $\mathbf{v}$ over the mesh as $\mathbf{v} = \max(1 - \lambda, 0)\,\mathbf{e}$, where $\|\mathbf{v}\|$ measures the amount of compression per unit of length.

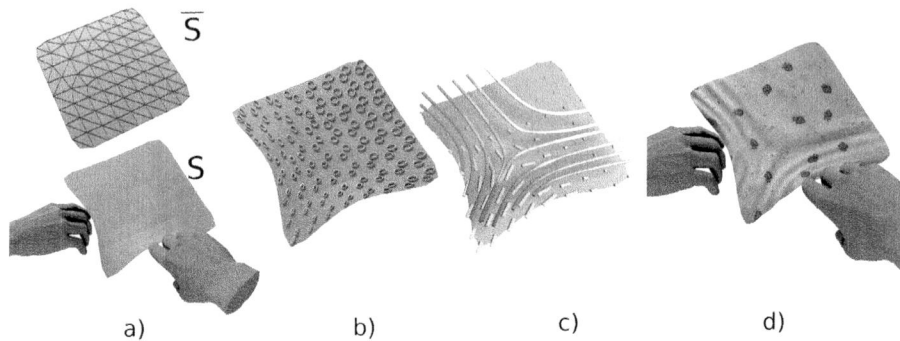

Fig. 3. Our active model for wrinkling cloth: a) Coarse input mesh and its initial, flat shape. b) Stretch tensor field. c) Wrinkle vector field **v** in green and the associated wrinkle curves. d) Final wrinkled mesh.

This allows to define wrinkle curves as streamlines of **v**: seed points are iteratively generated at vertices of maximal compression (while insuring that the distance between two of them remains larger than $2R_{\min}$) and streamlines are integrated in both direction from these points, until compression is smaller than a threshold.

4.2 Animating Fold Curves

Although this algorithm generates plausible wrinkle curves for a static frame (fig. 3d), it needs to be adapted to get temporal coherence during animation, i.e. prevent that small wrinkles appear and disappear or quickly slide over the cloth surface from frame to frame.

Firstly, curve seeds are tracked through time using a particle based approach: Seed positions from the previous frame are considered as potential seeds in the next frame, after being displaced toward the direction of the compression gradient. If the compression at the new location for the seed is still large enough, a wrinkle curve is generated there.

Secondly, the resulting wrinkle curve trajectories are smoothed over time to avoid discontinuities coming from the streamline integration process.

4.3 Wrinkle Geometry

The animated wrinkle curves now have to be used as deformers to generate the final wrinkled cloth geometry. Our previous work [2] described a solution for generating a fine folded mesh requiring re-triangulation. Here, we rather discuss how this solution can be modified to generate a displacement texture, easier to use in the context of real-time animation and directly applicable on GPU.

Wrinkle Parameters. The wrinkle shape is controlled by two parameters: a curvature radius $R(u)$ along the curve wrinkle and the wrinkle offset $\beta(u)$ determining the portion of the circular arc used for forming the wrinkle. We compute

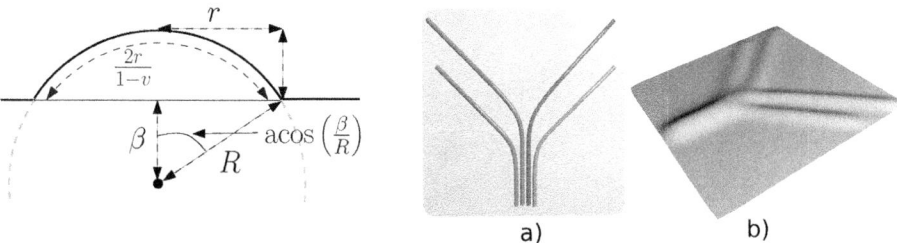

Fig. 4. Left: Relationship between width r, height h and the arc-length of a wrinkle generated at the offset distance β. Right: using wrinkle curves (a) as skeletons for an implicit deformer generates wrinkles which smoothly blend when they are close enough (b).

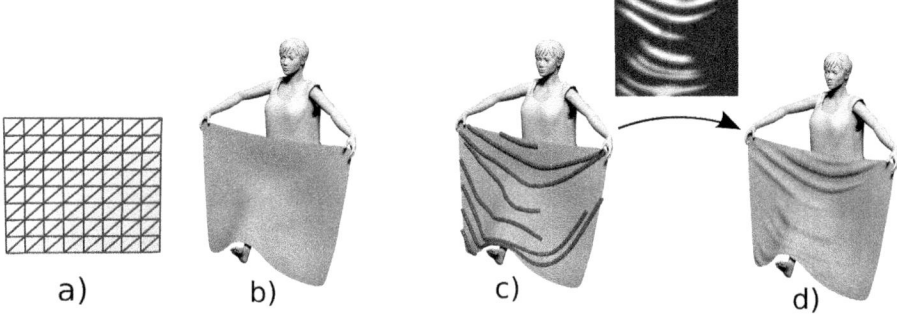

Fig. 5. Wrinkling a towel. a) Rest pose mesh. b) Deformed mesh using mass-spring system. c) Computed wrinkle curves. d) Final wrinkles generated using displacement texture map.

$R(u)$ from the norm of the wrinkle vector field as: $R(u) = (1 - 2/\pi)/\|\mathbf{v}(u)\|$. Once the radius is set, the wrinkle offset is computed such that the shrinkage with respect to the rest shape is minimized. Fig. 4(left) summarizes the geometric relationship between $\beta(u)$ and $\|\mathbf{v}(u)\|$.

Wrinkle Primitives. The wrinkles we generate should smoothly merge or split during animation, depending on the distance between them. We choose to model them as implicit deformers in order to handle seamlessly this behavior.

The wrinkle curves are used as skeletons generating a field function f, to which the wrinkle surface is the iso-surface $f(\mathbf{p}) = 1$. Among the implicit surface models, we use convolution surfaces for their ability to model smooth primitives along polylines: f is defined using convolution of the Cauchy kernel along the skeleton. This specific kernel provides a closed-form solution to the convolution integral (see [26] for details).

The final field function of the implicit deformer is computed by summing the fields generated by the individual wrinkle curves (which will smoothly blend them where needed) and adding an extra field representing the current, coarse

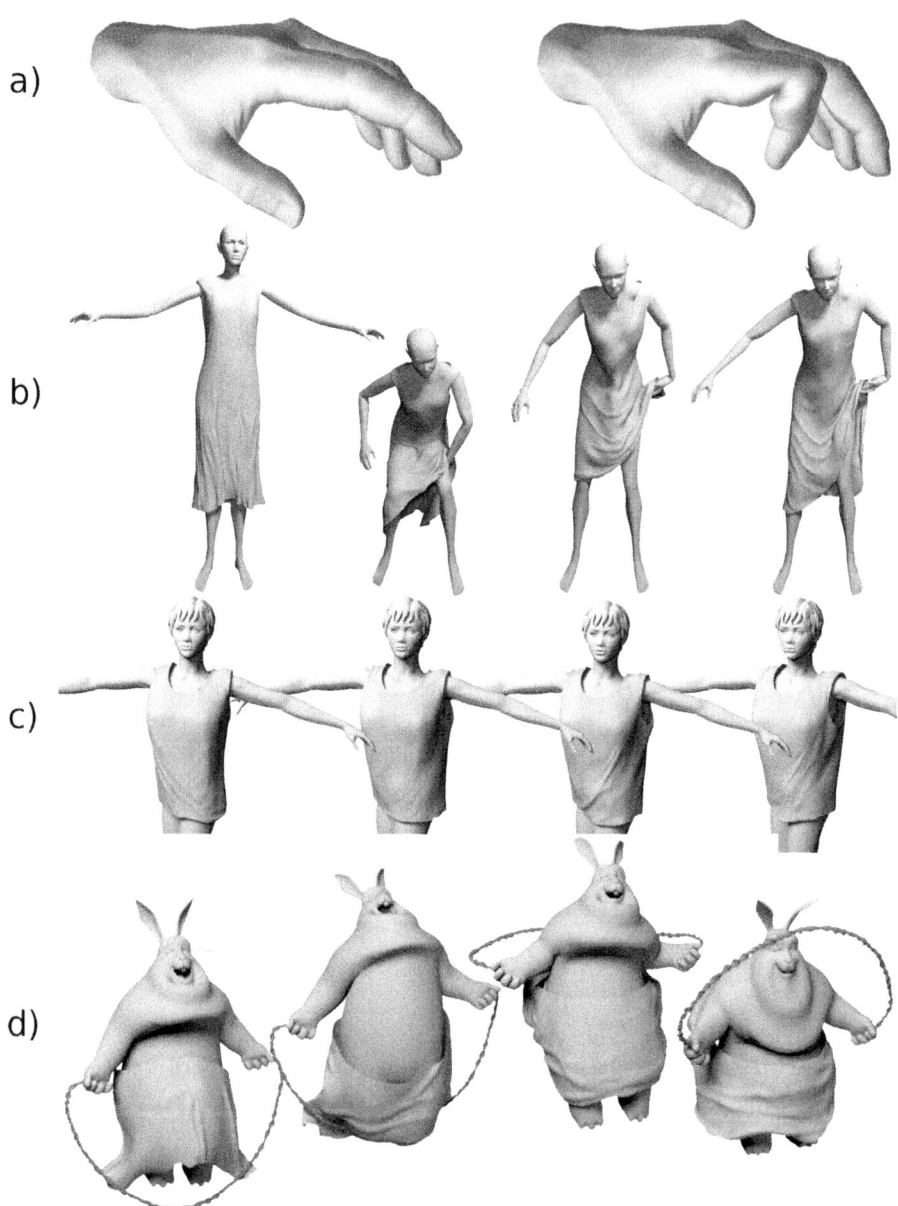

Fig. 6. a: Constant volume skinning of a human hand. Note the flesh bulges near the bent joint. b-d: Our wrinkles added on top of three coarse physically based animations of a human model and a bunny. Note the fine wrinkles added by our method where the coarse cloth model was shrinking.

cloth mesh as well, for the wrinkles to smoothly blend into the cloth surface. To do so in an efficient way, the field generated by the mesh is approximated by the field generated by the tangent plane at the mesh point which is the closest to each query point.

Finally, mesh deformation is expressed as a displacement texture map which can be stored on the cloth patterns (see fig. 5). To compute such a map, sampled positions on the original 3D mesh are projected along their normal direction onto the isosurface of isovalue 1. The amount of displacement is then stored on the texture map. Over-sampling is performed along the *wrinkle curve* location to ensure that the fine wrinkles are well captured by the texture.

5 Results and Conclusions

Our geometric models for constant volume and wrinkling cloth have been tested on a variety of animated characters such as a sitting elephant and a human hand for volume control and bending, twisting and jumping of dressed characters for the cloth (fig. 1 and 6). Fast physical simulation using Blender software was used to generate the input coarse cloth meshes.

No effort was made to optimize the code in the current version of the systems: the average frame rate for constant volume skinning is about 0.2 seconds per frame, while it is about 2 seconds per frame for our cloth wrinkles (when a refined mesh is generated for the wrinkled surface). However, the current results could be optimized. In particular, since computations are independent on each mesh vertex, GPU optimization should be possible for both methods and would bring a significant improvement. In addition, the volume correction part could also be made more efficient when the exact volume constraint is not required. In such a case, approximated partial volume computations for each bone could be used instead of computing the volume exactly on the gobal mesh.

In conclusion, the active geometric layers we have been advocating throughout this paper are efficient solutions for adding plausibility to an animation through geometric constraints. Their expression as an extra module added on top of a pre-existing animation makes them easy to tune, allows the use of intuitive parameters (such as profile curves for bulging flesh or min radius for cloth wrinkles) and insures that they remain compatible with a standard animation pipeline. We are currently experimenting with a third example of active geometric layer: some adapted convolution surface model used on top of a hair-guides simulation to animate volumetric, Manga-style hair with adequate wisps merging and splitting.

References

1. Rohmer, D., Hahmann, S., Cani, M.-P.: Exact volume preserving skinning with shape control. In: ACM SIGGRAPH/EUROGRAPHICS Symposium on Computer Animation (SCA), pp. 83–92 (August 2009)
2. Rohmer, D., Popa, T., Cani, M.-P., Hahmann, S., Sheffer, A.: Animation wrinkling: Augmenting coarse cloth simulations with realistic-looking wrinkles. To Appear in ACM Transactions on Graphics (TOG), Proceedings of ACM SIGGRAPH ASIA (December 2010)

3. Lewis, J., Cordner, M., Fong, N.: Pose space deformation: A unified approach to shape interpolation and skeleton-driven deformation. In: SIGGRAPH, pp. 165–172 (2000)
4. Kry, P., James, D., Pai, D.: Eigenskin: Real time large deformation character skinning in hardware. In: ACM SIGGRAPH/EUROGRAPHICS Symposium on Computer Animation (SCA), pp. 153–159 (2002)
5. Wang, R., Pulli, K., Popovic, J.: Real-time enveloping with rotational regression. To Appear in ACM Transactions on Graphics (TOG), Proceedings of ACM SIGGRAPH 26(3) (2007)
6. Wang, X., Phillips, C.: Multi-weight enveloping: Least-squares approximation techniques for skin animation. In: ACM SIGGRAPH/EUROGRAPHICS Symposium on Computer Animation (SCA), pp. 129–138 (2002)
7. Mohr, A., Gleicher, M.: Building efficient, accurate character skins from examples. ACM Transactions on Graphics (TOG), Proceedings of ACM SIGGRAPH 22(3) (2003)
8. Bloomenthal, J.: Medial-based vertex deformation. In: ACM SIGGRAPH/EUROGRAPHICS Symposium on Computer Animation (SCA), pp. 147–151 (2002)
9. Alexa, M.: Linear combination of transformations. ACM Transactions on Graphics (TOG), Proceedings of ACM SIGGRAPH 21(3) (2002)
10. Kavan, L., Collins, S., Zara, J., O'Sullivan, C.: Geometric skinning with approximate dual quaternion blending. ACM Transactions on Graphics (TOG) 27(4) (2008)
11. Larboulette, C., Cani, M.-P., Arnaldi, B.: Dynamic Skinning: Adding real-time dynamic effects to an existing character animation. In: Spring Conference on Computer Graphics, SCCG (2005)
12. Larboulette, C., Cani, M.-P.: Real-time dynamic wrinkles. In: Computer Graphics International, CGI (2004)
13. Angelidis, A., Singh, K.: Kinodynamic skinning using volume-preserving deformations. In: ACM SIGGRAPH/EUROGRAPHICS Symposium on Computer Animation (SCA), pp. 129–140 (2007)
14. Choi, K.-J., Ko, H.-S.: Stable but responsive cloth. ACM Transactions on Graphics (TOG), Proceedings of ACM SIGGRAPH 21(3) (2002)
15. English, E., Bridson, R.: Animating developable surfaces using nonconforming elements. ACM Transactions on Graphics (TOG), Proceedings of ACM SIGGRAPH 27(3) (2008)
16. Thomaszewski, B., Pabst, S., Strasser, W.: Continuum-based strain limiting. Computer Graphics Forum. Proocedings of EUROGRAPHICS 28(2) (2009)
17. Wang, H., Hecht, F., Ramanoorthi, R., O'Brien, J.: Example-based wrinkle synthesis for clothing animation. ACM Transactions on Graphics (TOG), Proceedings of ACM SIGGRAPH 29(4) (2010)
18. de Aguiar, E., Sigal, L., Treuille, A., Hodgins, J.: Stable spaces for real-time clothing. ACM Transactions on Graphics (TOG), Proceedings of ACM SIGGRAPH 29(4) (2010)
19. Decaudin, P., Juilius, D., Wither, J., Boissieux, L., Sheffer, A., Cani, M.-P.: Virtual grarments: A fully geometric approach for clothing design. Computer Graphics Forum. Proceedings of EUROGRAPHICS 25(3) (2006)
20. Muller, M., Chentanez, N.: Wrinkle meshes. In: ACM SIGGRAPH/EUROGRAPHICS Symposium on Computer Animation, SCA (2010)
21. Hadap, S., Bangerter, E., Volino, P., Magnenat-Thalmann, N.: Animating wrinkles on clothes. In: IEEE Proceedings on Visualization, pp. 175–182 (1999)

22. Kimmerle, S., Wacker, M., Holzer, C.: Multilayered wrinkle textures from strain. In: VMV, pp. 225–232 (2004)
23. Cutler, L., Gershbein, R., Wang, X., Curtis, C., Maigret, R., Prasso, L., Farson, P.: An art-directed wrinkle system for CG character clothing. In: ACM SIGGRAPH/EUROGRAPHICS Symposium on Computer Animation, SCA (2005)
24. Rohmer, D., Hahmann, S., Cani, M.-P.: Local volume preservation for skinned characters. In: Computer Graphics Forum, Proceedings of Pacific Graphics, vol. 27(7) (October 2008)
25. Elber, G.: Linearizing the area and volume constraints. Technical Report, TECHNION Israel (2000)
26. McCormack, J., Sherstyuk: Creating and rendering convolution surfaces. Computer Graphics Forum 17 (2001)

CAROSA: A Tool for Authoring NPCs

Jan M. Allbeck

Department of Computer Science,
Volgenau School of Information, Technology and Engineering,
George Mason University, Fairfax, VA 22030

Abstract. Certainly non-player characters (NPCs) can add richness to a game environment. A world without people (or at least humanoids) seems barren and artificial. People are often a major part of the setting of a game. Furthermore, watching NPCs perform and have a life outside of their interactions with the main character makes them appear more reasonable and believable. NPCs can also be used to move forward the storyline of a game or provide emotional elements. Authoring NPCs can, however, be very laborious. At present, games either have a limited number of character profiles or are meticulously hand scripted. We describe an architecture, called CAROSA (Crowds with Aleatoric, Reactive, Opportunistic, and Scheduled Actions), that facilitates the creation of heterogeneous populations by using Microsoft Outlook®, a Parameterized Action Representation (PAR), and crowd simulator. The CAROSA framework enables the specification and control of actions for more realistic background characters, links human characteristics and high level behaviors to animated graphical depictions, and relieves some of the burden in creating and animating heterogeneous 3D animated human populations.

Keywords: Non-player characters, Virtual crowds, Agent behaviors.

1 Introduction

As we journey through our day, our lives intersect with other people. We see people leaving for work or school, waiting for trains, meeting with friends, hard at work, and thousands of other activities that we may not even be conscious of. People create a rich tapestry of activity throughout our day. We may not always be aware of this tapestry, but we would definitely notice if it were missing, and it is missing from many games.

Most crowd simulations to date include only basic locomotive behaviors possibly coupled with a few stochastic actions. In the real world, inhabitants do not just wander randomly from location to location. They have goals and purposes that related to locations, objects, and activities. Military training simulations, virtual environments, games, and other entertainment enterprises, could benefit from varied, but controlled character behaviors to promote both plots and presence. Furthermore, the creation of these heterogeneous populations with contextual behaviors needs to be feasible. We do not want to demand that scenario authors be expert programmers. Ideally, they would be storytellers and creative visionaries. Likewise, defining these populations should not be an arduous, never ending process requiring levels of detail beyond what is immediately important to the scene author. An author should be allowed to concentrate on what is most critical to furthering the effectiveness of their virtual world and easily able to explore different possibilities.

In this paper we present a framework for creating and simulating what we term functional crowds. These are populations of virtual characters that inhabit a space instead of merely occupying it. The CAROSA (Crowds with Aleatoric, Reactive, Opportunistic, and Scheduled Actions) framework includes parameterized object and action representations, a resource manager, role definitions, and a simple calendar program. These components are vital to creating contextual behaviors in a heterogeneous population without overburdening the scenario author. To further increase richness, the CAROSA framework also includes four different types of actions: Scheduled actions arise from specified roles for individuals or groups. Reactive actions are triggered by contextual events or environmental constraints. Opportunistic actions arise from explicit goals and priorities. Aleatoric or stochastic actions are random but structured by choices, distributions, or parametric variations.

2 Related Work

As noted by Bates, in games as in life people are a part of setting [1]. Virtual characters can help establish context, provide emotional elements, and help drive a storyline, but to do so their behaviors need to be reasonable and appropriate. To be practical for a game, the mechanisms for authoring these characters need to be simple enough for non-programmers, but also provide enough freedom for designers to drive the story. In [11], Sanchez-Ruiz et al. present a framework for using ontologies and planning to automatically construct solutions that game designers can incorporate into *Behavior Trees*, but the results are still complex trees that can be difficult to manage and debug.

In many games and simulations, populations of non-player characters (NPCs) are programmed to follow a path and perhaps perform a few behaviors. They might also react to a certain set of stimuli, but they do not generally interact with many objects in the environment. They lack context, often only existing when in the players field of view [3].

Many entertainment companies, including producers of movies and games, use software such as Massive to build background characters [8]. Massive^{TM} provides mechanisms for creating and executing rules that govern the behaviors of characters. Creators can then replicate these behaviors and characters to produce very large populations. While creating and refining these rules still takes time and skill, the software makes construction of relatively homogeneous crowds much easier. There can even be some statistical variations in the models and the portrayals of the behaviors. While this is well suited to big battle scenes where the troops and their behaviors are all similar and object interactions are limited to

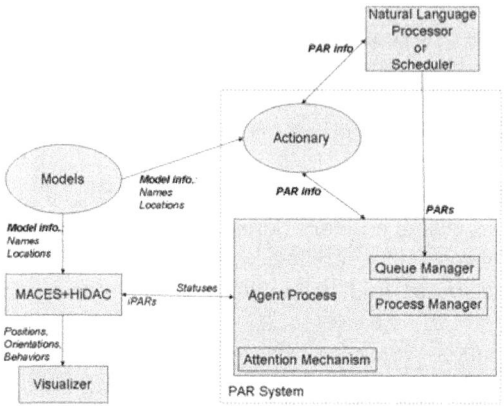

Fig. 1. System diagram

weapons attached to the characters, scenes that require functional, contextual characters are not feasible.

Much of the research in crowd simulations has been focused on maneuvering characters through a space while avoiding collisions with each other and the environment [10]. Other crowd simulation researchers have examined population diversification. McDonnell et al. have been studying how perceptible variations in character appearance and motion qualities are to human observers with the ultimate goal of simulating more visibly interesting, reasonable populations [9]. Lerner et al. use real world data to fit behaviors into pedestrian simulations [7]. In their work, character behaviors, such as two adjacent characters chatting, are based on the current spatial configuration of the characters in the environment. Other researchers have created variations in behavior by defining regions in the environment where certain behaviors would be displayed [4,6,12]. For example, a theater is labeled with sit and watch behaviors. However, within these regions behaviors are uniform. While these research enterprises have achieved more visually interesting populations, the character behaviors still lack larger purpose and context. Decision networks have been used for action selection [13], but they require crafting the prior probabilities of each action in context.

3 System Overview

The CAROSA framework includes many components common to game pipelines (See Figure 1). We are currently using the open source graphics engine, OGRE, for visualization. We are using HiDAC+MACES as an underlying crowd simulator [10]. HiDAC+MACES allows us to navigate characters from one location to another while avoiding collisions and providing us with notification when objects and other agents are perceived. Calls to playback animation clips also filter through HiDAC+MACES to the OGRE figures.

CAROSA's less common components include an action and object repository called the *Actionary*, *Agent Processes*, a *Resource Manager*, and a *Scheduler*. Scenario authors can use the *Scheduler* to schedule actions found in the *Actionary* and link them to locations or specific objects. These actions are, of course, also associated with the agents or characters that are to perform them. *Agent Processes* receive and process these actions, using the *Resource Manager* to allocate or bind any object parameters needed by the actions. The next few sections will describe some of these components in a bit more detail and discuss how they facilitate the creation of functional, heterogeneous populations.

4 Parameterized Representations

A key component of the CAROSA framework is the PAR system [2]. PARs include both parameterized actions and objects and mechanisms for processing them. The database of actions and objects, termed the Actionary, is a repository for general definitions or semantic descriptions. Both actions and objects are defined independent of a particular application or scenario. They are meant to be reusable building blocks for simulations. The Actionary currently contains more than 60 actions and nearly 100 object types.

We will concentrate on a few key aspects of PARs that help ground character behaviors in proper contexts. First, action definitions include *preparatory specifications*. These parameters provide a simple form of backward chaining that greatly reduces the burden on the scenario author. For example, in the simulation of a functional population, nearly every instantiated action is tied to an object participant (e.g. *sitting in a chair, using a computer, eating food, drinking coffee, etc.*). The character performing an action needs to walk to the associated object before the action can be performed. Requiring a game designer to specify these walk actions and other actions handled by preparatory specifications would more than double the amount of work required.

Like actions, the objects in the Actionary are stored in a hierarchy where children inherit parameter values from their parents. For objects, this hierarchy is driven by types (e.g. *Furniture, Person-Supporting-Furniture, Sofas, Chairs, etc.*). General action definitions, those that have not been instantiated for execution, then reference these types as object participant parameters. Hence, a general definition of *Sit* includes *Person-Supporting-Furniture*. Incorporating these types of commonsense items again lessens the work of scene authors who are not required to specify them. In the next section, we will discuss how the *Resource Manager* can be used to instantiate specific objects of the needed type. Our objects also include parameters such as sites that further specify how characters interact with them. They indicate, for example, where a character should stand to operate the object or how to grasp the object.

To increase the richness of simulations, we have extended the PAR representations to include different action types. Scheduled activities arise from specified roles for individuals or groups (See Section 6). Scheduled actions provide structure and control to simulations. Through scheduled actions, characters have purpose. Scheduled actions can be used to establish a daily routine or drive a storyline. For example, the game designer may schedule a meeting between two characters to show that they have an association or are a part of the same group (e.g., a gang or terrorist cell). We believe, however, that scheduled actions alone would result in characters that are too structured or robotic.

Reactive actions are triggered by contextual events or environmental constraints. Many of these behaviors arise from the underlying crowd simulator. For example, reactive behaviors include a character altering its heading or slowing to avoid collisions. Reactive actions, such as acknowledging someone as they pass in the hallway, are not handled by the crowd simulator. These reactions are specified and recognized in a ruled based *Attention Mechanism*. We believe that reactive actions help to add life to simulations. Perceiving and interacting with a dynamic environment indicates that the characters are not focused solely on achieving a specific goal. Reactive actions might also be used to create character flaws and emotional moments (e.g. mugging anyone walking alone, cowering in fear whenever they see a dog, avoiding police officers, etc).

Opportunistic actions arise from explicit goals and priorities. These need fulfilling behaviors are akin to the hill-climbing behavior choices of characters in the video game The Sims. While our opportunistic actions are similar, the implementation is different. In The Sims current proximity to resources is heavily weighted when choosing an action, and time is ignored entirely. We take into account time and future proximities. For example, a character may be working in his office and have a non-emergent energy need and a meeting to attend in a few minutes. The character could then attempt to address the need by stopping by the lunch room for a cup of coffee on the way to the meeting. This requires them to leave a couple of minutes early and to know that the lunch room has proximity to the path to the meeting room. Finding an object resource is essentially done through a depth-limited depth first search where the depth limit is based on the need level. As the need level increases, so does the distance the agent is willing to go out of their way to fulfill the need. In fact, as the need increases, so will the priority of the action that will fulfill it. If the need becomes great enough (higher priority than other actions), all other actions will be preempted or suspended. Opportunistic actions add plausible variability. Character begin each new run with random need levels and therefore fulfill their needs at different times creating emergent behaviors by reacting to the different stimuli they will encounter. These actions can also be used to establish character flaws and promote a storyline (e.g. characters that require large amounts of sleep, excitement, attention, alcohol, drugs, etc).

Aleatoric actions are random but structured by choices, distributions, or parametric variations. Take for example, working in an office. The aleatoric or stochastic nature of the behavior stems from what the sub-actions might be and their frequency of occurrence. For example, many professions entail working in an office for large portions of each day. Working in an office might include, using a computer, speaking on a telephone, and filing papers (See Figure 3(b)). These tasks are interspersed through out the day. A game designer could schedule each of these tasks during different segments of the day, but for most scenarios the exact timing of each sub-action is not important. The overarching *WorkInOffice* action should just look plausible. With aleatoric actions, the author need only specify *WorkInOffice* and the sub-actions will be performed in reasonable combinations. Each time a scenario is run, different combinations of actions will be chosen creating reasonable variations for the player encounter.

5 Resource Management

As we have stated, creating functional populations whose behaviors are contextual, involves associating actions with related objects. Making these associations by hand would overwhelm a game designer. We have implemented a *Resource Manager* that is used to automatically allocate objects.

Resources can be initialized from the environment definition using defined spaces and object model placement. This automated process also instantiates PAR objects and stores them under their corresponding types in the *Actionary*. Additionally it sets up location-content relationships. Object locations are set as the room that they are placed in and likewise the objects are listed as contents in those room objects. These relationships are then used to initiate resource management. There are different layers of resources. The objects in a room are resources of that room. Rooms themselves are also resources. A resource group is created for each room and all of the objects contained in that room are added as resources. Also, every room is added to the room resource group. Finally, all of the objects in the environment are added to a free objects list. This list is used to allocate objects for actions that do not have a location specified. For example, a housekeeper might be asked to clean anything. A random object would be chosen from the free objects list and used as a participant of the action. It is also possible to specify the type of object needed. This list is automatically updated as the Resource Manager allocates and de-allocates objects to agents. The entire resource allocation process itself is done automatically as a consequence of processing PAR actions and does not require game designer intervention.

Resources can be allocated according to type or a specific object instance can be requested. Furthermore, a preference function can be used to determine the best object to allocate to the character. This function argument could be used to indicate that particularly shady characters prefer to sit in the back corners of a restaurant, for example. It should also be noted that objects are not allocated to characters until the characters are at least in the target location or room of the action. While resources could be allocated at the time the PAR actions are assigned to the character, we feel this method leads to more natural behaviors including failures. For example, a character should have to walk to a room to know that there are no more chairs available.

6 Roles and Groups

Peoples functions or purposes through the course of a day are highly correlated to their roles: Students attend classes, Professors work in their offices and lecture,

Housekeepers clean, Researchers research, etc. Creating roles for game characters provides them focus and purpose. Defining groups of characters with common roles means reduced work for game designers.

Roles also give us a plausible starting point for enabling the assignment of default behaviors and locations that can lessen the load of an author. When they have nothing else to do, professors work in their offices, researchers conduct research in a lab, housekeepers clean, etc. This means that even if they have no scheduled activities, their behaviors will at least look plausible. Note that all of these actions, like most, require object participants. The professor needs an office to work in, the researchers need a lab desk, and the housekeeper needs something to clean. The question is how do we determine or allocate objects for default behaviors. Certainly we do not want to demand that a game designer assign such objects for every character in the population. We also do not want to just randomly choose objects whenever any of the default behaviors are to be performed. This would lead to inconsistent scenarios. Professors do not randomly choose offices to work in throughout a day. We have created a couple of mechanisms for automatically choosing appropriate object participants providing a more consistent context.

First, when defining a role, one can specify objects that the role should possess. For example, a professor should have an office. An author does not need to specify an instance of an object, though they are permitted. It is sufficient to specify a class like *Office*. Possessions are transitive. If you have an office, you also have the objects in the office. When a role has a possession specified, every character with that role is allocated (by the *Resource Manager*) a possession of that type during the initialization. Whenever a character initiates the performance of her default behavior, she first checks to see if an object of the type needed is located in her possessions. If it is, she uses that object. If it is not, the next method is tried.

This method also uses the association of object types with default actions. For example, researchers in our scenario need laboratory desks in order to do their research. As we did with professors, we could say that all researchers should possess a lab desk, but in a setting where resources are limited it might be more likely for researchers to take whatever desk is available. We do this by indicating that the action *Research* requires an object participant of the type *LabDesk*. When a researcher initiates this default action, an object of type *LabDesk* is allocated to her. If no object can be allocated, a failure is produced and the action will not be performed. This might lead to the character performing a different action or just standing still (or idling as a default action) until an appropriate object can be allocated.

A key aspect of this paper is facilitating the creation of heterogeneous populations. To achieve this, we need to provide a way to define groups as well as individuals. An author can name a group and provide the number of members in it, or an author can create individual characters and assign them to groups. In both cases, the information is stored in the *Actionary*. As needed, agents are created to fulfill the indicated membership numbers. If the name of a group happens to correspond to the name of a role, then all of the members of the group are automatically assigned that role and inherit all of the default behaviors and specifications for it. There can be more than one group per role and there may be groups that do not indicate a role. Entire groups can be assigned actions (of any type) as easily as a single individual.

7 Game Authoring

Beyond creating functional, heterogeneous crowds, our goal for the CAROSA framework was to facilitate authoring by non-programmers. Certainly to simulate a virtual environment, one will still require models and animations, but once a repository of these items has been created and a populated *Actionary* obtained and linked to them, a game designer should be able to use these building blocks.

When scheduling actions, an author only needs to specify the PAR action that should be performed, which character or group of characters should perform it, what objects or locations might participate in the action, and when it should be performed. In fact, the actions can either be simple PARs or complex lists of PARs composed together. A specific object participant can be specified, such as sit in *Chair_4* or a location can be given, such as sit in *ClassRoom_1*. If a location is specified, a chair instance will be allocated by the *Resource Manager* when the character arrives in the specified location. If the required resource cannot be allocated, a failure is reported to the *Agent Process* and the action is removed from the queue. In the future, a planner could be used to attempt to acquire the resource needed.

We are currently using the calendars of Microsoft Outlook®as a scheduling interface (See Figure 2(a)). It is a common, well-known software package that does not require users to be programmers. Calendars can be created for groups of characters or individuals as desired. Activities, locations, and times are then simply entered on the calendars. Then with a simple click of a button, a software package called GeniusConnect [5] links these calendars to tables in our MySQL database (i.e., the *Actionary*). These activities are read from the database and assigned to the appropriate character or characters by placing them on the appropriate *Action Queues* of the *Agent Processes*.

We have constructed simple custom Graphical User Interfaces (GUIs) for authoring reactive, opportunistic, and aleatoric actions (See Figure 2(b)). These GUIs are directly connected to the *Actionary*. The drop-down lists are populated from table entries in the database and submitting new actions of these types are written directly to database tables. These newly created actions can then also be referenced from Microsoft Outlook to, for example, schedule aleatoric actions.

The PAR representation of objects and actions provide semantic foundations that are referenced when authoring through any of these interfaces. The *Resource Manager* provides a means of filling in information and tying the behaviors to the environment. This allows a game designer to concentrate on aspects of the characters that are directly relevant to her storyline including the heterogeneity and functionality of the population. The precise nature of these interfaces is not

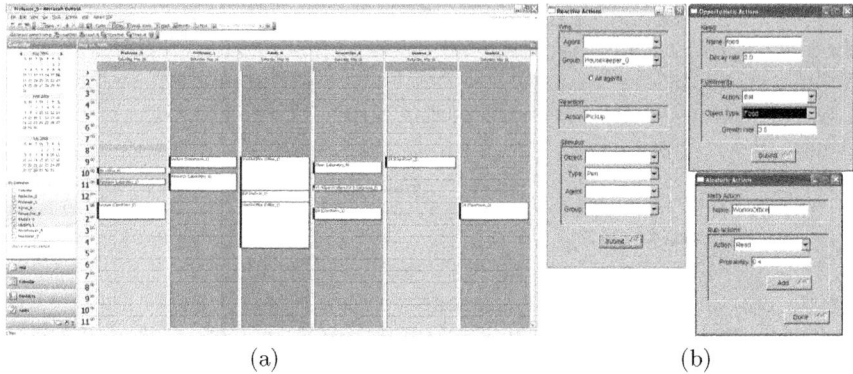

Fig. 2. (a) Microsoft Outlook®calendars are used as an interface for scheduling actions. (b) Custom GUI's for authoring Reactive, Opportunistic, and Aleatoric Actions.

important. What is important is that a limited, manageable amount of data is required to specify the behaviors of the characters. All of the required data could just as easily be specified in and then read from a spreadsheet or a simple text file.

8 Example Simulations

As an initial test-bed for the CAROSA framework, we simulated a university building environment. The environment is based on one floor of an engineering building at a university. It includes laboratories, classrooms, hallways, offices, a restroom, and a lounge (See Figure 3). The occupants of the building include professors, students, researchers, administrators, housekeepers, and maintenance personnel. Actions in this environment include working in an office, lecturing, attending class, lounging, cleaning, inspecting, researching, waving, eating, drinking, going to the restroom, picking up objects, as well as others. There is also collision-free locomotion.

Characters in the simulation adhere to their schedules as created by the scenario author through the *Scheduler*, but they also greet each other as they pass by and attend to their needs through opportunistic actions. If a portion of their day is unscheduled, the characters revert to their default behaviors which are in many cases aleatoric actions that provide ongoing reasonable behavior.

We ran many simulations of this environment and noted several emergent behaviors. For example, students tended to gather in the same areas because of the resources available and therefore tended to walk to classes together. Furthermore when groups were instructed to react to other group members by waving, students would greet each other before a class which seems like reasonable behavior. Students also tended to go directly for food and coffee after classes. Because need levels are currently all checked together, needs tend to be fulfilled at the same

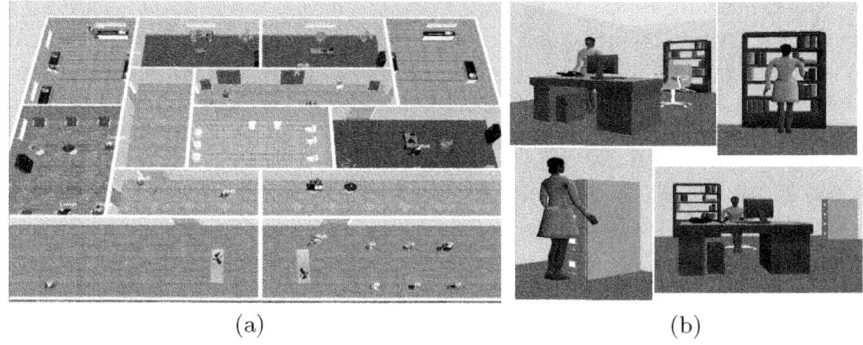

(a)   (b)

Fig. 3. Sample scenario based on activity in a university environment

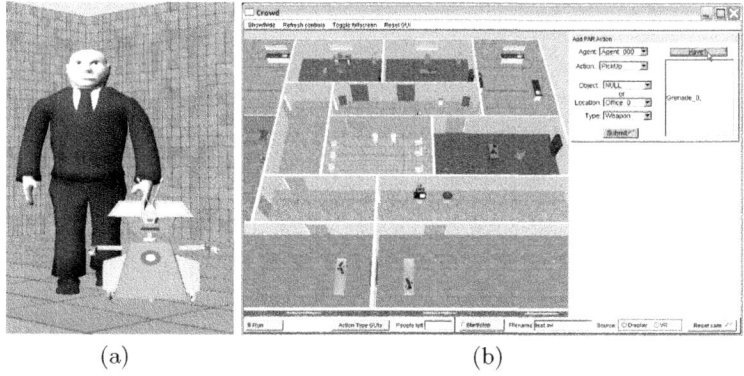

(a)   (b)

Fig. 4. Simulation of a robot, commander team tasked with clearing a building of weapons

time. For example, a character might go to the restroom and then to get coffee. Again this emergent behavior seems quite plausible.

We have also authored a scenario where a human commander and a robot team are tasked with clearing a building of weapons (See Figure 4(a)). This scenario includes searching the environment, picking up and dropping weapons, and alerting the commander to the presence of other beings. Reactions include alerting the commander to objects of certain types and other characters. For the robots, the main need is battery power, which is fulfilled by charging. Aleatoric actions can be used to vary the behaviors of the robots (including navigation routes) decreasing predictability that may be exploited by hostiles.

In addition to the interfaces for scheduling actions and authoring other action types, we constructed a GUI for giving the robot teams immediate high priority instructions. Through the interface depicted in Figure 4(b) a user chooses from drop-down menus the agent she is instructing, an action to be performed, and either an object participant or a location and object type. A PAR instance is then automatically created and added to the character's *Action Queue* with a very high priority. Having a high priority ensures that it will suspend or preempt any actions the robot is currently performing and start execution immediately.

9 Conclusions and Future Directions

In this paper, we have described the CAROSA framework that we have constructed to facilitate the creation of plausible, functional NPCs. Our characters do more than navigate from one location to another without colliding with each other or obstacles. Characters have roles that provide them purpose. Their behaviors are performed in proper context with the object participants. In fact, because actions reference PAR objects and not coordinates directly, the environmental setting can be completely rearranged without requiring additional work from the game designer. Objects can even be added or removed and the characters will still perform. Note, however, some actions may not be performed if there is a shortage of resources.

Through the use of the PAR representations, a resource manager, and definitions of roles and groups, we have a created a framework in which actions can be scheduled in forms analogous to the scheduling of real people. Game designers can then use these character behaviors to create more reasonable backdrops, motivate back-stories and drive storylines To add additional richness and more reasonable behaviors, we have also implemented reactive, opportunistic, and aleatoric actions that are equally easy to author and modify. These types of actions can be used to design flawed and emotional characters that provide additional depth and interest.

Additional development is still needed for the CAROSA framework. In particular, we would like to increase the scale of the populations. Our university simulation can run 30 characters in real-time on a standard PC. The most costly algorithm in the CAROSA framework is the calculation of opportunistic actions. Scheduling the fulfillment of a need can require searching through time for a gap in the character's schedule as well as through the environment to find a resource near a path the character will be traveling. We are considering caching these paths or locations or perhaps using stored way-points that act as road signs for resources.

Ideally, we would also like to include inverse kinematics for the characters. This would enable us to further showcase the PAR representation through more detailed object interactions. This would require additional semantic labeling of the objects in the environment (i.e. creation of sites), but objects can be reused in many simulations, so we believe the rewards would be worth the initial effort.

Finally, we would like to explore additional scenarios including outdoor environments. We are also working to build social and psychological characteristics into the characters. In particular, we are examining representing and simulating personalities and culture. As these characteristics are developed we would like to simulate their impact on character interactions and team behaviors such as cooperation, coordination, and competition. In particular, we would like to look at character communication in both verbal and non-verbal forms.

Acknowledgements

Partial support for this effort is gratefully acknowledged from the U.S. Army SUBTLE MURI W911NF-07-1-0216. We also appreciate donations from Autodesk.

References

1. Bates, B.: Game Design. Course Technology PTR (2004)
2. Bindiganavale, R., Schuler, W., Allbeck, J., Badler, N., Joshi, A., Palmer, M.: Dynamically altering agent behaviors using natural language instructions. In: Autonomous Agents, pp. 293–300. AAAI, Menlo Park (2000)
3. Brockington, M.: Level-of-detail ai for a large role-playing game. In: Rabin, S. (ed.) AI Game Programming Wisdom, pp. 419–425. Charles River Media, Inc., Hingham (2002)
4. DePaiva, D.C., Vieira, R., Musse, S.R.: Ontology-based crowd simulation for normal life situations. In: Proceedings of Computer Graphics International, pp. 221–226. IEEE, Stony Brook (2005)
5. GeniusConnect, http://www.geniusconnect.com/articles/products/2/3/ (last visited May 2009)
6. Lee, K.H., Choi, M.G., Hong, Q., Lee, J.: Group behavior from video: A data-driven approach to crowd simulation. In: ACM SIGGRAPH / Eurographics Symposium on Computer Animation, San Diego, pp. 109–118 (2007)
7. Lerner, A., Fitusi, E., Chrysanthou, Y., Cohen-Or, D.: Fitting behaviors to pedestrian simulations. In: Symposium on Computer Animation, pp. 199–208. ACM, New Orleans (2009)
8. Massive Software Inc.: 3d animation system for crowd-related visual effects, http://www.massivesoftware.com (last visited August 2010)
9. McDonnell, R., Micheal, L., Hernandez, B., Rudomin, I., O'Sullivan, C.: Eye-catching crowds: saliency based selective variation. In: ACM SIGGRAPH, pp. 1–10. ACM, New Orleans (2009)
10. Pelechano, N., Allbeck, J., Badler, N.: Virtual Crowds: Methods, Simulation, and Control. Synthesis Lectures on Computer Graphics and Animation. Morgan and Claypool Publishers, San Rafael (2008)
11. Sanchez-Ruiz, A., Llanso, D., Gomez-Martin, M., Gonzalez-Calero, P.: Authoring behaviour for characters in games reusing abstracted plan traces. In: Ruttkay, Z., Kipp, M., Nijholt, A., Vilhjalmsson, H. (eds.) IVA 2009. LNCS, vol. 5773, pp. 56–62. Springer, Heidelberg (2009)
12. Sung, M., Gleicher, M., Chenney, S.: Scalable behaviors for crowd simulation. Computer Graphics Forum 23(3), 519–528 (2004)
13. Yu, Q., Terzopoulos, D.: A decision network framework for the behavioral animation of virtual humans. In: Proceedings of ACM SIGGRAPH/Eurographics Symposium on Computer Animation, pp. 119–128. Eurographics Association, San Diego (2007)

BehaveRT: A GPU-Based Library for Autonomous Characters

Ugo Erra[1], Bernardino Frola[2], and Vittorio Scarano[2]

[1] Università della Basilicata, Potenza, Italy
ugo.erra@unibas.it
[2] Università di Salerno, Salerno, Italy
{frola,vitsca}@dia.unisa.it

Abstract. In this work, we present a GPU-based library, called BehaveRT, for the definition, real-time simulation, and visualization of large communities of individuals. We implemented a modular flexible and extensible architecture based on a plug-in infrastructure that enables the creation of a *behavior engine* system core. We used Compute Unified Device Architecture to perform parallel programming and specific memory optimization techniques to exploit the computational power of commodity graphics hardware, enabling developers to focus on the design and implementation of behavioral models. This paper illustrates the architecture of BehaveRT, the core plug-ins, and some case studies. In particular, we show two high-level behavioral models, picture and shape flocking, that generate images and shapes in 3D space by coordinating the positions and color-coding of individuals. We, then, present an environment discretization case study of the interaction of a community with generic virtual scenes such as irregular terrains and buildings.

1 Introduction

Autonomous character behavior simulation in computer graphics is a relatively new and complex area of research that is at the heart of modeling and simulation and behavioral and cognitive psychology. Within the last three decades, researchers in this area have attempted to model behaviors using simulation and visualization primarily for education and training systems. Today, behavior representation is also used in entertainment, commercials and non-educational simulations. The field of application ranges from crowd control [19] and evacuation planning [20] to traffic density [8] and safety [3].

The main goal is the simulation of autonomous virtual characters or agents to generate crowds and other flock-like coordinated group motion. In this type of simulation an agent is assigned a local behavioral model and moves by coordinating with the motion of other agents. The number of agents involved in such collective motion can be huge, from flocks of several hundred birds to schools of millions of fish. These simulations are attracting a large amount of interest [21] because, with appropriate modifications to interaction terms, they could be used for large number of systems, besides human crowds [1], in which local interactions among mobile elements scale to collective behavior, from cell aggregates, including tumors and bacterial bio-film, to aggregates of vertebrates such as migrating wildebeest. For this reason, we also use in this paper the term "individual" to indicate an element in the group.

Today, Graphics processing units (GPUs) are used not only for 3D graphics rendering but also in general-purpose computing because of increases in their price/performance ratio and hardware programmability and because of their huge computational power and speed. In particular, developments in programmability have significantly improved general accessibility thanks to a new generation of high-level languages simplifying the writing of a GPU program. In this scenario, the NVIDIA Compute Unified Device Architecture (CUDA) abstracts GPU as a general-purpose multithreaded SIMD (single instruction, multiple data) architectural model and offers a C-like interface for developers. However, developing an efficient GPU program remains challenging because of the complexity of this new architecture, which involves in-depth knowledge of resources available in terms of thread and memory hierarchy. Such knowledge remains a key factor in exploiting the tremendous computing power of the GPU, although manual optimizations are made time consuming, and it is difficult to attain a significant speed-up [15]. For this reason, several efforts have been devoted to exploit the processing power of GPUs without having to learn the programming tools required to use them directly. For instance, several framework based on the GPU can provides acceleration for the game physics simulation [6][11] and recently the GPU has been also exploited in artificial intelligence technology for path planning [2]. These works show that the level of interest in GPU-accelerated frameworks in video games is growing, and manufacturers are pushing new technologies in the GPUs.

In this work, we present BehaveRT as a flexible and extensible GPU-based library for autonomous characters. The library is written for CUDA language and enables the developer/modeler to implement behavioral models by using a set of plug-ins; the developer can also quickly prototype a new steering behavior by writing new plug-ins. The advantages of our framework are: (i) modelers can focus on specifying behavior and running simulations without in-depth understanding of CUDA's explicit memory hierarchy and GPU optimization strategies; (ii) the simulation performance exploits the computational power of the GPUs and allows massive simulation with high performance; (iii) CPU and GPU memory allocation of the data structures associated with agent population is hidden by BehaveRT, and these structures can be visualized in real time since agent data is already located on the GPU hardware.

The remainder of this paper is organized as follows: in section 2 we review previous relevant frameworks based on CPU and GPU approaches for agent-based simulation. In section 3, we describe the overall system. In section 4, we present the set of core plug-ins. Section 5, illustrates some case studies implemented using BehaveRT. Finally, section 6 concludes and discusses future work.

2 Related Work

Recently, researchers have begun to investigate the parallelism of GPUs in speeding up the simulation of large crowds. In this scenario, GPUs are used essentially for speeding up neighbors searches, which is an essential operation in agent-based simulation. Several approaches have been proposed. In [4], the authors use a GPU that enables the massive simulation and rendering of a behavioral model of a population of about 130 thousand at 50 frames per second. Similarly, the authors in [14] describe an efficient GPU implementation for real-time simulation and visualization. Their framework, called ABGPU, enables massive individual-based modeling underlying the graphical concepts of the GPU. In each case, these works focus more on exploiting GPU architecture for efficient implementation and less on employing a reusable software library.

OpenSteer is an open-source C++ library that implements artificial intelligence steering behaviors for virtual characters [13] for real-time 3D applications. Agents in OpenSteer react to their surroundings only by means of steering behaviors. Several such behaviors are implemented in OpenSteer, such as "walk along a path" or "align with a neighbor", corresponding to a steering vector computed from the current agent's state (position, orientation, velocity) and its local environment (nearby obstacles and neighboring agents, etc.). These simple behaviors may be combined into complex ones by combining the corresponding vectors. Some OpenSteer ideas have been adopted in BehaveRT, and we supply a plug-in with an interface similar to OpenSteer's.

SteerSuite is a flexible framework inspired by OpenSteerDemo component of Reynolds software. It provides functionality related to the testing, evaluating, recording, and infrastructure surrounding a steering algorithm [17].

3 Building a Behavior Engine

BehaveRT is a C++/CUDA library that enables the collective behaviors of a community of individuals by defining a customized *behavior engine*. Individuals belonging to the same community have the same characteristics, behave similarly and can interact with each other. In this paper, the term "behavior" is used to refer to the improvised actions of individuals.

The behavior engine is an entity containing *features* and *functionalities*. Features are data structures containing a number of elements equal to the number of individuals and representing their state, e.g., their position, orientation, velocity and so on. Functionalities are operations applied in a block to the whole community and can modify the features of individuals through behavior modification or can be used to manipulate features for particular operations such as determining all individuals that fall inside a range.

A behavior engine is assembled by means of an ordered set of plug-ins. The set of features and functionalities of the behavior engine is determined by those of each plug-in. Figure 1 shows an example of a behavior engine. The resulting behavior engine is able to manage a community in which each individual has a 3D

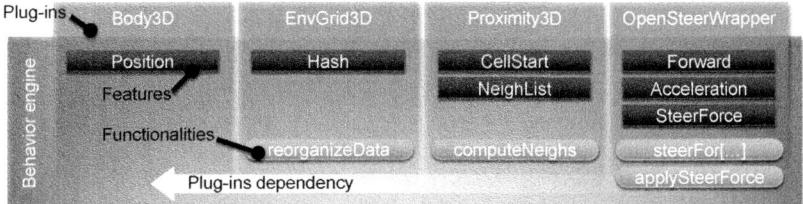

Fig. 1. Plug-in mixin. Body3D adds to the behavior engine one feature and no functionalities. EnvGrid3D adds one feature and a functionality for data reordering. Proximity3D covers related to neighborhood searching, such as neighborhood lists. The plug-in OpenSteerWrapper provides a set of functionalities that implements some steering behaviors. Each plug-in depends on the plug-in(s) to its left. For example, OpenSteerWrapper relies on the features and functionalities of Proximity3D, EnvGrid3D and Body3D.

position, recognize its neighboring individuals, and execute steering behaviors. These four plug-ins define the basic system core for the behavioral modeling of a community of individuals that can interact with each other.

3.1 Functionalities Implementation

Functionalities envelope one or more CUDA kernel calls. BehaveRT simplifies three different implementation phases (Figure 2 shows a comparison between a classical and BehaveRT approach). (*i*) Allocation: features are encapsulated and allocated simultaneously on both CPU and GPU. The library output is bind as OpenGL renderable object. (*ii*) Call: kernels call is simplified thanks to data structures encapsulation and allocation setups. (*iii*) Execution: kernels definition does not need to specify input and output features because they can be achieved from a GPU data structure (*FeaturesList*) automatically allocated and filled at allocation time. Figure 3 shows an implementation example, referring to *steerForSeparation* provided by the *OpenSteerWrapper* plug-in.

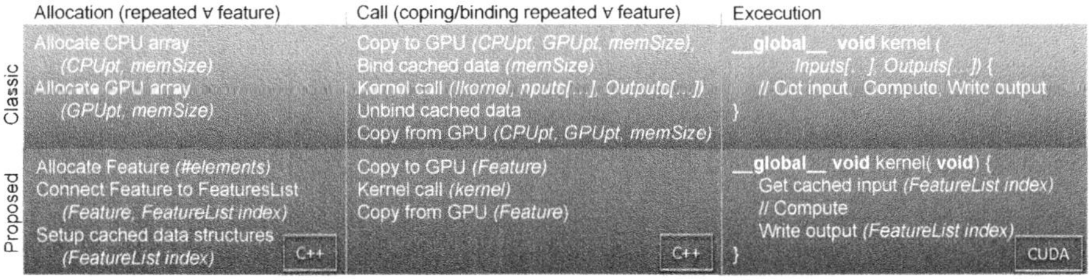

Fig. 2. Simplified representation of a functionality implementation – proposed approach compared to a classical approach. Items between brackets represent arguments.

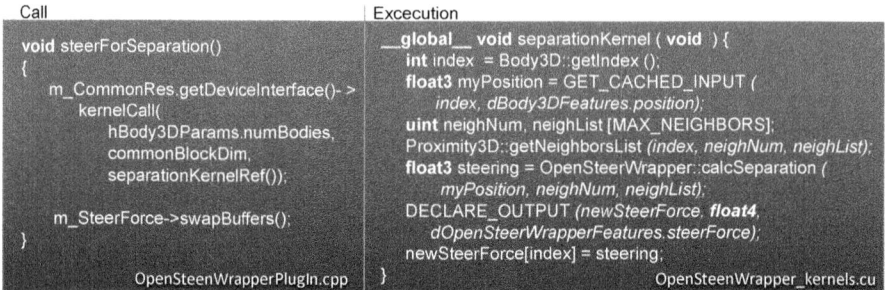

Fig. 3. Implementation of *steerForSeparation* functionality. Here are the last two phases shown in Fig. 2. On the right, the execution is simplified thanks to features and functionalities furnished by previous plugins, e.g., getNeighborsList.

3.2 Usage

The following code shows how to create the behavior engine shown in Fig.1. The set of plugins is specified by means of a mixin classes technique [18]. The BehaviorEngine class contains the attributes and methods of each plug-in class.

```
// Allocation phase
OpenSteerWrapperPlugIn < Proximity3DPlugIn < EnvGrid3DPlugIn <
      Body3DPlugIn >>>> behaviorEngine;
// Update - Call and execution phases
behaviorEngine.reorderData();         // EnvGrid3D functionality
behaviorEngine.computeNeighborhood(); // Proximity3D functionality
behaviorEngine.steerForSeparation();  // OpenSteerWrapper functionality
behaviorEngine.applySteeringForce( elapsedTime );
// Draw (using the current OpenGL context)
behaviorEngine.draw();  // Drawable3D functionality
```

4 Core Plug-Ins

This section describes the core plug-ins of BehaveRT. For each plug-in, we show its features and functionalities. These plug-ins will then be used to demonstrate a mixin plug-in to present some case studies of runnable behavior engines.

Body3D Plug-In. This plug-in provides a basis for mobile individuals in virtual environments. It offers only one feature, representing the 3D position of individuals. The plug-in manages a common radius that approximates the space occupied by each individual in the world, with a sphere. A more sophisticated version of this plug-in could define more features, such as a different shape for the space occupancy approximation, e.g., a parallelepiped or a cube. On the other hand, a simpler version of this plug-in could be defined, such as the 2D version Body2D.

EnvGrid3D Plug-In. The EnvGrid3D plug-in provides one feature and some functionalities for environment subdivision and data reorganization. Environment subdivision offers a static grid subdivision of the world in cubic cells of the same size and enables swift identification of all individuals within a given radius. Data reorganization enables the aggregation of individuals belonging to the same cell in continuous regions of the GPU memory. The data reorganization fulfills two roles. First, it simplifies the neighbors search provided by the Proximity3D plug-in. Second, the operation improves the performance of the memory bandwidth during execution of the functionalities of other plug-ins. Because the interaction of individuals necessitates access to data on neighboring individuals, features content referring to the same individual is read many times. We therefore used a cached region of the memory of the GPU, i.e., textures, to improve the performance of frequently repeated data reads.

Proximity3D Plug-In. The Proximity3D plug-in provides a neighbor search operation for each individual in the community. This operation is fundamental because each individual must take decisions only according to its neighbors, and so it must be able to pick out efficiently these individuals. In order to obtain interactive results, this phase requires a careful choice of graphics hardware resources to obtain superior performance results and avoid the bottlenecks of a massive simulation. The output of this process is a new feature containing neighbors lists. Neighbors lists are encoded into a GPU-suitable data type that allows fast neighborhood queries. Each individual looks for neighbors in the cell where it currently is and in adjacent cells. Because the plug-in EnvGrid3D enables the memory data reorganization of features content, the cell content look-ups do not need additional data structures. Groups of agents belonging to the same cell will be located in continuous regions of the GPU memory. A simple linear search starting from a proper index based on the cell hash function is then sufficient. The plug-in also offers an internal functionality that allows other plug-ins access to individuals' neighbor lists. Further implementation information, including performance and scalability evaluations, on this approach are discussed in [4,5].

OpenSteerWrapper Plug-In. This plug-in provides a set of steering behaviors similar to those described by Reynolds in [12] and implemented in OpenSteer [13]. Steering behaviors implemented in the plug-in are applied to the whole community of individuals, while those in Reynolds's OpenSteer are applied independently to each individual. In our implementation, behavior differentiation among individuals can be achieved only by means of data-driven conditions. According to Reynolds's implementation, each steering behavior yields a steering force, represented by a 3D vector. The sum of steering forces of many steering behaviors is applied to the 3D movement of each individual.

Features provided by the OpenSteerWrapper plug-in include Forward, SteerForce, and Acceleration. The plug-in offers a set of functionalities representing steering behaviors, including steerForSeparation, steerForCohesion, steerForAlignment, and steerForAvoidObstacle. It also provides functionality that applies steering force to individuals' positions (thus, modifying the Position feature provided by Body3D).

Drawable Plug-In. This plug-in provides by using a configurable rendering system a graphical representation of simulated individuals. Individuals can be rendered as simple points, billboards, or 3D models (by means of OpenGL geometry instancing). Positions and orientations are linked to VBOs and computed by ad-hoc OpenGL vertex and shader programs that directly generate a 3D representation of individuals.

5 Case Studies

This section introduces some case studies that applied to a large community of individuals generate impressive visual effects. Also we show how a community of individuals can interact with a generic environment. An ad-hoc BehaveRT plug-in, Shapes3D, provides picture and shape flocking by adding new features and functionalities on top of those described in the previous section. In both picture and shape flocking, individuals acts like a flock of bird by means of flocking behavior provided by the OpenSteerWrapper plug-in. In addition, each individual is associated with a target position and a target color. Per-individual target seeking is supplied by a functionality of the Shapes3D plug-in, while individual color managing is provided by the Drawable plug-in. The difference between picture and shape flocking is in the placement of target positions and the selection of target colors.

5.1 Picture Flocking

A community of individuals acting out picture flocking behavior recreates a 2D image in 3D space (Fig. 4a). Target positions are placed on a plane, and target colors are depicted from a picture in such a way as to associate each individual with a pixel of the picture. The Shape3D plug-in reads data from an input image file and initializes the target colors of individuals. When the simulation starts, individuals move toward the respective target positions, starting from random positions and moving in small groups by flocking. The Drawable3D plug-in also provides a smoothed translation of individuals' colors. For this reason, an individual attains the target color value only after several frames. When the simulation reaches a steady state, the result is a wavy image in space. This is due to the interaction between individuals that can never reach the exact target position because of inertia and separation behavior. As result, they simply wander near the target positions, aligned with and separated from their neighbors.

5.2 Shape Flocking

In shape flocking, behavioral target positions are placed on a 3D model surface. The Shapes3D plug-in reads a 3D model file and initializes the target positions to

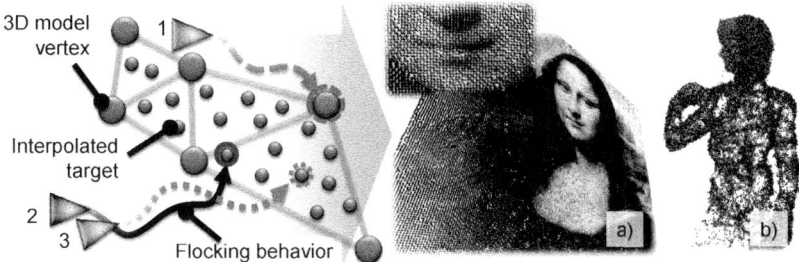

Fig. 4. A portion of the input 3D model and the path followed by three individuals during picture flocking (a) and shape flocking (b). The Shapes3D plug-in chooses target positions by randomly interpolating the positions of vertexes of an input 3D model. In this case, the target of individual 1 is a vertex, while targets of individuals 2 and 3 are interpolated points. Every individual has its own target position and moves toward that target exhibiting flocking behavior.

vertexes of the input 3D model. If the number of individuals is greater than the number of vertexes of the 3D model, Shapes3D interpolates the positions between vertexes. This generates a set of random points on the surface of each polygon of the 3D model (Fig. 4b). In the same way as picture flocking, each individual can be associated with a target color, depicted by the 3D model texture data. This characteristic is not currently implemented, but may be added to the Shapes3D plug-in initialization phase.

5.3 Shape Path Following

The 3D model can represent a path for the movement of individuals. Shape path-following behavior is an evolution of shape flocking. Only one characteristic differentiates these two behaviors: In shape path following, the target positions are not fixed. Indeed, target positions shift at every simulation step and effectively slide on the surface of the 3D model. The direction of the movement depends on the vertex order of the input model. Thus, given a set of ordered vertexes $v_1, v_2, ..., v_n$, the target position of agent j is the vertex v_i with $i = (j + \lfloor f * s \rfloor) \mod |G|$ where f is the frame counter, s is the global speed of the path begin followed, and G is the set of simulated individuals (Fig. 5).

5.4 Environment Interaction

Many applications using autonomous characters require a method that allows individuals to interact with their environment. Here we present a new simple technique for the interaction of a large number of individuals with irregular terrain and static objects. The resulting effect (Fig. 7b) and performances are quite similar to those described in [16], with the difference that our implementation does not use a global navigation system and it is not limited to a 2D ground level. Thus, is possible to populate multi-floor environments, such as buildings.

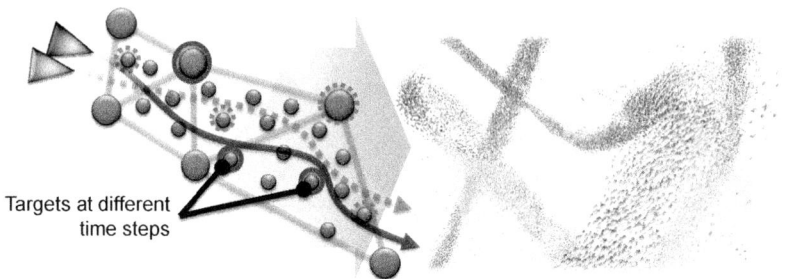

Fig. 5. The result of a community exhibiting shape path-following behavior, i.e., a flock of virtual birds moving on the surface of the input 3D model. At each step, the target positions change. They can be viewed as key points of a set of individual paths. The figure shows the paths of two individuals (continued and dashed) that wander around on the mesh of the 3D model.

An ad-hoc plug-in, Building3D, provides a feature that assigns a value to each cell of the environment grid (for simplicity we use the grid subdivision provided by the EnvGrid3D plug-in). A pre-computing phase samples the scene and checks for cells occupied by ground and other objects and assigns a binary value (empty/full) to a cell (Fig. 6a). This phase is similar to that described in [10] and [9] used for achieve navigation graphs from an approximate cell-decomposition of the navigable space. The purpose of the pre-computing phase is to achieve a discretized and simplified version of the scene. Individuals have access only to the information stored in their cell. The simulation phase is quick, because each individual has access to a restricted region of the memory shared among all individuals. By using modern commodity graphics hardware, the proposed technique permits the interaction of tens of thousands of individuals with a scene of medium complexity at interactive frame rates.

Additional Behaviors. Buildings3D provides the *floating behavior*, which allows individuals to float on irregular ground and objects. At each simulation step, an individual checks the value stored in its current cell and the value of its the future cell (Fig. 6b). Current and future cell information is used to modify the acceleration and the forward vector of characters (Acceleration and Forward are features provided by the OpenSteerWrapper plug-in). When an individual is not on the ground and the current cell is empty, then a downward factor is applied to the acceleration in order to simulate the gravitational force. When the future cell is full, impact with the ground is imminent. In this case an upward factor is applied to the acceleration (Fig. 6b) and the speed is reduced. However, there is a consequence to the speed reduction that occurs when individuals move up hills and mountains, in that individuals that reduce their speed become obstacles for approaching individuals. Through their separation behavior, approaching individuals steer to avoid these slowed individuals and move laterally. Owing to their alignment behavior, other individuals follow the group and avoid hills (Fig. 6c).

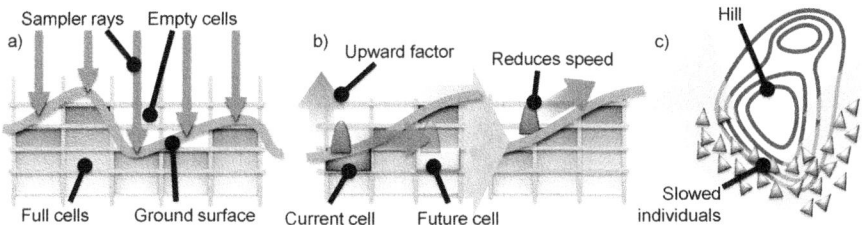

Fig. 6. (a) The pre-computing phase collects data on scene conformation. This scene is sampled with vertical rays. When a ray intersects the ground or an object, information are stored in the appropriate environment cell. (b) Floating behavior schema. Individuals check the current cell and future cell membership. In this example, the future cell data indicates the presence of the ground, and an upward force is applied to the character's acceleration. (c) An effect of the speed reduction in floating behavior.

Fig. 7. Environment interaction screenshots. The scene has been created with Ogre3D graphics engine [7]. Individuals are rendered in a native OpenGL phase added to the Ogre3D's rendering queue.

In addition to the floating behavior, individuals interact by executing Reynolds's cohesion, alignment, and a modified version of separation behavior. Unlike the original version, the modified version of separation does not allow sudden changes in the steering force. It merely reduces the speed of individuals when the (actual) separation force is opposite to the forward direction (cross product close to -1). In contrast, modified separation behavior increases their movement speed when the original separation force and forward direction are similar (cross product close to 1).

6 Conclusions and Future Work

In this paper, we have introduced BehaveRT, a library that enables the definition of collective behaviors for large communities of individuals. The library provides an extensible and flexible structure that subdivides the definition of behaviors and data structures among well-separated and reusable plug-ins.

We designed a set of core plug-ins implementing the basis of a simulation system of individual-based modeling. In order to show the extensibility of BehaveRT, we introduced picture and shape flocking, two high-level behaviors for generating particular visual effects by coordinating the positions and colors of individuals. We also saw how individuals can interact with a generic virtual scene by means of an environment discretization phase and the execution of floating behavior.

BehaveRT also exploits the power of commodity GPUs by using well-known reorganization memory techniques and simplifying the usage of high-speed memory regions in this type of hardware. In fact, on commodity graphics hardware (Nvidia 8800GTS 512 Mb RAM, CUDA v3.0), picture and shape flocking run at 15 FPS with 130 K individuals (billboard-based rendering). Environment interaction case study runs at 15 FPS with 100 K individuals with billboard-based rendering and 30 K individuals with geometry instancing (about 80 polygons per individual). These results, similar to those provided by cutting-edge specific GPU-based simulations systems [14][16], suggest that BehaveRT exploits the computational power of the GPUs enabling developers/modelers to assign local behavioral models more complex for massive simulation and rendering. As consequence, we believe that the GPU acceleration provided in BehaveRT is applicable to video games because, like graphics and physics processing on the GPU, individuals based simulation is driven by thousand of parallel computation.

In the short term BehaveRT will be released as open-source but future work will equip BehaveRT with several new features: (i) Multiple communities. Currently, more than one community can be initiated and simulated at the same time, but individuals of a community cannot interact with individuals of another community. Thus, how an inter-community interaction system can be created remains unanswered. (ii) Environment interaction. The current system version of the individuals' interaction with generic scenery is based on an environment discretization scheme that associates a binary value (empty/full) to each cell. An extension could be represented by a system similar to JPEG image compression. Rather than use binary values, a set of known patterns could be used. Each pattern then approximates the portion of the scene contained in the cell.

References

1. Azahar, M.A., Sunar, M.S., Daman, D., Bade, A.: Survey on real-time crowds simulation. In: Pan, Z., Zhang, X., El Rhalibi, A., Woo, W., Li, Y. (eds.) EDUTAINMENT 2008. LNCS, vol. 5093, pp. 573–580. Springer, Heidelberg (2008)
2. Bleiweiss, A.: Scalable multi agent simulation on the gpu. In: GPU Technology Conference (2009)
3. Courty, N., Musse, S.R.: Simulation of large crowds in emergency situations including gaseous phenomena. In: Proceedings of the Computer Graphics International 2005, CGI 2005, Washington, DC, USA, pp. 206–212. IEEE Computer Society, Los Alamitos (2005)
4. Erra, U., Frola, B., Scarano, V.: A GPU-based method for massive simulation of distributed behavioral models with CUDA. In: Proceedings of the 22nd Annual Conference on Computer Animation and Social Agents (CASA 2009), Amsterdam, the Netherlands (2009)

5. Erra, U., Frola, B., Scarano, V., Couzin, I.: An efficient GPU implementation for large scale individual-based simulation of collective behavior. In: International Workshop on High Performance Computational Systems Biology, pp. 51–58 (2009)
6. Havok. Havok physics, http://www.havok.com/
7. Junker, G.: Pro OGRE 3D Programming (Pro). Apress, Berkely (2006)
8. Loscos, C., Marchal, D., Meyer, A.: Intuitive crowd behaviour in dense urban environments using local laws. In: Proceedings of the Theory and Practice of Computer Graphics 2003, TPCG 2003, Washington, DC, USA, p. 122. IEEE Computer Society, Los Alamitos (2003)
9. Maim, J., Yersin, B., Pettre, J., Thalmann, D.: Yaq: An architecture for real-time navigation and rendering of varied crowds. IEEE Computer Graphics and Applications 29(4), 44–53 (2009)
10. Pettre, J., Laumond, J.-P., Thalmann, D.: A navigation graph for real-time crowd animation on multilayered and uneven. In: First Int'l Workshop Crowd Simulation (2005)
11. PhysX. PhysX physics, http://www.nvidia.com/object/physx_new.html
12. Reynolds, C.: Steering behaviors for autonomous characters (1999)
13. Reynolds, C.: Opensteer - steering behaviors for autonomous characters (2004), http://opensteer.sourceforge.net/
14. Richmond, P., Coakley, S., Romano, D.M.: A high performance agent based modelling framework on graphics card hardware with cuda. In: Proceedings of the 8th International Conference on Autonomous Agents and Multiagent Systems, Richland, SC, International Foundation for Autonomous Agents and Multiagent Systems, AAMAS 2009, pp. 1125–1126 (2009)
15. Ryoo, S., Rodrigues, C.I., Baghsorkhi, S.S., Stone, S.S., Kirk, D.B., Hwu, W.-m.W.: Optimization principles and application performance evaluation of a multithreaded GPU using CUDA. In: Proceedings of the 13th ACM SIGPLAN Symposium on Principles and Practice of Parallel Programming, PPoPP 2008, pp. 73–82. ACM, New York (2008)
16. Shopf, J., Oat, C., Barczak, J.: GPU Crowd Simulation. In: ACM SIGGRAPH Conf. Exhib. Asia (2008)
17. Singh, S., Kapadia, M., Faloutsos, P., Reinman, G.: An open framework for developing, evaluating, and sharing steering algorithms. In: Egges, A. (ed.) MIG 2009. LNCS, vol. 5884, pp. 158–169. Springer, Heidelberg (2009)
18. Smaragdakis, Y., Batory, D.S.: Mixin-based programming in c++. In: Butler, G., Jarzabek, S. (eds.) GCSE 2000. LNCS, vol. 2177, pp. 163–177. Springer, Heidelberg (2001)
19. Sung, M., Gleicher, M., Chenney, S.: Scalable behaviors for crowd simulation. Comput. Graph. Forum 23(3), 519–528 (2004)
20. Taaffe, K., Johnson, M., Steinmann, D.: Improving hospital evacuation planning using simulation. In: Proceedings of the 38th Conference on Winter Simulation, WSC 2006, pp. 509–515 (2006)
21. Thalmann, D., Musse, S.R.: Crowd Simulation. Springer-Verlag New York, Inc., Secaucus (2007)

Level of Detail AI for Virtual Characters in Games and Simulation

Michael Wißner, Felix Kistler, and Elisabeth André

Institute of Computer Science, Augsburg University,
86135 Augsburg, Germany
{wissner,kistler,andre}@informatik.uni-augsburg.de

Abstract. Following recent research that takes the idea of Level of Detail (LOD) from its traditional use in 3D graphics and applies it to the artificial intelligence (AI) of virtual characters, we propose our approach on such an LOD AI. We describe how our approach handles LOD classification and how we used it to simplify the simulation quality of multiple aspects of the characters' behavior in an existing application. Finally, we delineate how we evaluated our approach with both a performance and perception analysis and report our findings.

1 Introduction

In its classical meaning, Level of Detail (LOD) is a concept found in 3D games and simulations. It aims at reducing a scene's geometrical complexity and thus increasing the overall performance. To this end, each object's importance in the scene is determined, usually by measuring its distance from the camera and thus the player. The less important an object becomes for the player (i.e. the further away it is from his point of view), the fewer the number of polygons it is rendered with. This follows the rationale that at those distances, the human eye will not be able to spot any differences. Different LODs can be discrete, meaning they are switched as certain distance thresholds are crossed or they can be continuous and the number of polygons is updated dynamically. Recent research applies this traditional idea of LOD to virtual characters' artificial intelligence (AI) and the simulation of their behaviors, using the terms "Simulation LOD" or "LOD AI". Similar to the original approach, this is based on finding uninteresting or unimportant characters in the scene and reducing the simulation quality of their behavior accordingly. When building such an LOD AI for virtual characters, not all concepts can be simply transfered from the classical meaning of LOD. New questions arise, concerning how the different LODs should be defined and used: Is a character's distance from the player still a viable method to determine its LOD? Are there maybe situations in which a character might be interesting or important enough to show its complete behavior, regardless of where it is? Which aspects of the characters' behavior can be simplified and how? Will a player be able to spot these simplifications? Will the characters and their behavior still appear believable and consistent to the player?

In this work we put forward our approach to LOD AI and try to answer the above questions. We also report how we integrated and tested our system in two existing applications. The first application is Augsburg3D, a crowd simulation featuring 250 virtual characters in a reconstruction of the inner city of Augsburg. Aspects of the characters' behavior that are reduced by the LOD AI include path finding, animation and movement and the particle-based crowd simulation itself. The second application is the Virtual Beer Garden, a Sims-like simulation where characters' behavior is driven by certain needs which they try to satisfy through the interaction with other characters or smart objects. In addition to the previously mentioned crowd simulation related behaviors, this application also combines our LOD AI with these smart objects.

The novelty of our approach to a LOD AI lies in the many different types of behavior that can be reduced with it. Also, behaviors are not reduced during the behavior selection process but later, during behavior execution.

Performance tests and a user study show that the LOD-based behavior control increased the performance by a considerable amount while users barely noticed any difference in the characters' behavior.

The remainder of this paper is organized as follows: The next chapter introduces related work. Chapter 3 describes our LOD AI and how it was applied to the Augsburg3D application. Chapter 4 presents the results of the performance analysis and user perception study we conducted to evaluate our system and chapter 5 finally gives a conclusion and a short outlook on future work.

2 Related Work

Although they do not explicitly use the term "LOD" or reduce their characters' behavior simulation, the autonomous pedestrians described by Wei and Terzopoulos [1] are relevant for our work, as they display navigation and collision avoidance, interactions with other agents and the environment as well as dialogs. Patel et al. [2] describe dialog behavior for 'Middle Level of Detail Crowds', i.e. characters that are in the background but nevertheless important.

There is little previous research on the validation of crowd simulations with virtual characters through user perception: Pelechano et al. [3] describe how to validate crowds by making the user part of them and using the notion of presence. Peters and Ennis [4] showed short animations to users, asking them whether the groups in the animations were plausible.

Moving on to "Simulation LOD" or "LOD AI", Table 1 shows an overview of previous work on this topic, as well as our approach, characterized by the following criteria: LOD determination, LOD usage and actions performed by the AI.

Looking at the table, we see that our work offers a more flexible approach, with many different types behaviors involved, LOD determination based on distance and visibility and a rather large number of different LODs. Due to this flexibility, it can be used many different applications and scenarios.

Table 1. Comparison of different LOD approaches

Authors	LOD based on	Number of LODs	LOD applied to	AI behaviors
Chenney et al. [5]	potential visibility	2	updating movement	navigation, collision avoidance
O'Sullivan et al. [6]	distance	not specified	geometry, animations, collision avoidance, gestures and facial expressions, action selection	navigation, collision avoidance, complex dialogs with other agents
Brockington [7]	distance	5	scheduling, navigation, action selection in combat	navigation, collision avoidance, complex combat interactions
Niederberger and Gross [8]	distance and visibility	21	scheduling, collision avoidance, path planning, group decisions	navigation, collision avoidance, path planning
Pettré et al. [9]	distance and visibility	5	geometry, updating movement, collision avoidance, navigation	navigation, collision avoidance, path planning
Brom et al. [10]	simplified distance	4	action selection (with AND-OR trees), environment simplification	navigation, complex interactions with objects and other agents
Paris et al. [11]	distance	3	navigation, collision avoidance	path planning, navigation, collision avoidance
Lin and Pan [12]	distance	not specified	geometry, animations	locomotion
Osborne and Dickinson [13]	distance	not specified	navigation, flocking, group decisions	navigation
Our approach	distance and visibility	8	updating movement, collision avoidance, navigation, action execution	navigation, collision avoidance, desire-based interactions with agents and smart objects, dialogs

3 Level of Detail Based AI

Our LOD based AI for virtual characters is based on the Horde3D GameEngine (cf. [14]). The GameEngine is organized into different components such as a scene graph component, an animation component, a crowd simulation component, and a text to speech component. Because of this modular design, it can be simply enhanced by new components, as we did with our approach.

3.1 LOD Classification

The most important goal in classifying the different LODs is to keep their later usage as generic as possible. Because of that, the approach should offer a sufficient number of LODs. The current implementation classifies LOD in one of up to ten configurable levels. The main criterion for this classification is an entity's distance from the camera. Furthermore, if the entity is not visible, an additional and again configurable value, is added (as suggested by Niederberger and Gross). An entity's LOD configuration is placed in a sub node of its XML-entry for the GameEngine. This means that different entities can have different LOD configurations, adding to the applicability of the LOD AI to many different applications, scenarios and circumstances. As examples could be mentioned: the size and complexity of the environment, the camera perspective and its possible movements or specific restrictions given by the game logic. Figure 1 shows an example XML configuration. The attributes d0-d5 describe the discrete distances for the LOD levels in the game's environment, and "invisibleAdd" is the value added to the LOD in case the entity is not visible.

```
<AILOD d0="25" d1="40" d2="60" d3="90" d4="130" d5="200" invisibleAdd="2" />
```

Fig. 1. XML configuration for the LOD determination

In the LOD classification, all objects whose distance to the camera is lower than d0 are assigned a LOD value of 0. For all others, the following applies:

$$LOD(x) = d(x) + add(x)$$

where $d(x) = i$ for $d_i(x) \leq r(x) < d_{i+1}(x)$
with $r(x) = $ "distance of x to the camera"
with $d_i(x) = $ "discrete distance d_i as configured for x"

$$and\ add(x) = \begin{cases} invisibleAdd(x) & \text{if } x \text{ is invisible} \\ 0 & \text{otherwise} \end{cases}$$

As an example, let us assume the following: An entity's "invisibleAdd" is set to 2 and its distance to the camera is 100, which is between d3 and d4. So it is first given an LOD value of 3. If the entity should also be invisible as far as the current camera is concerned, the "invisibleAdd" is added, making the entity's LOD a total of 5.

The visibility determination takes place by testing bounding box of each entity that is subject to the LOD AI against the camera frustum. In that way almost visible agents are also counted as visible, in opposite to Niederberger and Gross. As most of the time only few entities fall into this category, the additional effort of this calculation is of little consequence, though.

Fig. 2. LOD classification. **Left:** Small section of the Augsburg3D environment, showing both visible and invisible agents at the same distance. **Right:** Complete scene, showing many agents with LOD 7.

If the GameEngine's occlusion culling is activated, this is also taken into account when determining an entity's final LOD: Entities occluded by others or obstacles are considered invisible and thus receive the "invisibleAdd" as well.

Figure 2 shows two examples of the LOD classification. Note that the transparent, white pyramid represents the frustum of the camera on which the LOD classification is based. Everything outside that frustum is invisible. All objects and characters are marked with colored circles according to their LOD value: LOD 0 is white and LODs 1-6 are incrementally darkening until black at LOD 7.

3.2 LOD Usage and Sample Applications

As the Horde3D GameEngine is separated into independent components, it seems logical to apply the Level Of Detail AI to each of them separately. Moreover, each component's LOD usage can be configured individually to suit the current application's needs.

Based on this, the following simplifications are applied to the characters' behavior so far:

- Path planning is simplified at higher LOD levels. Characters just select any route to their target, not necessarily the shortest one.
- Characters' movement is simulated differently: It ranges from a complete and continuous movement with full animations, over continuous movement without animations, to a more and more infrequent update of the movement (so the agents move with a jerk), and ends with the direct jump to the desired destination.
- Repulsive forces in the later discussed crowd simulation are ignored from a certain LOD on. As a result, agents walk through each other and through small obstacles.
- Characters' speech is only output up to a certain LOD.
- Animations are omitted at higher LODs.
- Characters lip-sync their dialogs only up to a certain LOD.
- Some parts of the characters' interactions are omitted at higher LODs.

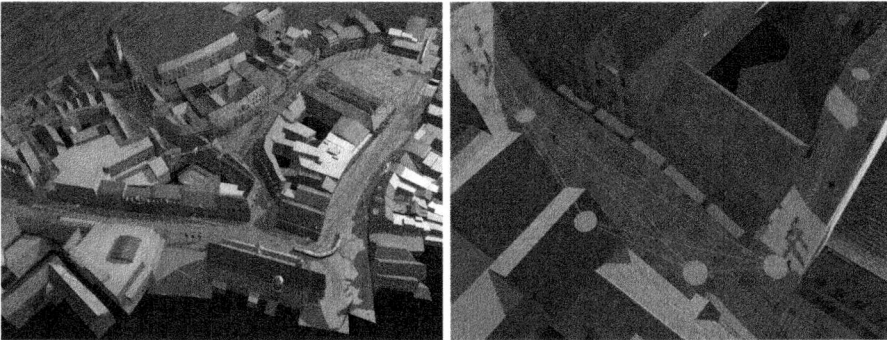

Fig. 3. Way points (blue) and the visibility graph (magenta) in the crowd simulation

Augsburg3D. Augsburg3D is a simulation of the inner city of Augsburg. The city is populated by virtual humans whose behavior is given by a crowd simulation (see [15]). This crowd simulation contains three basic elements: particles (the characters), obstacles and a net of way points. Using these way points, a visibility graph is generated, containing information about the reachability and distance among all way points. Figure 3 shows this for the whole scene (left) and in more detail (right).

Basically, the particles follow random paths between the way points. On their way, repulsive forces are applied between particles and obstacles on the one hand and between the particles themselves on the other hand. The way in these forces are applied in our crowd simulation is based on ideas by Heigeas et al. [16]. Figure 4 shows a character along with its three avoidance zones. As performance decreased with higher numbers of characters, we decided to apply level of detail to the characters' behavior. For that goal, the agents first get a LOD assigned as described in section 3.1. The discrete distances were defined as shown in Figure 1. Thus, all particles get a LOD between 0 and 5 according to the distance to the camera and a value of 2 is added in case of invisibility.

Now, two critical LOD values can be defined: One for the crowd simulation component and one for the movement animation component.

Fig. 4. Three avoidance zones d1 (blue), d2 (green) and d3 (red) in the crowd simulation

The latter is responsible for playing the characters' walk animations in an appropriate speed while they are moving. Further, it plays idle animations when the characters are standing around. If a character' LOD is above the defined critical value, walk and idle animations are simply skipped. Thus, the component adds no animation to the agent. The critical value for the movement animation component was set to 3 for Augsburg3D. That equates to a distance of 40 at invisibility or a distance of 90 at visibility, at which the agents in Augsburg3D are small enough, that the missing animation is not too conspicuous for the user.

The crowd simulation component applies three simplifications if its critical LOD value (also set to 3 for Augsburg3D) is reached:

1. The repulsive forces are dropped, so particles walk through small obstacles like lanterns or benches and also through other particles.
2. The path calculation is reduced. In that way, the start and end point is chosen by the first found possible way point.
3. The update rate for the particle movement is reduced. Therefore, every agent stores the time past since the last update. If an update occurs, the stored time is added to the current frame time to calculate the movement. As a consequence, the particles jump over longer distances at once after several frames, instead of changing their position incrementally after every frame. As the LOD calculation is only done if the character (or camera) has changed it's position, its LOD value is also calculated less frequently (at a static camera). The update rate $f_{update}(x)$ in updates per second for particle x with LOD value $lod(x)$ is defined as following:

$$f_{update}(x) = \begin{cases} \frac{1}{t(x)} & \text{if } lod(x) > criticalLOD(x) \\ \infty & \text{otherwise} \end{cases}$$

$$\text{where } t(x) = 0.55 \cdot lod'(x) - 0.5$$
$$\text{and } lod'(x) = lod(x) - criticalLOD(x)$$

With $criticalLOD(x) = 3$ that results in the following update rates:

LOD	0	1	2	3	4	5	6	7
f_{update}	∞	∞	∞	∞	20	$1,\overline{66}$	0.87	0.59

$f_{update}(x)$ always indicates the maximum update rate of x. Is the current frame rate below that value, the update rate is adequately smaller. Thus, ∞ is equivalent to an update in every frame. The time $t(x)$ until the next update is given by a simple linear function. The chosen gradient (0.55) and the y-intercept (0.5) ensure that, with a LOD one above the critical value, there is still a continuous movement with 20 updates per second. More complex functions did not provide an increased benefit in our experiments.

Additionally, if their LOD value is low enough, the agents also exhibit a new behavior to make the simulation more authentic: The pedestrians start simple conversations. For this matter, they search for a partner in their environment, that also has a sufficient small LOD. The two characters go toward each other, and the initiator starts a generated dialog as he reaches his partner. After they have exchanged some sentences, they both go their own ways again. To coordinate the different conversations, the particles allocate a so called interaction slot. There is only a limited number of those slots (six in the current configuration), so that the number of parallel dialogs in the simulation is limited as well. There is further a cool down timer (currently 5 to 15 seconds) for each agent to ensure he does not start a new conversation immediately after his last one. As the LOD value is dependent on the camera distance, the conversations only occur close to the camera, the only place where the observer really notices them. The actual speech output with lip synchronization only takes place if the LOD is equal or less than one. Otherwise, the current utterance is skipped and the agent just waits two seconds. The exact duration does not matter because the conversations consist of small talk which can be easily interrupted and resumed at any time without causing visible inconsistencies in the characters' behavior.

Virtual Beer Garden. The second application with LOD usage is the Virtual Beer Garden [17]. Details on the agents' artificial intelligence and how LOD is applied to it can be found in [18].

4 Evaluation

In order to evaluate the effectiveness of our approach, we compared two versions of Augsburg3D: one including our LOD system, and one without it. In the evaluation, we addressed the following two goals:

1. For the same number of agents, the performance of the application using the LOD system should be better, i.e. the frame rate should be higher.
2. The difference in the quality of the agents' behavior between the two versions should not be recognizable by a user.

To validate goal 1, we carried out a performance analysis of the Augsburg3D application. For goal 2, we conducted a web-based perception survey of both our applications.

4.1 Performance Analysis

We created two versions of the Augsburg3D application as described above. In both versions, we populated the scene with 50, 100, 150 and finally 250 agents. Each of these four configurations was run five times, for a total of 20 runs per version. Each run lasted 45 seconds and the camera position within the application was fixed for the entire time. We recorded average frame rates for each run and averaged those for all runs of a version. For comparison purposes, we also recorded another data set based off a third version which used the LOD system and the engine's occlusion culling. Figure 5 shows the results for all three versions. The machine we used is a Intel Core 2 Quad Q9550 with 2,83 Ghz, 4 GB Ram and a Nvidia Geforce GTS 250 with 512 MB Ram.

As we can see, the two versions using the LOD system have both an increased frame rate as well as a slower decline of the frame rate for larger numbers of agents. So, goal 1 can be seen as fulfilled.

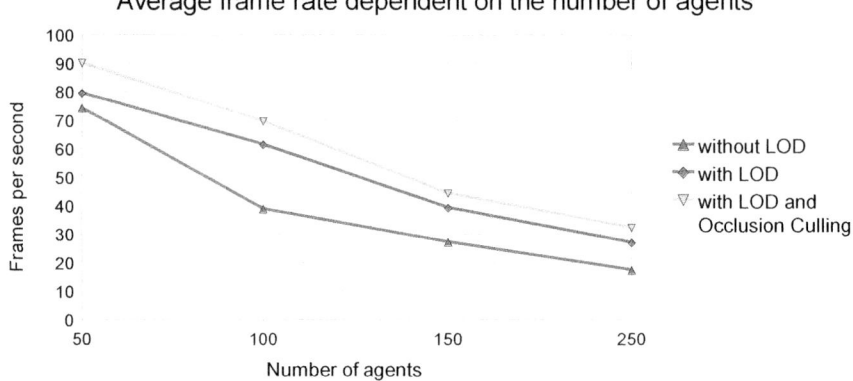

Fig. 5. Performance analysis

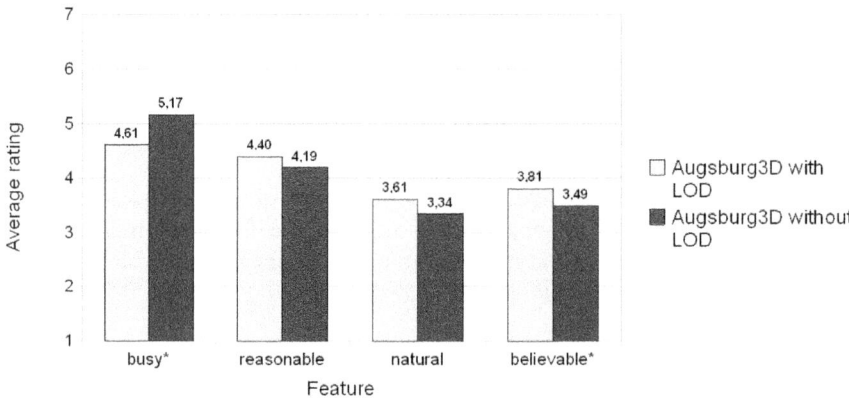

Fig. 6. Subjects' perception of Augsburg3D application, * indicates a significant difference

4.2 Perception Survey

For both the Augsburg3D as well as the Virtual Beer Garden application, we created two versions as described above. From each version we recorded a video of about 30 seconds. We created a web-based survey, asking the participants to watch the agents' behavior in the four videos and rate the following features of the scene on a 7-point Likert scale (1 to 7, 7 meaning full agreement): busy (in a positive sense, i.e. "alive"), reasonable, natural and believable. For each application, the order of the two videos was determined randomly, but the Augsburg3D videos were always shown first. Our study had 108 participants (69 male, 39 female) between ages 15 and 46. Figure 6 shows the mean values of their ratings for the Augsburg3D application.

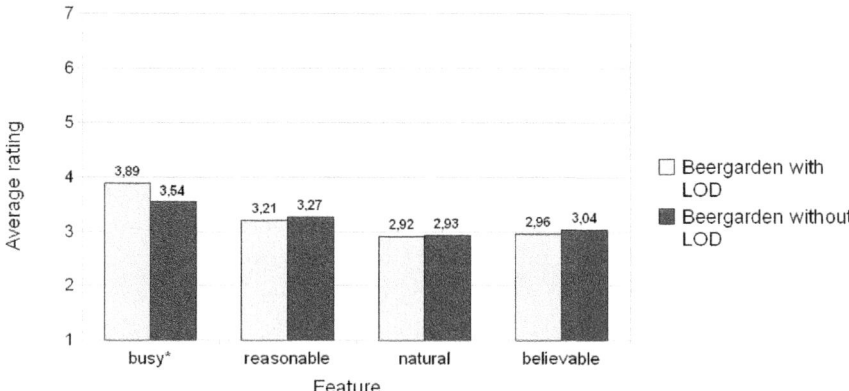

Fig. 7. Subjects' perception of Beer Garden application, * indicates a significant difference

It can be seen that subjects thought that the "normal" version was significantly busier, with mean values of 4.61 vs. 5.17 (t(107)=-4.599, p < 0.001). However, they rated the LOD-version slightly better regarding the other features, with the difference for "believable" also being significant with mean values of 3.81 vs. 3.49 (t(107)=2.040, p < 0.05). The effect size was medium for all four features.

Figure 7 shows the mean values of the subjects' ratings for the Virtual Beer Garden application.

Here, we get the exact opposite: Subjects rated the LOD-version busier, but thought that the normal version was better otherwise. Of these differences, only the one for "busy" was significant with mean values of 3.89 vs. 3.54 (t(107)=3.930, p < 0.001). The effect size was medium for "busy" and small for the other features.

Looking at these results with regards to goal 2, we have to state that the users noticed some differences. However, two of the three significant differences are actually in favor of our LOD system. So while there is definitely room for improvement regarding the busyness of Augsburg3D with the LOD system we see goal 2 as almost fulfilled.

5 Conclusion and Future Work

In this work we presented our idea of a Level of Detail AI for virtual characters based on discrete distances and visibility from a player's point of view. We applied it to an existing application and showed that many aspects of the characters' behavior can be simplified with our system due to its general approach. Applying the behavior reduction at the behavior execution level also provides us with high consistency during LOD changes.

An evaluation showed that our LOD AI considerably increased the application's performance. At the same time, users only barely observed changes in the characters' behavior (with the exception mentioned above), which suggests that our approach is viable but could also use some improvement.

Our ideas for future work include the following categories of improvement: First, the approach could be extended to additional of the characters' behaviors already present in the GameEngine, such as the gaze behavior system we described in [19]. Second, the usage of LOD in some components could be made more dynamic, especially regarding the animations: As of now, they are either fully played or not played at all. Here, it would be interesting to follow the approach described by [20] and create simplified animations that could be used in between. Finally, our LOD approach could also be adapted to other AI techniques used in games, such as HTN Planners, Hierarchical Finite State Machines or Behavior Trees.

References

1. Shao, W., Terzopoulos, D.: Autonomous Pedestrians. In: Proceedings of the 2005 ACM SIGGRAPH/EUROGRAPHICS Symposium on Computer Animation, SCA 2005, pp. 19–28. ACM Press, New York (2005)
2. Patel, J., Parker, R., Traum, D.: Simulation of Small Group Discussions for Middle Level of Detail Crowds. In: Army Science Conference (2004)
3. Pelechano, N., Stocker, C., Allbeck, J., Badler, N.: Being a Part of the Crowd: Towards Validating VR Crowds Using Presence. In: Proceedings of the 7th International Joint Conference on Autonomous Agents and Multiagent Systems, AAMAS 2008, Richland, International Foundation for Autonomous Agents and Multiagent Systems, pp. 136–142 (2008)
4. Peters, C., Ennis, C.: Modeling groups of plausible virtual pedestrians. IEEE Computer Graphics and Applications 29(4), 54–63 (2009)
5. Chenney, S., Arikan, O., Forsyth, D.A.: Proxy Simulations For Efficient Dynamics. In: Proceedings of EUROGRAPHICS 2001 (2001)
6. O'Sullivan, C., Cassell, J., Vilhjalmsson, H., Dingliana, J., Dobbyn, S., McNamee, B., Peters, C., Giang, T.: Levels of detail for crowds and groups. Computer Graphics Forum 21(4), 733–742 (2002)
7. Brockington, M.: Level-Of-Detail AI for a Large Role-Playing Game. In: AI Game Programming Wisdom, pp. 419–425. Charles River Media, Hingham (2002)
8. Niederberger, C., Gross, M.: Level-of-detail for cognitive real-time characters. The Visual Computer: Int. Journal of Computer Graphics 21(3), 188–202 (2005)
9. Pettré, J., de Heras Ciechomski, P., Maïm, J., Yersin, B., Laumond, J.P., Thalmann, D.: Real-time navigating crowds: scalable simulation and rendering. Comput. Animat. Virtual Worlds 17(3-4), 445–455 (2006)
10. Brom, C., Šerý, O., Poch, T.: Simulation Level of Detail for Virtual Humans. In: Pelachaud, C., Martin, J.-C., André, E., Chollet, G., Karpouzis, K., Pelé, D. (eds.) IVA 2007. LNCS (LNAI), vol. 4722, pp. 1–14. Springer, Heidelberg (2007)
11. Paris, S., Gerdelan, A., O'Sullivan, C.: CA-LOD: Collision Avoidance Level of Detail for Scalable, Controllable Crowds. In: Egges, A. (ed.) MIG 2009. LNCS, vol. 5884, pp. 13–28. Springer, Heidelberg (2009)

12. Lin, Z., Pan, Z.: LoD-Based Locomotion Engine for Game Characters. In: Hui, K.-c., Pan, Z., Chung, R.C.-k., Wang, C.C.L., Jin, X., Göbel, S., Li, E.C.-L. (eds.) EDUTAINMENT 2007. LNCS, vol. 4469, pp. 214–224. Springer, Heidelberg (2007)
13. Osborne, D., Dickinson, P.: Improving Games AI Performance using Grouped Hierarchical Level of Detail. In: Proc. of the 36th Annual Convention of the Society for the Study of Artificial Intelligence and Simulation of Behaviour (2010)
14. Horde3D GameEngine: University of Augsburg (2010), http://mm-werkstatt.informatik.uni-augsburg.de/projects/GameEngine/
15. Dorfmueller-Ulhaas, K., Erdmann, D., Gerl, O., Schulz, N., Wiendl, V., André, E.: An Immersive Game - Augsburg Cityrun. In: André, E., Dybkjær, L., Minker, W., Neumann, H., Weber, M. (eds.) PIT 2006. LNCS (LNAI), vol. 4021, pp. 201–204. Springer, Heidelberg (2006)
16. Heigeas, L., Luciani, A., Thollot, J., Castagné, N.: A Physically-Based Particle Model of Emergent Crowd Behaviors. In: Graphicon (2003)
17. Rehm, M., Endrass, B., Wißner, M.: Integrating the User in the Social Group Dynamics of Agents. In: Workshop on Social Intelligence Design (SID) (2007)
18. Kistler, F., Wißner, M., André, E.: Level of Detail based Behavior Control for Virtual Characters. In: Proc. of the 10th Int. Conf. on Intelligent Virtual Agents. Springer, Heidelberg (2010)
19. Wißner, M., Bee, N., Kienberger, J., André, E.: To See and to Be Seen in the Virtual Beer Garden - A Gaze Behavior System for Intelligent Virtual Agents in a 3D Environment. In: Mertsching, B., Hund, M., Aziz, Z. (eds.) KI 2009. LNCS, vol. 5803, pp. 500–507. Springer, Heidelberg (2009)
20. Giang, T., Mooney, R., Peters, C., O'Sullivan, C.: ALOHA: Adaptive Level of Detail for Human Animation: Towards a new Framework. In: EUROGRAPHICS 2000 Short Paper Proceedings, pp. 71–77 (2000)

Scalable and Robust Shepherding via Deformable Shapes[*]

Joseph F. Harrison, Christopher Vo, and Jyh-Ming Lien

Department of Computer Science, George Mason University,
4400 University Drive MSN 4A5, Fairfax, VA 22030 USA
{jharri1,cvo1,jmlien}@gmu.edu
http://masc.cs.gmu.edu

Abstract. In this paper, we present a new motion planning strategy for *shepherding* in environments containing obstacles. This instance of the *group motion control problem* is applicable to a wide variety of real life scenarios, such as animal herding simulation, civil crowd control training, and oil-spill cleanup simulation. However, the problem is challenging in terms of scalability and robustness because it is dynamic, highly underactuated, and involves multi-agent coordination. Our previous work showed that high-level probabilistic motion planning algorithms combined with simple shepherding behaviors can be beneficial in situations where low-level behaviors alone are insufficient. However, inconsistent results suggested a need for a method that performs well across a wider range of environments. In this paper, we present a new method, called DEFORM, in which shepherds view the flock as an abstracted deformable shape. We show that our method is more robust than our previous approach and that it scales more effectively to larger teams of shepherds and larger flocks. We also show DEFORM to be surprisingly robust despite increasing randomness in the motion of the flock.

Keywords: Motion planning, shepherding, simulation, group motion control, manipulation, deformation.

1 Introduction

Group motion control is the problem of moving a group of agents in coordination. One instance of this problem is *shepherding*, in which the objective is to herd a group of agents (e.g., a flock of sheep, crowd of people, etc.) using one or more "shepherd" agents (e.g., shepherds, riot police, etc.). The objective of shepherding is typically to guide the group to a goal, though other variants exist (e.g., escorting or protection, in which a "parent" tries to maintain separation between their "child" and a "stranger" [27]). A solution to the shepherding problem is typically given as a sequence of movements the shepherds can perform to guide the flock to the goal. This sequence may be the output of a high-level

---

[*] This work is partially supported by NSF IIS-096053.

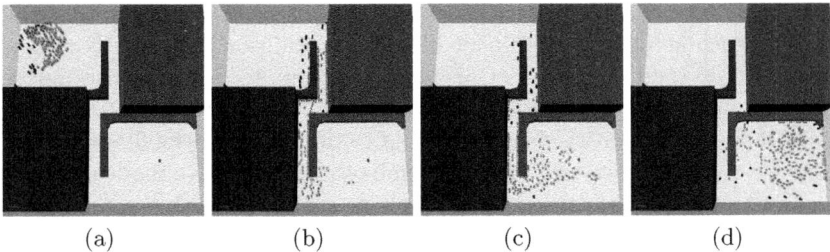

Fig. 1. A team of 15 shepherds successfully herds a flock with 200 members from the starting region in the top-left corner to the goal in the lower-right. Along the way, the flock must be coaxed through a narrow corridor. The flock are drawn in gray and the shepherds are drawn in black. We recommend the readers to view the attached video.

motion planner, the result of agent-based behaviors working independently or in coordination to achieve high-level commands, or some combination thereof.

Shepherding is applicable to a wide variety of real life scenarios, such as animal herding [7,1], civil crowd control [6,22], and micromanipulation [19]. Shepherding in these scenarios is very challenging (in terms of *scalability* and *robustness*) since it involves the underactuated control of a large number of dynamic agents whose trajectories may be unpredictable. While several works in robotics and computer animation have modeled the shepherding problem, none of them address shepherding for large groups (with tens to hundreds of members), shepherding in environments with obstacles, or how to handle uncertainty in the motion of the group.

Efficient group representation is necessary for any algorithm to be scalable and robust to uncertainty. Some naïve methods such as [23] and [24] represent the flock as individual agents which severely limits their scalability to large flocks. In this paper we present a simple approach called DEFORM which represents the flock as a *deformable blob* that can split and merge. This representation is more efficient than individual-based representations and approximates the flock more accurately than simpler representations such as discs or bounding boxes [25]. We will define the deformable blob and describe the strategies to manipulate the deformable blob in Section 4.

Despite its simplicity, DEFORM broadly outperforms our previous method (MAGB [25]), especially with larger flocks. Fig. 1 shows an example simulation where 15 shepherds successfully move a flock of 200 agents through a narrow passage to reach the goal. Previous methods rarely succeed in this scenario. DEFORM also shows impressive robustness vs. MAGB despite increasing randomness in flock behavior. Detailed experimental results will be presented in Section 5.

2 Related Work

We are interested in controlling large groups of active agents by providing external stimuli. There are several works along the vein of understanding the emergent behavior of simulated crowds when significant changes are made to the environment, such as adding movement barriers or additional agents. For example, Brenner et al. studied the effect of adding barriers to influence the movement of crowds in disaster scenarios for RoboCup [6]. Schubert and Suzić [22] used a genetic algorithm with agent simulation to determine optimal barrier deployments for controlling rioting crowds. Some other works have modeled the effect of adding agents with attractive or repulsive social forces [14] and agents with different roles (such as "leader" agents [3]). Yeh et al. also experimented with composite agents [27] which can exhibit different behaviors such as "guidance", using proxy agents as temporary obstacles.

Some experiments have also been performed to understand how to control living organisms using robotic agents. For example, Halloy et al. [11] showed that robotic agents can influence the collective behavior of a group of cockroaches through local interactions. Ogawa et al. [19] demonstrated the ability to track and manipulate Paramecium caudatum cells in a chamber using an electrical field. Some experiments have herded cows using "smart collars" which act as virtual fences [7,1]. When a cow approaches the virtual fence, the collar produces a noise which causes the cow to move away.

Explicit modeling of shepherding behaviors has also been studied in robotics and computer animation. In robotics, Schultz et al. [23] used a genetic algorithm to learn behaviors for a shepherding robot. Vaughan et al. [24] simulated and constructed a robot that shepherds a flock of geese in a circular environment. In computer animation, Funge et al. [10] simulated an interesting shepherd-like behavior in which a T. Rex chases raptors out of its territory. Potter et al. [20] studied a herding behavior using three shepherds and a single sheep in a simple environment. However, none of the aforementioned methods deal with large flocks or obstacle-filled environments.

In some existing approaches to shepherding, a bounding circle is used to model the flock. This is sufficient for relatively sparse workspaces but is excessively restrictive in environments with many obstacles or narrow corridors, or when the flock size is large. Using deformable object to model a coherent group is not new. Notably, Kamphuis and Overmars [13] used a hinged-box to represent a group of agents for navigating through an obstacle-filled environment. However, the deformable object in our work is modeled as a blob that can split and merge, and, to the best of our knowledge, our work is the first using a deformable representation for group motion control.

The general problem of planning motion for groups of agents is also relevant to our study. Recent work in the animation of group motion has shown that *Group Motion Graphs* [17] can be used to efficiently control small to medium-sized groups with a small computational load. In [16], using a similar data structure and a mesh-editing algorithm, a method to manipulate existing group motions was presented.

When multiple shepherds work together to move a flock, their movements must be coordinated to be effective. One common approach is to move the shepherds in formation [4]. Similar formations are commonly used by police and military personnel to disperse, contain, or block crowds and prevent riots [2]. In our own earlier work [18] we showed that shepherd formations can be used to effectively control a flock.

Recent work in the animation of group motion has shown that Group Motion Graphs [17] can be used to efficiently control small- to medium-sized groups with a small computational load.

In our most recent work on shepherding, [26], we presented a behavior-based shepherding method which uses the medial axis of the workspace as a roadmap, selecting intermediate milestones on its way to the goal. There, we referred to the method as "simulation only." In this paper, we use this method as a baseline for comparison, and for clarity, we refer to it hereafter as the "Medial Axis Graph-Based" (MAGB) approach.

Our work in [26] also used sample-based methods (RRT [15], EST [12], META-GRAPH) to find a sequence of intermediate milestones through free-space (not restricted to the medial-axis), and connected them with low-level shepherding behaviors. We found these approaches more effective than MAGB for certain types of environments (e.g., those requiring multiple changes of direction toward and away from the goal), but less effective on environments with multiple paths to the goal. The planning-based methods were also much less efficient, and when held to a fixed budget of simulation steps, were often unable to reach the goal.

3 Preliminaries

3.1 Definitions

A *flock* is a group of agents that move with some level of cohesion through the environment and away from *shepherds* which attempt to guide the flock toward a goal. We use F to denote the flock and S to refer collectively to the shepherds.

The configuration of an agent (shepherd or flock member) is represented by its position and velocity. Therefore, a group control problem with n flock members and m shepherds in 2-d will have a configuration space C in $4(n+m)$ dimensions. We refer to the joint configuration of S at time step t as $C_S(t)$ and F as $C_F(t)$. To simplify our notation, we will drop t whenever it does not cause any ambiguity. A *valid* configuration is one in which no agents are in collision with obstacles or other agents. Because n is usually much larger than m, the problem of shepherding is highly underactuated. Since this work focuses on controlling large flocks ($n > 10$), C is usually very high dimensional. A new representation, such as DEFORM, that can significantly reduce the dimensionality while still accurately approximate C is needed.

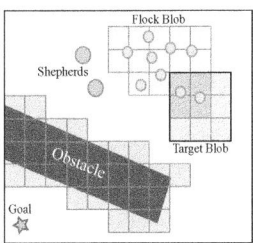

Fig. 2. We discretize the workspace to create an occupancy grid of obstacles and free space. The flock blob is the set of grid cells occupied by the flock; the target blob is the where the shepherds would like to move the flock en route to the goal.

We use the term *target* to denote an intermediate configuration toward which the shepherds attempt to steer the flock, and *steering point* to denote a point toward which a shepherd moves itself in order to do the steering.

The term *blob* in this paper refers to a subset of cells in the discretized grid of the environment. The *flock blob* is the area currently occupied by the flock and is denoted B_F. The target area to which the shepherds try to guide the flock is called the *target blob* and denoted B_T. The concept of blobs should be taken generally, as DEFORM may also be implemented using non-grid representations, such as polygons, contours, alpha-shapes, etc. Also, while blobs are often contiguous, they do not need to be.

We define the group-control problem as follows: Given $C_S(0)$ and $B_F(0)$, find a sequence of valid configurations $C_S(t)$ for $0 < t \leq 1$ so that all $B_F(t)$ are valid and $B_F(1) \in GC$, where GC is a set of user specified goal configurations.

3.2 Flock Behavior

The motion of our flock is similar to Reynolds' *Boids* model [21], in which each agent in the simulation steers itself according to three simple rules: *separation*, steering away to avoid colliding with other agents, *alignment*, steering toward the average heading of neighbors, and *cohesion*, steering toward the average position of neighbors. The *neighbors* of an agent are those that are located within its view, which is a fan-shaped region in front of the agent. The fan is defined by a viewing angle and distance. In our work, we consider agents with a viewing angle of 360° so their view depends only on proximity. We add the following rules to determine the motion of our flock: *avoidance* (steering away from nearby shepherds), *damping* (reducing speed over time when other forces are absent), and *entropy* (adjusting the direction randomly, defined in Eq. 1, and is usually set to zero).

4 Our Method: DEFORM

In this section, we describe the propose shepherding strategy, called DEFORM. We will first provide a sketch of the algorithm and the explain each step in detail for the rest of the section.

4.1 Overview of the Algorithm

DEFORM attempts to move the shepherds such that they guide the flock from the start configuration to the goal g. It does so by continually updating the flock blob (B_F), target blob (B_T), and steering points (s_{steer}). Algorithm 1 outlines the steps necessary.

4.2 Computing the Grid

We discretize the polygon-based environment to create an occupancy grid with cells as wide as the diameter of a flock member. We mark the cells as either

Algorithm 1. DEFORM(F, S, g)

Input: F (flock), S (shepherds), and g (goal)
while g is not reached **do**
    Compute the flock blob $B_F(t)$ of F at time t
    Determine the next target blob $B_T(t)$
    Find $s_{steer} \in C_S$ to morph B_F toward B_T
    Assign s_{steer} to S
    Update simulation

free or *in-collision* depending on whether they contain any part of an obstacle (see Fig. 2). For later use, we calculate a gradient outward from the goal, which provides the geodesic distance from each cell to the goal. Advantages of using grid-based representation also include efficiency and applicability to video games and mobile robots, in which environmental data may already be stored in bitmaps or occupancy grids.

4.3 Flock Blobs (B_F)

Given a flock configuration C_F, the corresponding flock blob B_F is the set of all grid cells occupied by members of the flock (see Fig. 3(a)). Since the grid size matches the diameter of the flock members, a one-member flock would have a flock blob with as few as 1 or as many as 4 cells. A large, tightly packed flock can fit within a blob with fewer cells than the size of the flock. Also note that a flock blob may contain multiple noncontiguous parts if the flock has split.

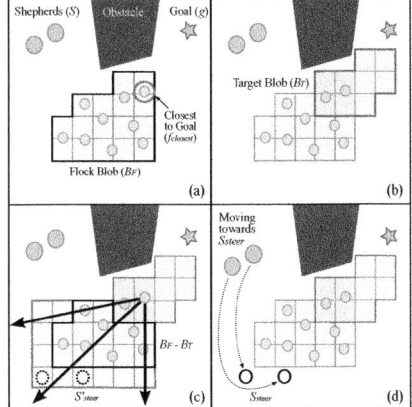

4.4 Target Blobs (B_T)

Given the current flock blob B_F, the target blob B_T is calculated as follows. First, let $f_{closest}$ be the member of the flock closest (in geodesic distance) to the goal. Next, grow B_T using free cells, starting with the 8-connected set and expanding in concentric rings around the cell containing $f_{closest}$. Stop when B_T contains as many cells as there are members of the flock. See Fig. 3(b) for an example.

Fig. 3. The process of choosing steering points: (a) calculate the flock blob and $f_{closest}$, (b) calculate the target blob, (c) take the boolean difference between flock blob and target blob and calculate potential steering points, (d) select steering points on the opposite side from the goal

This approach maximizes the thickness of the target blob, though different metrics could be chosen instead. We discuss some of them in Section 6.

4.5 Shepherds' Steering Points (s_{steer})

Given a flock blob B_F and a target blob B_T, we compute the set of potential steering points s'_{steer} and its subset $s_{steer} \subset s'_{steer}$ which get assigned to the shepherds. We begin by computing the boolean difference between B_F and B_T. Next, we select the points for s'_{steer}. We say a cell p is in s'_{steer} if (1) p is neighboring (using 8-connectivity) to at least a cell $c \in B_F - B_T$, (2) $p \notin (B_F \cup B_T)$, and (3) the geodesic distance from p to B_T is greater than from c to B_T.

Next we choose which points in s'_{steer} will be assigned to the m shepherds. If s'_{steer} contains fewer than m points, each will be given to a shepherd and the remaining shepherds will stay stationary. If there are more than m points in s'_{steer}, we partition s'_{steer} into m pie wedges centered at $f_{closest}$. In each pie wedge, the point in s'_{steer} farthest from the global goal is selected as a steering point. Fig. 3(c) shows an example.

After s_{steer} is computed, the steering points are assigned to the shepherds by forming a bipartite graph whose nodes are the points in s_{steer} and the shepherd positions, and whose edge weights are geodesic distances between them. The steering point assignment is then solved using a bipartite matching algorithm.

Once the steering points are matched to shepherds, the shepherds move toward their steering points during the next simulation step, as shown in Fig. 3(d).

5 Experiments

We used simulation to test the performance of DEFORM and compare it to our previous results using MAGB [18]. Each simulation begins with a starting and goal region for the flock. The flock is randomly scattered throughout the starting region and the shepherds are randomly placed around the perimeter of the flock. Providing different seeds for the random number generator gives us different initial conditions for each run. The shepherds and flock are modeled as 2-d discs for the collision detection.

We compared DEFORM against MAGB across 6 different test environments, shown in Fig. 4. For this paper, each experimental sample consists of 30 runs, and claims of statistical significance are based on 95% confidence. The performance of the shepherding strategies is measured by the *success rate* of herding the group to the goal within the limited amount of simulation time steps. Despite the simplicity of DEFORM, in our experiments, we have observed that DEFORM broadly outperforms MAGB in terms of scalability—the ability to steadily maintain high success rates despite an increasing simulation sizes (see details in Section 5.2). The DEFORM method also shows impressive robustness vs. MAGB in terms of maintaining 100% success rates despite increasing randomness in agent behavior (see details in Section 5.3).

5.1 Environments

We created several different environments to highlight different shepherding challenges (see Fig. 4). The "Empty" environment allows us to evaluate how well

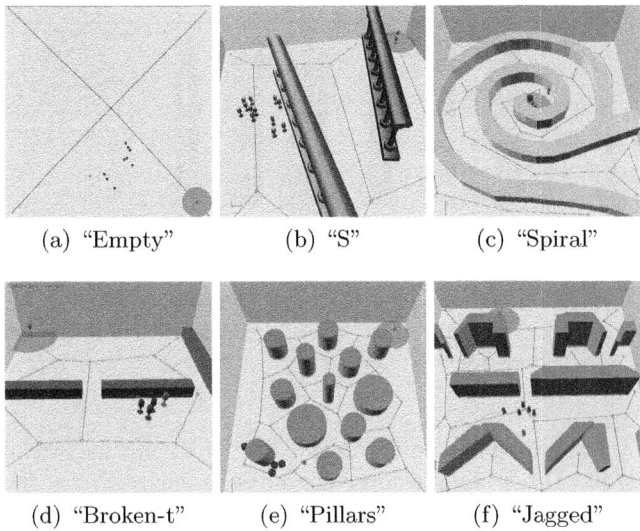

Fig. 4. Environments used in our experiments

a given shepherding approach can move the flock in the absence of obstacles. While it may seem trivially easy, it can be more difficult to move a flock through an open area than a down a narrow corridor. The "S" environment demonstrates a situation with simple local minima which require a shepherd to push the flock away from the goal temporarily. This environment is relatively easy for a motion planning algorithm, compared to the others. The "Spiral" environment requires repeated motions that move the flock closer to, then further from, the goal. This map is particularly difficult for sampling-based planning methods as it requires samples in each of several specific regions before it will be able to connect a path. The "Broken-t" environment presents a narrow passage in the middle of the map as well as another along the wall. The "Pillars" environment offers many different paths to the goal, which can be difficult for tree-based planning methods. The final environment, called "Jagged", also offers many different paths to the goal, but includes several areas where members of the flock may get stuck.

5.2 Experimental Results: Scalability

We presume that increasing the size of the flock should impact both the effectiveness and performance of the shepherding algorithm, and that more shepherds will be needed to move larger flocks to the goals in a timely manner. The *scalability* of these behaviors is defined by the ability for them to effectively herd increasingly large flocks to the goal within the given simulation time budget.

To test our hypotheses, we ran experiments with varying number of shepherds (2 and 4) and varying flock sizes (20 and 40) on each of the 6 test environments. The results of these experiments are shown in Figure 5. In this

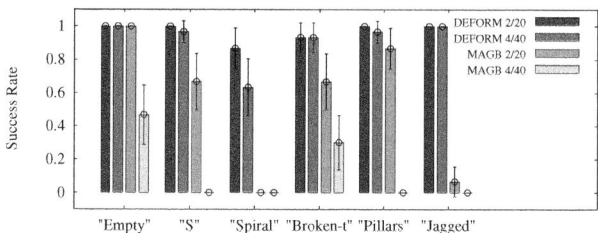

Fig. 5. Success rates on all environments using DEFORM and MAGB. "2/20" means 2 shepherds and a flock of size 20. "4/40" means 4 shepherds and a flock of size 40. Error bars indicate 95% confidence over 30 runs.

figure, note that across most environments, as the simulation size increases from a 2/20 to 4/40, MAGB significantly degrades in success rate, while DEFORM shows no significant degradation in success rate as the simulation size increases. To further investigate the scalability in the finer resolutions, we ran experiments with varying number of shepherds (1, 2, 3, and 4) and varying flock sizes (5, 10, 15, 20, and 25) on the "Broken-t" environments. We computed *success rates* as the proportions of sample runs where the algorithm successfully moved the flock to the goal. Table 1 shows example results for the "Broken-t" environment. For small flock sizes, DEFORM and MAGB perform similarly. However, as the size of the flock increases, MAGB shows significant decay in performance - it is clear that more shepherds are needed to control larger flocks using the MAGB behavior. On the other hand, DEFORM manages to achieve excellent results throughout (above 90% success rate) with no negative trend in performance up to 25 flock members.

We also ran experiments to test larger shepherd and flock sizes on the "S" environment. The results, shown in Fig. 6, show that DEFORM once again outperforms MAGB across the board, but especially with larger flock sizes.

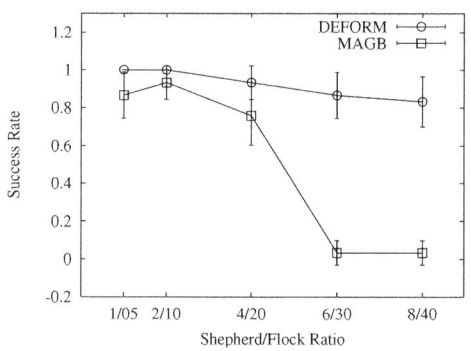

Fig. 6. Success rate on "S" Environment as flock size increases. Ratio is fixed at 1 shepherd per 5 sheep. Error bars are shown at 95% confidence over 30 runs.

5.3 Experimental Results: Robustness

We define *robustness* as the ability for the shepherds to control the flock in spite of unpredictable flock behavior. We modeled the unpredictable behavior

Table 1. Success Rates for DEFORM and MAGB with Varying Shepherd and Flock Sizes on "Broken-t"

		DEFORM				MAGB			
		# Shepherds				# Shepherds			
		1	2	3	4	1	2	3	4
Flock Size	5	0.93	1.00	0.90	0.83	0.90	1.00	0.93	0.97
	10	1.00	0.93	0.93	0.93	0.56	0.97	1.00	0.97
	15	1.00	1.00	1.00	0.93	0.47	0.80	0.77	0.83
	20	1.00	0.97	1.00	0.90	0.07	0.67	0.43	0.83
	25	1.00	1.00	0.97	0.97	0.00	0.40	0.33	0.43

by linearly mixing each flock agent's behavior with a random vector. We control this randomness by increasing and decreasing the magnitude of the random vector. More specifically, the randomness r is defined as:

$$r = \frac{|F_{rand}|}{|F_{rand}| + |F_{flock}|} , \qquad (1)$$

where F_{rand} and F_{flock} are the random force and the flocking force, respectively. Fig. 7 illustrates the results (with 95% confidence bars) of success rate versus increasing randomness. In these figures, randomness is shown as a ratio from 0.0 to 0.7 where 0.0 represents fully deterministic behavior and 0.7 represents the maximum randomness attempted. In general, DEFORM performs significantly better than MAGB with increasing flock randomness. An interesting effect is shown in Fig. 7 (top), where there is a bump in performance with increasing flock randomness for the MAGB behavior. We believe this occurs because the random perturbations help the flock pass through a narrow corridor like salt from a shaker.

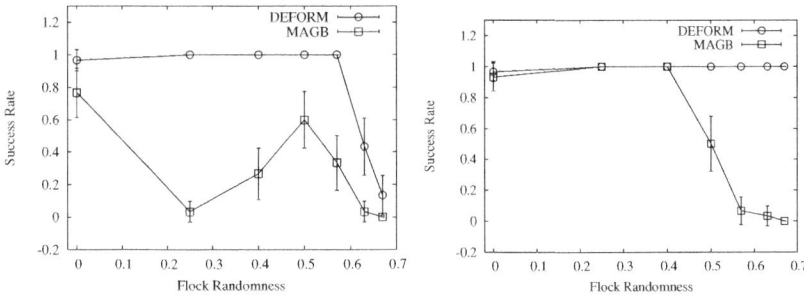

Fig. 7. (**left**) Success rate on "Broken-t" as flock motion becomes more random. (**right**) Success rate on "S". The numbers on the x-axis indicate the ratio of randomness to deterministic flocking motion. Error bars are shown at 95% confidence over 30 runs. In both experiments, two shepherds are used to herd 20 flock members.

6 Conclusion and Future Work

This paper introduces a new and powerful representation to the shepherding problem. Using a discretized, deformable blob to abstract the position of the flock members, our new DEFORM method performs shepherding more effectively than our previous best method [26], especially in terms of handling larger flock sizes and unpredictable flock behavior. Whereas our previous method relied on computing the medial axis of the workspace, the new method requires only an occupancy grid. This makes our new method more applicable to robotics scenarios where a medial axis is not available (i.e. when the workspace geometry is not known *a priori*).

There are some areas which deserve more attention. In the future, we'd like to try growing target blobs that favor metrics besides just "thickest". For example, it may be desired for target blobs to grow away from obstacles, towards planned waypoints, or along existing roadmap information. We are also further exploring the interaction between high level planning and these behaviors. For example, we would also like to apply the new deformable blob abstraction to the high level planning methods for shepherding that we have presented in [26], which combined shepherding behaviors with probabilistic motion planning algorithms such as EST [12], RRT [15], and our own, META-GRAPH.

As we mentioned in Section 3, DEFORM could be implemented to represent flock blobs using other geometric abstractions such as polygons, contours, or α-shapes [8,9]. α-shapes are particularly interesting in this regard since the value of α could be based on the flock's sensing range to prevent the shepherds from disturbing the flock unnecessarily. The Iterative Closest Point (ICP) algorithm [5] could then be used to compute successive transformations of the flock. However, to be efficient, a new variation would be needed that exploits the temporal and spatial coherence of the flock's motion.

References

1. Anderson, D.M.: Virtual fencing–past, present and future. The Rangeland Journal 29, 65–78 (2007)
2. Applegate, R.: Riot control: materiel and techniques. Stackpole Books (1969)
3. Aubé, F., Shield, R.: Modeling the effect of leadership on crowd flow dynamics. In: Sloot, P.M.A., Chopard, B., Hoekstra, A.G. (eds.) ACRI 2004. LNCS, vol. 3305, pp. 601–621. Springer, Heidelberg (2004)
4. Balch, T., Arkin, R.: Behavior-based formation control for multirobot teams. IEEE Trans. Robot. Automat. 14(6), 926–939 (1998)
5. Besl, P.J., McKay, N.D.: A method for registration of 3-d shapes. IEEE Trans. Pattern Anal. Mach. Intell. 14(2), 239–256 (1992)
6. Brenner, M., Wijermans, N., Nussle, T., de Boer, B.: Simulating and controlling civilian crowds in robocup rescue. In: Bredenfeld, A., Jacoff, A., Noda, I., Takahashi, Y. (eds.) RoboCup 2005. LNCS (LNAI), vol. 4020, Springer, Heidelberg (2006)
7. Butler, Z., Corke, P., Peterson, R., Rus, D.: Virtual fences for controlling cows. In: Robotics and Automation. In: Proceedings of IEEE International Conference on ICRA 2004, April-1- May, vol. 5, pp. 4429–4436 (2004)

8. Edelsbrunner, H., Kirkpatrick, D.G., Seidel, R.: On the shape of a set of points in the plane. IEEE Trans. Inform. Theory IT- 29, 551–559 (1983)
9. Edelsbrunner, H., Mücke, E.P.: Three-dimensional alpha shapes. ACM Trans. Graph. 13(1), 43–72 (1994)
10. Funge, J., Tu, X., Terzopoulos, D.: Cognitive modeling: Knowledge, reasoning and planning for intelligent characters. In: Computer Graphics, pp. 29–38 (1999)
11. Halloy, J.: Colleagues: Social Integration of Robots into Groups of Cockroaches to Control Self-Organized Choices. Science 318(5853), 1155–1158 (2007)
12. Hsu, D., Latombe, J.C., Motwani, R.: Path planning in expansive configuration spaces. Int. J. Comput. Geom. & Appl., 2719–2726 (1997)
13. Kamphuis, A., Overmars, M.: Motion planning for coherent groups of entities. In: Proc. IEEE Int. Conf. Robot. Autom. (ICRA), vol. 4, pp. 3815–3822 (2004)
14. Kirkland, J., Maciejewski, A.: A simulation of attempts to influence crowd dynamics. In: IEEE International Conference on Systems, Man and Cybernetics, vol. 5, pp. 4328–4333 (2003)
15. Kuffner, J.J., LaValle, S.M.: RRT-Connect: An Efficient Approach to Single-Query Path Planning. In: Proc. of IEEE Int. Conf. on Robotics and Automation, pp. 995–1001 (2000)
16. Kwon, T., Lee, K.H., Lee, J., Takahashi, S.: Group motion editing. ACM Trans. Graph. 27(3), 1–8 (2008)
17. Lai, Y.C., Chenney, S., Fan, S.: Group motion graphs. In: Proceedings of the 2005 ACM SIGGRAPH/Eurographics Symposium on Computer Animation, SCA 2005, pp. 281–290. ACM, New York (2005)
18. Lien, J.M., Bayazit, O.B., Sowell, R.T., Rodriguez, S., Amato, N.M.: Shepherding behaviors. In: Proc. IEEE Int. Conf. Robot. Autom. (ICRA), pp. 4159–4164 (2004)
19. Ogawa, N., Oku, H., Hashimoto, K., Ishikawa, M.: Microrobotic visual control of motile cells using high-speed tracking system. IEEE Transactions on Robotics 21(4), 704–712 (2005)
20. Potter, M.A., Meeden, L., Schultz, A.C.: Heterogeneity in the coevolved behaviors of mobile robots: The emergence of specialists. In: IJCAI, pp. 1337–1343 (2001)
21. Reynolds, C.W.: Flocks, herds, and schools: A distributed behaviroal model. In: Computer Graphics, pp. 25–34 (1987)
22. Schubert, J., Suzic, R.: Decision support for crowd control: Using genetic algorithms with simulation to learn control strategies. In: Military Communications Conference, MILCOM 2007, pp. 1–7. IEEE, Los Alamitos (October 2007)
23. Schultz, A.C., Grefenstette, J.J., Adams, W.: Robo-shepherd: Learning complex robotic behaviors. In: Robotics and Manufacturing: Recent Trends in Research and Applications, vol. 6, pp. 763–768. ASME Press (1996)
24. Vaughan, R.T., Sumpter, N., Henderson, J., Frost, A., Cameron, S.: Experiments in automatic flock control. J. Robot. and Autonom. Sys. 31, 109–117 (2000)
25. Vo, C., Harrison, J.F., Lien, J.M.: Behavior-based motion planning for group control. In: Proceedings of the IEEE International Conference on Intelligent Robots and Systems (IROS), St. Louis Missouri (2009) (to appear)
26. Vo, C., Harrison, J.F., Lien, J.M.: Behavior-based motion planning for group control. In: Proceedings of the IEEE/RSJ International Conference on Intelligent Robots and Systems (2009)
27. Yeh, H., Curtis, S., Patil, S., van den Berg, J., Manocha, D., Lin, M.: Composite agents. In: Gross, M., James, D. (eds.) Proceedings of EUROGRAPHICS / ACM SIGGRAPH Symposium on Computer Animation (2008)

Navigation Queries from Triangular Meshes

Marcelo Kallmann

University of California, Merced

Abstract. Navigation meshes are commonly employed as a practical representation for path planning and other navigation queries in animated virtual environments and computer games. This paper explores the use of triangulations as a navigation mesh, and discusses several useful triangulation–based algorithms and operations: environment modeling and validity, automatic agent placement, tracking moving obstacles, ray–obstacle intersection queries, path planning with arbitrary clearance, determination of corridors, etc. While several of the addressed queries and operations can be applied to generic triangular meshes, the efficient computation of paths with arbitrary clearance requires a new type of triangular mesh, called a Local Clearance Triangulation, which enables the efficient and correct determination if a disc of arbitrary size can pass through any narrow passages of the mesh. This paper shows that triangular meshes can support the efficient computation of several navigation procedures and an implementation of the presented methods is available.

Keywords: Path planning, reactive behaviors, navigation, crowd simulation.

1 Introduction

Navigation meshes are commonly used as a representation for computing navigation procedures for autonomous characters. The term navigation mesh however has been broadly used to refer to any type of polygonal mesh able to represent walkable areas in a given environment, and no specific attention has been given to establishing properties and algorithms for specific types of navigation meshes. This paper explores the use of triangulated meshes and summarizes several operations, queries and properties which are useful for the implementation of navigation strategies for autonomous agents.

One main advantage of relying on triangulations is that the obtained triangular cell decomposition of the environment has $O(n)$ cells, where n is the number of segments used to describe the obstacles in the environment. As a result, spatial processing algorithms will depend on n, which is related to the complexity of the environment (the number of edges to describe obstacles) and not to the size (or extent) of the environment. Therefore triangulated meshes are in particular advantageous for representing large environments where uniform grid–based methods become significantly slower.

The algorithms described in this paper target several navigation queries needed in applications with many agents in complex and large environments. There are two main types of obstacles which have to be addressed: 1) static objects are those describing the environment, they typically do not move over time but it is acceptable that they change their position from time to time (like a door which can be open or closed), and

2) dynamic objects are those which are continuously in movement, as for example the agents themselves.

Following the most typical approach taken with navigation meshes, the presented triangulations are considered to only represent static objects. Even if dynamic updates of obstacles are possible, specific approaches (based on triangulations or not) for handling dynamic objects will usually be more efficient. This paper addresses the use of triangulations for handling these issues and also for computing several additional navigation queries, such as ray–obstacle intersections and path planning with clearance (see Figure 1). The discussed data structures and algorithms have been implemented and are available from the author's web site[1].

Fig. 1. Examples of several paths computed with arbitrary clearances in different environments

2 Related Work

Triangulations are powerful representation structures which have been used in different ways for the purpose of computing navigation queries. Triangulations have in particular been used for extracting adjacency graphs for path planning from given environments. A variety of approaches have been devised, as for example to automatically extract roadmaps considering several layers (or floors) [16], and to hierarchically represent environments with semantic information and agents of different capabilities [19].

Most of the previous work on the area has however focused on specific application goals, and not on the underlying algorithms and representations. For instance, one main drawback of reducing the path planning problem to a search in a roadmap graph is that the obtained paths will still need to be smoothed. In addition, it also becomes difficult to compute paths with other useful properties, such as being optimal in length and having a given clearance from obstacles.

The most popular approach for computing the geometric shortest path among polygonal obstacles defined by n segments is to build and search the *visibility graph* [3,17] of

---

[1] http://graphics.ucmerced.edu/software.html

the obstacles, what can be achieved in $O(n^2)$ time [21, 24]. The shortest path problem is however $O(n \log n)$ and optimal [11] and near-optimal [20] algorithms are available following the *continuous Dijkstra* paradigm. However, in particular when considering arbitrary clearances from obstacles, it is difficult to achieve efficient algorithms which are suitable for practical implementations. The *visibility–voronoi complex* [27] is able to compute paths in $O(n^2)$ time and is probably the most efficient implemented approach to compute paths addressing both global optimality and arbitrary clearance. It is possible to note that the use of dedicated structures is important and planar meshes have not been useful for computing globally-optimal geometric shortest paths.

Nevertheless, navigation meshes remain a popular representation in practice, and recent works have started to address properties and algorithms specifically for them. My previous work of 2003 [14] addressed the insertion and removal of constraints in a Constrained Delaunay Triangulation (CDT), and showed that environments can be well represented by CDTs for the purpose of path planning. The implementation developed in [14] has been used by other researchers [4] and significant improvements in performance were reported in comparison to grid–based methods. Extensions to the original method for handling clearance have also been reported [4], however without correctly solving the problem. In a recent publication [13], I have showed how arbitrary clearances can be properly addressed with the introduction of a new triangulation called a *Local Clearance Triangulation* (LCT), which after a precomputation of $O(n^2)$, is able to compute *locally shortest* paths in $O(n \log n)$, and even in $O(n)$ time, achieving high quality paths very efficiently. A summary of this approach is given in Section 7.

Other efficient geometric approaches are also available. In particular the approach based on corridor maps [7, 8] is also able to efficiently achieve paths with clearance. One fundamental advantage of using triangulated meshes is that the environment is already triangulated, and thus channels or corridors do not need to be triangulated at every path query according to given clearances.

Path planning is not the only navigation query needed for the simulation of locomotion in complex environments. Handling dynamic agents during path execution is also important and several approaches have been proposed: elastic roadmaps [6], multi agent navigation graphs [26], etc. Avoidance of dynamic obstacles has also been solved in a reactive way, for instance with the use of *velocity obstacles* [2, 5]. Hardware acceleration has also been extensively applied [12] for improving the computation times in diverse algorithms. Although these methods are most suitable for grid–based approaches, methods for efficiently computing Delaunay triangulations using GPUs have also been developed [22].

Among the several approaches for computing navigation queries, this paper focuses on summarizing techniques which are only based on triangulations.

3 The Symedge Data Structure for Mesh Representation

The algorithms presented in this paper require a data structure able to represent planar meshes and to encode all adjacency relations between the mesh elements in constant time. The data structure used here follows the adjacency encoding strategy of the quad–edge structure [9] and integrates adjacency operators and attachment of information per element similarly to the half–edge structure [18]. The obtained data structure is called *Symedge*, and its main element represents an oriented edge which is always symmetrical to the other oriented edge adjacent to the same edge. Oriented edges in this representation are hence called symedges, and each one will always be adjacent to only one vertex, one edge, and one face.

Each symedge keeps a pointer to the next symedge adjacent to the same face, and another pointer to the next symedge adjacent to the same vertex. The first pointer is accessed with the $nxt()$ operator and the second with the $rot()$ operator, since it has the effect of rotating around the adjacent vertex. These operators rely on a consistent counter-clockwise orientation encoding. In addition, three optional pointers are stored in each symedge for quick access to the adjacent vertex, edge and face elements, which are used to store user–defined data as needed. These pointers are accessed with operators $vtx()$, $edg()$, and $fac()$, respectively. Figure 2-left illustrates these operators.

Note that the two described primitive adjacency operators are enough for retrieving all adjacent elements of a given symedge in constant time and additional operators can be defined by composition. For instance operator $sym()$ is defined as $sym() = nxt() \to rot()$ for accessing the symmetrical symedge of a given symedge s. Inverse operators are also defined: $pri() = nxt()^{-1} = rot() \to sym()$, and $ret() = rot()^{-1} = sym() \to nxt()$. Also note that $sym()^{-1} = sym()$. Figure 2 illustrates all the mentioned element retrieval and adjacency operators.

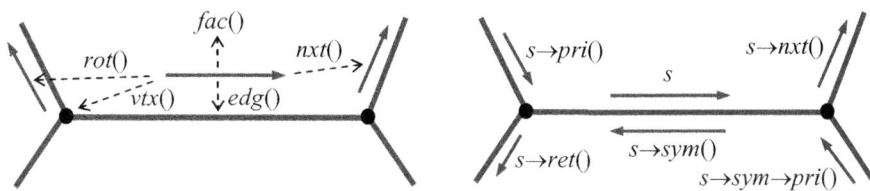

Fig. 2. Pointers are stored per symedge for fast retrieval of adjacent information (left), and several adjacency operators are defined for accessing any adjacent symedge in constant time (right)

The described symedge structure includes many additional utilities useful for the construction of generic meshes. In particular, construction operators are also included as a safe interface to manipulate the structure and Mäntylä's Euler operators [18] are implemented as the lowest–level interface. In a previous work [15], the mentioned *simplified quad–edge* structure is equivalent to the symedge structure described here. The benchmark performed in this previous work indicates that the symedge structure is among the fastest ones for describing general meshes. Although the algorithms discussed here mainly rely on triangulated meshes, using a generic structure has several advantages, in particular for the correct description of intermediate meshes during operations, and for the correct description of generic outer borders. The algorithms described in this paper have been implemented using the described symedge data structure, which is therefore also included in the available implementation.

4 Mesh Construction and Maintenance

Let $S = \{s_1, s_2, ..., s_n\}$ be a set of n input segments describing the polygonal obstacles in a given planar environment. Segments in S may be isolated or may share endpoints forming closed or open polygons. The input segments are also called constraints, and the set of all their endpoints is denoted as $\mathcal{P}$.

A suitable navigation mesh can be obtained with the Constrained Delaunay Triangulation (CDT) of the input segments. Let T be a triangulation of $\mathcal{P}$, and consider two arbitrary vertices of T to be visible to each other only if the segment connecting them does not intercept the interior of any constraint.

Triangulation T will be the CDT of S if 1) it enforces the constraints, i.e., all segments of S are also edges in T, and 2) it respects the Delaunay Criterion, i.e., the circumcircle of every triangle t of T contains no vertex in its interior which is visible from all three nodes of t.

The first step to build a CDT therefore consists of identifying the segments delimiting obstacles in a given environment. Sometimes these segments will already be available, but very often designers will specify them by hand. One main difficulty in the process is that most CDT implementations will require a clean input segment set. Instead, the implemented solution chooses to handle self–intersections, overlaps, and duplicated vertices automatically as constraints are inserted in the triangulation. For example, Figure 3 shows the segments modeled by a designer to represent the walls of an apartment. The segments were intuitively organized in rectangles but with several intersections and overlapping parts. Nevertheless a correct CDT can still be obtained.

The employed corrective incremental insertion of constraints is described in a previous work [14]. Note that the alignment problem is in particular important. For instance if two adjacent walls do not precisely share common vertices, a non–existing gap will be formed. In order to automatically detect and fix such possible gaps, the whole CDT construction uses a user–provided ϵ value and performs two specific corrective tests for each new segment inserted in the CDT: 1) if the distance between a new vertex and an existing one is smaller than ϵ, the new vertex is not inserted and instead the

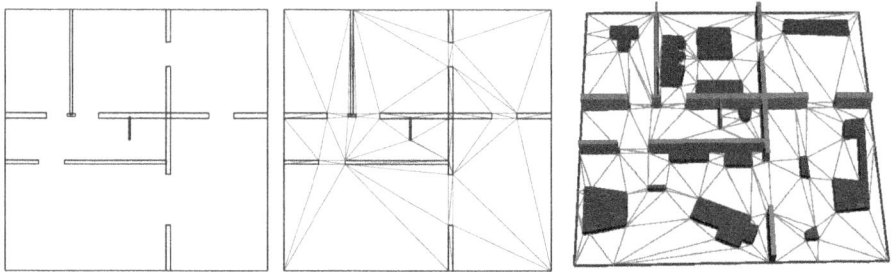

Fig. 3. Given the input set S of segments delimiting a given environment (left), $CDT(S)$ provides a suitable navigation mesh for several queries (center). Note that included validity tests [14] are able to automatically handle overlapping and intersecting constraints. Additional obstacles in the environment can also be incrementally inserted in the CDT as needed (right).

existing one becomes part of the input segment currently being inserted, and 2) if the distance between a new vertex and an already inserted segment (a constrained edge in the current CDT) is smaller than ϵ, the vertex is projected to the segment and its insertion precisely subdivides the segment in two collinear sub–segments. This ϵ–based approach for cleaning the input data also represents a way to improve the robustness of the geometric algorithms involved during the CDT construction. However robustness cannot be guaranteed for all types of input sets only using these two tests. Still, it seems to be possible to extend the approach to handle any possible situation. Note that this corrective approach for robustness is fundamentally different than addressing robustness only in the involved geometric computations, which is the usual approach in CDT implementations targeting mesh generation for finite element applications [23].

Given that the input segment set $\mathcal{S}$ can be correctly handled, the mesh obtained with $CDT(\mathcal{S})$ will always well conform to the obstacles. If only closed obstacles are represented, each triangle of the mesh will be either inside or outside each obstacle. Note that it is also possible to represent open polygons or simple segments, and therefore the representation is flexible to be used in diverse situations (see Figure 7). The obtained $CDT(\mathcal{S})$ is suitable for all operations described in this paper, except for the determination of paths with clearance, which will require additional properties leading to the introduction of the Local Clearance Triangulation $LCT(\mathcal{S})$, as discussed in Section 7.

Obstacles can also at any point be removed and re–inserted in $CDT(\mathcal{S})$. This is possible by associating with each inserted segment an id which can be later used to identify segments to be removed. The correct management of ids is described in detail in [14]. Removal of segments only involves local operations, however if the segment is long and connects to most of other segments in the triangulation, the removal may be equivalent to a full re–triangulation of the environment. In any case, the ability to efficiently update the position of small obstacles is often important. For example, Figure 3-right shows the apartment environment with additional obstacles representing some furniture. The ability to update the position of furniture or doors as the simulation progresses is important in many situations.

5 Agent Placement and Avoidance

Once the mesh representation of the environment is available, one particular problem which often appears is to efficiently place several agents in valid initial locations. Usual approaches will often make use of spatial subdivisions to keep track of the agents already inserted in different parts of the environment, in order to locally verify the validity of new agents being inserted. An efficient approach only based on the maintenance of a Delaunay triangulation of the agents as they are inserted is also possible.

Let r be the radius of the circle representing the agent location and let T be a Delaunay triangulation (of points) being built. First, T is initialized with (usually four) points delimiting a region containing the entire environment, with a margin of $2r$ space. Candidate locations are then sampled at random or following any scripted distribution strategy. Each candidate location p is only inserted in T if no vertices of T are closer to p than $2r$. The triangle t containing p is then determined by efficient point location routines [14, 23], and all edges around t are recursively visited for testing if their vertices respect the distance of $2r$ from p. Adjacency operators are used to recurse from the seed edges (the edges of t) to their neighbor edges, and marking of visited vertices will avoid overlaps. When all edges closer to p than $2r$ are visited with no illegal vertices found, then p is inserted as a new vertex of the triangulation and a new agent of radius r can be safely inserted at p without intersection with other agents. If the environment also has obstacles, an additional similar recursive test is performed to check if the circle centered at p with radius r does not intersect any constrained edge represented in the navigation mesh of the environment. See Figure 4 for examples.

Agents of different sizes can also be handled by storing the size of each agent in the vertices of the triangulation and performing vertex–specific distance tests during the recursive procedures described above. Alternatively, the recursive test can be avoided by first inserting each candidate point as a new vertex v in T, and then if an adjacent vertex v is too close, v is removed from T and a new candidate location is processed.

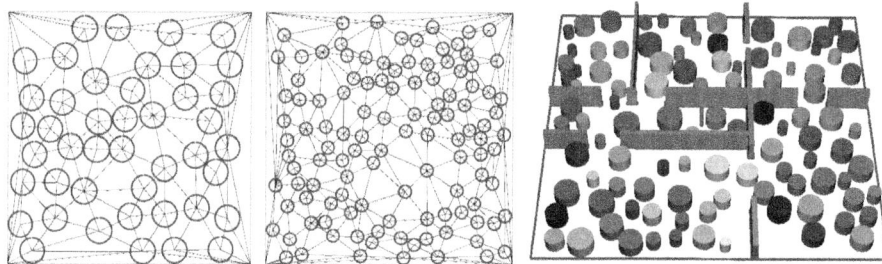

Fig. 4. The placement of non-overlapping agents with a given radius can be efficiently performed with a Delaunay triangulation tracking the inserted locations. Each location is also tested against the obstacles represented in the navigation mesh in order to achieve valid placements in the given environment.

After agents are correctly placed in valid locations navigation modules can then take control of the agents. The described Delaunay placement strategy can also be extended to efficiently perform collision avoidance strategies between the agents. One typical scenario is when agents are following given free paths in respect to the static environment while reactively avoiding the other agents on the way. For that, each agent needs to know the location of the closest agents around it at all times. The vertices of the initial Delaunay triangulation used to place the agents can then be updated as the agents move, such that each agent can quickly query the location of all agents around it. The key element of this strategy is to efficiently update the vertices of the triangulation. Fortunately there are several known algorithms able to track the position of moving vertices and only perform topological changes (of $O(n \log n)$ cost) when needed [1].

6 Ray–Obstacle Intersection Queries

Another important class of navigation queries is related to the simulation of sensors. Sensors are useful in a number of situations: for simulating laser sensors attached to robotic agents, for obtaining a simplified synthetic vision module, for querying visibility length along given directions, for aiming and shooting actions, etc. Figure 5 shows the example of a generic ray–obstacle intersection query. In this example a ray direction is given, and the ray query can be computed as follows. First, the edge e_0 first crossing the ray is determined by testing among the three edges of the triangle containing the ray source point. Then, the other edges on the next triangle adjacent to e_0 are tested for intersection and the next intersection edge e_1 is determined. The process continues until a given number of constrained edges are crossed or until a given ray length is reached. In most cases only the first crossing is important, but the algorithm can compute any number of crossings, as showed in Figure 5. Several extensions can be easily designed, for example for covering a cone sector, or a full circular region around the agent.

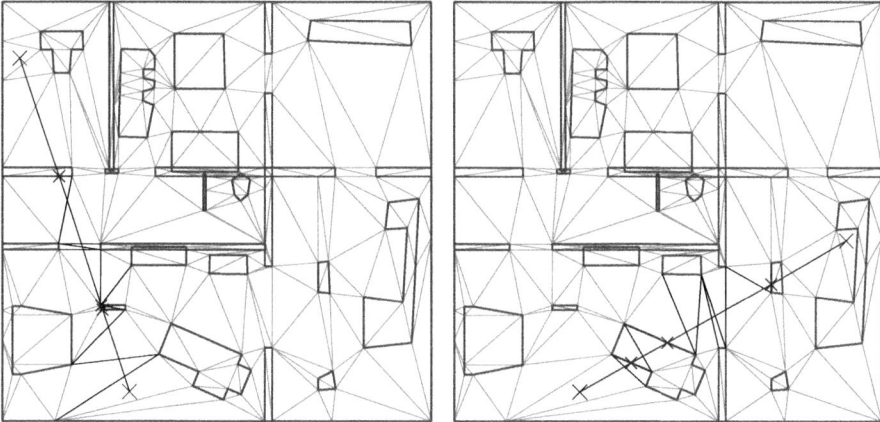

Fig. 5. In both examples, the illustrated ray–obstacle intersection query starts at the marked bottom location and identifies the first three obstacle intersections. All traversed edges are marked in black and the final top location represents the length of the query.

7 Path Planning and Paths with Clearance

Although $CDT(S)$ is already able to well represent environments, an additional property is required for enabling the efficient computation of paths with arbitrary clearance. This property is called the *local clearance property* [13] and will guarantee that only local clearance tests are required during the search for paths with clearance. Its construction starts with the $CDT(S)$, and then refinement operations are performed until the local clearance property is enforced for all triangle traversals in the mesh. The obtained mesh is a *Local Clearance Triangulation* (LCT) of the input segments.

Once $T = LCT(S)$ is computed, T can be efficiently used for computing free paths of arbitrary clearance. Let $\mathbf{p}$ and $\mathbf{q}$ be two points in $\mathbb{R}^2$. A non–trivial free path between $\mathbf{p}$ and $\mathbf{q}$ will cross several triangles sharing unconstrained edges, and the union of all traversed triangles is called a *channel*. A path of r clearance is called *locally optimal* if 1) it remains of distance r from all constrained edges in T and 2) it cannot be reduced to a shorter path of clearance r on the same channel. Such a path is denoted π_r, and its channel C_r. Note that a given path π_r joining two points may or not be the globally shortest path. If no shorter path of clearance r can be found among all possible channels connecting the two endpoints, the path is then the *globally optimal* one.

The key issue for finding a path π_r is to search for a channel C_r which guarantees that there is enough clearance in all traversed triangles. A graph search over the adjacency graph of the triangulation is then performed, starting from the initial triangle containing **p**, and until reaching **q**. For each triangle traversed during this search, a precomputed clearance value will determine if that single triangle traversal is guaranteed to have clearance r. The refinement operations performed to build a LCT will guarantee that each traversal can be locally tested for clearance and thus enabling the precomputation of two clearance values per edge for testing the clearance of all possible triangle traversals. Figure 6 shows a typical problem which can occur in $CDTs$ but which will not occur in $LCTs$. Note that if a channel C_r is not found, the goal is not reachable.

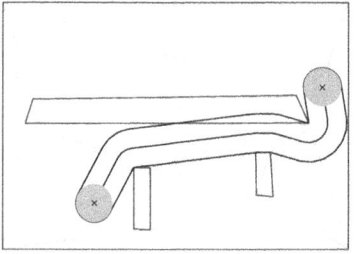 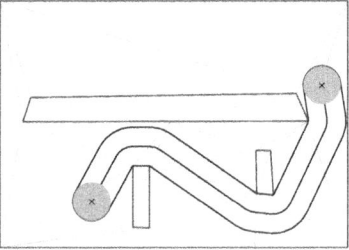

Fig. 6. Local clearance tests in CDT's cannot guarantee the correct clearance determination of paths (left). The corresponding LCT (right) will always lead to free paths with correct clearances.

Once a channel C_r of arbitrary clearance is found, its locally optimal path π_r can be computed in linear time in respect to the number of triangles in the channel. This is achieved with an extended *funnel algorithm* [10] handling clearances, which is detailed in [13].

The result is a flexible and efficient approach for path planning. The LCT can be precomputed in $O(n^2)$, and then paths of arbitrary clearance can be retrieved in $O(n \log n)$ by using a standard A* search (as implemented in [13]), or even in $O(n)$ time as the generated structure is suitable for the application of linear time planar search algorithms [25]. Figures 1 and 7 show several examples.

8 Determination of Corridors and Extensions

Note that the search for free channels (with or without clearances) during the described path planning procedure automatically determines free corridors around the computed paths. A channel C is the union of all traversed triangles and therefore the boundary of the channel will be a polygon representing a corridor containing the path. See Figure 8 for an example. Figure 8 also illustrates the computation of *extra clearances*, which deform the path in order to achieve higher clearance than the minimum required. Extra

Navigation Queries from Triangular Meshes   239

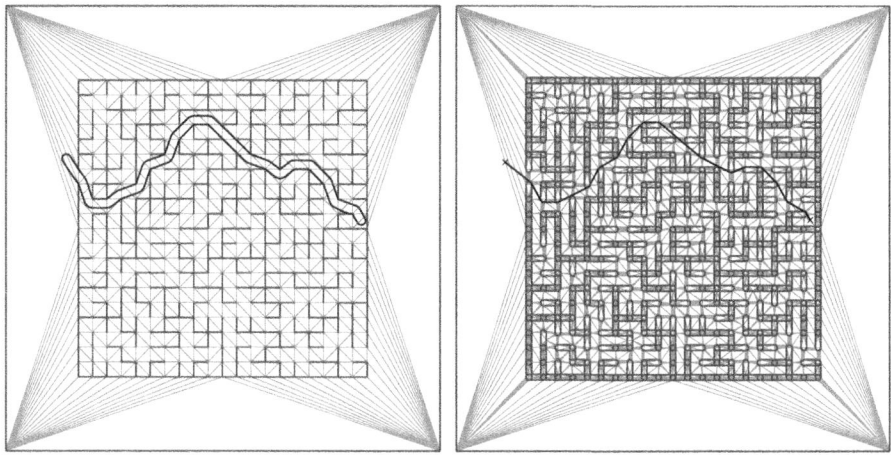

Fig. 7. The LCT of the input segments is required for computing paths with arbitrary clearance (left). Alternatively, if the clearance is constant, the environment can be inflated and paths without clearance can be extracted from the CDT of the inflated input segments (right). Also note that triangulated meshes can well represent environments described by input segments which do not form closed obstacles (left).

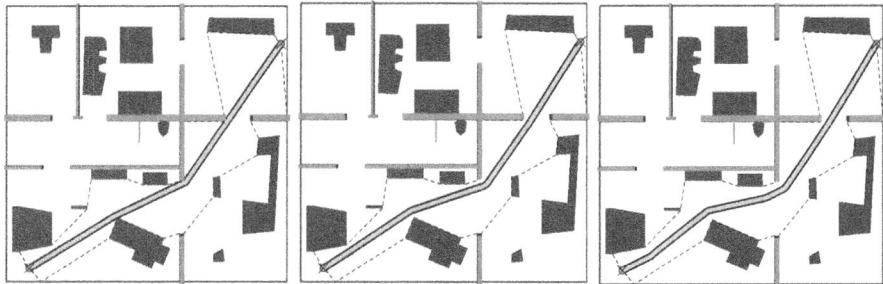

Fig. 8. Three paths with same minimum clearance but with extra clearances of 0, 0.45, and 0.9

clearances can be computed with post-optimization of obtained paths and can model a variable range of locomotion behaviors, from attentive in passages with minimum clearance to safe navigation in higher clearance areas.

Many other extensions can be devised. For instance the corridor search procedure can be optimized (significantly in certain cases) by introducing a smaller connectivity graph which excludes the triangles inside corridors, which are those that have two constrained edges and thus have only one way of being traversed. Hierarchical representations of several levels (common in grid–based approaches) can also be translated to triangle meshes.

Note that efficient path planning queries are also important for decision modules. For example, the ability to query goal reachability with different clearances and to compute lengths of obtained paths may be important for deciding which target locations to visit first in case of several choices. Many other uses of the proposed methods exist. For instance, the handling of intersections and overlaps in the input segment set can be used to perform Boolean operations with obstacles, what is useful for optimizing the navigation mesh. Finally, one important extension in many applications is the ability to model uneven terrains. Due to their irregular decomposition nature, triangulations are well suited for the representation of terrains, however each geometric test in the described procedures would have to be generalized for handling non-planar surfaces.

9 Final Remarks

This paper presented several triangulation–based methods for the efficient computation of diverse navigation queries for autonomous agents. With the growing research activity in the area and the appearance of several new development tools, triangulation–based navigation meshes can be expected to become increasingly popular.

References

1. Albers, G., Mitchell, J.S., Guibas, L.J., Roos, T.: Voronoi diagrams of moving points. Internat. J. Comput. Geom. Appl. 8, 365–380 (1992)
2. Berg, J., Lin, M., Manocha, D.: Reciprocal velocity obstacles for real-time multi-agent navigation. In: Proceedings of the International Conference on Robotics and Automation, ICRA 2008 (2008)
3. De Berg, M., Cheong, O., van Kreveld, M.: Computational geometry: algorithms and applications. Springer, Heidelberg (2008)
4. Demyen, D., Buro, M.: Efficient triangulation-based pathfinding. In: Proceedings of the 21st National Conference on Artificial Intelligence, AAAI 2006, pp. 942–947. AAAI Press, Menlo Park (2006)
5. Fiorini, L.P., Shiller, Z.: Motion planning in dynamic environments using velocity obstacles. International Journal of Robotics Research 17(7), 760–772 (1998)
6. Gayle, R., Sud, A., Andersen, E., Guy, S.J., Lin, M.C., Manocha, D.: Interactive navigation of heterogeneous agents using adaptive roadmaps. IEEE Transactions on Visualization and Computer Graphics 15, 34–48 (2009)
7. Geraerts, R.: Planning short paths with clearance using explicit corridors. In: Proceedings of the IEEE International Conference on Robotics and Automation, ICRA 2010 (2010)
8. Geraerts, R., Overmars, M.H.: The corridor map method: a general framework for real-time high-quality path planning: Research articles. Computer Animation and Virtual Worlds 18(2), 107–119 (2007)
9. Guibas, L., Stolfi, J.: Primitives for the manipulation of general subdivisions and the computation of voronoi. ACM Trans. Graph. 4(2), 74–123 (1985)
10. Hershberger, J., Snoeyink, J.: Computing minimum length paths of a given homotopy class. Computational Geometry Theory and Application 4(2), 63–97 (1994)
11. Hershberger, J., Suri, S.: An optimal algorithm for euclidean shortest paths in the plane. SIAM Journal on Computing 28, 2215–2256 (1997)
12. Hoff III, K.E., Culver, T., Keyser, J., Lin, M., Manocha, D.: Fast computation of generalized voronoi diagrams using graphics hardware. In: Proceedings of the Sixteenth Annual Symposium on Computational Geometry (2000)
13. Kallmann, M.: Shortest paths with arbitrary clearance from navigation meshes. In: Proceedings of the Eurographics / SIGGRAPH Symposium on Computer Animation, SCA (2010)

14. Kallmann, M., Bieri, H., Thalmann, D.: Fully dynamic constrained delaunay triangulations. In: Brunnett, G., Hamann, B., Mueller, H., Linsen, L. (eds.) Geometric Modeling for Scientific Visualization, pp. 241–257. Springer, Heidelberg (2003) ISBN 3-540-40116-4
15. Kallmann, M., Thalmann, D.: Star vertices: A compact representation for planar meshes with adjacency information. Journal of Graphics Tools 6(1), 7–18 (2001)
16. Lamarche, F.: TopoPlan: a topological path planner for real time human navigation under floor and ceiling constraints. Computer Graphics Forum 28 (03 2009)
17. Lozano-Pérez, T., Wesley, M.A.: An algorithm for planning collision-free paths among polyhedral obstacles. Communications of ACM 22(10), 560–570 (1979)
18. Mäntylä, M.: An introduction to solid modeling. Computer Science Press, Inc., New York (1987)
19. Mekni, M.: Hierarchical path planning for situated agents in informed virtual geographic environments. In: Proceedings of the 3rd International ICST Conference on Simulation Tools and Techniques, SIMUTools 2010, pp. 1–10 (2010)
20. Mitchell, J.S.B.: Shortest paths among obstacles in the plane. In: Proceedings of the Ninth Annual Symposium on Computational Geometry, SCG 1993, pp. 308–317. ACM, New York (1993)
21. Overmars, M.H., Welzl, E.: New methods for computing visibility graphs. In: Proceedings of the fourth Annual Symposium on Computational Geometry, SCG 1988, pp. 164–171. ACM Press, New York (1988)
22. Rong, G., Tan, T.-s., Cao, T.-t.: Computing two-dimensional delaunay triangulation using graphics hardware. In: Proceedings of the Symposium on Interactive 3D Graphics and Games, I3D (2008)
23. Shewchuk, J.R.: Triangle: Engineering a 2D Quality Mesh Generator and Delaunay Triangulator. In: Lin, M.C., Manocha, D. (eds.) FCRC 1996 and WACG 1996. LNCS, vol. 1148, pp. 203–222. Springer, Heidelberg (1996); from the First ACM Workshop on Applied Computational Geometry
24. Storer, J.A., Reif, J.H.: Shortest paths in the plane with polygonal obstacles. J. ACM 41(5), 982–1012 (1994)
25. Subramanian, S., Klein, P., Klein, P., Rao, S., Rao, S., Rauch, M., Rauch, M.: Faster shortest-path algorithms for planar graphs. Journal of Computer and System Sciences, 27–37 (1994)
26. Sud, A., Andersen, E., Curtis, S., Lin, M.C., Manocha, D.: Real-time path planning in dynamic virtual environments using multiagent navigation graphs. IEEE Transactions on Visualization and Computer Graphics 14, 526–538 (2008)
27. Wein, R., van den Berg, J.P., Halperin, D.: The visibility–voronoi complex and its applications. In: Proceedings of the Twenty-First Annual Symposium on Computational Geometry, SCG 2005, pp. 63–72. ACM, New York (2005)

Motion Parameterization with Inverse Blending

Yazhou Huang and Marcelo Kallmann

University of California, Merced

Abstract. Motion blending is a popular motion synthesis technique which interpolates similar motion examples according to blending weighs parameterizing high-level characteristics of interest. We present in this paper an optimization framework for determining blending weights able to produce motions precisely satisfying multiple given spatial constraints. Our proposed method is simpler than previous approaches, and yet it can quickly achieve locally optimal solutions without pre-processing of basis functions. The effectiveness of our method is demonstrated in solving two classes of problems: 1) we show the precise control of end-effectors during the execution of diverse upper-body actions, and 2) we also address the problem of synthesizing walking animations with precise feet placements, demonstrating the ability to simultaneously meet multiple constraints and at different frames. Our several experimental results demonstrate that the proposed optimization approach is simple to implement and effectively achieves realistic results with precise motion control.

Keywords: Motion parameterization, character animation, walk synthesis, spatial constraints.

1 Introduction

Keyframe animation and motion capture represent popular approaches for achieving high-quality character animation in interactive applications such as in 3D computer games and virtual reality systems. In particular, motion blending techniques [15, 17, 18] provide powerful interpolation approaches for parameterizing pre-defined example animations according to high-level characteristics. While direct blending of motions is able to produce fast and realistic motions, it remains difficult to achieve blendings and parameterizations able to precisely satisfy generic spatial constraints. We show that with an optimization approach we are able to always solve spatial constraints when possible, and usually less example motions are required to cover the spatial variations of interest.

Our method models each spatial constraint as an objective function whose error is to be minimized. The overall multi-objective inverse blending problem is solved by optimizing the blending weights until a locally optimal solution is reached. Solutions can be found in few milliseconds and no pre-computation of basis functions is needed. The method is therefore suitable for interactive applications and several results running in real-time are presented.

While previous work has addressed the maintenance of spatial properties in a single motion interpolation step [11, 15, 18], we focus on optimizing blending weights until best meeting multiple spatial constraints. Our approach has the additional flexibility of modeling spatial constraints with objective functions which are independent of the abstract space used by the motion blending. Generic spatial constraints can be handled and Inverse Kinematics problems can also be solved based on motion blending. We have in particular recently applied the presented framework to an interactive virtual reality training application [5], and the obtained results were very effective.

This paper demonstrates our methods in three scenarios: pointing to objects, pouring water and character locomotion. The spatial constraints of inverse blending are modeled differently for each scenario. As a result, our interactive motion modeling framework allows animators to easily build a repertoire of realistic parameterized human–like motions (gestures, actions, locomotion, etc) from examples which can be designed by hand or collected with motion capture devices.

2 Related Work

Several approaches to motion interpolation have been proposed involving different techniques such as: parameterization using Fourier coefficients [22], hierarchical filtering [4], stochastic sampling [23], and interpolation based on radial basis functions (RBFs) [17]. Our motion blending framework is closely related to an extension of the verbs and adverbs system which performs RBF interpolation to solve the Inverse Kinematics (IK) problem [18]. RBFs can smoothly interpolate given motion examples and the types and shapes of the basis functions are optimized in order to better satisfy the constraint of reaching a given position with the hand.

More generically, spatial properties such as feet sliding or hand placements are well addressed by the geostatistical interpolation method [15], which computes optimal interpolation kernels in accordance with statistical observations correlating the control parameters and the motion samples. Another approach for improving the maintenance of spatial constraints is to adaptively add pseudo–examples [11, 18] in order to better cover the continuous space of the constraint. This random sampling approach however requires significant computation and storage in order to meet constraints accurately and is not suited for handling several constraints.

In all these previous methods spatial constraints are only handled as part of the motion blending technique employed, i.e., by choosing sample motions which are close to the desired constraints and then using the abstract interpolation space to obtain motion variations which should then satisfy the constraints. Another possible technique sometimes employed is to apply Inverse Kinematics solvers in addition to blending [6, 18], however risking to penalize the obtained realism.

The work on mesh–based IK [20] does address the optimization of blending weights for the problem of blending example meshes. However, although our approach is simple and intuitive, no specific previous work has specifically analyzed and reported results applied to skeletal motion, and in particular also simultaneously solving multiple constraints and at different frames.

The Scaled Gaussian Process Latent Variable Model (SGPLVM) [9] provides a more specific framework targeting the IK problem which optimizes interpolation kernels specifically for generating plausible poses from constrained curves such as positional trajectories of end–effectors. The approach however focuses on maintaining constraints described by the optimized latent spaces. Although good results are obtained, constraints cannot be guaranteed to be precisely met.

The presented approach can be seen as a post–processing step for optimizing a given set of blending weights, which can be initially computed by any motion blending technique. We demonstrate in this paper that our optimization framework is able to address any kind of constraints without even the need of specifying an abstract parameterization space explicitly. Only error functions for the spatial constraints are necessary in order to optimize the blending weights using a given motion interpolation scheme.

We also apply our method for parameterization of locomotion, which is a key problem in character animation. Many methods have been previously proposed for finding optimal solutions for path following [12, 14], for reaching specified locomotion targets [10, 19], or also for allowing interactive user control [1, 14, 21]. Most of these works combine motion blending techniques with motion graphs [2, 3, 8, 12, 19] and can then generate different styles of locomotion and actions. Different than these methods, we give specific attention to precisely meet specified feet placements. The geostatistical interpolation method [15] reduces feet sliding problems but still cannot guarantee to eliminate them. Other precise feet placement techniques [7, 13] are available, however not based on a generic motion blending approach.

In conclusion, diverse techniques based on motion blending are available and several of these methods are already extensively used in commercial animation pipelines for different purposes. Our work presents valuable experimental results demonstrating the flexibility and efficiency of a simple optimization framework for solving inverse blending problems involving multiple spatial constraints. Our approach is effective and easy to implement, and was first described in [5], where it was applied to an interactive virtual reality training application. The formulation is again described here for completeness purposes, and then several extensions and new results are presented to effectively model diverse parameterized upper–body actions, and also to model parameterized walking animations with precise footstep placements.

3 Inverse Blending

Given a set of similar and time–aligned motion examples, we first employ a traditional RBF motion interpolation scheme to compute an initial set of blending weights. These weights are then optimized to meet a given constraint C.

Each motion M being interpolated is represented as a sequence of poses with a discrete time (or frame) parameterization t. A pose is a vector which encodes the root joint position and the rotations of all the joints of the character. Rotations are encoded with exponential maps but other representations for rotations (as quaternions) can also be used. Each constraint C is modeled with a function $e = f(M)$, which returns the error evaluation e quantifying how far away the given motion is from satisfying constraint C. We first select the k motions which are the ones best satisfying the constraints being solved. For example, in a typical reaching task, the k motion examples having the hand joint closest to the target will be selected. The k initial motions are therefore the ones with minimal error evaluations. The initial set of blending weights w_j, $j = \{1, \ldots, k\}$, are then initialized with a RBF kernel output of the input $e_j = f(M_j)$. We constrain the weights to be in interval $[0, 1]$ in order to stay in a meaningful interpolation range and we also normalize them to sum to 1. Any kernel function can be used, as for example kernels in the form of $exp^{-\|e\|^2/\sigma^2}$. In this work we do not attempt to optimize kernel functions in respect to the constraints [15, 17] and instead we will optimize the weights independently of the interpolation method. Our interpolation scheme computes a blended motion with:

$$M(w) = \sum_{j=1}^{k} w_j M_j, \quad w = \{w_1, \ldots, w_k\}. \tag{1}$$

In order to enforce the given constraint C, our goal is to find the optimal set of blending weights w, which will produce the minimum error e^* as measured by the constraint error function f:

$$e^* = min_{w_j \in [0,1]} \ f\left(\sum_{j=1}^{k} w_j M_j\right). \tag{2}$$

The presented formulation can easily account for multiple constraints by combining the error metric of the given constraints in a single weighted summation. In doing so we introduce two more coefficients for each constraint $C_i, i = \{1, \ldots, n\}$: a normalization coefficient n_i and a prioritization coefficient c_i. The goal of n_i is to balance the magnitude of the different constraint errors. For example, positional constraints depend on the used units and in general have much larger values than angular values in radians, which are typically used by orientation errors. Once the normalization coefficients are set, coefficients $c_i \in [0, 1]$ can be interactively controlled by the user in order to vary the influence (or priority) of one constraint over the other.

The result is essentially a multi–objective optimization problem, with the goal being to minimize a new error metric composed of the weighted sum of all constraints' errors:

$$f(M(w)) = \sum_{i=1}^{n} (c_i \ n_i \ f_i \ (M(w))). \tag{3}$$

Independent of the number of constraints being addressed, when constraints are fully satisfied, $e^* \to 0$.

4 Action Parameterization with Inverse Blending

Figure 1 presents several examples obtained for parameterizing upper–body actions. Different types of spatial constraints were used. Constraint C_{pos} is a 3-DOF positional constraint which requires the end–effector (hand, finger tip, etc) to precisely reach a given target location. Constraint C_{line} is a 2-DOF positional constraint for aligning the hand joint with a given straight line in 3D space. Constraint C_{ori} is a 1 to 3 DOFs rotational constraint which requires the hand to comply to a certain given orientation. Note that all these constraints are only enforced at one given frame of the motion. Constraints can also be combined in order to allow additional ways of parameterizing motions. For

Fig. 1. This figure presents several results obtained by inverse blending. A C_{line} constraint is used for precisely pointing to distant objects (a) and for pouring exactly above the teapot (c). Positional C_{pos} and rotational C_{ori} constraints are used for pin–pointing a button on the telephone (b). Note that the finger tip precisely reaches the button, and the hand orientation matches that of the shown x-axis. Constraints C_{pos} and C_{ori} are also used for achieving precise grasping motions (d).

example by combining C_{line} and C_{ori}, precise grasping targets can be achieved (Figure 1-d), and different hand orientations can be obtained when pin–pointing buttons (Figure 1-b).

5 Parameterized Character Locomotion with Inverse Blending

This section demonstrates our inverse blending framework for generating parameterized walking motions with precise control of feet placements for character navigation.

First, two sets of motion examples are prepared with clips obtained from motion capture. The first set $\boldsymbol{R}_m$ consists of right foot stepping motions. Each example motion $M_r \in \boldsymbol{R}_m$ represents one full step forward with the right foot while the left foot remains in contact with floor as the support foot, see Figure 2-a. The second set of example motions $\boldsymbol{L}_m$ is prepared in the same way but containing stepping examples for left foot.

The motions in both sets contain many variations, e.g. step length, step direction, body orientation, velocity, root joint position, etc. These should well span the variations of interest, which are to be precisely parameterized by inverse blending. Figure 2-b illustrates how some of the motions in our database look like. No alignment of the motions is needed and the variations will actually be explored by the inverse blending optimization in order to reach any needed alignments on–line. We also mirror the example motions in both sets in order to guarantee the same number of examples (and variations) are available in each set.

As we are interested in precisely controlling each footstep location during walking, the length and direction of each step is parameterized while the support foot contact on the floor is maintained. Let θ^e be the orientation of the support foot at the starting frame of one stepping motion example (rotational constraint). Let v^s and v^e be vectors encoding the position of the stepping foot in respect to the support foot at the start and at the end frames respectively (positional constraints). Figure 2-c illustrates these parameters for one left step. Each stepping motion of interest is then specified as a function of these parameters with $M(v^s, v^e, \theta^e)$, which will be obtained by inverse

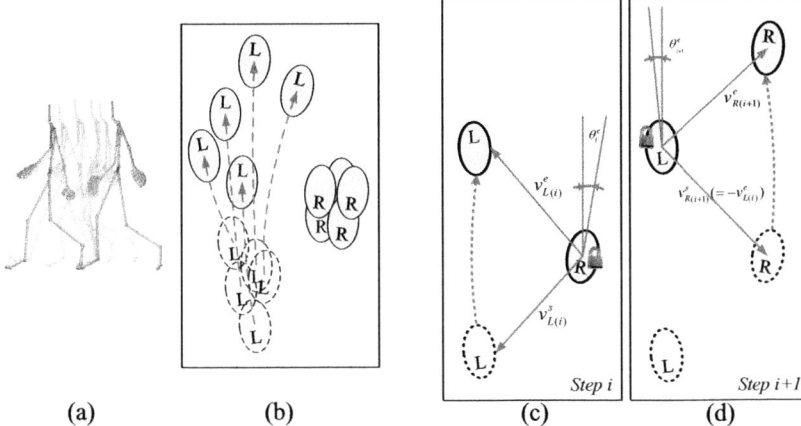

Fig. 2. (a) shows snapshots of one right foot stepping motion from R_m example set. (b) is a top–down footprint view of several left foot stepping motions used in L_m example set. The footprints are from diverse walk segments and do not need to be aligned. (c) shows the constraints for computing step i: θ^e is the rotational constraint for the support foot (with lock icon), v^s and v^e are positional constraints for the stepping foot at the start and end of the motion respectively. (d) shows the constraints for step $i + 1$, which immediately follows step i.

blending procedures based only on the stepping examples available in the R_m and L_m motion sets.

With the given constraints v^s, v^e, θ^e described above, the process for obtaining $M(v^s, v^e, \theta^e)$ is composed of 4 phases, as described in the next paragraphs.

Phase 1: the set of blending weights w^s is computed by inverse blending such that the stepping foot respects the positional constraint v^s at the start frame t^s. As these weights are computed to meet constraints v^s we use the notation $w^s(v^s)$ for the obtained blending weights.

Phase 2: we solve a new inverse blending problem for determining the blending weights w^e at the end frame t^e in order to meet two constraints: the positional constraint v^e for the stepping foot and the rotational constraint θ^e for the support foot. Therefore the obtained weights $w^e(v^e, \theta^e)$ produce an end posture with the stepping foot reaching location v^e, while the support foot respects the orientation specified by θ^e.

Phase 3: the average lengths l_{avg} of the example motions in phase 1 and 2 is used to time–align the blended motions. The blending weights used to produce the required stepping motion is finally obtained as a function of the frame time t, such that $w(t)$ $=interp(w^s(v^s), w^e(v^e, \theta^e), t)$, $t \in [0, l_{avg}]$. The interpolation function $interp$ employs a smooth in and out sine curve and each of the motions are time warped to l_{avg} in order to cope with variations of step lengths and speeds in the used motion set. The final parameterized stepping motion is then obtained with $M(v^s, v^e, \theta^e) = w(t)\Sigma_{i=1}^{k} M_i$. This process is illustrated in Figure 3.

Phase 4: this phase consists of a velocity profile correction [24] in order to maximally preserve the overall quality of the original motions since several blending operations have been performed at this point. For that we select the root joint velocity profile of the motion example which gave most contribution in the inverse blending procedures. The time parameterization of $w(t)$ is then adapted on the fly in order to obtain a motion with the root joint velocity profile matching the selected reference profile. Figure 3 bottom–right exemplifies the root increment against frames during a two–step walking sequence showing how root velocity changes over time. This has been proven to well preserve the quality of the obtained results.

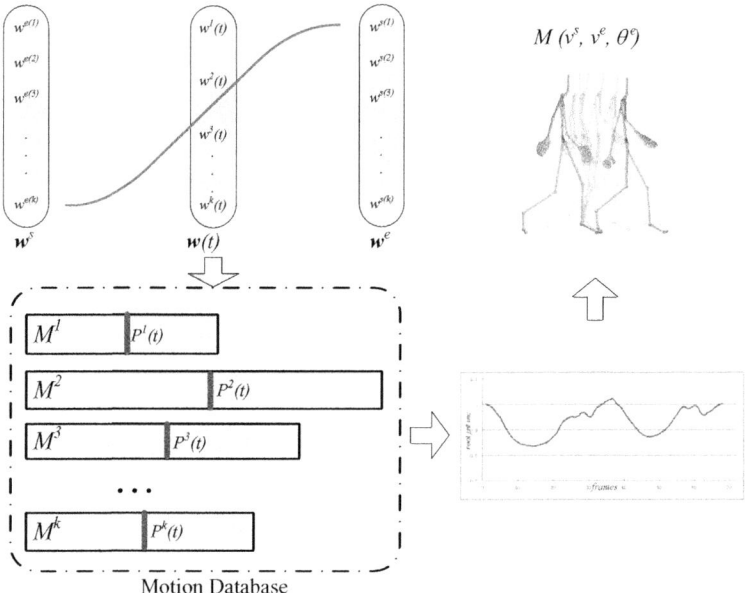

Fig. 3. The motion examples selected by the inverse blending $M^i (i = 1 \sim k)$ are blended with interpolated weights $w(t)$ which ensure spatial constraints both at the start and end frame of the motions. Velocity profiles are adapted on–line to preserve the original quality and realism.

The procedure described above is applied each time a stepping motion has to be generated. For producing stepping motions for the right foot, $M_R(v^s, v^e, \theta^e)$ is obtained by using the example motions from $\boldsymbol{R}_m$. Left foot stepping motions $M_L(v^s, v^e, \theta^e)$ are similarly obtained using examples from $\boldsymbol{L}_m$. As a result a walking sequence achieving precise feet placements at each step can be obtained with the following concatenation of alternating steps: $M_L(v_1^s, v_1^e, \theta_1^e)$, $M_R(v_2^s, v_2^e, \theta_2^e)$, $M_L(v_3^s, v_3^e, \theta_3^e)$, $\cdots$.

During each stepping motion in the sequence above, the character is translated at every frame to make sure the support foot does not slide on the floor (i.e. its location and orientation are maintained), this will essentially make the character walk forward. At the end of each stepping, the support foot changes, and its location and orientation are updated, ready for the following step. With this, the common problem of feet–sliding is here eliminated.

When computing each stepping motion, we make constraint v^s_{i+1} equal to $-v^e_i$ from the previous step (see Figure 2-c and 2-d), for smooth transition between step i and step $i + 1$. The negation appears because the support foot and stepping foot are swapped from step i to $i + 1$.

Figure 4 shows the end posture P^e_L (thick red line) of the left step $M_L(v^s_i, v^e_i, \theta^e_i)$ and the start posture P^s_R (thick green line) of the right step $M_R(v^s_{i+1}, v^e_{i+1}, \theta^e_{i+1})$. With $v^s_{i+1} = -v^e_i$, inverse blending generates postures P^e_L and P^s_R matching the constraints and with body postures which are very close to each other. The small difference in the body posture is handled by smoothly concatenating the stepping motions with a brief ease–in blending period from M_L going into M_R, achieving a smooth overall transition.

In the examples presented in this paper we have used only six stepping motion examples in each motion set, and yet the described inverse blending procedure can precisely control each foot placement within a reasonable range. If larger databases are used, a wider range for the constraints can be specified. Figure 5 shows several results obtained by our real–time walking control application. Several animation sequences are also available in the video accompanying this paper.

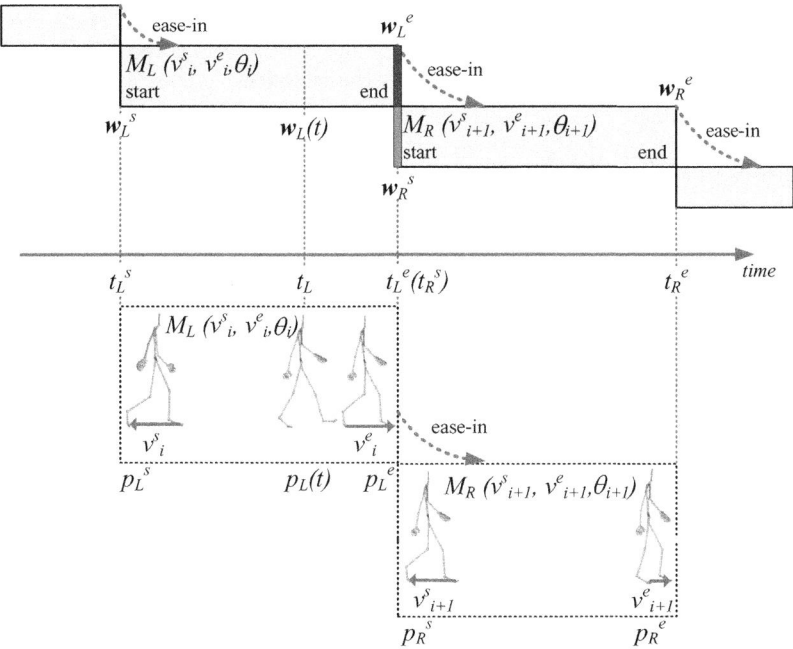

Fig. 4. This figure illustrates the generation of $M_L(v_n)$, and the concatenation of $M_L(v_n)$ and $M_R(v_{n+1})$ for achieving the final walking synthesis

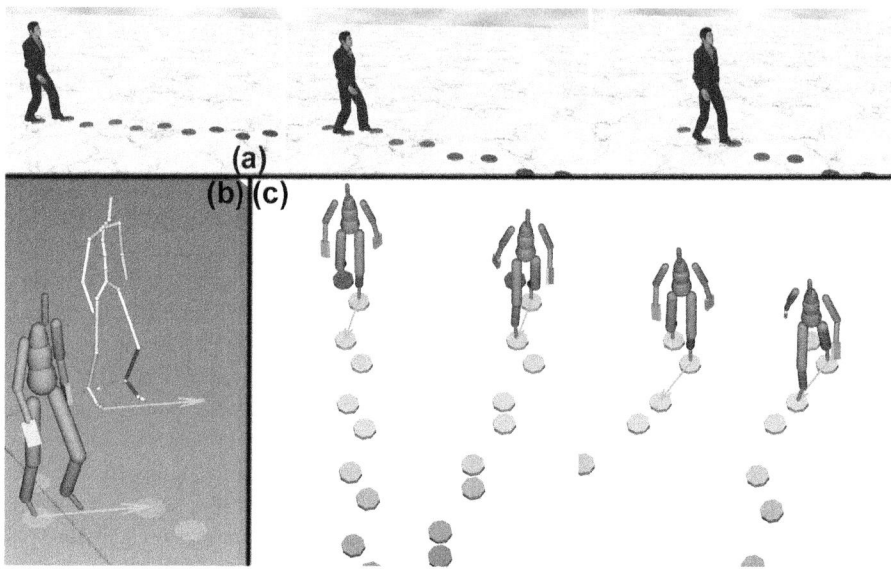

Fig. 5. The figure presents several snapshots of obtained walking motions where each step precisely meets the given feet targets (a and c). The targets for each step can be adjusted on the fly achieving a controllable locomotion generator with precise feet placements. The generation of the stepping motions is illustrated in the lower-left image (b), where the gray skeleton shows the inverse blending solution for the left foot, prior to concatenation.

6 Results and Discussion

With suitable example motions in a given cluster, inverse blending can produce motions exactly satisfying given spatial constraints and fast enough for real–time applications. Several of the figures in this paper illustrate the many experiments successfully conducted in different scenarios. To evaluate the performance of our method, a reaching task was designed to measure the errors produced by our method against a single RBF interpolation, with the 16 reaching motion database from Mukai et.al [15]. A total of 144 reaching targets (shown as yellow dots in Figure 6, each specifying a 3-DOF C_{pos} constraint) were placed evenly on a spherical surface within reach of the character. The end locations of the hand trajectory in 16 example motions are shown as gray dots.

For each reaching target we first apply standard RBF interpolation alone to generate a reaching motion and record the final hand position where the character actually reaches. These 144 final positions were used to construct a mesh grid, which is shown in Figure 6-a. Each triangle on the mesh is colored in respect to the average errors from its vertices, representing the distance error between the final hand positions and their corresponding reaching targets. We then use our inverse blending optimization to perform the same tasks, and the constructed mesh is shown in Figure 6-b. The reaching motions generated by inverse blending can precisely reach most of the targets. Errors measured are practically zero across most of the mesh, and increase only at the boundary of the surface. In this specific task, the radius of the spherical surface was set to $80cm$, and both methods used eight example motions from the database ($k = 8$) for computing each reaching task.

Additional experiments were performed by varying the radius of the spherical surface to be $65, 70, 75, 80, 85$ and $90cm$. Again a total of 144 reaching targets were generated on the spherical surfaces, covering a large volume of the workspace. These spherical surfaces are shown in Figures 6-c and 6-d. The constructed meshes by inverse blending are shown in Figure 6-d, and the results obtained with the RBF interpolation are shown in Figure 6-c. It is possible to note that the inverse blending optimization produces a smooth mesh very well approximating the yellow dots, and the errors produced by our method are clearly much lower, with most areas in pure blue.

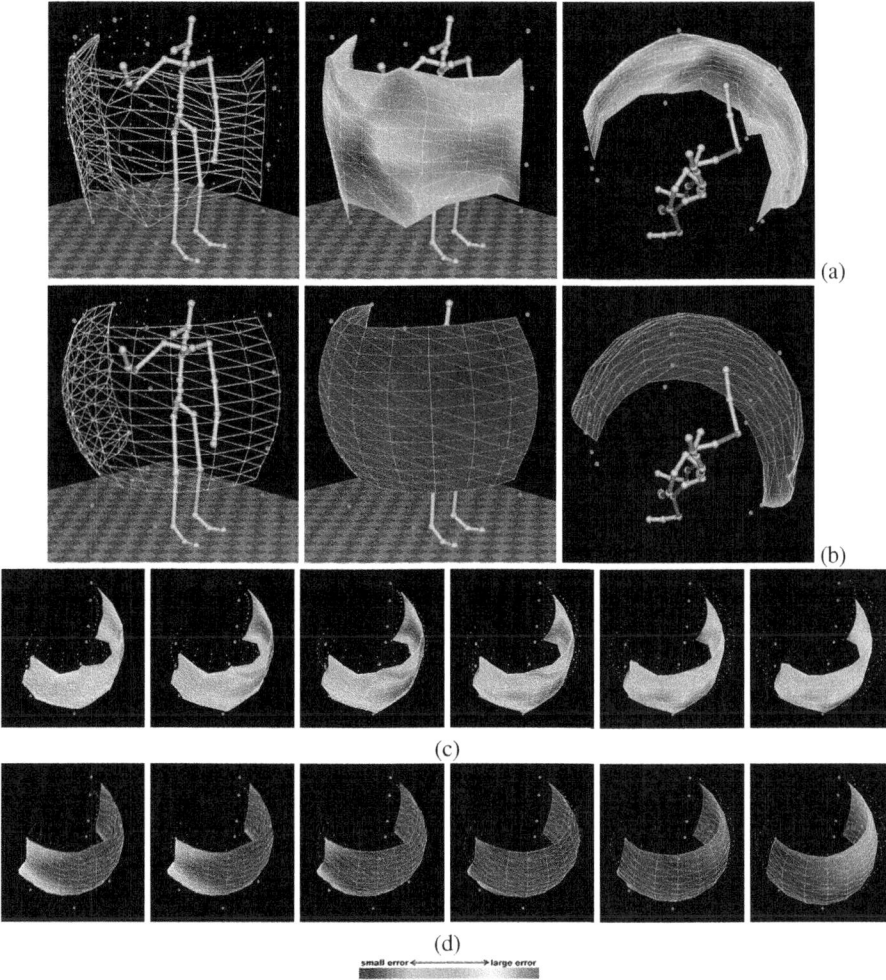

Fig. 6. Error evaluations. The meshes constructed by a standard RBF interpolation (a and c) result in much larger errors than by our inverse blending optimization (b and d), which most often produces no error.

Using standard optimization techniques [16] our inverse blending problems could be solved under 1.16 ms of computation time on average, with worse cases taking 2.29 ms (with a non–optimized single core code on a Core 2 Duo 2.13 GHz). Three scenarios (character performing pointing, pouring water and walking with precise feet placements) were used for this evaluation, with each scenario solving 5000 inverse blending problems towards random placements.The approach is therefore suitable for real–time applications, and in addition, it does not require pre–processing of the motion database, making it suitable for systems interactively updating the motion database (as in [5]).

In terms of limitations, two main aspects have to be observed. First, the ability of enforcing constraints greatly depends on the existing variations among the motion examples being blended. The number of needed example motions also depend on the size of the target volume space. For example, our walk generator can produce good results with only 6 stepping example motions (6 for left foot stepping, mirrored to become 12 for both feet) due to great variations available in the motions. However more example motions are typically needed to well cover a large reaching or pointing volume, and we have used 35 example motions in some cases. The second limitation, which is related to the first, is that the computational time required for finding solutions will also depend on the quality and number of motion examples (the k value). However, as shown in our several examples, these limitations are easy to address by appropriately modeling example motions, and balancing the coverage vs. efficiency trade–off specifically for each action being modeled.

7 Conclusions

We have presented an optimization framework for satisfying spatial constraints with motion blending. Our approach is simple and can handle any type of spatial constraints. Several different actions (pointing, grasping, pouring, and walking) were successfully modeled and parameterized with precise placement of end–effectors. Our inverse blending framework has therefore shown to be a simple and powerful tool for achieving several useful motion parameterizations. We believe that our overall framework can significantly improve the process of modeling full–body motions for interactive characters.

Acknowledgments. This work was partially supported by NSF Awards IIS-0915665 and CNS-0723281, and by a CITRIS seed fund.

References

1. Abe, Y., Liu, C.K., Popović, Z.: Momentum-based parameterization of dynamic character motion. In: 2004 ACM SIGGRAPH/EUROGRAPHICS Symposium on Computer Animation, SCA 2004, pp. 173–182. Eurographics Association, Aire-la-Ville (2004)
2. Arikan, O., Forsyth, D.A.: Synthesizing constrained motions from examples. Proceedings of SIGGRAPH 21(3), 483–490 (2002)
3. Arikan, O., Forsyth, D.A., O'Brien, J.F.: Motion synthesis from annotations. ACM Transaction on Graphics (Proceedings of SIGGRAPH) 22(3), 402–408 (2003)
4. Bruderlin, A., Williams, L.: Motion signal processing. In: SIGGRAPH 1995, pp. 97–104. ACM Press, New York (1995)

5. Camporesi, C., Huang, Y., Kallmann, M.: Interactive motion modeling and parameterization by direct demonstration. In: Proceedings of the 10th International Conference on Intelligent Virtual Agents, IVA (2010)
6. Cooper, S., Hertzmann, A., Popović, Z.: Active learning for real-time motion controllers. ACM Transactions on Graphics (SIGGRAPH 2007) 26(3) (August 2007)
7. Coros, S., Beaudoin, P., Yin, K.K., van de Pann, M.: Synthesis of constrained walking skills. ACM Trans. Graph. 27(5), 1–9 (2008)
8. Gleicher, M., Shin, H.J., Kovar, L., Jepsen, A.: Snap-together motion: assembling run-time animations. In: Proceedings of the Symposium on Interactive 3D Graphics (I3D), pp. 181–188. ACM Press, New York (2003)
9. Grochow, K., Martin, S., Hertzmann, A., Popović, Z.: Style-based inverse kinematics. ACM Transactions on Graphics (Proceedings of SIGGRAPH) 23(3), 522–531 (2004)
10. Heck, R., Gleicher, M.: Parametric motion graphs. In: Proc. of the 2007 Symposium on Interactive 3D Graphics and Games, I3D 2007, pp. 129–136. ACM Press, New York (2007)
11. Kovar, L., Gleicher, M.: Automated extraction and parameterization of motions in large data sets. ACM Transaction on Graphics (Proceedings of SIGGRAPH) 23(3), 559–568 (2004)
12. Kovar, L., Gleicher, M., Pighin, F.H.: Motion graphs. Proceedings of SIGGRAPH 21(3), 473–482 (2002)
13. Kovar, L., Schreiner, J., Gleicher, M.: Footskate cleanup for motion capture editing. In: Proceedings of the ACM SIGGRAPH/EUROGRAPHICS Symposium on Computer Animation (SCA), pp. 97–104. ACM Press, New York (2002)
14. Kwon, T., Shin, S.Y.: Motion modeling for on-line locomotion synthesis. In: Proceedings of the 2005 ACM SIGGRAPH/EUROGRAPHICS Symposium on Computer Animation, SCA 2005, pp. 29–38. ACM Press, New York (2005)
15. Mukai, T., Kuriyama, S.: Geostatistical motion interpolation. In: ACM SIGGRAPH, pp. 1062–1070. ACM Press, New York (2005)
16. Press, W.H., Teukolsky, S.A., Vetterling, W.T., Flannery, B.P.: Numerical Recipes: The Art of Scientific Computing, 3rd edn. Cambridge Univ. Press, New York (2007)
17. Rose, C., Bodenheimer, B., Cohen, M.F.: Verbs and adverbs: Multidimensional motion interpolation. IEEE Computer Graphics and Applications 18, 32–40 (1998)
18. Rose III, C.F., Sloan, P.P.J., Cohen, M.F.: Artist-directed inverse-kinematics using radial basis function interpolation. Computer Graphics Forum (Proceedings of Eurographics) 20(3), 239–250 (2001)
19. Safonova, A., Hodgins, J.K.: Construction and optimal search of interpolated motion graphs. In: ACM SIGGRAPH 2007, p. 106. ACM, New York (2007)
20. Sumner, R.W., Zwicker, M., Gotsman, C., Popović, J.: Mesh-based inverse kinematics. ACM Trans. Graph. 24(3), 488–495 (2005)
21. Treuille, A., Lee, Y., Popović, Z.: Near-optimal character animation with continuous control. In: ACM SIGGRAPH 2007 Papers, SIGGRAPH 2007, p. 7. ACM Press, New York (2007)
22. Unuma, M., Anjyo, K., Takeuchi, R.: Fourier principles for emotion-based human figure animation. In: SIGGRAPH 1995, pp. 91–96. ACM Press, New York (1995)
23. Wiley, D.J., Hahn, J.K.: Interpolation synthesis of articulated figure motion. IEEE Computer Graphics and Applications 17(6), 39–45 (1997)
24. Yamane, K., Kuffner, J.J., Hodgins, J.K.: Synthesizing animations of human manipulation tasks. In: ACM SIGGRAPH 2004, pp. 532–539. ACM, New York (2004)

Planning and Synthesizing Superhero Motions

Katsu Yamane[1,2] and Kwang Won Sok[2]

[1] Disney Research, Pittsburgh
4615 Forbes Ave. Suite 420, Pittsburgh, PA 15213, USA
[2] Carnegie Mellon University
kyamane@disneyresearch.com
http://www.disneyresearch.com/

Abstract. This paper presents an approach to planning and synthesizing collision-free motions of characters with extreme physical capabilities, or superheroes, using a human motion database. The framework utilizes the author's previous work on momentum-based motion editing, where the user can scale the momentum of a motion capture sequence to make more or less dynamics motions, while maintaining the physical plausibility and original motion style. In our new planning framework, we use a motion graph that contains all possible motion transitions to list the candidate motion segments at each planning step. The planner then computes the momentum scale that should be applied to the original motion segment in order to make it collision-free. Experimental results demonstrate that the planning algorithm can plan and synthesize motions for navigating through a challenging environment, using a relatively small motion capture data set.

Keywords: Planning, Motion Graphs, Momentum Editing, Motion Capture.

1 Introduction

Recently, motion and performance capture have been a popular way to create character animations in films and games because human motion data contain rich expressions and styles which even talented animators may spend a long time to create by hand. On the other hand, reusing human motion data is not straightforward because a captured motion is only an instance of the behavior performed by a specific subject in a specific environment.

Therefore, many algorithms have been developed for adapting motion capture data to various characters and constraints. Earlier work in this area includes retargetting motions to new characters [3] and interpolating example motions to satisfy new constraints [8]. As large-scale motion capture databases became available, the concept of *motion graph* was implemented by many groups [1,4,5,9]. Motion graphs describe the possible transitions among poses, and can be used to plan motions under various constraints and objectives by employing graph search techniques. More recently, many researchers combined motion capture data with physical simulation to make characters respond to external disturbances [10,2,7,6].

Unfortunately, most of the work utilizing human motion data assumes human-like kinematic structure and physical capability, and therefore it is not straightforward to extend these approaches to characters with different physical capabilities. Consider an example where a character has to move to the other side of a valley. If the character is limited to human capability, it will have to climb down to the bottom of the valley and climb up. If the character can make extremely wide jump, on the other hand, it can simply jump over the valley. However, a motion database created from human motions would not have such data because the jump is beyond human capability.

In this paper, we present a framework for planning and synthesizing *superhero* motions from a human motion database. We consider characters with the ability to exert large joint torques and endure large impact forces, while performing the motions in human-like styles. We apply our previous work on momentum-based motion editing [11], where an animator can directly scale the momentum to obtain more dynamic motions. Because the method only considers the physical consistency in terms of the center of mass motion, it is possible to synthesize highly dynamic motions while maintaining the overall physical plausibility.

As a case study, we consider the problem of planning collision-free motions that navigate through environments with large obstacles where the character may have to make extremely high jumps. We describe a method for determining the momentum scale that allows the character to clear the obstacles and land on a flat surface with a particular height. The planner also utilizes the concept of motion graphs to enumerate possible motion transitions.

This paper is organized as follows. We first introduce the momentum-based motion editing technique in Section 2. We then describe the motion planning and synthesis algorithm in Section 3. Section 4 shows several experimental results, followed by the concluding remarks in Section 5.

2 Momentum-Based Motion Editing

The original momentum-based motion editing technique [11] takes a user-specified momentum scale as input and synthesizes a new motion whose momentum matches the scaled one. A problem occurs when the user also wants to maintain the final position because the velocity changes due to the momentum scaling also changes the center of mass (COM) trajectory. Also when the character is in flight, the COM motion should obey the conservation of momentum or the gravitational acceleration, which will not be the case with the new trajectory.

We solve these issues by modifying the duration of the motion. For this purpose, we introduce the concept of *normalized dynamics* where the time axis is normalized and can be modified by simply changing the time scale. In this section, we derive the normalized dynamics formulation using the vertical linear direction as an example. The formulations for the other linear directions are exactly the same without the gravity. It is also straightforward to extend to angular momentum if we consider the rotation around a single axis. See [11] for a detailed discussion.

2.1 Normalized Dynamics

The total external force f, the total linear momentum P, and the COM velocity $\dot{z}$ are related by

$$f = \dot{P} + mg \qquad (1)$$
$$P = m\dot{z} \qquad (2)$$

where m is the total mass of the character and g is the gravity acceleration.

We derive the *normalized dynamics* by normalizing the time t by a scaling factor c. We denote the normalized time, or *phase*, by $\tau = t/c$. Using c, the time derivative is converted to the phase derivative by

$$\frac{d}{dt} = \frac{1}{c}\frac{d}{d\tau}. \qquad (3)$$

Equations (1) and (2) are converted to the phase space as

$$\phi = \frac{dP}{d\tau} + \mu\gamma \qquad (4)$$
$$P = \mu\frac{dz}{d\tau} \qquad (5)$$

where the normalized quantities are defined as

$$\phi = cf, \ \mu = m/c, \ \gamma = c^2 g. \qquad (6)$$

Equations (4) and (5) yield the equations to calculate the linear momentum and position:

$$P(\tau) = P_0 - \mu\gamma\tau + \int_0^\tau \phi(\sigma)d\sigma \qquad (7)$$
$$z(\tau) = z_0 + \frac{1}{\mu}\int_0^\tau P(\sigma)d\sigma \qquad (8)$$

where we have assumed that c is constant during the integration period and therefore μ and γ are also constant. In fact, we can assume that the phase is equivalent to the time in the original motion, in which case we have $c = 1$.

2.2 Force and Momentum Scaling

We now consider scaling the external force by a constant scaling factor s, i.e.,

$$\hat{f}(t) - mg = s(f(t) - mg). \qquad (9)$$

Note that the difference from gravitational force is scaled.

If we simply plug the new external force into the standard equation of motion, the new COM trajectory would be uniquely determined. With the normalized dynamics, however, we can still modify the trajectory by adjusting the time scale from c to another constant value $\hat{c}$, which is equivalent to changing the duration of the motion. With the new time scale, the normalized mass and gravity become $\hat{\mu}$ and $\hat{\gamma}$ accordingly. The normalized force is scaled as

$$\hat{\phi}(\tau) - \hat{\mu}\hat{\gamma} = s(\phi(\tau) - \mu\gamma). \tag{10}$$

The new momentum in the phase space becomes

$$\begin{aligned}\hat{P}(\tau) &= \hat{P}_0 - \hat{\mu}\hat{\gamma}\tau + \int_0^\tau \hat{\phi}(\sigma)d\sigma \\ &= \hat{P}_0 + s\int_0^\tau (\phi(\sigma) - \mu\gamma)d\sigma \\ &= \hat{P}_0 + s(P(\tau) - P_0) \\ &= sP(\tau)\end{aligned} \tag{11}$$

where we have used Eq.(7) to eliminate the integration term, and the relationship $\hat{P}_0 = sP_0$ to yield the final result. Eq.(11) means that the momentum itself is scaled by the same ratio as the external force. We can then obtain the new COM trajectory by

$$\begin{aligned}\hat{z}(\tau) &= \hat{z}_0 + \frac{1}{\hat{\mu}}\int_0^\tau \hat{P}(\sigma)d\sigma \\ &= \hat{z}_0 + \hat{c}s(z(\tau) - z_0)\end{aligned} \tag{12}$$

where we have used Eq.(8) to eliminate the integration term.

Eq.(12) shows that the COM trajectory can be controlled by two free variables $\hat{c}$ and s. In our previous work [11], we let the user choose s and the system determined $\hat{c}$ based on constraints such as the final position. In this paper, the planner will determine both $\hat{c}$ and s based on the environment constraints.

When there is no external force, the motion during the flight phase is determined by the initial momentum and time scale. Substituting $\phi = 0$ to Eq.(7) and then using Eq.(8), we obtain the COM trajectory as

$$z(\tau) = z_0 + \hat{c}v_0\tau - \frac{1}{2}g\hat{c}^2\tau^2 \tag{13}$$

where $v_0 = P_0/m$. Eq.(13) represents the parabolic trajectory with initial velocity v_0 and constant acceleration $-g$.

3 Planning and Synthesis

We employ the momentum-based motion editing technique to plan and synthesize physically plausible, collision-free motions. More specifically, we consider the problem of navigating through an environment with many large obstacles that are difficult to go around or climb over for humans. Motion planning using a human motion database does not work for such environments.

Our approach is to allow the planner to modify the clips in the motion database by scaling their momentum. Because the planner will have a much wider variety of motions to choose from, it has a larger chance of finding a collision-free motions.

The planning algorithm is based on A* (A-star) search [12]. The planning algorithm uses a motion graph that contains short motion segments and transitions among them to obtain a list of possible segments at each step of the planning. For a given motion segment and character location, the planner then computes the momentum scale to make the motion collision-free in that particular location in the environment using our momentum-based motion editing technique. When the planning is finished, we can obtain an optimal, collision-free motion represented as a series of scaled motion segments.

3.1 Motion Graph

We construct a motion graph from multiple motion capture sequences. Although many techniques for automatic construction of motion graphs have been developed [1,4,5], we build the motion graph manually to have better control of the contact states at each node.

After cleaning up the captured motions, we first segment the motion data such that each node starts at the touchdown of a foot and ends at the touchdown of the other foot. The node can therefore be divided into contact and flight phases as shown in Fig. 1.

We then define the possible transitions among motion segments. Obviously, the contact state at the end of a segment m_1 should match that at the beginning of another segment m_2 in order for the transition from m_1 to m_2 to be feasible. We may also enforce similarities between the final pose of m_1 and the initial pose m_2. In our implementation, we simply consider the contact state to determine whether a transition is possible.

Finally, we form a motion graph by connecting nodes representing the motion segments and by edges representing all possible transitions. We also add special empty segments labeled INITIAL and FINAL to specify the segments that can be used for starting and ending the whole sequence. Figure 2 shows an example of motion graphs constructed as described above.

In practice, we can further populate the graph by adding motions synthesized by mirroring or the momentum scaling technique. For example, we can synthesize slower and faster running segments by reducing or increasing the linear momentum along the forward direction, even if we have only captured one running sequence.

3.2 Motion Planning with Momentum Scaling

Notations. Before describing the algorithm in detail, we first introduce the notations we will use in the algorithm description.

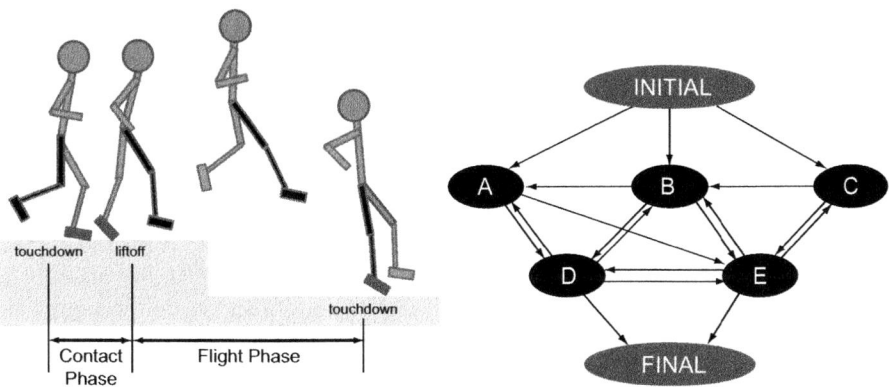

Fig. 1. Contact and flight phases in a motion segment. The red blocks indicate the links in contact.

Fig. 2. An example of motion graph

For clarity, we represent the position and orientation of the root joint of the character by a single variable R and call it *root configuration*. We specify the start and goal of the planning by a pair of root configurations.

We define the following methods operating on a motion segment m:

- $m.NextSegments()$ returns a list of motion segments that can be directly transitioned from segment m. In the example shown in Fig. 2, $D.NextSegments()$ returns $\{A, B, E, FINAL\}$.
- $m.FinalRootConfig(R)$ returns the final root configuration when the entire segment is transformed so that the initial root configuration matches R.

Our planning framework follows the standard A* search algorithm [12]. During the search, the configuration of the character is represented as a state x. A state contains the information on the motion segment $x.motion$ performed by the character, as well as the initial and final root configurations $x.R_I$, $x.R_F$. We define the following methods operating on a state x:

- $x.CalcScale()$ computes a pair of (possibly time-varying) time and momentum scales, $x.c$ and $x.s$ respectively, that make $x.motion$ collision-free. This process may fail when, for example, a negative scale is required.
- $x.ApplyScale()$ computes a new root trajectory based on the time and momentum scales, and returns the final root configuration.
- $x.SetAscendant(x')$ sets another state x' as the ascendant of x so that we can trace the solution after a goal state is found.

A* algorithm uses a priority queue Q that includes all active states sorted in the ascending order of $C(x) + G(x)$, where $C(x)$ and $G(x)$ are the cost-to-come and underestimated cost-to-go of state x respectively. These costs are discussed in detail later. We define the following methods operating on a queue Q:

- $Q.Add(x)$ adds state x to Q and sort the states in the ascending order of $C(x) + G(x)$, while $Q.Remove(x)$ removes state x from Q.
- $Q.GetFirst()$ returns the first state in Q, i.e., the state with the minimum total cost.

Planning Algorithm. Algorithm 1 shows the general planning framework. Each iteration starts by finding the state x in the search tree with the minimum total cost (line 4). We then consider all transitions available from the motion segment m associated with x (the for loop starting from line 7). We attempt to compute the time and momentum scales (line 10) and if successful, we apply the scales and add x to the queue. This iteration is repeated until we find a state whose final root configuration is sufficiently close to the goal, or the priority queue becomes empty. In the latter case, the search fails.

Cost Functions. The cost-to-come at a state x, $C(x)$, represents the cost to move from the start configuration to x. Our cost-to-come of state x' is

$$C(x') = C(x) + D(x'.R_I, x'.R_F) + w_s S(x'.c, x'.s) \qquad (14)$$

where x is the predecessor of x', $D(R_1, R_2)$ computes the distance between two root configurations R_1 and R_2, and $S(c, s)$ takes the time and momentum scales as inputs and computes

$$S(c, s) = (1 - c)^T (1 - c) + (1 - s)^T (1 - s) \qquad (15)$$

where $\mathbf{1}$ is a column vector of appropriate size whose elements are all 1. The term $D(*)$ prioritizes shorter paths, and $S(*)$ prioritizes scales closer to one.

The cost-to-go is the cost that will be accumulated until the path reaches the goal. Because this cost is unknown until a feasible path is found, we have to estimate the value. In particular, if we always underestimate the cost-to-go, then it is guaranteed that we can find the optimal path in terms of the cost function (14). In our implementation, we use $G(x') = D(x'.R_F, R_G)$ which is obviously always smaller than the actual cost.

3.3 Automatic Scaling of Momentum

The $CalcScale()$ is the core of the planning algorithm. The function is responsible for automatically determining the momentum and time scales to make the character's motion collision-free.

In our implementation, we only scale the linear momentum along the vertical axis because this is the most critical axis to jump over tall obstacles. We also assume that the center of mass trajectory during the contact phase is unchanged, and only the flight phase section is modified to avoid the obstacles.

When $CalcScale()$ is called for a new state x' with motion segment k, we first transform the entire k so that the initial root configuration matches $x'.R_I$, which is identical to the final root configuration of the predecessor state. We then

Algorithm 1. Motion Planning with Momentum Scaling

Require: start and goal configurations R_S and R_G
1: create state x_0 with $x_0.R_I = x_0.R_F = R_S$ and $x_0.motion = \phi$
2: $Q.Add(x_0)$
3: **repeat**
4:    $x \leftarrow Q.GetFirst()$
5:    $Q.Remove(x)$
6:    $m \leftarrow x.motion$
7:    **for all** $k \in m.NextSegments()$ **do**
8:       create new state x'
9:       $x'.motion \leftarrow k$, $x'.R_I \leftarrow x.R_F$
10:      **if** $x'.CalcScale()$ is successful **then**
11:          $x'.R_F = x'.ApplyScale()$
12:          **if** $x'.R_F$ is close to R_G **then**
13:             **return** x'
14:          **end if**
15:          $Q.Add(x')$
16:          $x'.SetAscendant(x)$
17:      **end if**
18:    **end for**
19: **until** Q is empty
20: **return** NULL

check if the transformed k makes any collision with the environment during the contact phase. If a collision is found, $CalcScale()$ returns a failure state because we have assumed that the motion during the contact phase is unchanged, and hence we cannot resolve the collision.

If the contact phase is collision-free, we then look at the flight phase to compute the momentum and time scales. Let v_0 denote the original vertical COM velocity at the beginning of the flight phase. As already shown in Eq.(13), the COM trajectory in normalized time during the flight phase is uniquely determined by the initial COM velocity v_0. Here we consider the case where the momentum has been scaled by a factor s by the end of the contact phase, and therefore the initial velocity of the flight phase is now $\hat{v}_0 = sv_0$. In this case, the vertical COM trajectory will be

$$z(\tau) = z_0 + \hat{c}sv_0\tau - \frac{1}{2}g\hat{c}^2\tau^2 \ (0 \leq \tau \leq T) \tag{16}$$

where T is the duration of the flight phase. We determine $\hat{c}$ and s so that

- The lowest point on the character is above the topmost surface at $0 \leq \tau < T$.
- The character makes contact with the topmost surface at $\tau = T$.

We discretize the normalized time axis with a fixed time step $\Delta\tau$ into $N+1$ frames. The normalized time of i-th frame is $\tau_i = i\Delta\tau$ $(i = 0, 1, \ldots N)$ with $\tau_N = T$. We obtain the height of the COM and the lowest point of the character at frame i, $Z(i)$ and $L(i)$ respecively, in the original motion. We also obtain the height of the topmost surface, $H(i)$, at the lowest point of the character.

By making the initial root configuration match the final root configuration of the predecessor state, we can assume that

$$z(0) = H(0) + Z(0) - L(0) \stackrel{\triangle}{=} z_0 \quad (17)$$

holds. The first constraint translates to

$$z(\tau_i) > H(i) + Z(i) - L(i) \stackrel{\triangle}{=} \hat{z}_i \quad (18)$$

and the second constraint to

$$z(T) = H(N) + Z(N) - L(N) \stackrel{\triangle}{=} z_N. \quad (19)$$

Plugging Eq.(19) into (16), we obtain

$$\hat{c}sv_0 T - \frac{1}{2}g\hat{c}^2 T^2 = z_N - z_0 \quad (20)$$

which leads to

$$\hat{c}s = \frac{z_N - z_0 + g\hat{c}^2 T^2/2}{v_0 T}. \quad (21)$$

Because both $\hat{c}$ and s have to be positive, a negative $\hat{c}s$ value means that we cannot use the motion segment for this particular location of the environment.

We can eliminate s from Eq.(16) by plugging Eq.(21) into Eq.(16) as

$$z(\tau) = z_0 + \frac{\tau}{T}(z_N - z_0) + \frac{1}{2}g\hat{c}^2 \tau(T - \tau) \quad (22)$$

Constraint (18) yields the condition on c as

$$\frac{1}{2}g\tau_i(T - \tau_i)\hat{c}^2 > \hat{z}_i - z_0 - \frac{\tau_i}{T}(z_N - z_0). \quad (23)$$

Note that the coefficient of $\hat{c}^2$ in the left-hand side of Eq.(23) is always positive. Therefore, if the right-hand side of Eq.(23) is negative, the character can clear the surface with any positive $\hat{c}$. Otherwise, $\hat{c}$ has to satisfy Eq.(23) to avoid the collision at frame i. By taking the maximum of the minimum $\hat{c}$ that satisfies Eq.(23) at every frame, we obtain the minimum $\hat{c}$ that makes the entire segment collision-free. This $\hat{c}$ is used as the time scale during the flight phase. Once the time scale is computed, the momentum scale s can be obtained by Eq.(21).

If the right-hand side of Eq.(23) is negative for all frames, we assume $\hat{c} = s$ to avoid excessive change to the motion segment. In this case, $\hat{c}$ and s become the square root of the left-hand side of Eq.(21).

Finally, we determine the time and momentum scales during the contact phase. We already know that the momentum scale at the end of the contact phase has to be the s obtained above. We also know the momentum scale at the end of the previous state, s_0. In our implementation, we obtain the momentum scale at the i-th frame of the contact phase, s_i by linearly interpolating between s_0 and s to make the momentum continuous. On the other hand, the time scale at the i-th frame is determined by $c_i = 1/s_i$. This choice of c_i keeps the COM trajectory unchanged as shown in Eq.(12).

4 Results

4.1 Environment Model

We assume a simple environment with a horizontal ground and multiple objects, and the character is able to step on any top-most horizontal surface on either the ground or an object. The height of the topmost surface is obtained by rendering the top view of the scene in parallel projection and obtaining the z-buffer value at the pixel corresponding to the character location. In the experiment, we use an environment model with six obstacles of various sizes randomly placed in a field of 20m×20m.

4.2 Motion Graph

The motion graph for the experiment consists of behaviors including running, jumping forward, jumping up, jumping down, turning left, and turning right. Each motion clip is segmented as described in Section 3.1.

We also create more segments by manipulating the original motion segments. We create a slow running by scaling the momentum in the forward direction by 0.7, and wide jumps by scaling the momentum by 1.5. A mirrored version of each jump motion is also added. Running and jumping motions with different width and feet are essential to adjust the footstep locations to the environment.

Figure 3 illustrates the motion graph used for the experiment. The thick arrows indicate that every segment on one end can transition to all segment(s)

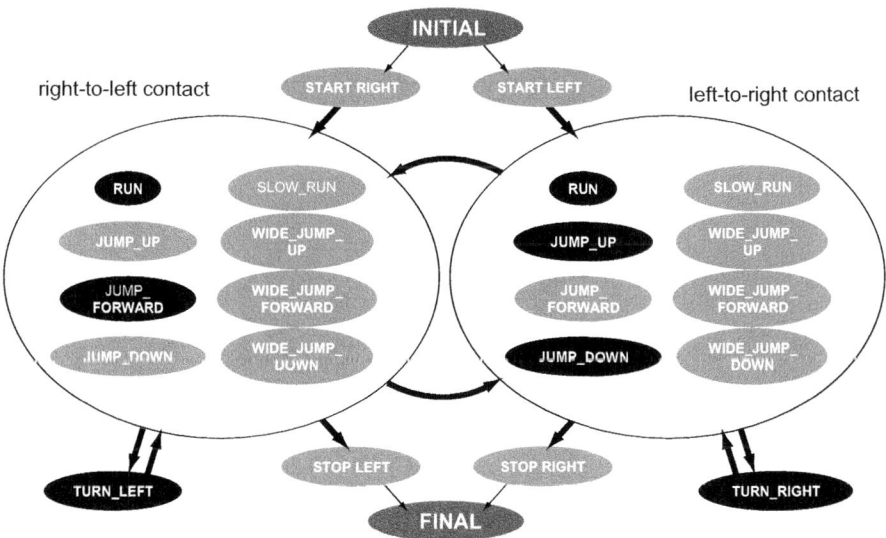

Fig. 3. The motion graph used for the experiment

Fig. 4. Snapshots from a planned motion

Fig. 5. Closeups of jumping up and down

on the other end. The green (light gray) segments are synthesized from captured motion sequences by mirroring and/or momentum scaling.

4.3 Planned Motions

The planner takes about 10 minutes to plan a motion that navigates from the start to the goal in about 18 seconds as shown in 4. Figure 5 shows closeups of some examples of the jumping up and down motions.

5 Conclusion

This paper presented a framework for planning and synthesizing motions of characters with extreme physical capability based on a human motion database.

The planner utilizes the momentum-based motion editing technique to modify the original captured motions to make collision-free motions.

This technique will be useful in interactive games where the environment model is created on the fly, because it may be impossible to make collision-free motions only with the original motion clips contained in the database.

Acknowledgements. The authors would like to thank Justin Macey for his assistance in capturing and cleaning up the motion capture data used in the experiment.

References

1. Arikan, O., Forsyth, D.A.: Synthesizing Constrained Motions from Examples. ACM Transactions on Graphics 21(3), 483–490 (2002)
2. Da Silva, M., Abe, Y., Popović, J.: Interactive simulation of stylized human locomotion. ACM Transactions on Graphics 27(3), 82 (2008)
3. Gleicher, M.: Retargetting Motion to New Characters. In: Proceedings of SIGGRAPH 1998, Orlando, FL, pp. 33–42 (1998)
4. Kovar, L., Gleicher, M., Pighin, F.: Motion graphs. ACM Transactions on Graphics 21(3), 473–482 (2002)
5. Lee, J., Chai, J., Reitsma, P.S.A., Hodgins, J.K., Pollard, N.S.: Interactive Control of Avatars Animated With Human Motion Data. ACM Transactions on Graphics 21(3), 491–500 (2002)
6. Lee, Y., Kim, S., Lee, J.: Data-driven biped control. ACM Transactions on Graphics 29(4), 129 (2010)
7. Muico, U., Lee, Y., Popović, J., Popović, Z.: Contact-aware nonlinear control of dynamic characters. ACM Transactions on Graphics 28(3) (2009)
8. Rose, C., Cohen, M., Bodenheimer, B.: Verbs and Adverbs: Multidimentional Motion Interpolation. IEEE Computer Graphics and Applications 18(5), 32–40 (1998)
9. Safonova, A., Hodgins, J.: Interpolated motion graphs with optimal search. ACM Transactions on Graphics 26(3), 106 (2007)
10. Sok, K., Kim, M., Lee, J.: Simulating biped behaviors from human motion data. ACM Transactions on Graphics 26(3) (2007)
11. Sok, K., Yamane, K., Lee, J., Hodgins, J.: Editing dynamic human motions via momentum and force. In: Eurographics/ACM SIGGRAPH Symposium on Computer Animation (2010)
12. LaValle, S.M.: Planning Algorithms. Cambridge University Press, New York (2006)

Perception Based Real-Time Dynamic Adaptation of Human Motions

Ludovic Hoyet[1], Franck Multon[1,2], Taku Komura[3], and Anatole Lecuyer[1]

[1] IRISA - INRIA Bretagne Atlantique, Bunraku Team, Rennes, France
[2] M2S - Mouvement Sport Santé University Rennes 2, France
[3] IPAB, University of Edinburgh, Scotland
{ludovic.hoyet,franck.multon,Anatole.Lecuyer}@irisa.fr,tkomura@ed.ac.uk

Abstract. This paper presents a new real-time method for dynamics-based animation of virtual characters. It is based on rough physical approximations that lead to natural-looking and physically realistic human motions. The first part of this work consists in evaluating the relevant parameters of natural motions performed by people subject to various external perturbations. According to this pilot study, we have defined a method that is able to adapt in real-time the motion of a virtual character in order to satisfy kinematic and dynamic constraints, such as pushing, pulling and carrying objects with more or less mass. This method relies on laws provided by experimental studies that enable us to avoid using complex mechanical models and thus save computation time. One of the most important assumption consists in decoupling the pose of character and the timing of the motion. Thanks to this method, it is possible to animate up to 15 characters at 60Hz while dealing with complex kinematic and dynamic constraints.

Keywords: Motion adaptation, physics, interaction, perception-based approach, virtual human.

1 Introduction

The use of motion capture data is now widely spread in computer animation but the data are specific to a given environment, to given constraints and to the morphology of the actor. Three main types of techniques were developed to enable the use of such captured motions under new constraints: solving constraints, designing controllers or using huge data structures, such as motion graphs. While most of the techniques offer some control of the virtual human, one of the most difficult problem is to generate natural motions. It is a key issue for applications involving real-time interactions with a user, such as video games, virtual reality and serious games. In such applications, naturalness has a strong effect on the believability of the simulation and thus the feeling of being involved in the experiment (named Presence in virtual reality). If a lot of techniques were developed to handle physical laws (part of the naturalness of the resulting motion), it is still an open problem for real-time applications. Recently, new motion control techniques based on captured motions have been presented to handle physically correct interactions using simulation [1,2,3] or biomechanical knowledge [4]. Designing such controllers is a key point as physically-valid motions may look unnatural. Thus, these controllers have to select a natural motion among all the physically-valid ones which is still difficult to handle. In this paper, we propose to use knowledge provided by experimental sciences such as biomechanics or neurosciences to drive the simulation to a motion that is supposed to be natural.

More than naturalness, the way a motion is perceived by a user is the key point in many applications. A paradox is that some animations may look natural while the physical laws are not completely satisfied, such as animating cartoon characters. In recent works [5] we have evaluated the sensitivity of users to perceive the physical properties of objects manipulated by virtual humans. In this paper, we describe how to use this knowledge to develop an efficient method to adapt the motion subject to physical perturbations. We also show that the velocity profile is a key biomechanical factor that conveys relevant information about the dynamic status of the motion. The method developed in this paper has been applied to pushing, pulling, carrying and walking motions subject to continuous external perturbations, such as adding masses or pushing continuously the character. The resulting method is able to simulate numerous characters in real-time as it requires only few computation time.

The paper is organized as follow. Section 2 presents related work addressing the problem of dynamic animation of virtual characters. A dedicated biomechanical experiment is described in section 3, while Section 4 provides some details about the resulting method. Results obtained with this new method are presented in Section 5 for a selection of motions. Finally, Section 6 concludes and gives some perspectives to this work.

2 Background

Motion synthesis using controllers is one of the most popular method to handle the dynamics of human motion [6,7,8]. However, the controller is specialized for one type of motion and can be difficult to tune to handle new situations. Moreover, it can lead to unnatural or robotics-like motions. To overcome this limitation, Shiratori et al. [4] created controllers for trip recovery responses that may occur during walking using biomechanical literature and motion capture data (as the work of Zordan and Hodgins [9]). However, such controllers are still complex to design for a wide variety of motions while keeping naturalness.

An alternative consists in adapting an existing motion by solving concurrently kinematic and dynamic constraints through spacetime optimization [10,11]. One of the key points here is to design the relevant cost function, as it has direct influence on computed motion. This type of method is not limited to simple systems but can handle complex skeletons [12]. Some simplifications have been introduced to save computation time, such as using biomechanical knowledge to guide the otimization process [13], modeling the joint trajectories with parametric curves [14] or working in reduced spaces thanks to Principal Component

Analysis data reduction [15]. Despite these efforts spacetime optimization is still time demanding and is difficult to apply to interactive animations. Moreover the design of the cost function is still debated.

As generating natural motions is a challenge, some authors have proposed to use databases of captured motions to drive a virtual human subjects to sudden external perturbations [1,16,17]. When the perturbation occurs, the method searches into the database for the most relevant reaction while dealing with the dynamic state of the system. Simplified models (such as an inverted pendulum [18]) or concatenation of retimed clips [19] have been introduced recently for specific motions, such as aerial ones. However, these techniques rely on the quality and dimension of the database to handle more or less complex situations.

On the opposite, some authors have proposed to adapt a unique motion to satisfy the physical laws. The most famous method is called Dynamic Filters [20,21,22] and consists in a per-frame adaptation process. Given an edited motion capture input, the dynamics filter keeps the feet planted on the ground while filtering impossible forces and torques. A Kalman filter is generally used [20] to predict the next physically correct posture by simulating the system over a short time step. The predicted posture is then filtered to get closer to the original motion. Although this method enables real time dynamic correction, the parameters used to design the Kalman filter are difficult to tune. Other dynamic filters tried to adjust uncorrect poses (unbalanced motions [23,3] or unfeasible forces and torques applied on unactuated joints [21]) thanks to optimization [22,3] or local joint adjustments ([21,23]).

All the above methods aim at generating a physically-valid motion which requires high computation time and sometimes complex dynamic models. Past studies [24,19,25] have shown that users are able to detect physically unfeasible motions or identified the maximum error tolerance of physical simulation for complex scenarios [26]. Other perceptual studies [27,28] have demonstrated the capability of subjects to determine the physical properties of manipulated objects by simply picking information in the kinematic data of the performer. However, these studies also demonstrated that this perceptual skill is not very accurate. We thus wondered if it was possible to take advantage of this inaccuracy in order to simplify the motion adaptation process. To test this hypothesis, we have developed a two-steps process:

1. estimating the accuracy of people to perceive the physical properties of manipulated objects by simply watching a virtual performer,
2. designing a simplified dynamic filter that neglects some terms in the dynamic equations and that uses biomechanical knowledge to compute a natural pose.

3 Physics in Natural Motions

Runeson and Frykholm [28] showed that human beings are able to evaluate the mass of lifted objects just by watching the motion of a performer. Bingham [27] even showed that these dynamic properties are perceived through kinematic data of motions. In the same way, we have demonstrated that people can recognize

Fig. 1. Subject lifting a 6kg dumbbell

the heaviest mass carried by a virtual human while only watching its motion [5]. Hence, modifying joint angles of a virtual human may affect the way people perceive the dynamic properties of the environment. We have shown that people can perceive differences of 3Kg (almost 50%) for lifting motions of dumbbells ranging from 2Kg to 10Kg. Although this accuracy is limited a specific motion, this result clearly shows that it does not seem needed to perform accurate computation of physical models for such cases.

In this paper we describe a pilot study to determine the kinematic parameters that seem relevant for determining the dynamic status of human motion. Thanks to this study it should be possible to determine how these relevant kinematic data changes according to dynamic constraints. It could be helpful for developing operators that would directly adapt global variables without the use of complex physical model. To this end, we recorded both video sequences and motion capture data of a subject carrying a dumbbell placed over a table in front of him. The subject was asked to carry masses ranging only from 2 to 10kg determine if kinematic parameters significantly change even for such small variation in dynamic constraints. The subject performed 10 times each motion.

To determine which are the most relevant kinematic parameters that could be used for perceiving such dynamic properties, we carried-out statistical analyzes. We tested the following parameters: trajectory and velocity of hand, elbow and shoulder. To evaluate if there exists significant differences between two lifting motions t_1 and t_2 using these various parameters, we defined the following distance metrics:

$$e(t_1, t_2) = \left\| \sqrt{\frac{1}{N} \sum_{i=1}^{N} (t_1(i) - t_2(i))^2} \right\|$$

where N corresponds to a sampling of the time-independent trajectories, and $\|.\|$ to the euclidean norm as points on trajectories t_i are in the three dimensional space. This root mean square error has been used by an one-way ANOVA statistical test for each parameter listed above. Significance was set to $p < 0.05$ (after correction $p < 0,00625$). Each set of motions associated with mass i was compared to the motions of mass j. It leads to 72 tests for each parameter. The results are summarized in Table 1.

Despite the number of significant differences provided in this table, our results show that a difference of only 1-kg in the dumbbell mass leads to a significant difference in hand and elbow velocities (for more than 80% of the cases). As masses used for the study ranged from 2 to 10kg, it almost corresponds to at least a 10% of the maximum lifted mass. This result should be verified for heavier masses and for other types of movements. In [5] this difference was estimated to

Table 1. Number of significant differences between all the combinations of lifting motions with various masses, according to the parameter used in the distance metrics

Parameter	# of sign. diff.	Parameter	# of sign. diff.
hand trajectory	43/72 (59.7%)	hand velocity	60/72 (83.3%)
elbow trajectory	47/72 (66.3%)	elbow velocity	62/72 (86.1%)
shoulder trajectory	43/72 (59.7%)	shoulder velocity	45/72 (62.5%)

3Kg when studying the perception of users watching captured motions of weight lifting. All these results tend to demonstrate that even if people can perceive modifications in the velocity profile due to changes in dynamic constraints, this perceptual skill is not so accurate or may focus on wrong variables. We have used this knowledge to develop an efficient method to adapt the motion of a virtual human subjects to various physical constraints.

4 Real-Time Dynamic Adaptation of Human Motions

We propose a new method for adapting a motion when new external forces are exerted, such as pushing, pulling or carrying objects with various masses. If we consider the global system of the virtual human, Newton laws give:

$$\begin{cases} \mathbf{P} + \mathbf{GRF} + \sum_j \mathbf{F}_j = m\mathbf{a} \\ \mathbf{OG} \times \mathbf{P} + \mathbf{OCoP} \times \mathbf{GRF} + \sum_j \mathbf{OA_j} \times \mathbf{F}_j = \dot{\mathbf{H}}_{/O} \end{cases} \quad (1)$$

where $\mathbf{P} = m\mathbf{g}$ is the weight force acting on the character applied at its center of mass G, $\mathbf{GRF}$ is the ground reaction force applied at the position of the center of pressure CoP, $\mathbf{a}$ is the acceleration of G and $\dot{\mathbf{H}}_{/O}$ is the derivative of the angular momentum of the character computed at O. The character is also subject to j external forces $\mathbf{F}_j$ applied respectively at the position A_j. Adding forces $\mathbf{F}_j$ involves changes in all the other terms to satisfy this equation. If the virtual human has to push a 40kg-object 1m far from his center of mass, $\mathbf{OA_j} \times \mathbf{F}_j$ is almost 400Nm which is very important. To avoid solving nonlinear differential equations we propose to divide the solving process into two steps: first solving the system in the static situation and then adapting the velocity profile (as it has been identified to be relevant in the previous experiment). In addition to this process applied during the interaction, it is necessary to adapt the motion that is performed before this interaction (named the preparation phase).

4.1 Static Correction of a Posture at Interaction Time

Hence, firstly, we assume that the terms involving acceleration could be ignored:

$$\begin{cases} \mathbf{P} + \mathbf{GRF} + \sum_j \mathbf{F}_j = 0 \\ \mathbf{OG} \times \mathbf{P} + \mathbf{OCoP} \times \mathbf{GRF} + \sum_j \mathbf{OA_j} \times \mathbf{F}_j = 0 \end{cases} \quad (2)$$

Based on this assumption, the system is much simpler to solve. However it remains an infinity of solutions (i.e. posture adaptations) that satisfy the above equations. Only some of them may look natural. To select one of them, we apply some knowledge provided by the biomechanics and neurosciences literature.

Given a posture q subjects to forces $\mathbf{F}_j$ not present in the original motion, it is possible to correct q to ensure static balance using the above Equation 2. Biomechanical knowledge is used to guide this process to a natural solution.

Hof [29] has identified three strategies for humans to naturally maintain balance. First, humans move their CoP inside the base of support to compensate small perturbations. However, the CoP quickly reaches the limit of the base of support. Thus, humans switch to a second strategy to compensate these perturbations: they use counter rotation of segments around the ankle to compensate strong perturbations (ankle strategy). Then, the last strategy consists in applying external forces (such as holding a bar) or changing the base of support. In this method, we propose to successively apply these strategies to mimic the natural behavior of humans to maintain balance.

When the CoP runs out of the base of support, the first strategy consists in correcting the posture by rotating the body around the ankle. Using the angular momentum of Equation 2, we compute $\mathbf{C}$ for posture q

$$\mathbf{C} = \mathbf{OG} \times \mathbf{P} + \mathbf{OCoP} \times \mathbf{GRF} + \sum_j \mathbf{OA_j} \times \mathbf{F}_j \quad (3)$$

Regarding Equation 2, if the posture is statically balanced, $\mathbf{C}$ should be null. As we do not have access to a captured value of the CoP, we use the CoP trajectory of the original motion (computed with static equation 2 for original motion without external forces). However, when new forces $\mathbf{F}_j$ are applied on a posture, this equation usually does not hold anymore. An additional torque is introduced by each force $\mathbf{F}_j$. In that case, we search for the ankle rotation θ_{ankle} that ensures a null value of $\mathbf{C}$. The relation between $\mathbf{C}$ and θ_{ankle} can be linearized using the Jacobian matrix $\mathbf{J}$

$$\Delta C = J \Delta \theta_{ankle} \quad (4)$$

By inversing Equation 4, it is possible to compute a linear approximation of θ_{ankle}. However, as dynamics is not linear, it is necessary to iterate around the solution until convergence. In most of the cases, a few iterations are sufficient to obtain θ_{ankle} which leads to a null value of $\mathbf{C}$. This corresponds to the computation of the static configuration respecting as much as possible the static CoP position of the original motion. Figure 2 shows two examples of posture correction by rotating around the ankle. The shadowed character represents the original captured motion (without external forces), while the apparent character corresponds to a static correction with additional external forces (respectively pushing 30kg and 80kg). In these examples, pushing was done using two arms and the external force was equally divided between both arms.

When this strategy is not enough to preserve balance, as suggested by Hof [29], the character has to adapt his base of support. It consists in adapting the feet location prior to the interaction, as described in the next subsection.

(a) (b)

Fig. 2. Shadowed character: original posture where the character is pushing nothing (0kg). Apparent character: adapted pose using the static solver a) pushing 30kg cupboard and b) pushing 80kg cupboard.

4.2 Dynamic Correction of the Motion

The experiment described in Section 3 has shown that the velocity profile of some body parts conveys a lot of information about the dynamics of the motion. Thus, only modifying the global pose of the character without considering time is not enough to take this information into account. To this end, we have applied a time-warping algorithm to the motion, as proposed in previous works [19]. Hence, once the interaction has begun, the system changes the posture and then retimes the current action.

4.3 Preparation to the Interaction Phase

The preparatory phase to a new upcoming dynamic constraint is divided in three phases. Detecting a new constraint in a near future corresponds to an upcoming event that will modify the character's actions. As the dynamic properties of the object are generally different from those used during the motion capture session, it is necessary to modify the motion to take these new properties into account. When dealing with the manipulation of virtual objects, it mainly corresponds to adding new forces exerted on the virtual character ($\mathbf{F}_j$ in equation 2).

When a constraint is detected at the time of interaction t_i, our solver starts by computing a correct static posture ensuring balance with the given forces at t_i, as described in Section 4.1. After having observed several behaviors during the motion capture session used for this paper, we assume that the supporting foot is generally close to the projection of the center of mass on the ground at interaction time. Based on this empirical knowledge, the static pose at t_i is also modified to ensure that the supporting foot remains close to this projection in order to maintain balance. This is performed by simply changing the support foot location and applying inverse kinematics [30].

Then, as the posture at t_i is different than the one corresponding to the motion capture data, the sequence of footprints generally has to be modified. It is necessary to compute the new foot sequences to reach the wanted configuration (second step). Two cases can occur:

1. Several steps exist between the current time and the time of interaction (i.e. the motion involves displacements): it is possible to modify the existing footprint sequence to reach the required configuration,
2. There is no displacement or the location of the next interaction is too close to the current position: it is necessary to generate new steps to reach the required configuration.

Once the new foot sequences are computed, the preparatory motion is adapted to reach the target posture while respecting the new foot sequences.

5 Results

We have applied the results of our method to various motions involving interactions with different dynamic properties. Figure 3 shows the current posture (shadowed characters) when an interaction is detected in the near future. The apparent character represents the posture after adaptation at interaction time. Figure 4 compares the results of our method to experimental data. Compared to

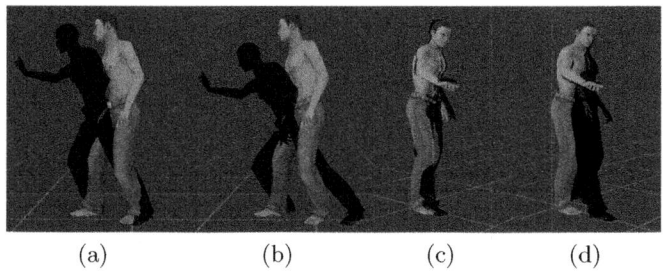

(a)   (b)   (c)   (d)

Fig. 3. Apparent character: current posture of the simulation. Shadowed character: corrected posture at the time of interaction to handle the physical constraint. Motions: pushing with an additional force of a) 50N and b) 150N, and pulling with an additional force of c) 50N and d) 150N.

Table 2. Comparison of the method for different scenarios (Motion × Additional force). A: time needed to compute the correct posture for the time of interaction. B: mean time needed for the foot sequence modification/generation of one frame. C: mean time needed for the dynamic correction of one frame and D: mean number of iterations needed to converge toward a correct static posture with the method of section 4.1.

Motion	Additional force	A	B	C	D
Push right hand	20N	0.23ms	0.005ms	0.27ms	1
	100N	0.24ms	0.005ms	0.35ms	2
	200N	0.26ms	0.005ms	0.51ms	2.15
Pull right hand	20N	0.24ms	0.005ms	0.43ms	1.92
	100N	0.24ms	0.005ms	0.40ms	2
	200N	0.24ms	0.005ms	0.42ms	1.99

the original sequence (pulling or pushing nothing), the resulting motion seems to be more adapted to reacting to an external force of 200N. For the pulling scenario, the actor decided to perform a completely different strategy and this preliminary result should be confirmed by further experiments.

Table 2 presents computation time for the different phases (standard laptop Dualcore 2.33GHz CPU, 2GB RAM and NVidia Quadro FX 2500M GC). Column A presents the time needed to compute the correct interaction posture and the new foot sequences in reaction to the upcoming interaction. Column B and C present respectively the average time for correction of postures in the preparatory phase (to satisfy the new feet positions and prepare the character for interaction) and during the interaction phase. Column D presents the number of iterations for the linearization of the static equations of Section 4.1. For the implementation presented in this paper, the maximum computation time for one frame is always under 1ms per character. Thus, it is possible to animate up to 15 characters with dynamic interactions at 60Hz, without visualization.

To evaluate the error committed during this process, we have compared the trajectory of the CoP (thanks to Equation 1) for the original and the corrected motion (see Figure 5). In this figure, the curve in the middle represents the original motion without changes, the curve on top one depicts the trajectory of the CoP if we consider 200N additional force without adaptation, and the curve in the bottom one stands for the final adapted motion. By definition, the

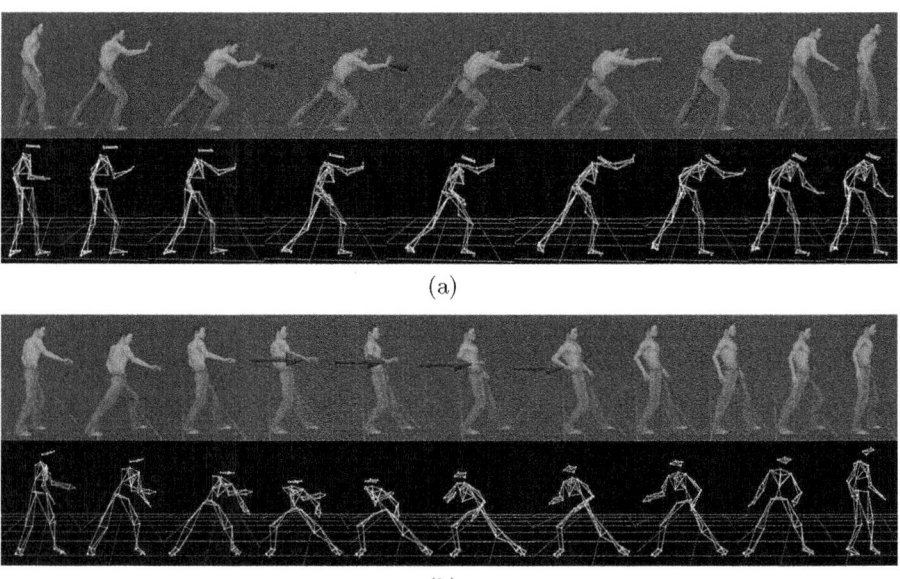

(a)

(b)

Fig. 4. Motions subject to a 200N external perturbation corrected with our method compared to motions interacting with a similar intensity captured on a real subject for a) pushing and b) pulling

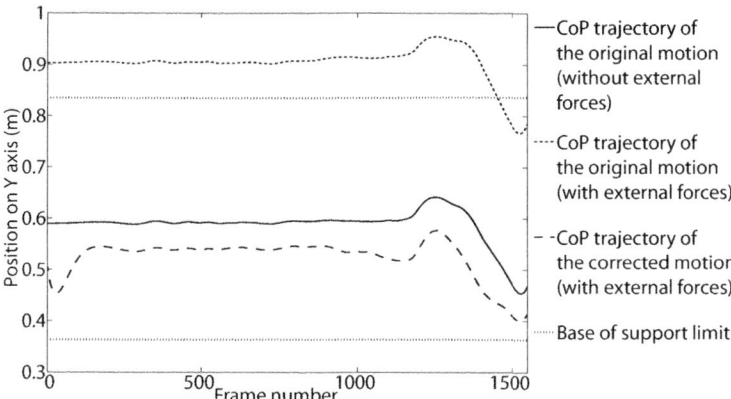

Fig. 5. Trajectory of the CoP for pulling with an external force of 200N. Original motion without taking external force into account (middle), taking this force into account but without adaptation (top) and final adapted motion (bottom).

trajectory of the CoP is restricted to remain inside the base of support (depicted with black dot lines). Figure 5 clearly shows that the CoP of the adapted motion remains close to the CoP of the original one, inside the base of support.

6 Conclusion

The first contribution of this paper is to identify the relevant kinematic parameters that convey information about the dynamic constraints of the environment. To this end, we have carried-out a specific experiment based on carrying dumbbells with various masses. The preliminary results show that the velocity profile is clearly affected by the external load carried by the subject. Hence, we could assume that this change in the velocity profile could be perceived by people who are watching the scene. Past works [5] have shown that this perception skill is not very accurate, which enabled us to make some simplifications of the simulation process. All these results are preliminary and should be verified for various motions, with a wider range of external loads and various types of performers.

Based on these preliminary results we have proposed a two-steps process to adapt the motion to physical perturbations, such as pushing, pulling, or carrying various masses. This process is based on the assumption that the pose and the velocity profile can be adapted separately. This assumption obviously leads to inaccuracies but we assume that it remains acceptable from the perception point of view, as suggested by the previous results [5]. Hence, the first step consists in solving the problem by considering the system as static (null accelerations). At this stage, the base of support and the global orientation of the character are adapted, as suggested in the biomechanical literature to maintain balance [29]. If necessary, either the few preliminary footprints are adapted (if the task follows displacements) or the system generates new steps (if no displacement occur) to ensure balance (by adapting the base of support).

The second step consists in retiming the motion, as suggested in [19]. This is an important issue as it is supposed to convey lot of information about the dynamic status of the system. The results show that computation time enables us to animate dozens of characters at 30Hz on a common PC. Other methods, such as physical simulation with controllers or spacetime constraints would certainly not reach this performance. Now, the key question is to determine if people can perceive the errors induced by our method. As a perspective we thus wish to carry-out perceptual experiments to check this point.

Using the perception skills of subjects to design appropriate simulation methods is a promising approach for the future. The studies and the methods reported in this paper are clearly preliminary but offer promising results. They should be improved by further experiments and biomechanical analyzes.

Acknowledgments

This work was partially funded by Rennes Metropole through the "Soutien la mobilité des doctorants" project.

References

1. Zordan, V.B., Majkowska, A., Chiu, B., Fast, M.: Dynamic response for motion capture animation. ACM Trans. Graph. 24(3), 697–701 (2005)
2. Yin, K., Loken, K., van de Panne, M.: Simbicon: simple biped locomotion control. ACM Trans. Graph. 26(3), 105 (2007)
3. Macchietto, A., Zordan, V., Shelton, C.R.: Momentum control for balance. ACM Trans. Graph. 28(3) (2009)
4. Shiratori, T., Coley, B., Cham, R., Hodgins, J.K.: Simulating balance recovery responses to trips based on biomechanical principles. In: Proceedings of SCA 2009, pp. 37–46 (2009)
5. Hoyet, L., Multon, F., Komura, T., Lecuyer, A.: Can we distinguish biological motions of virtual humans? Perceptual study with captured motions of weight lifting. To appear in the Proceedings of VRST 2010 (2010)
6. Hodgins, J.K., Wooten, W.L., Brogan, D.C., O'Brien, J.F.: Animating human athletics. In: Proceedings of SIGGRAPH 1995, pp. 71–78 (1995)
7. Faloutsos, P., van de Panne, M., Terzopoulos, D.: Composable controllers for physics-based character animation. In: Proceedings of SIGGRAPH 2001, pp. 251–260 (2001)
8. Faloutsos, P., van de Panne, M., Terzopoulos, D.: The virtual stuntman: dynamic characters with a repertoire of autonomous motor skills. Computers and Graphics 25(6), 933–953 (2001)
9. Zordan, V.B., Hodgins, J.K.: Motion capture-driven simulations that hit and react. In: Proceedings of SCA 2002, pp. 89–96 (2002)
10. Witkin, A., Kass, M.: Spacetime constraints. In: Proceedings of SIGGRAPH 1988, pp. 159–168 (August 1988)
11. Cohen, M.F.: Interactive spacetime control for animation. SIGGRAPH Comput. Graph. 26(2), 293–302 (1992)

12. Liu, C.K., Popović, Z.: Synthesis of complex dynamic character motion from simple animations. In: Proceedings of SIGGRAPH 2002, pp. 408–416 (2002)
13. Liu, C.K., Hertzmann, A., Popović, Z.: Learning physics-based motion style with nonlinear inverse optimization. ACM Trans. Graph. 24(3), 1071–1081 (2005)
14. Fang, A.C., Pollard, N.S.: Efficient synthesis of physically valid human motion. In: Proceedings of SIGGRAPH 2003, pp. 417–426 (2003)
15. Safonova, A., Hodgins, J.K., Pollard, N.S.: Synthesizing physically realistic human motion in low-dimensional, behavior-specific spaces. ACM Trans. Graph. 23(3), 514–521 (2004)
16. Komura, T., Ho, E.S.L., Lau, R.W.H.: Animating reactive motion using momentum-based inverse kinematics: Motion capture and retrieval. Comput. Animat. Virtual Worlds 16(3-4), 213–223 (2005)
17. Arikan, O., Forsyth, D.A., O'Brien, J.F.: Pushing people around. In: Proceedings of SCA 2005, pp. 59–66 (2005)
18. Mitake, H., Asano, K., Aoki, T., Salvati, M., Sato, M., Hasegawa, S.: Physics-driven multi dimensional keyframe animation for artist-directable interactive character. Comput. Graph. Forum 28(2), 279–287 (2009)
19. Majkowska, A., Faloutsos, P.: Flipping with physics: motion editing for acrobatics. In: Proceedings of SCA 2007, pp. 35–44 (2007)
20. Yamane, K., Nakamura, Y.: Dynamics filter - concept and implementation of online motion generator for human figures. In: Proceedings of the 2000 IEEE International Conference on Robotics and Automation, pp. 688–695 (2000)
21. Pollard, N., Reitsma, P.: Animation of humanlike characters: Dynamic motion filtering with a physically plausible contact model. In: Proc. of Yale Workshop on Adaptive and Learning Systems (2001)
22. Tak, S., Ko, H.S.: A physically-based motion retargeting filter. ACM Trans. Graph. 24(1), 98–117 (2005)
23. Shin, H.J., Kovar, L., Gleicher, M.: Physical touch-up of human motions. In: Proceedings of the 11th Pacific Conference on Computer Graphics and Applications, PG 2003, p. 194 (2003)
24. O'Sullivan, C., Dingliana, J., Giang, T., Kaiser, M.K.: Evaluating the visual fidelity of physically based animations. ACM Trans. Graph. 22(3), 527–536 (2003)
25. Reitsma, P.S.A., O'Sullivan, C.: Effect of scenario on perceptual sensitivity to errors in animation. ACM Trans. Appl. Percept. 6(3), 1–16 (2009)
26. Yeh, T.Y., Reinman, G., Patel, S.J., Faloutsos, P.: Fool me twice: Exploring and exploiting error tolerance in physics-based animation. ACM Trans. Graph. 29(1) (2009)
27. Bingham, G.P.: Kinematic form and scaling: further investigations on the visual perception of lifted weight. Journal of Experimental Psychology. Human Perception and Performance 13(2), 155–177 (1987)
28. Runeson, S., Frykholm, G.: Visual perception of lifted weight. Journal of Experimental Psychology: Human Perception and Performance 7, 733–740 (1981)
29. Hof, A.L.: The equations of motion for a standing human reveal three mechanisms for balance. J. Biomech. 40, 451–457 (2007)
30. Kulpa, R., Multon, F., Arnaldi, B.: Morphology-independent representation of motions for interactive human-like animation. Computer Graphics Forum, Eurographics 2005 Special Issue 24, 343–352 (2005)

Realistic Emotional Gaze and Head Behavior Generation Based on Arousal and Dominance Factors

Cagla Cig[2], Zerrin Kasap[1], Arjan Egges[2], and Nadia Magnenat-Thalmann[1]

[1] MIRALab, University of Geneva, Switzerland
[2] Department of Information and Computing Sciences, Universiteit Utrecht, The Netherlands
{ccig,egges}@cs.uu.nl, {zerrin.kasap,thalmann}@miralab.ch

Abstract. Current state-of-the-art virtual characters fall far short of characters produced by skilled animators in terms of behavioral adequacy. This is due in large part to the lack of emotional expressivity in physical behaviors. Our approach is to develop emotionally expressive gaze and head movement models that are driven parametrically in real-time by the instantaneous mood of an embodied conversational agent (ECA). A user study was conducted to test the perceived emotional expressivity of the facial animation sequences generated by these models. The results showed that changes in gaze and head behavior combined can be used to express changes in arousal and/or dominance level of the ECA successfully.

Keywords: Emotional gaze and head behavior, expressive animation, virtual humans, facial animation, natural motion simulation.

1 Introduction

One of the ultimate goals of the research on ECAs is to build believable, trustworthy, likeable, human- and life-like virtual characters. This involves, amongst other things, having the character display the appropriate signs of a changing mood, a recognizable personality and a rich emotional life. One of the many ways of conveying these signs is through emotionally expressive gaze and head behavior. Gaze and head movements need to be appropriately modeled and advantageously included in realistic facial animations of human-like models in order to more effectively and completely mimic human behavior. Despite the numerous models of gaze and head behavior in ECAs, there currently has been very little exploration of how changes in emotional state affect changes in the manner of gaze and head behavior.

We employ a parametric approach based on behavioral sciences and human physiology to model lifelike gaze and head behavior. The first criterion is the realism of the animations, thus appropriate models are chosen separately for both gaze and head movement generation. Second, both behavior generators are linked with the emotional state machine developed at MIRALab [11], where mood is represented with Mehrabian's Pleasure-Arousal-Dominance (PAD) Temperament Model [14]. Arousal and dominance dimensions are used to drive the parameters of the gaze and head movement models.

2 Related Work

To date, only a few works have explicitly considered an ECA capable of revealing its emotional state through the manner of its gaze and/or head behavior. We will first discuss previous work on modeling gaze, then we will discuss previous work on modeling head movements.

Fukayama et al. [6] used an animated pair of eyes to display affective signals to observers. Queiroz et al. [17] have proposed a parametric model for automatically generating emotionally expressive gaze behavior. The authors manually annotated scenes from Computer Graphics movies with labels describing emotional content and gaze behavior and utilized this behavioral database to create their parametric model. The major shortcoming of their work results from the fact that overall emotional expressivity is limited by the predefined set of emotions and implemented gaze behaviors. Their system does not simulate the internal dynamics of emotions and can only express six different types of basic emotions. This shortcoming has been observed in a majority of emotionally expressive gaze and head movement models.

Lance et al. have presented the Gaze Warping Transformation [12], a method of generating emotionally expressive head and torso movements during gaze shifts that is derived from the difference between human motion data of emotionally neutral and emotionally expressive gaze shifts. Busso et al. [2] employed a technique that entails quantizing motion-captured head poses in a finite number of clusters and building a Hidden Markov Model (HMM) for each of these clusters based on the prosodic features of the accompanying speech. Additionally, different sets of HMMs are built and trained for each emotional category (i.e. sad, happy, angry and neutral). In the synthesis step, the most likely head movement sequence is generated based on prosodic features and emotional content of the input speech. The results of their user study have shown that perceived emotion was different from the intended one in some cases; therefore their model is not able to completely preserve the emotional content of the input signal.

3 Emotionally Expressive Gaze and Head Movement Models

For realistic gaze and head movement generation, our approach is to develop emotionally expressive gaze and head movement models that are driven parametrically in real-time by the instantaneous mood of an ECA, where mood is represented with Mehrabian's PAD Temperament Model [14]. This way, emotional expressivity is tied to the character's internal emotional state rather than triggered by a limited set of discrete emotional labels. Pleasure/displeasure level relates to positivity or negativity of the emotional state, arousal/non-arousal shows the level of physical activity and mental alertness and dominance/submissiveness indicates the feeling (or lack) of control. Evidence for the relationship of gaze and/or head movements to the display of pleasure is relatively limited when compared with the arousal and dominance dimensions; therefore our models are parameterized continuously based on arousal and dominance dimensions only. In addition to this, our models are enhanced by the integration of custom designed animation files. This way, they benefit both from the richness of offered detail provided by the animation files and also from the advantage of being able to generate flexible animations according to high-level parameters provided by the parametric approach. In this section, we will first detail the parametric approach, then the integration of custom designed animation files and lastly our emotionally expressive gaze and head movement models.

3.1 Parametric Approach

The parameters used to drive the emotionally expressive gaze model can be classified into two categories. The first category is related to blinking and the only parameter utilized is *blink rate*, which is measured by the number of blinks per minute. The second category is related to gaze aversion and the related choice of parameters has been inspired by the works of Lee *et al.* [13] and Fukayama *et al.* [6]. Noting that saccades are rapid movements of both eyes from one gaze position to another, implications of the gaze aversion-related parameters are listed as follows:

- *Saccade duration* – amount of time taken to complete a saccade
- *Inter-saccade interval* – amount of time that elapses between the termination of one saccade and beginning of the next one
- *Saccade velocity* - average rate of change of position of the eyes
- *Gaze points while averted*– region that delineates the gaze points in gaze-averted mode (controlled by manipulating the *saccade direction* and *saccade magnitude* parameters)
- *Total amount of gaze aversion* – ratio between the total amount of time spent during eyes in gaze-averted mode and the total amount of time spent during direct-gaze mode

The parameters used to drive the emotionally expressive head movement model are identical to the aversion related parameters utilized in the emotionally expressive gaze model except that in this context, saccades are rapid movements of the head from one position to another.

3.2 Custom Designed Animation Files

The parametric gaze and head movement models are enhanced by the integration of custom designed MPEG-4 Facial Animation Parameter (FAP) files [9]. One of these files is utilized for generating idle gaze and head behavior animation while the other two are for generating dominant and submissive gaze and head behavior animations, respectively.

Aside from improving the realism of the resulting gaze and head behavior animations, the custom designed FAP files are also employed for determining the directions of the saccades for both the eyes and the head during gaze aversion. Basically, each animation file specifies particular patterns of gaze and head behavior and the frequencies of the animation files being played are linked to the instantaneous dominance level by a probabilistic approach. This approach can be seen as a simple motion graph [10]. There are three distinct FAP animation files forming the nodes of the motion graph (see Fig. 1). These files specify values for the orientation of the eyeballs and the head. Each node is connected to every other node including itself and allows traversal to and from the connected nodes. Emotional changes in the dominance dimension trigger an immediate traversal from the current node to the future one.

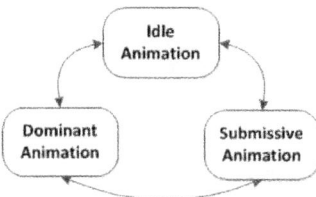

Fig. 1. Simple motion graph built from the three animation files employed in our research

3.3 Emotionally Expressive Gaze Model

In this section, we give an overview of the parameters used for the related dimensions of the PAD Temperament Model and the patterns of change for each parameter.

Arousal. Arousal can be linked to both blink-related and aversion-related parameters. Blink rate is increased during states of anxiety or tension, but decreased during concentrated thinking or visual attention [8,1]. Additionally, there is a strong relationship between gaze aversion and arousal. For example, embarrassment is an emotion with a high level of arousal and it has been shown that people exhibit less eye contact when asked embarrassing questions than when asked plain question [4]. Gaze aversion has also been used for detecting deception which invokes emotions with high levels of arousal. Some researchers claim that one reason we avert our gaze is to reduce heightened physiological arousal caused by prolonged mutual gaze [5]. The subset of gaze model parameters utilized in our model for displaying arousal/non-arousal and the corresponding relationships can be seen in Fig. 2.

Gaze Model Parameters	Relationship with Arousal
Blink Rate	Direct Proportion
Saccade Duration	Inverse Proportion
Inter-saccade Interval	Inverse Proportion
Saccade Velocity	Direct Proportion
Saccade Magnitude	Direct Proportion
Total Amount of Gaze Aversion	Direct Proportion

Fig. 2. Overview of the high-level relationship between gaze model parameters and arousal

Dominance. The relationship between gaze and dominance has been suggested by various studies that show changes in the gaze direction or frequency of gaze aversion as a function of dominance. The majority of the research conducted in this area suggests that gaze aversion serves as a sign of submissiveness that subordinates display to dominant individuals whereas staring at another person is a dominance gesture [3]. In our gaze model, direct eye contact in the form of an unwavering gaze is used to express dominance while gaze aversion is used to express submissiveness. Fukayama *et al.* [6] conducted a user study and concluded that low gaze position induces a lower dominance impression than the other positions (i.e. random/upward/downward/lateral gaze). Based on this observation, in our gaze model, it was decided to restrict the positioning of the gaze points while averted to the lower part of the eye by modifying the *saccade direction* parameter accordingly. For this purpose, the aforementioned custom designed submissive animation file is utilized. The frequency of submissive animation file being played is linked to dominance level by a probabilistic approach which will be detailed later under the section presenting the integration of the dominance dimension to our head movement model. The subset of gaze model parameters utilized in our model for displaying dominance/submissiveness and the corresponding relationships can be seen in Fig. 3.

Gaze Model Parameters	Relationship with Dominance
Saccade Direction	Probabilistic Approach
Saccade Magnitude	Inverse Proportion
Total Amount of Gaze Aversion	Inverse Proportion

Fig. 3. Overview of the high-level relationship between gaze model parameters and dominance

3.4 Emotionally Expressive Head Movement Model

In this section, we give an overview of the parameters used for the related dimensions of the PAD Temperament Model and the patterns of change for each parameter.

Arousal. While there has been little work on how arousal/non-arousal is specifically related to head movements, velocity has been shown to be an indicator of arousal [16]. Lance et al. [12] have found that low arousal head movement animations are 80% slower than high arousal head movement animations. The subset of gaze model parameters utilized in our model for displaying arousal/non-arousal and the corresponding relationships can be seen in Fig. 4.

Head Movement Model Parameters	Relationship with Arousal
Saccade Duration	Inverse Proportion
Inter-saccade Interval	Inverse Proportion
Saccade Velocity	Direct Proportion
Saccade Magnitude	Direct Proportion
Total Amount of Head Aversion	Direct Proportion

Fig. 4. Overview of the high-level relationship between head movement model parameters and arousal

Dominance. Performing appeasement and dominance gestures is quite frequent, especially in primates. For instance, a subordinate chimpanzee greets dominant ones by crouching and bowing its upper body almost to the ground whereas a more dominant one adopts a vertical erect posture and raises its body hairs to increase in stature. Head behavior in humans is also a component of dominance expression homologous to those of animals. The study of Mignault et al. [15] shows that a bowed head is perceived as submissive and displaying inferiority emotions whereas when a head is raised, the face is perceived as more dominant and displaying greater superiority emotions. Lance et al. [12] have found that low dominance head pitch is 25 degrees lower than high dominance head pitch. In our head movement model, a raised head is used to express dominance while a bowed head is used to express submissiveness. The aforementioned custom designed animation files are used to control the relationship between dominance/submissiveness and *saccade direction* parameter. One of these files is for generating an animation sequence consisting mainly of bowed head behavior that suggests submissiveness whereas the other is for generating an animation sequence consisting mainly of raised head behavior that suggests dominance. The frequencies of submissive and dominant head behavior animations being played are linked to dominance level by a probabilistic approach. The details of the probabilistic occurrence model can be seen in Fig. 5.

On the other hand, the relationship between dominance/submissiveness and *saccade magnitude* parameter depends on which animation file is currently being played. If the idle animation file is being played, the relationship is inversely proportional in order to allow for a more direct eye contact. This is also the case for the submissive animation file, however for a different reason. As the dominance value is decreased, *total amount of head aversion* is increased by increasing *saccade magnitude* in order to emphasize the head movements specified in the animation file that mainly consists of bowed head behavior. On the other hand, the opposite case applies to the dominant animation file. As the

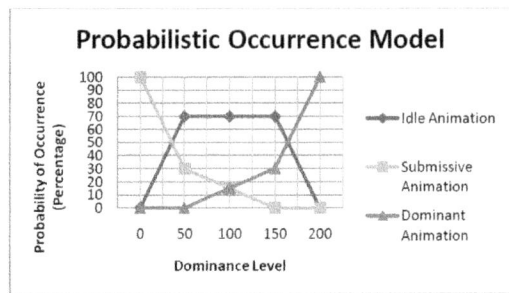

Fig. 5. Probabilistic occurrence model showing the relationship between dominance level and percentage of occurrence for the animation files

dominance value is increased, *total amount of head aversion* is increased by increasing *saccade magnitude* in order to emphasize the head movements specified in the animation file that mainly consists of raised head behavior. The subset of head movement model parameters utilized in our model for displaying dominance/submissiveness and the corresponding relationships can be seen in Fig. 6.

Head Movement Model Parameters	Relationship with Dominance
Saccade Direction	Probabilistic Approach
Saccade Magnitude	Inverse Proportion
Total Amount of Head Aversion	Inverse Proportion

Fig. 6. Overview of the high-level relationship between head movement model parameters and dominance

See Fig. 7 for example frames taken from the facial animation sequences generated by our emotionally expressive gaze and head movement models.

4 User Study

A user study was conducted to test the perceived emotional expressivity and naturalness of the facial animation sequences generated by our gaze and head movement models.

4.1 Participants

The study included 30 participants (21 males and 9 females) aged 21-46 with an approximate average age of 26 years and all selected from the students or staff of the University of Utrecht. 28 participants had no to little experience with affective computing and 26 participants had no to little experience with robots.

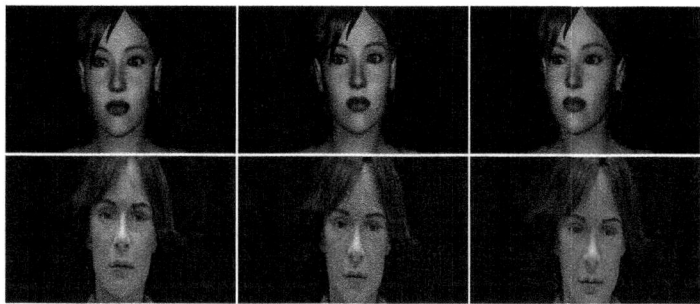

Fig. 7. Virtual and robotic MIRALab-ECAs expressing maximum (above) and minimum (below) levels of dominance

4.2 Design and Procedure

The experiment consisted of two sections, one for the virtual ECA and another for the robotic ECA. Moreover, each section was composed of two parts corresponding to two different tasks. A within-subjects design was employed, thus each participant completed both tasks of both sections. The transfer of learning effects was lessened by varying the order in which the tasks were tackled. The tasks basically involved the participants watching videos on a monitor and rating the emotional state of the ECA in the videos. The videos were recorded using a Sony DCR-VX2100E MiniDV digital camcorder. The participants watched the videos on a 1440x900 17" Toshiba TruBrite WXGA TFT High Brightness display monitor that was located approximately 55 cm in front them.

In the first part of the user study, test runs were generated by varying the expression type (gaze, head, combined), emotional dimension (arousal, dominance) and emotional level (-100, -50, +50, +100) variables. For the robotic ECA, using gaze or head behavior alone for emotional expressivity was not tested and thus the value of the expression type variable was set to *combined* for all test runs. During each test run, the participants were shown 2 videos consisting of the ECA performing emotionally neutral and emotionally expressive nonverbal behaviors, respectively. Each video was 10 seconds long. At the end of each test run, the participants were presented with a row of 6 SAM figures [7] illustrating arousal/dominance (see Fig. 8).

The participants were instructed that the leftmost figure corresponded to the first video and they were asked to circle the figure that they thought best corresponded to the second video among the remaining figures on the right. The leftmost figure was always the figure corresponding to neutral arousal/dominance level and participants were asked to rate the arousal/dominance level in the second video relative to the first video. The hypothesis under inspection was whether changes in gaze and/or head behavior generated by our gaze and head movement models can be used to express changes in arousal or dominance levels of the ECA successfully.

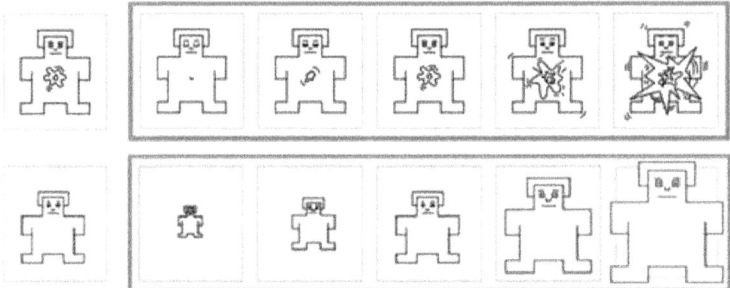

Fig. 8. SAM figures utilized for evaluating the perceived emotional expressivity in terms of arousal (above) and dominance (below) dimensions

The second part of the user study consisted of 4 test runs: one for positive arousal and positive dominance, one for positive arousal and negative dominance, one for negative arousal and positive dominance and lastly, one for negative arousal and negative dominance. During each test run, the participants were shown a video of 30 seconds duration. Each video consisted of the ECA performing emotionally expressive nonverbal behaviors. In each test run, after watching the video, the participants were presented with two rows of SAM figures corresponding to arousal and dominance levels, respectively. The participants were asked to circle one figure from each row of 5 figures that they thought best corresponded to the video. Following this, they were asked to circle the main emotion that they thought was expressed in the video from a group of 8 emotional descriptors listed as follows: *anxious, bored, dependent, disdainful, docile, exuberant, hostile and relaxed*. The only major difference between the first and second parts of the user study regarding the SAMs system was the number of emotional dimensions evaluated in a test run. In the first part, emotional dimensions were tested separately whereas they were tested simultaneously in the second part. Simultaneous testing made it possible to inspect the emotional expressivity of the gaze and head movement models in a more practical setup as well as the possible relationships between arousal and dominance dimensions in the context of emotional expressivity.

Lastly, at the end of each section of the user study, the participants were asked to rate the following aspects of the animation sequences using a 5-point Likert scale: naturalness, contribution of gaze behavior to emotional expressivity, contribution of head behavior to emotional expressivity and lastly, synchrony and consistency between gaze and head behaviors.

4.3 Results

In the first part of the current research, Wilcoxon Signed-Rank Test was employed for the statistical analysis of the results. The important results are listed as follows:

- For the virtual ECA, changes in gaze behavior can be used to express changes in arousal level of the ECA successfully (Z = -4.411, p = 0,000). However, changes in gaze behavior alone cannot be used to express changes in dominance level of the ECA successfully (Z = -0.451, p = 0.652).
- For the virtual ECA, changes in head behavior can be used to express changes in arousal (Z = -3.428, p = 0.001) and dominance (Z = -4.612, p = 0.000) levels of the ECA successfully.
- For the virtual ECA, changes in gaze and head behavior combined can be used to express changes in arousal (Z = -4.819, p = 0.000) and dominance (Z = -4.719, p = 0.000) levels of the ECA successfully.
- For the robotic ECA, changes in gaze and head behavior combined can be used to express changes in arousal (Z = -4.608, p = 0.000) and dominance (Z = -4.813, p = 0.000) levels of the ECA successfully.

In the second part of the current research, Wilcoxon Signed-Rank Test was employed for the statistical analysis of the results regarding the SAMs system. On the other hand, for the part of the statistical analysis regarding the emotional descriptors, Pearson's Chi-Square Test was employed. It is important to note that the chance of determining the correct descriptor by a random guess was 25% since in each test, 2 adjectives among the set of 8 were considered as correct. This value was used as the expected value during Pearson's Chi-Square Tests. The important results are listed as follows:

- Changes in gaze and head behavior combined can be used to express changes in arousal/dominance level of the virtual/robotic ECA successfully. This holds true also when the emotional level of the ECA is being changed simultaneously in the other emotional dimension.
- Changes in gaze and head behavior combined can cause the participants to perceive nonexistent changes in arousal/dominance level of the virtual/robotic ECA. This undesirable effect was experienced more often with the virtual ECA than with the robotic ECA.
- Gaze and head behavior combined could convey the correct descriptor of the emotion expressed by the ECA in 50% and 75% of the cases for the virtual and robotic ECAs, respectively.

Lastly, the important results regarding the overall impressions of the participants are listed as follows:

- A Wilcoxon Signed-Ranks Test showed that perceived naturalness of the virtual and robotic ECAs differed significantly (Z = -3.389, p = 0.001). Virtual ECA was found to be more natural than the robotic ECA.
- A Wilcoxon Signed-Ranks Test showed that perceived contribution of gaze and head behaviors differed significantly for both ECAs (Z = -2.758, p = 0.006 for the virtual ECA and Z = -3.081, p = 0.002 for the robotic ECA). In both cases, head behavior was found to have a bigger contribution to emotional expressivity in the resulting facial animation sequences than gaze behavior.

5 Conclusions and Discussion

For realistic gaze and head movement generation, we developed emotionally expressive gaze and head movement models and conducted a user study to test the perceived emotional expressivity and naturalness of the facial animation sequences generated by these models. The most important results of the user study regarding the perceived emotional expressivity are listed as follows. For the virtual ECA, changes in gaze behavior can be used to express changes in arousal level of the ECA successfully. Moreover, for the virtual ECA, changes in head behavior can be used to express changes in arousal/dominance level of the ECA successfully. Lastly, for both the virtual and robotic ECAs, changes in gaze and head behavior combined can be used to express changes in arousal/dominance level of the ECA successfully. This holds true also when the emotional level of the ECA is being changed simultaneously in the other emotional dimension.

Virtual ECA was found to be significantly more natural than the robotic ECA. However, unexpected cases regarding the participants perceiving nonexistent changes in arousal/dominance levels occurred more often with the virtual ECA than with the robotic ECA. In addition to this, participants performed the task regarding determining the correct descriptor of the emotion expressed by the ECA more successfully with the robotic ECA than with the virtual ECA. This can be explained with the uncanny valley effect. Robotic ECA was found to be less natural but emotionally more expressive than the virtual ECA.

The contribution of the gaze model to the overall emotional expressivity was found to be significantly less than the contribution of the head movement model. In addition to this, gaze behavior generated by the current model did not suffice to express changes in dominance level of the virtual ECA successfully. Therefore, more research should be conducted on the display of emotions with gaze behavior (especially in the dominance dimension) and findings should be incorporated into the emotionally expressive gaze model. For instance, parameters related with pupil diameter and eye openness can be integrated into the gaze model since both factors are expected to change in accordance with the changes in emotional levels.

For both models, gaze aversion-related parameters play a major part in specifying the relationships between nonverbal behaviors and emotional states. However, modeling gaze aversion without using an eye tracking mechanism is not very meaningful since for instance, a gaze shift intended for gaze aversion can result in an unintended direct stare and vice versa depending on the relative locations of the ECA and the interacting user. Therefore, integration of an eye tracking mechanism can possibly improve the expressivity of the gaze and head movement models. Moreover, the effects of this improvement can be particularly observable in the case of gaze conveying changes in dominance levels.

Acknowledgements

This study is partly funded by the EU Projects 3DLife (IST-FP7 247688). We would like to thank Nedjma Cadi-Yazli for the preparation of the animation files which was crucial for this work and Maher Ben Moussa for his contributions on the gaze and head deformation library.

References

1. Argyle, M.: Bodily Communication. Methuen, London (1998)
2. Busso, C., Deng, Z., Neumann, U., Narayanan, S.: Learning Expressive Human-Like Head Motion Sequences from Speech. In: Data-Driven 3D Facial Animations, pp. 113–131. Springer, New York (2007)
3. Dovidio, J.F., Ellyson, S.L.: Patterns of visual dominance behavior in humans. In: Power, Dominance, and Nonverbal Behavior, pp. 129–149. Springer, New York (1985)
4. Exline, R., Gray, D., Schuette, D.: Visual behavior in a dyad as affected by interview content and sex of respondant. Journal of Personality and Social Psychology 1, 201–209 (1965)
5. Field, T.: Infant gaze aversion and heart rate during face-to-face interactions. Infant Behaviour and Development 4, 307–315 (1981)
6. Fukayama, A., Ohno, T., Mukawa, N., Sawaki, M., Hagita, N.: Messages embedded in gaze of interface agents - impression management with agent's gaze. In: Proceedings of the ACM CHI 2002 Conference on Human Factors in Computing Systems Conference, pp. 41–48. ACM Press, Minnesota (2002)
7. Grimm, K., Kroschel, K.: Evaluation of natural emotions using self assessment manikins. In: Proceedings of the IEEE Workshop on Automatic Speech Recognition and Understanding, pp. 381–385. IEEE Press, Mexico (2005)
8. Harris, C.S., Thackray, R.I., Shoenberger, R.W.: Blink rate as a function of induced muscular tension and manifest anxiety. Perceptual and Motor Skills 22, 155–160 (1966)
9. International Standards Office. ISO/IEC IS 14496-2 Information technology - Coding of audio-visual objects - Part 2: Visual. ISO, Geneva (1999)
10. Kovar, L., Gleicher, M., Pighin, F.: Motion graphs. In: Proceedings of ACM SIGGRAPH 2002, vol. 21, pp. 473–482 (2002)
11. Kasap, Z., Moussa, M.B., Chaudhuri, P., Magnenat-Thalmann, N.: Making Them Remember-Emotional Virtual Characters with Memory. IEEE Computer Graphics and Applications 29, 20–29 (2009)
12. Lance, B., Marsella, S.C.: Emotionally Expressive Head and Body Movement During Gaze Shifts. In: Pelachaud, C., Martin, J.C., André, E., Chollet, G., Karpouzis, K., Pelé, D. (eds.) IVA 2007. LNCS (LNAI), vol. 4722, pp. 72–85. Springer, Heidelberg (2007)
13. Lee, S.P., Badler, J.B., Badler, N.I.: Eyes alive. ACM Transactions on Graphics 21, 637–644 (2002)
14. Mehrabian, A.: Analysis of the big-five personality factors in terms of the pad temperament model. Australian Journal of Psychology 48, 86–92 (1996)
15. Mignault, A., Chaudhuri, A.: The Many Faces of a Neutral Face: Head Tilt and Perception of Dominance and Emotion. Journal of Nonverbal Behavior 27, 111–132 (2003)
16. Paterson, H.M., Pollick, F.E., Sanford, A.J.: The Role of Velocity in Affect Discrimination. In: Proceedings of the Twenty-Third Annual Conference of the Cognitive Science Society, pp. 756–761. Lawrence Erlbaum Associates, London (2001)
17. Queiroz, R.B., Barros, L.M., Musse, S.R.: Providing expressive gaze to virtual animated characters in interactive applications. Computers in Entertainment (CIE) 6, 1–23 (2008)

Why Is the Creation of a Virtual Signer Challenging Computer Animation?

Nicolas Courty and Sylvie Gibet

Université de Bretagne Sud, Laboratoire VALORIA, Bâtiment Yves Coppens,
F-56017 Vannes, France

Abstract. Virtual signers communicating in signed languages are a very interesting tool to serve as means of communication with deaf people and improve their access to services and information. We discuss in this paper important factors of the design of virtual signers in regard to the animation problems. We notably show that some aspects of these signed languages are challenging for up-to-date animation methods, and present possible future research directions that could also benefit more widely the animation of virtual characters.

1 Introduction

Signed languages (SL), defined as visual languages, were initially intended to be a mean of communication between deaf people. They are entirely based on motions and have no written equivalent. They constitute full natural languages, driven by their own linguistic structure. Accounting for the difficulties of deaf to read text or subtitles on computers or personal devices, computer animations of sign language improve the accessibility of those media to these users [27,5,20,9]. The use of avatars to this purpose allows to go further the restrictions of videos, mostly because the possibilities of content creation with avatars are far more advanced, and because avatars can be personalized along with the user's will. They also allow the anonymity of the interlocutor.

However, animating virtual signers has revealed to be a tedious task [17], mostly for two reasons: *i)* our comprehension of the linguistic mechanisms of signed languages are still not fully achieved, and computational linguistic software may sometimes fail in modeling particular aspects of SL *ii)* animation methodologies are challenged by the complex nature of gestures involved in signed communication. This paper focuses on this second class of problems, even though we admit that in some sense those two aspects are indissociable.

In fact, signs differ sensibly from other non-linguistic gestures, as they are by essence multichannel. Each channel of a single sign (those being the gestures of the two arms and the two hands, the signer's facial expressions and gaze direction) conveys meaningful information from the phonological level to the discourse level. Moreover, signs exhibit a highly spatial and temporal variability that can serve as syntactic modifiers of aspect, participants, etc. Then, the combination in space and time of two or more signs is also possible and sometimes mandatory

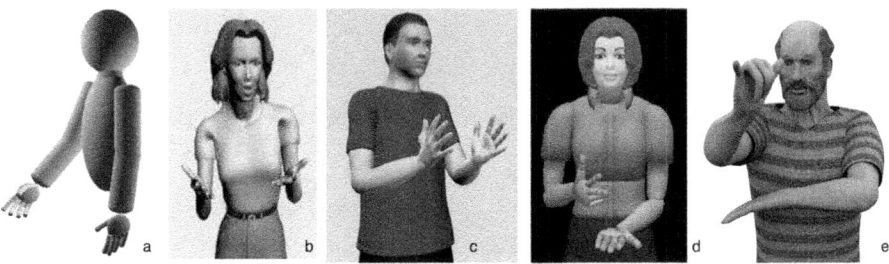

Fig. 1. Some virtual signers classified in chronological order: (a) the GESSYCA system [10] (b) Elsi [8] (c) Guido from the eSign european project [20] (d) the virtual signer of the City University of New-Yord [17] (e) Gerard [2]

to express concisely ideas or concepts. This intricate nature is difficult to handle with classical animation methods, that most of the time focus on particular types of motions (walk, kicks, etc.) that do not exhibit a comparable variability and subtleties.

The remainder of the paper is organized as follows: a brief state-of-the-art presents some existing virtual signers and the two aspects of sign generation: procedural and data-driven methods (Section 2), then challenges in the production of signs are exposed (Section 3) and finally a collection of unresolved virtual character animation problems are presented (Section 4).

2 Existing Virtual Signers

We first begin by reviewing some of the technologies used to animate virtual signers. Figure 1 presents in chronological order some existing virtual signers.

2.1 Descriptive and Generative Methods

Several gesture taxonomies have already been proposed in [19] and [28], some of which rely on the identification of specific phases that appear in co-verbal gestures and sign language signs [22]. Recent studies dedicated to expressive gesture rely on the segmentation and annotation of gestures to characterize the spatial structure of a sign sequence, and on transcribing and modeling gestures with the goal of later re-synthesis [21].

Studies on sign languages formed early description/transcription systems, such as [33] or [31]. More recently, at the intersection of linguistics and computation, gestures have been described with methods ranging from formalized scripts to a dedicated gestural language. The BEAT system [4], as one of the first systems to describe the desired behaviors of virtual agents, uses textual input to build linguistic features of gestures to be generated and then synchronized with speech. Gibet et al. [10] propose a gesture synthesis system based on a quantified description of the space around the signer; using the HamNoSys [31] sign language notation system as a base, the eSign project has further designed

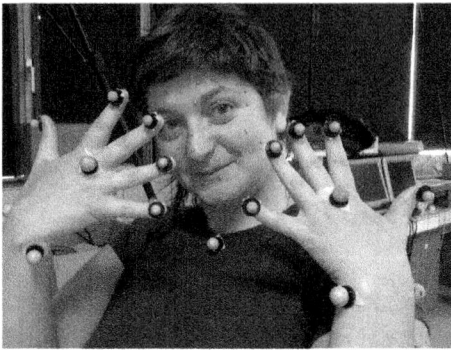

Fig. 2. Photo of the motion capture settings in the Signcom project

a motion specification language called SigML [7]. Other *XML*-based description languages have been developed to describe various multimodal behaviors, some of these languages are dedicated to conversational agents behaviors, as for example MURML [24], or describe style variations in gesturing and speech [30], or expressive gestures [11]. More recently, a unified framework, containing several abstraction levels has been defined and has led to the XML-based language called BML [36], which interprets a planned multimodal behavior into a realized behavior, and may integrate different planning and control systems.

Passing from the specification of gestures to their generation has given rise to a few works. Largely, they desire to translate a gestural description, expressed in any of the above-mentioned formalisms, into a sequence of gestural commands that can be directly interpreted by a real-time animation engine. Most of these works concern pure synthesis methods, for instance by computing postures from specification of goals in the 3D-space, using inverse kinematics techniques, such as in [10], [35], [23]. Another approach uses annotated videos of human behaviors to synchronize speech and gestures and a statistical model to extract specific gestural profiles; from a textual input, a generation process then produces a gestural script which is interpreted by a motion simulation engine [29].

Alternatively, data-driven animation methods can be substituted for these pure synthesis methods. In this case the motions of a real signer are captured with different combinations of motion capture techniques. Since it is not possible to record every possible sentences, new strategies are to be devised in order to produce new utterances, The next paragraph presents an example of a fully data-driven approach.

2.2 An Example of a Full Data-Driven Approach: The Signcom Project

An example of a full data-driven virtual signer is given by the Signcom project, which aims at improving the quality of the real-time interaction between real humans and avatars, by exploiting natural communication modalities such as

Why Is the Creation of a Virtual Signer Challenging Computer Animation? 293

Fig. 3. Screenshots of the virtual signer "Sally" from the Signcom project

gestures, facial expressions and gaze direction. Based on French Sign Language (FSL) gestures, the real human and the virtual character produce statements towards their interlocutor through a dialog model. The final objective of the project consists in elaborating new ways of communication by recognizing FSL utterances, and synthesizing adequate responses with a 3D avatar. The motion capture system uses Vicon MX infrared camera technology to capture the movements of our LSF informants at frame rates of 100 Hz. The setup was as follows: 12 motion capture cameras, 43 facial markers, 43 body markers, and 12 hand markers. In order to replay a complete animation, several post operations are necessary. First, the fingers' motion were reconstructed by inverse kinematics, since only the fingers' end positions were recorded. In order to animate the face, cross-mapping of facial motion capture data and blendshapes parameters was performed [6]. This technique allows to animate directly the face from the raw motion capture data once a mapping has been learned. Finally, since no eye gazes were recorded during the informants performance, an automatic eye gazing systems was designed. Figure 3 gives some illustrations of the final virtual signer "sally" replaying captured motions. A corpus annotation was also conducted. Annotations expand on the mocap data by identifying each sign type with a unique gloss, so that each token of a single type can be easily compared. Other annotations include grammatical and phonological descriptions.

From recorded FSL sequences, multichannel data are retrieved from a dual-representation indexed database (annotation and mocap data), and used to generate new FSL utterances [2], in a way similar to [1]. At that time, the final system is currently under evaluation with native LSF signers.

3 Challenges in Sign Production

Though data-driven animation methods significantly improve the quality and credibility of animations, there are nonetheless several challenges to the reuse of motion capture data in the production of sign languages. Some of them are presented in the following.

Spatialization of the content. As sign languages are by nature spatial languages, forming sign strings requires a signer to understand a set of highly spatial and temporal grammatical rules and inflection processes unique to a sign language. We can separate plain signs that do not use space semantically (like the American Sign Language sign HAVE which does not make any notable use of space other than which is necessary for any sign) from signs that incorporate depiction. This second group of signs includes the strongly iconic signs known as depicting verbs (or classifiers), which mimic spatial movements, as well as size-and-shape specifiers, which concern static spatial descriptions.

Moreover, indicating signs like indicating verbs and deictic expressions require the signer to interface with targets in the signing space by effecting pointing-like movements towards these targets. Indicating verbs include such signs as the LSF sign INVITER, in which the hand moves from the area around the invited party toward the entity who did the inviting . Depending on the intended subject and object, the initial and final placements of the hand vary greatly within the signing space. Deixis, such as pronouns, locatives, and other indexical signs are often formed with a pointed index finger moving toward a specific referent, though other hand configurations have been reported in sign languages, such as American Sign Language.

Small variations can make big semantic differences. Sign languages require precision and rapidity in their execution, but at the same times imperfection in the realization of the signs or bad synchronization can change the semantic content of the sentence. We give here some challenging elements in the execution of signs:

- **Motion precision.** The understandability of signs require accuracy in the realization of the gestures. In particular in finger spelling the degree of openness of a fingers leads to different letters. Some of fhe different hand shapes used in FSL only differ by the positions of one finger or by the absence or not of a contact. This calls for a great accuracy in the capture and animation processes.
- **spatio-temporal aspects of the gestures.** The sign language being a language with highly spatio-temporal components, the question of timing and dynamics of gesture is crucial. In fact, three elements are of interest for a sign: first, the spatial trajectory of the hands are rather important. They do not only constitute transitions in space between two key positions, but may be constituent of the sign. This raises the problem of the coding of this trajectory. Second, synchronization of the two hands is a major component,

and usually hands do not have to this regard a symmetric role, In the case of PAS D'ACCORD (not agree), the index start from from the forehead and meets the other index in front of the signer. The motion of the second hand is clearly synchronized on the first hand. Third, the dynamics of the gesture (acceleration profile along time) allows the distinction between two significations. An example is the difference between the signs CHAISE (chair) and S'ASSEOIR (to sit), which have the same hands configurations, the same trajectories in space, but different dynamics. Let us finally note that the dynamics of contacts between the hand and the body (gently touching or striking)is also relevant.
- **facial expressions and non manual elements.** While most of the description focus on the hands configuration and their motions, important non manual elements should also be taken into account, like shoulder motions, head swinging, changes in gazes or facial mimics. For example, the gaze can be used either to recall a particular object of the signing space, or either directed by the dominating hand (like in the sign LIRE, to read, where the eyes follow the motion of fingers). In the case of facial mimics, some facial expressions may serve as adjectives (for instance inflated cheeks will make an object big, while wrinkled eyes would make it thin) or indicate wether the sentence is a question (raised eyebrows) or an affirmation (frowning). It is therefore very important to preserve these informations in the facial animation.

4 Unresolved Animation Problems

Regarding the different requirements exposed in the previous Section, several unresolved computer animation problems are presented here. Those problems are not particularly exclusive to the animation of virtual signers, and can address more widely general virtual character animation problems.

High frequency full body and facial motion capture. Signs are by nature very dexterous and quick gestures, that involve at the same time several modalities (arms, hands, body, gaze and facial expressions). Capturing accurately all these channels with an appropriate frequency (> 100 Mhz) actually pushes motion capture equipment to their very limits. It could be argued that splicing methods such as [26] would allow to capture independently the different modalities, and then combine them during a post process phase. However, the temporal synchronization issues raised by this method seem hard to alleviate. Moreover, asking the signer to perform alone the facial expressions corresponding to given sentences is also out of reach, since most of the facial mimics are generally done unconsciously. A parallel could be drawn with non-verbal communication: could we ask someone to perform accompanying gestures of an unspoken discourse ? Finally, new technologies such as surface capture [32], that captures simultaneously geometry and animation, are very attractive, but yet the resolution is not sufficient to capture the body and the face with an adequate precision, and only very few methods exist to manipulate this complex data in order to produce new animations.

Expressivity filtering. As seen in the previous Section, the spatio-temporal variability of signs can be used as adjectives, or in a more general way, to inflect the nature of a sentence and enhance the global expressivity of the virtual signer. It has been shown [15] that temporal alignment methods [13] can be efficiently used to change the style and expressivity of a captured sentence. Nevertheless, big variations in style are can not only obtained by changing the timing of gestures, but most often by the change of spatial trajectories, and sometimes may inflect the entire sentence. Most of existing methods that build statistical models [37] of gestures may fail for this purpose, mostly because the style transfer is encoded by higher level linguistic rules, and because pure signal approaches are insufficient to model this variability.

Advanced motion retargeting. Most of the actual motion retargeting techniques focus on the adaptation of motion to changing the physical conditions of the motion [34] or more frequently kinematic constraints [18,25,12] through the use of inverse kinematic techniques. In the case of sign language the spatial relations between the fingers and the arms or the head are key elements for the comprehension of the discourse and should be preserved in the retargeting process. To this end, the recent work of Ho and colleagues [16] is really attractive, provided that the important relation between limbs could be preserved by their methods. Yet Its application to sign language synthesis remains to be explored. Whereas interaction with the floor or objects in the environment lead to hard constraints which lead to difficult optimization problems and procedures, constraints in sign language may be more diffuse or expressed qualitatively (e.g. "the thumb should touch the palm of the hand"). Algorithms dealing with such fuzzy or high level constraints could be extremely interesting, both numerically (more degrees of freedom while optimizing) and from a usability point of view. Finally, since arms motions are involved, a planing phase may also be required to avoid self collisions. Combined inverse kinematics and planing algorithms could be used [3], as well as more recent hybrid approaches [38]. Yet, real time algorithms for this class of problems remain to be found.

Multichannel combinations. As exposed in [2], the possibility of building new signed utterances by composing selectively pre-exisiting elements of a corpus data is possible. In this option, not only the spatial coherency should be preserved, but as well the channel's temporal synchronization:

- *spatial coherency.* Sign language allows to combine different gesture with different meanings at the same time, thus providing several information in a minimum of gestures. This combination differs from the classical blending approaches which mix motions together to produce new ones [1], as far as topological constraints should be preserved in the composition process. An example is given in Figure 4, where the same pose indicates at the same

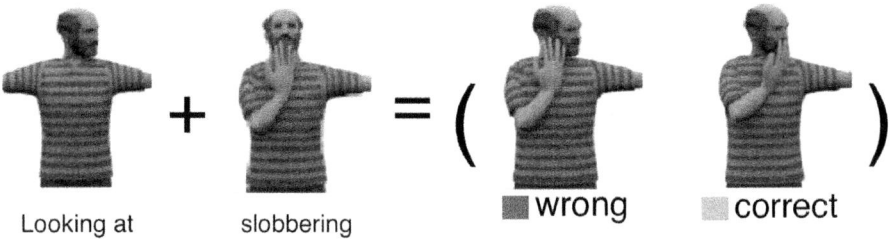

Fig. 4. Combination of two signs ("looking" and "slobbering")

time that a dog is looking at (first sign) something while slobbering (sign 2). If both signs were to be recorded independently, a naive blending operation would fail because the hand would not anymore be located in front of the mouth. Moreover, as exposed in the previous Section, every spatialized gestures should be retargeted with respect to the current signing space. This brings us back to the problem of advanced motion retargeting, but also clearly reveals that the combination process should be driven by more abstract definition, possibly of linguistic nature.

- *temporal synchronization.* It is likely that the different motion elements have not the same duration. The consequent problem is twofold: *i)* a common timeline has to be found, eventually as the result of a combinatorial optimization, or driven by linguistic rules. Up to our knowledge though, no existing model of sign language describe such temporal rules or model the synchronization of the different channels *ii)* once a correct time plan has been devised, the temporal length of the motion chunks has to be adapted, while preserving the dynamic of the motions. To this end, time warping techniques can be used [13]. However, inter channels synchronizations may exist (for example between the hand and the arm motions [14]). Those synchronization schema can be extracted from analysis, but the proper way to introduce this empirical knowledge in the synthesis process has not been explored yet.

5 Conclusion

We examined in this article the different challenges posed by the animation of virtual agent communicating in sign language. While data-driven animation techniques clearly lead to the best natural results, a lot of improvements are still mandatory to fulfill the requirements of sign languages. Among others, capture techniques and retargeting algorithms are severely challenged by the complex spatial and temporal schemas involved in signs. In parallel, those improvements should accompany progresses in the modeling of sign language, which is in itself a critical issue. In a second step, the usability and acceptability of virtual signers to the community of deaf people should also be evaluated thoroughly, notably through the help of native signers. Though those issues have recently attracted the attention of several research groups, a lot remain to be done before signing avatars can be used in our everyday environments.

References

1. Arikan, O., Forsyth, D., O'Brien, J.: Motion synthesis from annotations. ACM Trans. on Graphics 22(3), 402–408 (2003)
2. Awad, C., Courty, N., Duarte, K., Le Naour, T., Gibet, S.: A combined semantic and motion capture database for real-time sign language synthesis. In: Ruttkay, Z., Kipp, M., Nijholt, A., Vilhjálmsson, H.H. (eds.) IVA 2009. LNCS (LNAI), vol. 5773, pp. 432–438. Springer, Heidelberg (2009)
3. Bertram, D., Kuffner, J., Dillmann, R., Asfour, T.: An integrated approach to inverse kinematics and path planning for redundant manipulators. In: Int. Conf. on Robotic and Automation, ICRA 2006, pp. 1874–1879 (2006)
4. Cassell, J., Sullivan, J., Prevost, S., Churchill, E.F.: Embodied Conversational Agents. MIT Press, Cambridge (2000)
5. Chiu, Y.H., Wu, C.H., Su, H.Y., Cheng, C.J.: Joint optimization of word alignment and epenthesis generation for chinese to taiwanese sign synthesis. IEEE Trans. on Pattern Analysis and Machine Intelligence 29(1), 28–39 (2007)
6. Deng, Z., Chiang, P.i.-Y., Fox, P., Newmann, U.: Animating blendshape faces by cross-mapping motion capture data. In: Proc. of the 2006 Symp. on Interactive 3D Graphics and Games, Redwood City, California, pp. 43–48 (March 2006)
7. Elliott, R., Glauert, J., Jennings, V., Kennaway, J.: An overview of the sigml notation and sigml signing software system. In: Workshop on the Representation and Processing of Signed Languages, 4th Int'l Conf. on Language Resources and Evaluation (2004)
8. Filhol, M., Braffort, A., Bolot, L.: Signing avatar: Say hello to elsi? In: Proc. of Gesture Workshop 2007, Lisbon, Portugal, LNCS. Springer, Heidelberg (June 2007)
9. Fotinea, S., Efthimiou, E., Caridakis, G., Karpouzis, K.: A knowledge-based sign synthesis architecture. Universal Access in the Information Society 6(4), 405–418 (2008)
10. Gibet, S., Lebourque, T., Marteau, P.F.: High level specification and animation of communicative gestures. Journal of Visual Languages and Computing 12, 657–687 (2001)
11. Hartmann, B., Mancini, M., Pelachaud, C.: Implementing expressive gesture synthesis for embodied conversational agents. Gesture in Human-Computer Interaction and Simulation 3881, 188–199 (2006)
12. Hecker, C., Raabe, B., Enslow, R., DeWeese, J., Maynard, J., van Prooijen, K.: Real-time motion retargeting to highly varied user-created morphologies. ACM Trans. on Graphics 27(3), 1–11 (2008)
13. Héloir, A., Courty, N., Gibet, S., Multon, F.: Temporal alignment of communicative gesture sequences. Computer Animation and Virtual Worlds 17, 347–357 (2006)
14. Héloir, A., Gibet, S.: A qualitative and quantitative characterisation of style in sign language gestures. In: Sales Dias, M., Gibet, S., Wanderley, M.M., Bastos, R. (eds.) GW 2007. LNCS (LNAI), vol. 5085, pp. 122–133. Springer, Heidelberg (2009)

15. Héloir, A., Kipp, M., Gibet, S., Courty, N.: Evaluating data-driven style transformation for gesturing embodied agents. In: Prendinger, H., Lester, J.C., Ishizuka, M. (eds.) IVA 2008. LNCS (LNAI), vol. 5208, pp. 215–222. Springer, Heidelberg (2008)
16. Ho, E., Komura, T., Tai, C.-L.: Spatial relationship preserving character motion adaptation. ACM Trans. on Graphics 29(4), 1–8 (2010)
17. Huenerfauth, M., Zhao, L., Gu, E., Allbeck, J.: Evaluation of american sign language generation by native asl signers. ACM Trans. Access. Comput. 1(1), 1–27 (2008)
18. Choi, K.j., Ko, H.s.: On-line motion retargetting. Journal of Visualization and Computer Animation 11, 223–235 (2000)
19. Kendon, A.: Human gesture. In: Tools, Language and Cognition, pp. 43–62. Cambridge University Press, Cambridge (1993)
20. Kennaway, J.R., Glauert, J.R.W., Zwitserlood, I.: Providing signed content on the internet by synthesized animation. ACM Trans. Comput. Hum. Interact. 14(3), 15 (2007)
21. Kipp, M., Neff, M., Kipp, K., Albrecht, I.: Toward natural gesture synthesis: Evaluating gesture units in a data-driven approach. In: Pelachaud, C., Martin, J.-C., André, E., Chollet, G., Karpouzis, K., Pelé, D. (eds.) IVA 2007. LNCS (LNAI), vol. 4722, pp. 15–28. Springer, Heidelberg (2007)
22. Kita, S., van Gijn, I., van der Hulst, H.: Movement phase in signs and co-speech gestures, and their transcriptions by human coders. In: Wachsmuth, I., Fröhlich, M. (eds.) GW 1997. LNCS (LNAI), vol. 1371, pp. 23–35. Springer, Heidelberg (1998)
23. Kopp, S., Wachsmuth, I.: Synthesizing multimodal utterances for conversational agents. Journal Computer Animation and Virtual Worlds 15(1), 39–52 (2004)
24. Kranstedt, A., Kopp, S., Wachsmuth, I.: MURML: A Multimodal Utterance Representation Markup Language for Conversational Agents. In: Falcone, R., Barber, S.K., Korba, L., Singh, M.P. (eds.) AAMAS 2002. LNCS (LNAI), vol. 2631. Springer, Heidelberg (2003)
25. Kulpa, R., Multon, F., Arnaldi, B.: Morphology-independent representation of motions for interactive human-like animation. Comput. Graph. Forum 24(3), 343–352 (2005)
26. Majkowska, A., Zordan, V.B., Faloutsos, P.: Automatic splicing for hand and body animations. In: Proc. of the 2006 ACM SIGGRAPH/Eurographics Symposium on Computer Animation, SCA 2006, pp. 309–316 (2006)
27. Marshall, I., Safar, E.: Grammar development for sign language avatar-based synthesis. In: Proc. of the 3rd Int. Conf. on Universal Access in Human-Computer Interaction, UAHCI 2005 (2005)
28. McNeill, D.: Hand and Mind - What Gestures Reveal about Thought. The University of Chicago Press, Chicago (1992)
29. Neff, M., Kipp, M., Albrecht, I., Seidel, H.-P.: Gesture modeling and animation based on a probabilistic re-creation of speaker style. ACM Transactions on Graphics 27(1), 1–24 (2008)
30. Noot, H., Ruttkay, Z.: Variations in gesturing and speech by gestyle. Int. J. Hum.-Comput. Stud. 62(2), 211–229 (2005)
31. Prillwitz, S., Leven, R., Zienert, H., Hanke, T., Henning, J.: Hamburg Notation System for Sign Languages - An Introductory Guide. University of Hamburg Press (1989)
32. Starck, J., Hilton, A.: Surface capture for performance-based animation. IEEE Computer Graphics and Applications 27(3), 21–31 (2007)

33. Stokoe, W.C.: Semiotics and Human Sign Language. Walter de Gruyter Inc., Berlin (1972)
34. Tak, S., Ko, H.-S.: A physically-based motion retargeting filter. ACM Tra. On Graphics 24(1), 98–117 (2005)
35. Tolani, D., Goswami, A., Badler, N.: Real-time inverse kinematics techniques for anthropomorphic limbs. Graphical Models 62(5), 353–388 (2000)
36. Vilhalmsson, H., Cantelmo, N., Cassell, J., Chafai, N.E., Kipp, M., Kopp, S., Mancini, M., Marsella, S., Marshall, A.N., Pelachaud, C., Ruttkay, Z., Thorisson, K., van Welbergen, H., van der Werf, R.J.: The behavior markup language: Recent developments and challenges. In: Pelachaud, C., Martin, J.-C., André, E., Chollet, G., Karpouzis, K., Pelé, D. (eds.) IVA 2007. LNCS (LNAI), vol. 4722, pp. 99–111. Springer, Heidelberg (2007)
37. Wang, J.M., Fleet, D.J., Hertzmann, A.: A multifactor gaussian process models for style-content separation. In: Proc. of Int. Conf. on Machine Learning, ICML (June 2007)
38. Zhang, L., Lin, M.C., Manocha, D., Pan, J.: A hybrid approach for simulating human motion in constrained environments. Computer Animation and Virtual Worlds 21(3-4), 137–149 (2010)

Realtime Rendering of Realistic Fabric with Alternation of Deformed Anisotropy*

Young-Min Kang

Dept. of Game Engineering, Tongmyong University
Nam-gu, Busan 608-711, Korea
ymkang@tu.ac.kr
http://ugame.tu.ac.kr/ymkang/wiki

Abstract. In this paper, an efficient method is proposed to produce photorealistic images of woven fabrics without material data such as the measured BRDFs. The proposed method is applicable both to ray tracer based offline renderers and to realtime applications such as games. In order to enhance the realism of cloth rendering, researchers have been utilizing the measured data of surface reflectance properties. Although the example-based approaches drastically enhance the realism of virtual fabric rendering, those methods have serious disadvantage that they require huge amount of storage for the various reflectance properties of diverse materials. The proposed method models the reflectance properties of woven fabric with alternating anisotropy and deformed MDF(microfacet distribution function). The experimental results show the proposed method can be successfully applied to photorealistic rendering of diverse woven fabric materials even in interactive applications.

Keywords: Fabric rendering, alternating anisotropy, deformed anistropy, realtime rendering.

1 Introduction

The goal of physically-based rendering is to produce photorealistic images. The goal can be achieved by modeling the surface reflection properties of virtual objects. In this paper, a new procedural approach is proposed for photorealistic rendering of woven fabric as shown in Figure 1. Fabric appearance is important in industrial applications of computer graphics in the textile, garment, and fabric care industries. There are two main types of fabrics namely, knitwear and woven fabrics. Our goal is to produce photorealistic images of woven fabrics without empirical data such as the measured BRDFs. The physically plausible rendering results can be obtained when the accurate reflection models is found. However, the actual objects in real world usually have complex reflection properties which

---

* This research was supported by the MKE(Ministry of Knowledge Economy), Korea, under the ITRC(Information Technology Research Center) support program supervised by the NIPA(National IT Industry Promotion Agency) (NIPA-2010-(C-1090-1021-0006)).

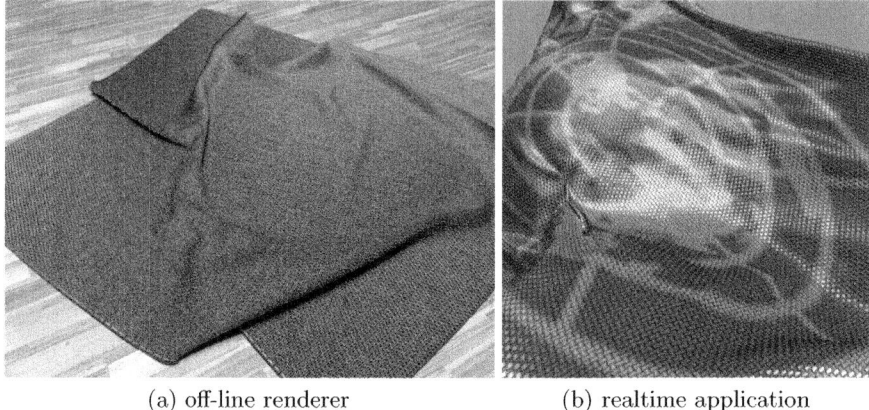

(a) off-line renderer (b) realtime application

Fig. 1. The results of the proposed method

cannot be easily modeled. The method proposed in this paper produces realistic image of virtual woven fabric by procedurally modeling the alternating anisotropic reflection on the fabric surface and deforming the microfacet distribution function(MDF) in accordance with the weave patterns and yarn-level structure.

2 Related Work

Fabrics can be categorized into two classes. One of them is knitted fabric and the other is woven fabric. There have been various research efforts to model and render the knitted fabric[7,8,9,10,15,19]. However, those methods cannot be directly employed for rendering woven fabrics because those methods usually focus on the fluffy texture and the knitted structures.

In this paper, we exploit the microfacet model for modeling the surface reflection property of fabric. The microfacet model was first proposed by Torrance and Sparrow[18], and then introduced to graphics literature by Cook and Torrance[5].

Yasuda et al. proposed a shading model for woven fabric by applying anisotropic reflectance according to the yarn direction[20]. This approach is one of the earliest efforts for plausible fabric rendering with procedural techniques. Their method is based on the microfacet model proposed by Blinn and Newell[11]. Although this method is the first approach to woven fabric rendering with weave-based reflectance model, it is not capable of rendering the close-up scene where weave patterns are visible.

Adabala et al. proposed a woven fabric rendering method that can be applied to both distant and close-up observations of woven surface[2,1]. Their method is based on the microfacet model proposed by Ashikhmin and Shirley[3,4], and utilizes horizon map proposed by Sloan et al.[17]. The distant viewing of the method is the generalized model of the satin fabric rendering in [3]. The major advantage of the method is that it can render various weave patterns and represent the spatially varying reflectance based on the weave patterns. However, this method focuses on variety of weave patterns and treated the light reflection on the yarn surface somewhat lightly.

Photorealistic rendering of woven fabric requires spatially varying anisotropic reflectance model. Some researchers tried to capture the spatially varying BRDF (SVBRDF) for realistic representation of the fabric material [6,13,14]. However, capturing the SVBRDF requires expensive devices and huge amount of storage. Wang et al proposed a SVBRDF measuring techniques using data captured from a single view[12]. This method made it possible to capture SVBRDF with low-cost measuring device by synthesizing microfacet model based on measured partial normal distribution functions. Although this method can reproduce photorealistic image of woven fabric, it still requires huge amount of storage for SVBRDF. Moreover, one needs to measure all kinds of fabric which will be possibly used in rendering.

Sattler et al. employed BTF(bidirectional texture function) proposed by Dana et al.[6] to render photorealistic woven fabric[16]. However, this method also suffers from the common disadvantages of example-based approach. The measured data requires huge amount of storage, and the reflectance property of rendered fabric cannot be easily controlled.

Zinke and Weber proposed bidirectional fiber scattering distribution function(BFSDF) for physically plausible rendering of dielectric filament fiber[21]. However, the BFSDF is 8 dimensional function so that it cannot be easily dealt. They introduced various approximation techniques to reduce the dimensionality of the scattering function. The disadvantage of this method is that this method requires geometry for the fiber. Therefore, an extremely complex geometry is required to represent a woven fabric object.

3 Woven Fabric Rendering

In this section, we will introduce a procedural techniques that represent the surface reflectance of woven fabric. The proposed method is based on microfacet model and the microfacet distribution function is deformed in order to express the peculiar reflectance of woven fabric.

3.1 Alternating Anisotropy and Weave Control

The reflectance property of microfacet model is dominated by the MDF (microfacet distribution function) $D(\omega_h)$ which gives the probability that the normal vector of a microfacet is oriented to ω_h. Ashikhmin et al. proposed an anisotropic reflectance model as follows[3]:

$$D(\omega_h) = \frac{\sqrt{(e_x + 1)(e_y + 1)}}{2\pi}(\omega_h \cdot \mathbf{n})^{e_x \cos^2 \phi + e_y \sin^2 \phi} \quad (1)$$

where e_x and e_y are the exponents for the distribution function for controlling the anisotropy of the reflectance, and ϕ denotes the azimuthal angle of the half vector

ω_h. One can easily control the anisotropic reflectance by changing the parameters e_x and e_y. These two parameters control the shape of the specular lobe. Since ω_h is a unit direction vector, it can be represented as $(\omega_{h.x}, \omega_{h.y}, \sqrt{1 - \omega_{h.x}^2 - \omega_{h.y}^2})$

Woven fabric has weft and warp yarns. Because the yarns are oriented in different directions, the reflectance anisotropy is alternating according to the yarn direction. Therefore, we employed alternating anisotropy for woven fabric, and the anisotropy is determined by the underlying weave patterns. We can easily alternate the anisotropy by swapping the parameters e_x and e_y. First we determine whether the sampled point is weft yarn, warp yarn, or inter-yarn gap. We then apply different distribution functions D_{weft} and D_{warp} for weft yarns and warp yarns respectively as follows:

$$D_{weft}(\omega_h) = \frac{\sqrt{(e_x+1)(e_y+1)}}{2\pi}(\omega_h \cdot \mathbf{n})^{e_x \cos^2 \phi + e_y \sin^2 \phi} \quad (2)$$

$$D_{warp}(\omega_h) = \frac{\sqrt{(e_x+1)(e_y+1)}}{2\pi}(\omega_h \cdot \mathbf{n})^{e_x \sin^2 \phi + e_y \cos^2 \phi}$$

The method proposed in this paper is based on the anisotropy that is determined by weave patterns. Therefore, it is very important to define the weave pattern and determine whether the sampled point is on a weft yarn or a warp yarn. Once the yarn direction is determined, the alternating MDFs described in Eq. 2 are applied.

(a) 2D MDF (b) Anisotropic reflectance (c) Alternating Anisotropy

Fig. 2. Alternating anisotropic reflectance

Figure 2 demonstrates the effect of the the alternating anisotropy. The distribution function can be easily visualized in 2D space where two axis are aligned with $\omega_{h.x}$ and $\omega_{h.y}$ as shown in Figure 2 (a). Figure 2 (b) shows the simple anisotropic reflectance and (c) shows the result when the anisotropy is alternated along the yarn direction by applying the distribution functions shown in Eq. 2.

The MDFs shown in Eq. 2 are in fact the same because we can easily rotate one MDF to coincide with another. This concept is described in Figure 3.

3.2 Deformed MDF for Woven Surface Representation

The alternating anisotropy can be efficiently and effectively utilized for describing the woven fabric reflectance. However, the alternating anisotropy cannot represent

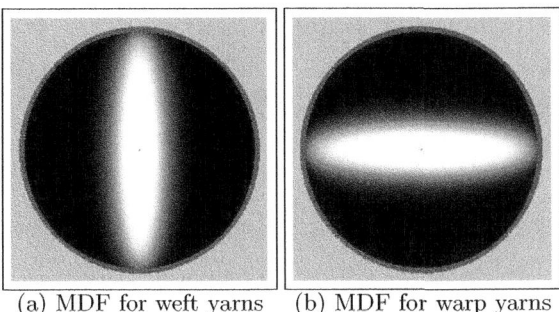

(a) MDF for weft yarns   (b) MDF for warp yarns

Fig. 3. MDF Rotation for Alternating Anisotropy

the bumpy surface of woven fabric. In order to produce realistic bumpy surface caused by woven structure, we used deformed MDFs rather than the original base MDFs shown in Eq. 2.

If we can rotate the MDF in order to represent the alternating anisotropic reflectance on the woven surface as shown in Figure 3, we can also deform the MDF to represent the different reflectance on the bumpy surface.

The result shown in Figure 2 assumes that the most probable normal direction of the sampled point is exactly the same as the normal given by underlying geometry. However, our goal is to procedurally generate the bumpy woven surfaces defined by weave patterns. The normal vectors on woven surface should be perturbed according to the defined weave patterns.

In order to perturb the normal vector at the sampled point, we need to know how far the sampled point is from the axis of the yarn The offsets in weft and warp yarns are denoted as σ^v and σ^u respectively. These offsets range from -1 to 1. The offset value is 0 when the sampled point is on the axis of the yarn, and 1 or -1 at the both ends of the yarn.

Let us denote the perturbed normal as $\tilde{N}$ while the original normal given by mesh data is denoted as N. In the surface coordinate system (tangent space), N is always represented as $(0, 0, 1)$. Therefore, the perturbed normal $\tilde{N}$ can be easily denoted as $(\Delta x, \Delta y, \sqrt{1 - \Delta x^2 - \Delta y^2})$. The basic idea of this paper is that we can easily obtain the approximate reflectance at the sampled point with the perturbed normal by deforming the base MDF. The deformation translates the center of the distribution function with the amount of $(\Delta x, \Delta y)$.

Figure 4 shows the basic concept of the deformed MDF. Figure 4 (a) and (c) show the base MDFs, and (b) and (d) show the corresponding deformed MDFs with the amount of $(\Delta x, \Delta y)$.

Figure 5 (a) shows the cross section of a perfectly circular warp yarn. As shown in the figure, the sampled point p should be rendered as if it is located at the displaced point p'. We can easily notice that the x component of the perturbed normal is proportional to the offset ratio σ^u which ranges from -1 to 1. Therefore, the MDF of the sampled point on a warp yarn can be described by deforming the base MDF with $(c_y \sigma^u, 0)$ where c_y denotes the yarn curvature

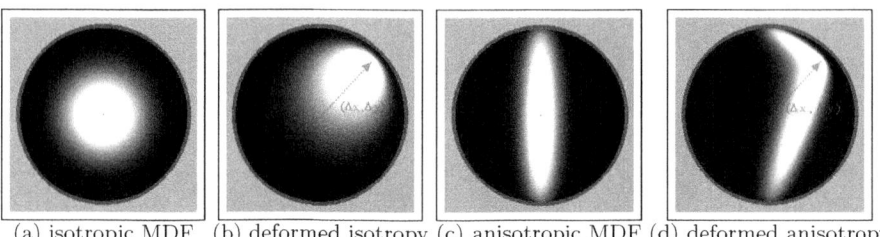

(a) isotropic MDF (b) deformed isotropy (c) anisotropic MDF (d) deformed anisotropy

Fig. 4. Visualization of MDF deformation

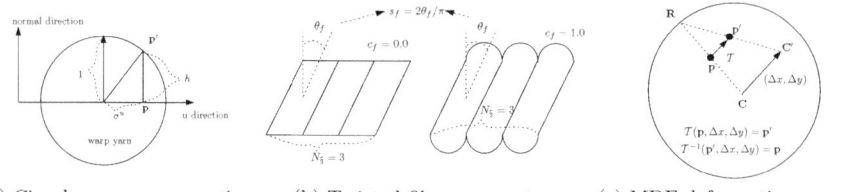

(a) Circular yarn cross section   (b) Twisted fiber parameters   (c) MDF deformation concept

Fig. 5. Bumpy cloth model and reflectance deformation

control parameter ranging from 0 to 1. We can similarly model the MDF of a sampled point on a weft yarn with the deformation amount of $(0, c_y\sigma^v)$.

A single yarn is usually the collection of twisted fibers. Figure 5 (b) shows the parameters for controlling the reflectance of a yarn based on the twist. θ_f is the slope angle of the twist, and c_f is the curvature control parameter for twisted fiber. For simplicity, we denoted $2\theta_f/\pi$ as s_f to control the slope, and then s_f ranges from 0 to 1. As shown in the figure, the twisted fiber is flat when c_f is zero, and curvature increases as c_f increases. The parameter $N_§$ denotes the number of twists within a weave element space. We can easily notice that the twisted structure affects the y components of MDF deformations for warp yarns while it affects the x components for weft yarns. For the warp yarns, the deformation caused by twisted fiber is $(0, c_f(2\mathbf{fract}(N_§(v_w - s_f\sigma^v)) - 1)$ where the function $\mathbf{fract}(x)$ returns $x - \lfloor x \rfloor$. The MDF deformation for weft yarns caused by the twisted structure can also similarly described. Therefore, the MDF deformation on woven surface can be finally described as follows:

$$(\Delta x, \Delta y) \tag{3}$$
$$for\ weft : (c_f(2\mathbf{fract}(N_§(u_w - s_f\sigma^u)) - 1, c_y\sigma^v)$$
$$for\ warp : (c_y\sigma^u, c_f(2\mathbf{fract}(N_§(v_w - s_f\sigma^v)) - 1)$$

It is obviously inefficient to compute the deformed MDF for every sampled point. Instead, we used only the original base MDF $D(w)$ shown in Eq. 2 and transformed coordinate. Figure 5 (c) shows the MDF deformation concept. The value at the point $\mathbf{p}$ will be moved to $\mathbf{p}'$ in the deformed MDF. We can define a function that transforms the point **p** to **p'** as $\mathcal{T}$. Then we can easily obtain the inverse function $\mathcal{T}^{-1}$. Let **R** be the intersection point of the circumference of the MDF and the ray from the deformed center **C'** passing through the deformed point **p'**. We can easily notice that the deformed point and the original point **p** are related as follows:

$$\mathcal{T}^{-1}(\mathbf{p'}, \Delta x, \Delta y) = \mathbf{p'} - \frac{|\mathbf{R} - \mathbf{p'}|}{|\mathbf{R} - \mathbf{C'}|}(\Delta x, \Delta y) \qquad (4)$$

In order to compute the reflectance intensity at a sampled point where the normal is perturbed with the amount $(\Delta x, \Delta y)$, we simply apply usual microfacet based procedural BRDF with $D(\mathcal{T}^{-1}(\omega_h, \Delta x, \Delta y))$.

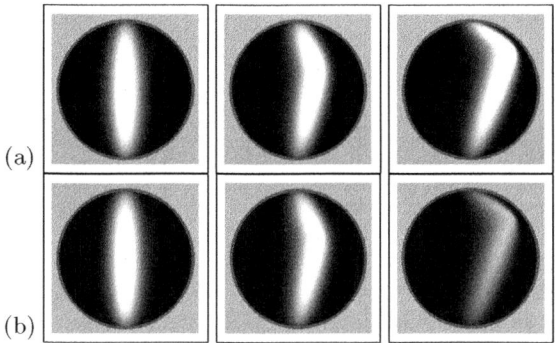

Fig. 6. Approximate ambient occlusion for MDF deformation: (a) simple MDF deformations, (b) MDF deformations with approximate ambient occlusion effects

Although the deformed MDFs of the sampled points on the woven surface effectively perturb the reflectance on the surface, the sampled points are still on a flat plane. Therefore the self-shadowing by the bumpy structure of the woven surface cannot be represented with deformed MDFs. In this paper, we approximated the ambient occlusion to increase the realism of rendered woven fabrics. Instead of casting rays in every direction from the surface, we simply approximated the ambient occlusion with the amount of the MDF deformation. Therefore, the reflected intensity was scaled by $(1 - \sqrt{\Delta x^2 + \Delta y^2})$. Figure 6 shows the effect of the approximate ambient occlusion. The row (a) shows the simple MDF deformation while (b) shows the deformed MDF with the consideration of the approximate ambient occlusion.

4 Experiments

We implemented the proposed method for both physically based renderer and realtime application.

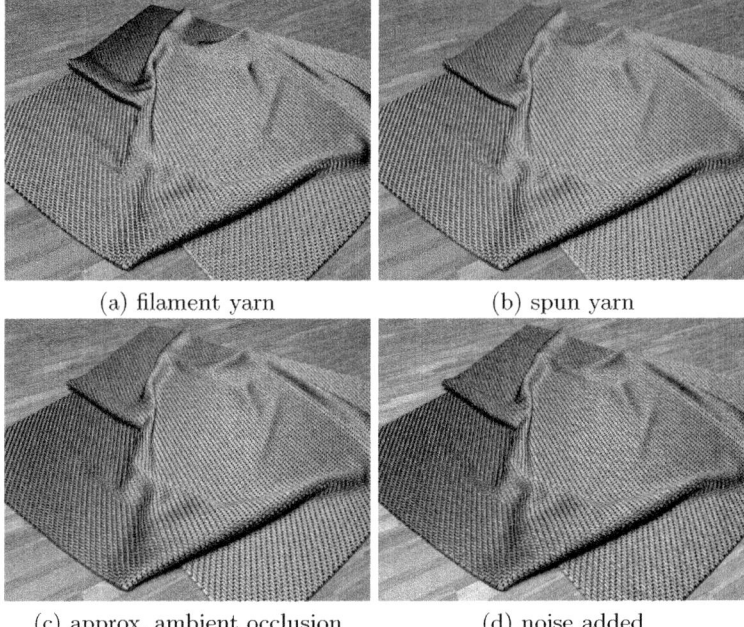

(a) filament yarn (b) spun yarn

(c) approx. ambient occlusion (d) noise added

Fig. 7. The effect of the proposed techniques: (a) MDF deformation without twisted structure ($e_x = 100, e_y = 0.1, c_y = 0.5$), (b) MDF deformation with twisted structure ($c_f = 0.75, N_\S = 4$), (c) approximate ambient occlusion applied to MDF, and (d) realism enhanced with noise

Figure 7 shows the effect of the proposed techniques when applied to ray tracing based renderer. Figure 7 (a) shows the result when MDF was deformed to represent the curved surface of the cylinder-shaped filament yarns while (b) shows the result when the twisted fiber structure is also considered. As shown in the figure, the spun yarn model reduces the specularity, and brightens and smoothens the shaded area on the woven surface. The result shown in Figure. 7 (c) shows the effect of the approximate ambient occlusion and (d) shows the effect of the thread-wise noise.

Figure 8 compares the real fabric images and rendered images of virtual fabric. As shown in the figure, the proposed method can produce plausible virtual fabric objects.

The proposed method can be easily implemented with GPU programming languages and applied to realtime applications. Figure 9 demonstrates the proposed method can effectively render the woven surface even in realtime applications. The results were obtained by implementing a hardware shader with OpenGL shading language and the spatially varying reflectance on the woven surface is plausibly expressed. Table 1 show the comparison result when the time expenses taken by proposed rendering method and other rendering methods is compared with the cost of OpenGL default Gouraud shading. As shown in the table, the

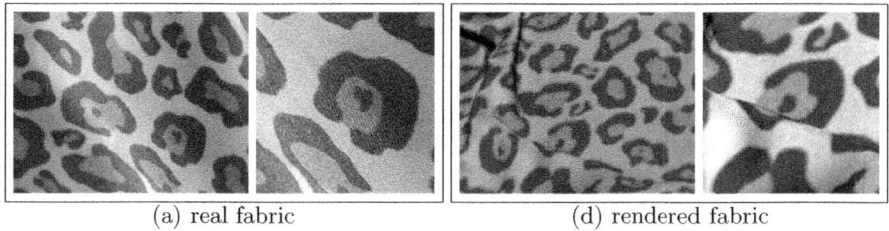

(a) real fabric　　　　　　　　　　(d) rendered fabric

Fig. 8. The comparison of real fabric and rendered virtual fabric: (a) real fabric image, (b) virtual fabric rendered with the proposed method

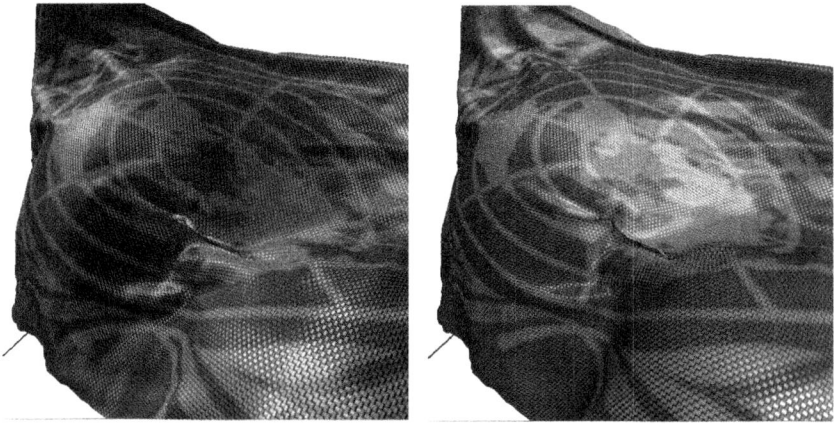

Fig. 9. Spatially varying reflectance according to the light movement in a realtime application

proposed method is efficient enough to be used in realtime applications. For the experiment, the proposed method was test on Mac OS X with 2.26 GHz intel core 2 CPU, 2G 1067 Mhz DDR3 RAM, and NVIDIA GeForce 2400M Graphics Hardware.

Table 1. Time expenses of rendering methods are compared with OpenGL default Gouraud shading

	Gouraud shading	Per-pixel lighting	Alternating anisotropy	Proposed method
Time Cost	1	1.415	1.785	1.846

Figure 10 shows the various rendering results by changing the distance between the observation camera and the fabric objects. As shown in the figure, the proposed method can effectively generate realistic fabric in arbitrary distances both in realtime applications and in offline renderers.

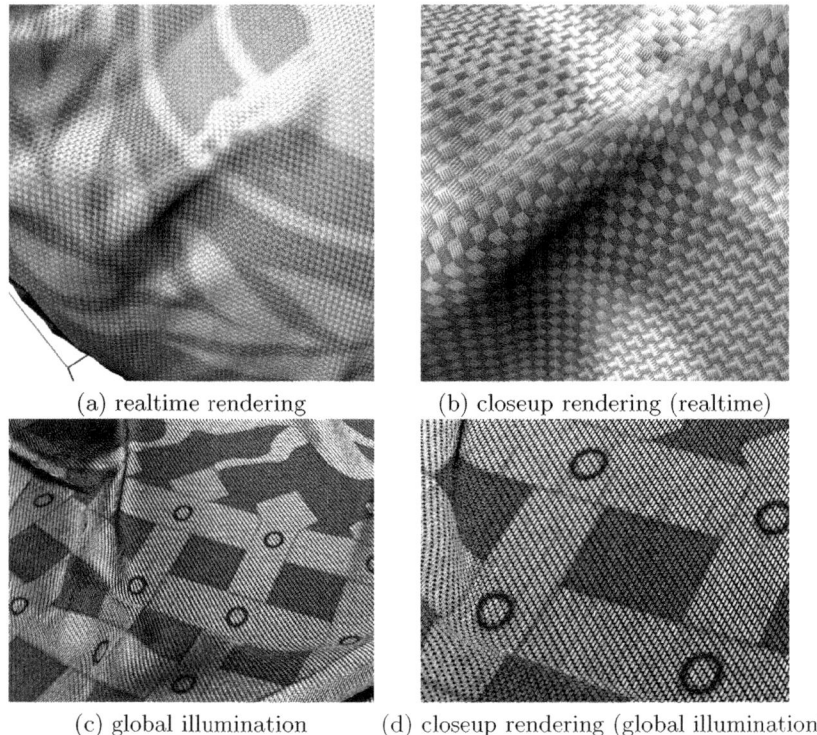

Fig. 10. Rendering of woven surface with varying observation distance: (a) and (b) realtime rendering results, (c) and (d) offline rendering with global illumination

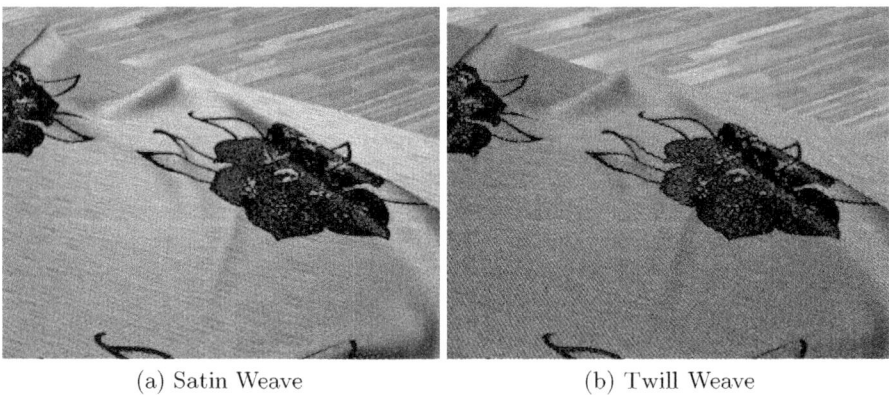

Fig. 11. The rendering results according to the weave patterns with $e_x = 20$ and $e_y = 2$: (a) satin fabric ($n_w = \infty$) (b) twill fabric ($n_w = 2, n_\pi = 1, n_\sigma = 1$)

Figure 11 demonstrates the effect of weave pattern control. As shown in the figure, we can effectively generate various fabric reflectance by simply control the weave pattern parameters, n_w and n_σ. Figure 11 (a) shows the rendering result with satin weave while (b) is rendered with twill weave pattern.

5 Conclusion

In this paper, a procedural approach to photorealistic fabric rendering is proposed. The proposed method reproduces natural reflectance of woven fabric by alternating the anisotropy. Moreover, the proposed method generates the bumpy illusion on the woven fabric by deforming the microfacet distribution function(MDF). The proposed method takes into account the weave patterns and twisted fiber structure. The twisted fiber model and realism enhancement of the proposed method drastically increase the rendering quality.

The experimental results show that the proposed method can be successfully employed for photorealistic rendering of diverse woven fabric materials even in realtime applications such as games and VR systems. The proposed method was successfully implemented with OpenGL shading language and produced realistic cloth image in a realtime rendering application.

Since the reflectance properties of the woven surface was procedurally modeled, the observation of the virtual fabric can be performed in arbitrary distance. The experimental results show that the proposed method always provides plausible rendering results regardless of the observation distance.

Even the simple BRDF data usually requires heavy data so that an extremely large amount of storage will be required to express various material. Even worse, the spatially varying BRDF (SVBRDF) is essential for plausible representation of woven fabric, and the SVBRDF data requires far larger amount of storage. However, the proposed method efficiently renders plausible woven fabric objects without any expensive measured data.

The efficiency and the procedural approach of the proposed method enable interactive applications such as games to employ the rendering techniques to express realistic woven fabric.

References

1. Adabala, N., Magnenat-Thalmann, N., Fei, G.: Real-time rendering of woven clothes. In: Proceedings of the ACM Symposium on Virtual Reality Software and Technology, pp. 41–47 (2003)
2. Adabala, N., Magnenat-Thalmann, N., Fei, G.: Visualization of woven cloth. In: Proceedings of the 14th EUROGRAPHICS Workshop on Rendering. ACM International Conference Proceeding Series, vol. 44, pp. 178–185 (2003)
3. Ashikhmin, M., Premoze, S., Shirley, P.: A microfacet-based brdf generator. In: Proceedings of ACM SIGGRAPH 2000, pp. 65–74 (2000)
4. Ashikhmin, M., Shirley, P.: An anisotropic phong brdf model. Journal of Graphics Tools 5(2), 25–32 (2002)

5. Cook, R.L., Torrance, K.E.: A reflectance model for computer graphics. Computer Graphics, ACM SIGGRAPH 1981 Conference Proceedings 15(3), 307–316 (1981)
6. Dana, K.J., Nayar, S.K., Van Ginneken, B., Koenderink, J.J.: Reflectance and texture of real-world surfaces. ACM Transactions on Graphics 18(1), 1–34 (1999)
7. Daubert, K., Lensch, H.P.A., Heindrich, W., Seidel, H.P.: Effcient cloth modeling and rendering. In: Proc. 12th Eurographics Workshop on Rendering, Rendering Techniques 2001, pp. 63–70 (2001)
8. Daubert, K., Seidel, H.P.: Hardwarebased volumetric knitwear. Computer Graphics Forum EUROGRAPHICS 2002 Proceedings 21, 575–584 (2002)
9. Gröller, E., Rau, R., Strasser, W.: Modeling and visualization of knitwear. IEEE Transactions on Visualization and Compute Graphics, 302–310 (1995)
10. Gröller, E., Rau, R., Strasser, W.: Modeling textile as three dimensional texture. In: Proc. 7th EUROGRAPHICS Workshop on Rendering, Rendering Techniques 1996, pp. 205–214 (1996)
11. Blinn, J., Newell, M.: Texture and reflection in computer generated images. Communication of ACM 19(10), 542–547 (1976)
12. Wang, J., Zhao, S., Tong, X., Synder, J., Guo, B.: Modeling anisotropic surface reflectance with example-based microfacet synthesis. ACM Transactions on Graphics (SIGGRAPH 2008) 27(3), 41:1–41:9 (2008)
13. Lawrence, J., Ben-Artzi, A., Decoro, C., Matusik, W., Pfister, H., Ramamoorthi, R., Rusinkiewicz, S.: Inverse shade trees for non-parametric material representation and editing. ACM Transactions on Graphics 25(3), 735–745 (2006)
14. McAllister, D.K., Lastra, A.A., Heidrich, W.: Efficient rendering of spatial bi-directional reflectance distribution functions. In: Proceedings of the 17th EUROGRAPHICS/ SIGGRAPH Workshop on Graphics Hardware (EGGH 2002), pp. 79–88 (2002)
15. Meissner, M., Eberhardt, B.: The art of knitted fabrics, realistic and physically based modeling of knitted patterns. Computer Graphics Forum (EUROGRAPHICS 1998 Proceedings), 355–362 (1998)
16. Sattler, M., Sarlette, R., Klein, R.: Efficient and realistic visualization of cloth. In: Proceedings of the 14th EUROGRAPHICS Workshop on Rendering, EGRW 2003, pp. 167–177 (2003)
17. Sloan, P.-P., Cohen, M.F.: Interactive horizon mapping. In: Proceedings of the Eurographics Workshop on Rendering Techniques 2000, pp. 281–286 (2000)
18. Torrance, K.E., Sparrow, E.M.: Theory for off-specular reflection from roughened surfaces. Journal of Optical Society of America 57(9) (1967)
19. Xu, Y., Lin, Y.C.S., Zhong, H., Wu, E., Guo, B., Shum, H.: Photorealistic rendering of knitwear using the lumislice. In: Proceedings of SIGGRAPH 2001, pp. 391–398 (2001)
20. Yasuda, T., Yokoi, S., Toriwaki, J., Inagaki, K.: A shading model for cloth objects. IEEE Computer Graphics and Applications 12(6), 15–24 (1992)
21. Zinke, A., Weber, A.: Light scattering from filaments. IEEE Transactions on Visualization and Computer Graphics 13(2), 342–356 (2007)

Responsive Action Generation by Physically-Based Motion Retrieval and Adaptation

Xiubo Liang[1], Ludovic Hoyet[2], Weidong Geng[1], and Franck Multon[2,3,*]

[1] State Key Lab of CAD&CG, Zhejiang University, 310027 Hangzhou, China
[2] Bunraku project, IRISA, Campus de Beaulieu, 35042 Rennes, France
[3] M2S, University Rennes2, av. Charles Tillon, 35044 Rennes, France
{liangxiubo,gengwd}@zju.edu.cn,
{lhoyet,Franck.Multon}@irisa.fr

Abstract. Responsive motion generation of avatars who have physical interactions with their environment is a key issue in VR and video games. We present a performance-driven avatar control interface with physically-based motion retrieval. When the interaction between the user-controlled avatar and its environment is going to happen, the avatar has to select the motion clip that satisfies both kinematic and dynamic constraints. A two-steps process is proposed. Firstly, it selects a set of candidate motions according to the performance of the user. Secondly, these candidate motions are further ranked according to their capability to satisfy dynamic constraints such as balance and comfort. The motion associated with the highest score is finally adapted in order to accurately satisfy the kinematic constraints imposed by the virtual world. The experimental results show that it can efficiently control the avatar with an intuitive performance-based interface based on few motion sensors.

Keywords: Motion sensors, motion retrieval, physical constraints, motion adaptation, virtual human, avatar.

1 Introduction

With traditional gaming interfaces, players can only play games by keyboard, mouse, joystick or steering wheel, which are not intuitive enough to control the avatars in the virtual world. Expressive control interfaces can obviously improve the end-user experience in interactive VR applications. Such interfaces usually translate user's intentions accurately to the motion of the virtual character in a natural manner. Recently, new input devices (e.g. Wii Remote) have been employed to implement performance-based animation interfaces, which take advantage of human perceptual capabilities in order to present user's motion authoring intention in native ways [1,2,3]. One of the key technical points behind such kind of systems is content-based motion retrieval, which has been widely

---

* Corresponding author.

investigated by a lot of researchers [4,5,6]. Most previous works aimed at solving kinematic constraints, which consist of selecting the motion clips that best matches the query by numerically or logically defined similarity. However, using such metrics, it's almost impossible to distinguish some similar motions which are actually quite different in terms of the ever-changing dynamic constraints in the virtual world. For instance, in the grasping scenario for some VR applications, users can only perform a rough grasp motion without the exact sense of object weight. The system should generate the responsive actions that are physically correct. To the best of our knowledge, this problem of physically-based motion retrieval hasn't been investigated by any researchers yet. In this paper, we present a novel approach to creating responsive character animation using motion sensors based on the performer's movements (see Figure 1). The goal of our research is to develop a motion retrieval and adaptation method which meets the physical constraints in the virtual environment.

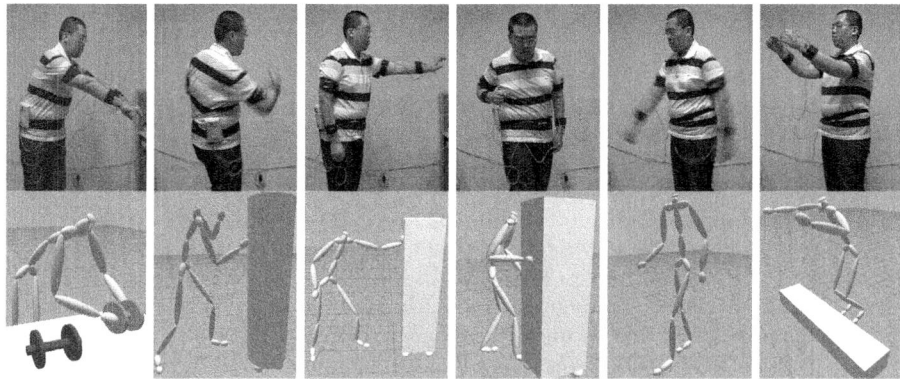

Fig. 1. A user wearing a few motion sensors controls the avatar by his performance. From left to right: grasping, punching, pushing, pulling, walking and jumping.

Motion sensor is one of the most popular devices that can capture versatile movements by measuring acceleration and orientation. Although capturing high quality motions reliably is still very hard for them, they are powerful enough to get rough movements conveniently. However the motion of the user cannot be directly imposed to the avatar because the virtual environment is not the same than the real world in which the user if moving. For example, while the user is pushing nothing with the hands, the avatar has to push virtual objects with various masses. If the movement of the avatar is not adapted it may lead to visually unrealistic motions; the avatar may move heavy objects without visible efforts. Section 2 presents a set of related works addressing two separated problems: performance-driven animation of an avatar and satisfying dynamic constraints. Section 3 describes an overview of the proposed method and the remaining parts of the paper give detailed information of its various components.

2 Related Work

Performance-driven animation interfaces are good choices for natural interaction with a virtual world. Various types of devices have been used to interact with a virtual human such as video [1] or accelerometers [2,7] or other types of sensors [8]. However, these methods mostly focused on driving the virtual human with kinematic data and generally did not take physical constraints into account. One can cite the work of [3] who developed a Wiimote-based interface for the locomotion of a mechanical model of character. Indeed, one of the most common method used to ensure the physical correctness of computer animations is to employ dynamic models driven by controllers [9,10,11]. Recent advances in this field have demonstrated the capabilities of designing robust locomotion controllers [12,13] but it is still difficult to generalize to any kind of natural motion. Coupling controllers and motion capture data were successfully applied to adapt clips to physical constraints [14]. Dynamic filters [15,16,17] have also been introduced to convert a physically infeasible reference motion into a feasible one for the given human figure. It has been applied to ensuring balance and comfort of the resulting motions [18] but it's difficult to adjust the parameters embedded in such methods. Non-linear optimization has been widely used in the past to solve both kinematic and dynamic constraints in a unified spacetime optimization [19,20,21]. But, these methods are based on non-linear optimization which leads to heavy computation and may fall into local minimums when the resulting motion is too far from the original one. An alternative of using this non-linear optimization consists of computing a displacement map that takes the physical constraints into account [22] which decreases computation time and complexity.

These previous methods are based on modifying an imposed motion clip. It is mostly applicable in the neighborhood of the original motion clip. However, constraints may lead to completely different motor strategies, such as carrying light or very heavy objects. Methods based on adapting a unique motion may fail to address this problem. In the specific case of reacting to hits and punches, it's possible to sequentially blend passive physical simulation with a relevant motion clip [23] or to select the most appropriate keyframes into a database [24]. But these methods have only been applied to navigation task. Reinforcement learning can also address this type of problem by designing controllers based on reusing and blending several clips together [25,26]. However this approach requires a lot of examples for each motion and for each dynamic constraint. On the opposite, Ishigaki et al. [27] used a small database of examples to recognize some key features of the users' actions and to control a simplified physical model. However, it cannot handle a wide variety of complex motions, such as manipulation tasks. In this paper, we propose a method that consists in using more general criteria such as comfort and balance directly in the motion retrieval process. The resulting method is illustrated with pushing, pulling and carrying motions with variable physical constraints.

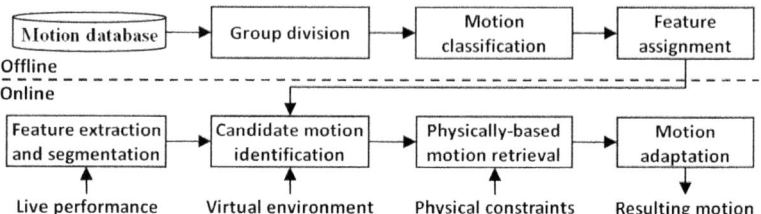

Fig. 2. The schematic block diagram of the system

3 Overview

Our system takes orientation readings as input, and then applies motion recognition and adaptation based on pre-captured database. The schematic block diagram of the system is given in Figure 2. Offline, motions in the database are divided into a set of groups (e.g. grasping, punching etc). Then the motions in a group are further classified into a set of classes (e.g. lifting left object with right hand). At last, each class is annotated with a predefined geometric feature sequence which will serve as the template for motion identification. Online, the user puts the motion sensors at the specified position on his body and performs some actions. The system segments the sensing data continuously and generates a geometric feature stream which is used to search the database for the most appropriate motion classes (see section 4). The algorithm only searches in the motion groups the candidates which could have the correct interaction with the object in the virtual world. When a class of candidate motions are identified, they will be further ranked according to their property to satisfy the dynamic constraints (e.g. balance and comfort), as described in section 5. Finally, the motions with the highest score will be further adapted with a fast inverse kinematics solver to accurately meet the kinematic constraints (see section 6). Some results and validations are proposed in section 7.

4 Motion Sensing and Action Identification

Human motion can be driven by the relative orientation of each joint to its parent [8]. The configuration of motion sensors (Xsens MTx) on human body are shown in Figure 3(a). If the user is only interested in the motions of upper or lower part, he can simply use the sensors on the trunk and limbs (5 sensors are needed in that case). With such few sensors, the system can only acquire rough motions performed by the user. To make matters worse, there are always noise distortions in the data readings. In order to recognize the rough and noisy motions, we propose an algorithm based on geometric features. Motivated by dance movement notation language of Labanotation, we developed our feature extraction mechanism based on subspace division. However, unlike the religious dance motion, the same interaction movement performed by different users in VR applications exhibits much more spatio-temporal variance. In order to cover the variance, we don't use as many subspaces as being used in Labanotation. The local space of each joint is only divided into six parts (See Figure 3(b)). If the feature symbol in a frame is different from the one in its previous frame, a new segment is generated. When several joints are taken into account together, the overall feature sequence will be the conjunction of the feature sequences of all the joints (See Figure 3(c)). In this way, a motion is segmented and translated into a feature sequence represented by a set of subspace symbols. Therefore, the problem of motion identification is converted into the problem of symbol matching. If a predefined feature sequence template in the database is matched to a subsequence in the continuously performed motion sequence, a class of motion is identified. Just like the process the choreographer choreographs dance motions using Labanotation, we select some key postures for each motion class and assign the feature sequence generated from the key postures to the class as the template for motion identification. In order to improve efficiency and reduce error, each object in the virtual environment is connected to one or more groups. When the distance between the virtual character and some object is less than a threshold, the system thinks the user is going to interact with the object. The candidate motion is only searched in the associated groups. Note that the "moving" group should be connected to all objects, so that the character can move away when the user finishes the interaction and performs a moving motion.

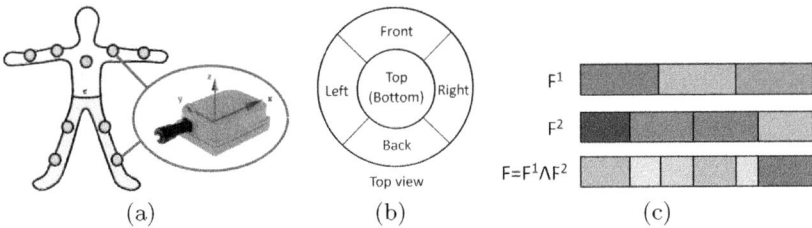

Fig. 3. (a) The configuration of motion sensors on human body. (b) The subspace division of each joint. (c) The conjunction of two feature sequences $F = F^1 \wedge F^2$.

Let $FS = (d_1, d_2, ..., d_T)$ be the feature stream computed from the real-time sensing data where T is the number of segments. Let $PS = \{P^i | P^i = (f_1^i, f_2^i, ..., f_{m_i}^i), 1 \leq i \leq n\}$ be the set of candidate motion classes, where m_i is the number of key features in class P^i, n is the total number of classes in PS, $d_j, f_j^i \in S^r$ where S is the set of subspace symbols and r is the number of joints. The problem of motion identification is transformed into the problem of finding a sub-sequence $SS = (d_{x_1}, d_{x_2}, ..., d_{x_k})$ in FS which satisfies:

$$SS = P^i \Leftrightarrow \begin{cases} d_{x_1} = f_1^i \\ d_{x_2} = f_2^i \\ ... \\ d_{x_k} = f_{m_i}^i \end{cases}, where\ 1 \leq x_1 < x_2 < ... < x_k \leq T \quad (1)$$

When a new segment d_{T+1} is detected from the sensed data, it will be pushed back to FS. The system then looks over if there is a sub-sequence in FS which matches P^i. If so, all the elements before the last occurrence of d_{x_k} in FS will be removed and the motions of this class will be fed to the physically-based motion retrieval module. If not, the system will wait for a new segment d_{T+2}. It's possible that some motion classes are assigned with the same feature sequence. In that case, we need to further select the best motion class. Principle component analysis (PCA) is employed to deal with this problem. The positions and velocities of the wrist and elbow joints (ankle and knee joints are also used if legs are involved) are extracted from all the motions in the jth class to form a matrix $M_j = [V_1, V_2, ..., V_N]$, where V_i is the data of the ith frame and N is the total number of frames in the class. Let C_j be the matrix composed of the first m components of M_j. When transforming M_j to the low dimension space with C_j and reconstructing the data back with the transpose matrix C_j^T, there will be a reconstruction error $e_j = \left| M_j - C_j^T \times (C_j \times M_j) \right|^2$. The class with the smallest error will be selected.

5 Physically-Based Motion Retrieval

Several physical factors may have an influence on the motion naturally selected by humans in complex tasks. Among these factors, many authors agree to state that balance and comfort are very relevant in human motion control. These variables have been used in many approaches of robotics and computer animation.

5.1 Physical Constraints

In many domains, dynamic balance is defined by ensuring that the Zero Momentum Point (ZMP) remains inside the base of support. ZMP could be calculated with three main models: the inverted pendulum model which is widely used for locomotion, the particle model in which inertia is neglected and the articulated model assuming that humans could be modeled as connected rigid bodies [28]. The particle model is widely used as it's a good compromise between computation complexity and accuracy. In this model, each body part is modeled by its center of mass. The equation to compute ZMP(denoted $c(c_x, 0, c_z)$) is as follows:

$$c_x = \frac{\sum_{i=1}^{n}(m_i(\ddot{y}_i+g)x_i - m_i(y_i-c_y)\ddot{x}_i)}{\sum_{i=1}^{n} m_i(\ddot{y}_i+g)}, c_z = \frac{\sum_{i=1}^{n}(m_i(\ddot{y}_i+g)z_i - m_i(y_i-c_y)\ddot{z}_i)}{\sum_{i=1}^{n} m_i(\ddot{y}_i+g)} \quad (2)$$

The parameters of the dynamic model are set to those of the virtual human and not of the actor. The key idea is to be able to manage various mass repartitions and not limited to those of the actor. If the human-like figure is carrying an object whose mass is m_o and whose center of mass is (x_o, y_o, z_o), the equation becomes:

$$c_x^O = \frac{N_x + m_o(\ddot{y}_o+g)x_o - m_o(y_o-c_y)\ddot{x}_o}{D + m_o(\ddot{y}_o+g)}, c_z^O = \frac{N_z + m_o(\ddot{y}_o+g)z_o - m_o(y_o-c_y)\ddot{z}_o}{D + m_o(\ddot{y}_o+g)} \quad (3)$$

where N_x, N_z are the numerators of c_x, c_z respectively and D is the denominator of c_x, c_z. In a more general manner, the human-like figure may be subject to external forces $F_{ext}(F_x, F_y, F_z)$ (in addition to gravity and ground reaction force) exerted at point $P_{force}(P_x, P_y, P_z)$ on the body. In that case, equation 2 becomes:

$$c_x^F = \frac{N_x - c_y \times F_x - P_y \times F_x - P_x \times F_y}{D - F_y}, c_z^F = \frac{N_z - c_y \times F_z - P_y \times F_z - P_z \times F_y}{D - F_y} \quad (4)$$

If several forces are added to the system, this equation can be easily generalized. All these equations lead to computing positions and accelerations of center of mass of each body part based on motion capture data. As there is much noise, a low-pass filter is preliminary applied to the computed data.

To maximize comfort, we wish to minimize the torques at some key joints. The joint forces $F_{j->i}$ and torques $\tau_{j->i}$ are computed using the Newton's equations for the key joint i:

$$\begin{aligned} F_{j->i} &= m \times \ddot{x}_i - \sum_{k \neq j}(F_{k->i}) - W_i \\ \tau_{j->i} &= \dot{H}_{O_{ij}} - O_{ij}O_{ik} \times F_{k->i} - O_{ij}x_i \times W_i - \tau_{k->i} \end{aligned} \quad (5)$$

where x_i is the ith center of mass, $F_{k->i}$ and $\tau_{k->i}$ are the force and torque exerted by segment k on segment i, W_i is the weight of segment i, O_{ij} is the contact point between segment i and j and $\dot{H}_{O_{ij}}$ is the angular momentum. Note that this algorithm cannot be used directly to calculate joint torques for joint chains with loops (e.g. feet on the ground, hands connected by an object). We deal with this case in an idealized way: divide the object into two seperated parts with the same weight which will be lifted the two hands accordingly.

5.2 Retrieval Algorithm

The typical idea in the field of content-based motion retrieval is the template matching algorithm with a wide range of variants. keeping in mind the idea of physically-based motion retrieval, we need the templates for the physical constraints which are the ground truth ZMPs and torques according to the physical environment in which the actor did the motion capture. When doing motion retrieval, we compute the distances between the ground truth templates and runtime ZMP and torque curves with virtual environment. See Figure 4 for an example of the ZMP (Figure 4(a)) and torque (Figure 4(b)) curves of a grasping motion with different object weights. Note that the curves with 10kg object are ground truth whereas the others are the curves when grasping objects of other weights in the virtual environment with the same motion. Let $c_t(M_i), \tau_t(M_i)$ be the ZMP and torque template of motion M_i and $c_r(M_i), \tau_r(M_i)$ be the runtime ZMP and torque curve with virtual environment. The ZMP distance $D_c(M_i)$ and torque distance $D_\tau(M_i)$ between them are computed using typical Dynamic Time Warping algorithm: $D_c(M_i) = DTW(c_t(M_i), c_r(M_i))$ and $D_\tau(M_i) = DTW(\tau_t(M_i), \tau_r(M_i))$. However, for some motions, e.g. pushing or punching, it's not easy to measure the pysical properties such as the contact forces. Thus, we cannot compute the ground truth templates. In this case, we employ another metric: $D_c(M_i)$ is computed as the average distance between ZMP and the support area during the interaction phase. The feet of the virtual character are modeled as rectangles. An ordered convex hull is computed thanks to the eight vertexes. The resulting convex hull is supposed to be the support area. Then the distance between ZMP and the support area is the shortest length between ZMP and the convex hull (See Figure 4(c)). $D_\tau(M_i)$ is simply computed as the average value of runtime torques $\tau_r(M_i)$.

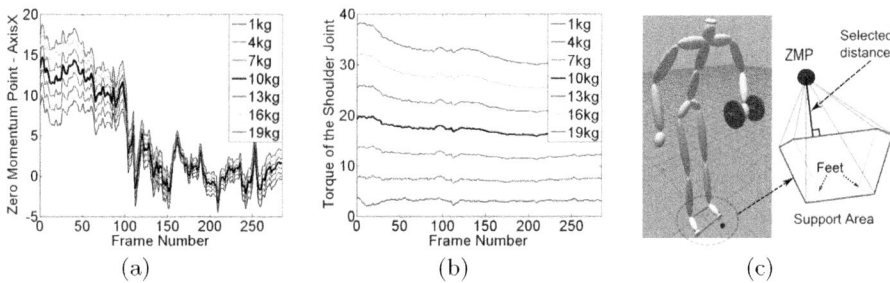

Fig. 4. Distance calculation for a left hand grasping motion. (a) The ZMP curves of the grasping motion with objects of different weights. (b) The torque curves of the left shoulder joint. (c) The distance from ZMP to the support area in a frame.

After $D_c(M_i)$ and $D_\tau(M_i)$ are computed, we can sort all the candidate motions according to these two scores. First, the score $S_c(M_i)$ based on $D_c(M_i)$ and the score $S_\tau(M_i)$ based on $D_\tau(M_i)$ are calculated as follows:

$$S_c(M_i) = \frac{1}{D_c(M_i)} / \sum_{j=1..n} \frac{1}{D_c(M_j)}, \quad S_\tau(M_i) = \frac{1}{D_\tau(M_i)} / \sum_{j=1..n} \frac{1}{D_\tau(M_j)} \quad (6)$$

Then the two scores are integrated with corresponding weights ω_c and ω_τ to compute a final score: $S(M_i) = \omega_c \times S_c(M_i) + \omega_\tau \times S_\tau(M_i)$. The motion with the highest final score is selected to be further adapted.

6 Real-Time Motion Adaptation

The retrieved motions should be adapted to the ever-changing virtual environment. We develop our real-time constraint solver based on the work of Kulpa et al. [29]. The main idea is to divide the human body into a set of sub-groups (arms, legs, head and trunk); then trying to solve the constraints with analytical method for limbs; if it fails, the traditional Cyclic Coordinate Descent (CCD) algorithm is employed to adapt other body parts. The main problem for the analytical solution of arms (the same for legs) is how to compute the position of elbow given the location of wrist. In shoulder's local coordinates, the position of elbow $P_0(0, -Y_0, -Z_0)$ can be calculated with Equation 7 (See Figure 5(a)).

Supposing P_e is the elbow's position to be sought, P_s is the shoulder's position, R_x is the matrix which rotates the normal of the half plane containing the shoulder, elbow and wrist joints to x direction and R_z is the matrix which rotates the limb vector $\boldsymbol{L}$ to the z direction, then $P_e = P_s + R_x^{-1} * R_z^{-1} * P_0$.

$$Z_0 = (L_1^2 - L_2^2 + L^2)/(2 \times L), \quad Y_0 = \sqrt{L_1^2 - Z_0^2} \qquad (7)$$

Note that the above solution assumes that the half plane's normal is known in advance which is actually not. [29] gets the normal from a reference motion. However, we can't always have a proper reference motion in hand. Therefore, we propose a way to automatically estimate it. When the wrist moves to a new position, the normal of the half plane may also change. But it must lie in the plane vertical to the limb vector. We then project the original normal vector to that plane. The projected vector is selected as the new normal (See Figure 5(b)). In this way, as much as information is maintained as the ground truth motion is changed as less as possible. See Figure 5(c) for an example of motion adaptation.

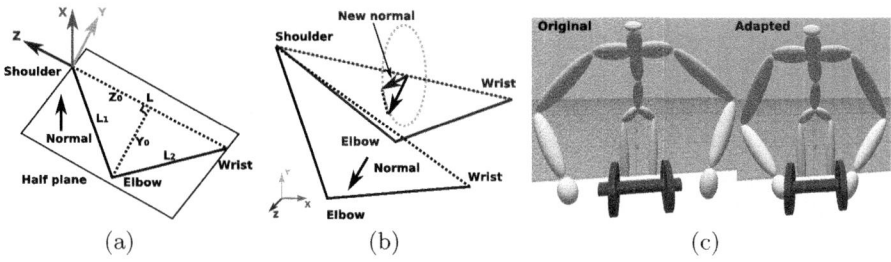

Fig. 5. Constraint solver. (a) The half plane containing the arm in shoulder's local coordinates. (b) The best new normal of the new half plane. (c) An adaptation example.

7 Experimental Results

To evaluate the responsive action generation system, we captured a series of motions: grasping, punching, pushing, pulling light, medium and heavy objects, walking and jumping. 154 motion clips in the BVH format with a 160Hz frame-rate are used to create the database, which are divided into four groups (grasping, punching, pushing-pulling and moving) and classified into 34 classes (lifting/putting left/middle/right object with left/right/both hands; hack/hook/ straight/swing punching with left/right hand; pushing/pulling with left/right/ both hands; walking; jumping). The total time of the database is about 12 minutes. Note that this paper doesn't focus on motion blending and other techniques used to generate a continuous sequence. We first tested the recognition rate of the overall algorithm. We count a match as correct if both the motion class and the physical reaction to the virtual environment are correct, i.e., the user performs a pushing motion with his left hand to move a heavy object, the recognized motion should be a left hand heavy pushing motion. The recognition

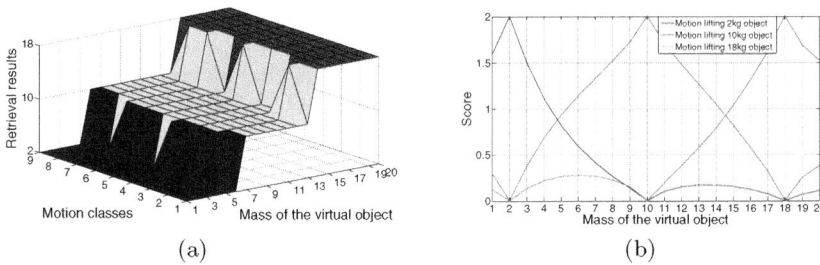

Fig. 6. Validation of motion retrieval. (a) The retrieval results for 9 classes of grasping motions when changing the mass of the virtual object. (b) The score curves for the grasping motions of class 4 in (a).

rate was greater than 90% with our database. The system can work on a larger database without the need of adding new subspaces if the motions are carefully selected into different motion groups. Then we validated the physical retrieval algorithm with the grasping example. When the virtual object's mass is set to 1-6, 7-13 and 14-20, the motions of grasping 2kg, 10kg, 18kg objects are selected respectively (Figure 6(a)). The detailed score curves for motions of class 4 in Figure 6(a) are illustrated in Figure 6(b). Figure 7 shows several retrieval and adaptation results in terms of different physical constraints.

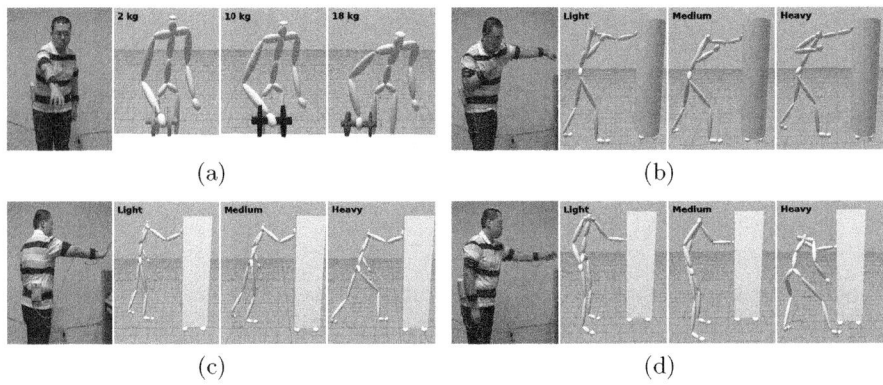

Fig. 7. In each sub-figure, from left to right: the performed action, the selected and adapted motions interacting with light, medium and heavy objects. (a) A grasping motion. (b) A punching motion. (c) A pushing motion. (d) A pulling motion.

8 Conclusion and Future Work

Our main contribution consists in introducing dynamic variables in the motion retrieval process. We have demonstrated that the avatar selects the most relevant motion of the database for a unique given performance of the user. For example, the "pulling a heavy mass" motion is actually selected when the object manipulated by the avatar is heavy. One has to notice here that the selection is not simply based on comparing the masses pulled in the original clip and the current situation. This method is very promising for many applications involving real-time measurements of the user's performance, such as video games, serious games, training in VR... In all these applications, the naturalness of the motion is strongly linked to the dynamic constraints imposed by the virtual environment.

Because the method mainly relies on motion retrieval, the resulting motion may not perfectly satisfy the mechanical laws. Although the choice is the best one among the proposed motions, the balance status of this motion may be affected. In the future it could be interesting to add adaptation facilities in order to perfectly satisfy the physical constraints. In the current version, only kinematic constraints have been considered. Another drawback is that the current system requires some manual tuning. Some automatic feature extraction methods will be developed to overcome this problem. As a perspective, we wish also to process the data in the real-time flow instead of segmented isolated clips. However, it is a complex task as the algorithm would have to select the most relevant motion with only a partial knowledge about the current actions of the user.

Acknowledgments

This work has been partially funded by ANR through the SignCom project, and INRIA through the "Bird" Associate Team fundings. It was also partly supported by NSFC 60633070 and 60773183, NCET-07-0743, and PCSIRT 065218.

References

1. Chai, J., Hodgins, J.: Performance animation from low-dimensional control signals. ACM Trans. on Graph. 24(3), 686–696 (2005)
2. Liang, X., Li, Q., Zhang, X., Zhang, S., Geng, W.: Performance-driven motion choreographing with accelerometers. Computer Animation and Virtual Worlds 20(2-3), 89–99 (2009)
3. Shiratori, T., Hodgins, J.: Accelerometer-based user interfaces for the control of a physically simulated character. ACM Trans. on Graph. 27(5), 1–9 (2008)
4. Liu, F., Zhuang, Y., Wu, F., Pan, Y.: 3d motion retrieval with motion index tree. Computer Vision and Image Understanding 92(2-3), 265–284 (2003)
5. Keogh, E., Palpanas, T., Zordan, V., Gunopulos, D., Cardle, M.: Indexing large human motion databases. In: Proceedings of the 30th International Conference on Very Large Data Bases, pp. 780–791 (2004)
6. Muller, M., Roder, T., Clausen, M.: Efficient content-based retrieval of motion capture data. ACM Trans. on Graph. 24(3), 677–685 (2005)
7. Slyper, R., Hodgins, J.: Action capture with accelerometers. In: Proceedings of ACM SIGGRAPH/Eurographics Symposium on Computer Animation, pp. 193–199 (2008)
8. Liang, X., Zhang, S., Li, Q., Pronost, N., Geng, W., Multon, F.: Intuitive motion retrieval with motion sensors. In: Proceedings of Computer Graphics International, pp. 64–71 (2008)

9. Hodgins, J., Wooten, W., Brogan, D., O'Brien, J.: Animating human athletics. In: Proceedings of ACM SIGGRAPH 1995, pp. 71–78 (1995)
10. Hodgins, J., Pollard, N.: Adapting simulated behaviors for new characters. In: Proceedings of ACM SIGGRAPH 1997, pp. 153–162 (1997)
11. Wooten, W.L., Hodgins, J.: Animation of human diving. Computer Graphics Forum 15, 3–13 (1996)
12. Yin, K., Loken, K., van de Panne, M.: Simbicon: Simple biped locomotion control. ACM Trans. on Graph. 26(3), 105 (2007)
13. Sok, K.W., Kim, M., Lee, J.: Simulating biped behaviors from human motion data. ACM Trans. Graph. 26(3), 107 (2007)
14. Zordan, V., Hodgins, J.: Motion capture-driven simulations that hit and react. In: Proceedings of ACM SIGGRAPH/Eurographics Symposium on Computer Animation, pp. 89–96 (2002)
15. Tak, S., Song, O.Y., Ko, H.S.: Motion balance filtering. Computer Graphics Forum 19(3), 437–446 (2000)
16. Yamane, K., Nakamura, Y.: Dynamic filters - concept and implementations of online motion generator for human figures. IEEE Trans. on Robotics and Automation 19(3), 421–432 (2003)
17. Tak, S., Ko, H.: A physically-based motion retargeting filter. ACM Trans. on Graph. 24(1), 98–117 (2005)
18. Tak, S., Song, O.Y., Ko, H.S.: Spacetime sweeping: a interactive dynamic constraints solver. In: Proceedings of IEEE Computer Animation, pp. 261–270 (2002)
19. Witkin, A., Kass, M.: Spacetime constraints. In: Proceedings of ACM SIGGRAPH, pp. 159–168 (1988)
20. Sofonova, A., Hodgins, J., Pollard, N.: Synthesizing physically realistic human motion in lowdimensional, behavior-specific spaces. ACM Trans., on Graph. 23(3), 514–521 (2004)
21. Jain, S., Ye, Y., Liu, K.: Optimization-based interactive motion synthesis. ACM Trans. on Graph. 28(1), 10:1–10:12 (2009)
22. Shin, H., Kovar, L., Gleicher, M.: Physical touch-up of human motions. In: Proceedings of Pacific Graphics 2003, pp. 194–203 (2003)
23. Zordan, V., Majkowska, A., Chiu, B., Fast, M.: Dynamic response for motion capture animation. ACM Trans. on Graph. 24, 697–701 (2005)
24. Mitake, H., Asano, K., Aoki, T., Marc, S., Sato, M., Hasegawa, S.: Physics-driven multi dimensional keyframe animation for artist-directable interactive character. Computer Graphics Forum 28(2), 279–287 (2009)
25. Treuille, A., Lee, Y., Popović, Z.: Near-optimal character animation with continuous control. ACM Trans. Graph. 26(3), 7:1–7:7 (2007)
26. Cooper, S., Hertzmann, A., Popović, Z.: Active learning for real-time motion controllers. ACM Trans. on Graph. 26(3), 5 (2007)
27. Ishigaki, S., White, T., Zordan, V.B., Liu, C.K.: Performance-based control interface for character animation. ACM Trans. on Graph. 28(3), 61:1–61:8 (2009)
28. Kajita, S.: Humanoid Robot, Ohmsha, Japan (2005)
29. Kulpa, R., Multon, F., Arnaldi, B.: Morphology-independent representation of motions for interactive human-like animation. Computer Graphics Forum 24, 343–352 (2005)

Visibility Transition Planning for Dynamic Camera Control

Thomas Oskam[1], Robert W. Sumner[2], Nils Thuerey[1], and Markus Gross[1,2]

[1] ETH Zurich, Switzerland
[2] Disney Research Zurich, Switzerland

Abstract. We present a real-time camera control system that uses a global planning algorithm to compute large, occlusion free camera paths through complex environments. The algorithm incorporates the visibility of a focus point into the search strategy, so that a path is chosen along which the focus target will be in view. The efficiency of our algorithm comes from a visibility-aware roadmap data structure that permits the precomputation of a coarse representation of all collision-free paths through an environment, together with an estimate of the pair-wise visibility between all portions of the scene. Our runtime system executes a path planning algorithm using the precomputed roadmap values to find a coarse path, and then refines the path using a sequence of occlusion maps computed on-the-fly. An iterative smoothing algorithm, together with a physically-based camera model, ensures that the path followed by the camera is smooth in both space and time. Our global planning strategy on the visibility-aware roadmap enables large-scale camera transitions as well as a local third-person camera module that follows a player and avoids obstructed viewpoints. The data structure itself adapts at runtime to dynamic occluders that move in an environment. We demonstrate these capabilities in several realistic game environments.

The Application of MPEG-4 Compliant Animation to a Modern Games Engine and Animation Framework

Chris Carter, Simon Cooper, Abdennour El Rhalibi, and Madjid Merabti

School of Computing & Mathematical Sciences, Liverpool John Moores University,
Byrom Street, Liverpool, L3 3AF
{C.J.Carter@2007,S.Cooper@2003,a.elrhalibi@,
m.merabti@}ljmu.ac.uk

Abstract. The MPEG-4 standards define a technique for 3D facial and body model animations (FAPS / BAPS respectively), as seen in animation systems such as Greta. The way this technique works is in contrast to the set of animation techniques currently used within modern games technologies and applications, which utilize more advanced, expressive animation systems such as Skeletal, Morph Target and Inverse Kinematics. This paper describes an object oriented, Java-based framework for the integration and transformation of MPEG4 standards-compliant animation streams known as Charisma. Charisma is designed for use with modern games animation systems; this paper illustrates the application of this framework on top of our Java / OpenGL-based games engine framework known as Homura.

Keywords: MPEG-4, FAPS, Facial Animation, Skinning, Homura, Charisma.

1 Introduction

The virtual, synthetic representation of human characters has been a prominent research challenge for many years. Their applications are many and varied: within real-time interactive applications such as games; the construction of Embodied Conversation Agents (ECAs) [1]; animated movies and television shows; telecommunication systems and many more. In order to construct a believable representation, it is crucial that the virtual character must be able to express the many facets which the human face can convey: emotional displays, speech synthesis, gaze, gestures and must be achieved in a natural, expressive manner in order to achieve believability. One such method for the construction of virtual "Talking Heads" [2] is the application of the Motion Picture Expert Group (MPEG) ISO standardized virtual character (VC) animation specification, commonly known as a Facial and Body Animation (FBA) stream [3], which forms part of the wider MPEG-4 standards whose focus is primarily concerned with audio-visual coding. Whilst the MPEG-4 animation standard has been widely utilized within academia, it has been superseded by more modern, fluid techniques such as morphable animation [4] and the use of blend shapes [5]. Moreover, the per-vertex manipulation of feature points is in contrast to the skeletal and inverse-kinematic [6] animation solutions commonly used in modern real-time graphics and games engines and the asset production pipelines utilized by such engines.

As such, we feel there is a need to provide an MPEG-4 compliant animation solution which utilizes skeletal animation in a manner that is compatible with the aforementioned pipeline and engines, whilst interoperating seamlessly with alternative animation techniques. In this paper, we describe a technique for combining MPEG-4 complia nt facial animation with skeletal-based animations, in order to improve the fidelity and smoothness of the movements within the animation, as well as simplifying the animation pipeline. This work has resulted in the production of an animation API known as Charisma. The rest of this paper is organized as follows: Section 2 describes the current state of the art and MPEG-4 standards for facial animation systems; Section 3 describes the Charisma API and the mechanisms employed to generate the animations; Section 4 describes the rigging process, it's role within the framework and integration with an artist's pipeline using industry-standard modelling software; Section 5 illustrates a real-world use-case of the API within a games engine context; finally, in Section 6, we conclude the paper and discuss our future work.

2 Related Work

Over the last twenty years, a large amount of research has been conducted in regards to the virtual synthesis of human characters. In this section we focus upon techniques for facial modelling and facial animation. In order to qualify the state of research within the field, Ersotelos et al. conducted a survey into the state-of-the-art techniques [7], which will be covered in section 2.1. Meanwhile, the creation of applications and frameworks which adhere to the MPEG-4 FBA specification have also been prominent in the last few years, such as the open source X-Face [8] framework and the ECA application, Greta [9]. Therefore, section 2.2 details the typical MPEG-4 approach to modelling and animation, whilst section 2.3 describes existing implementations of the MPEG-4 specification.

2.1 State of the Art Facial Modelling and Animation

This section provides an overview into the current research conducted in regards to both facial modelling and facial animation techniques.

Modelling the virtual character can be cumbersome and time consuming process. Typically a character is modelled as a polygon mesh; whilst alternative techniques exist such as multi-layer modelling which include underlying structures such as the muscle and skeletal formations, these techniques are still considered too computationally expensive, particularly in the context of real-time applications [7]. There are several approaches to constructing a modelled character [7]: *Standard Geometric Face Modelling*, an approach typically used in animated feature films and games; *Generic Model Individualization* (GMI), where a specific model is constructed through the deformation of an existing model, such as those constructed from video streams and 3D face scanners; and *Example-based Face Modelling* (EFM), where a face is constructed with the desired facial features through the linear combinations of an existing face model collection, an example of this is the morphable facial modelling technique described by Blanz and Vetter [4]. Since our work is concerned with the efficient real-time application of facial animation and integration within game engine and game-related asset pipelines, we have adopted the standard geometric face modelling technique, using a single layer polygon mesh. In relation to facial animation, there are three prominent classifications for the techniques employed [7]: A *Simulation-based* approach, in which synthetic facial movements are generated through the mimicking of facial muscle contractions, whose application is prominent in the field of medical-based visualization. The *Performance-Driven* approach, where facial actions are typically applied to a face model, from sources such as video recordings and performance capture. The MPEG-4 standard falls into this classification, but can also be driven through procedural means. The third classification is the *Blend Shape-based* approach, which typically creates a set of desired animations through the combination of a number of existing face models. There are various levels of granularity for blend shapes, such as interpolation between models, interpolation between images, Blanz and Vetter's [4] morphable model approach, and a facial segmentation approach, where the face is divided into small regions, which are manipulated and interpolated between poses. The latter approach has become increasingly prevalent within the games industry. In this case, facial models are "rigged" with segmentations based on skeletal joints/ bones, blend shape objects, or a hybrid combination of both. The Face rig is then manipulated through the artist's tools or through automated techniques such as motion and performance capture. Examples of this case can be found through visual outsourcing specialists such as Image Metrics [10], whose proprietary technique has been integrated with commercial game engine solutions such as Emergent's Gamebyro engine. In the context of facial animation, our approach is to adopt techniques from the real-time face segmentation blend-shape approach and apply them to MPEG-4 based techniques at the lowest level of integration. Whilst blend shape objects have been used to great effect within the film industry (e.g. the Lord of the Rings films' Gollum character was comprised of 946 blend shape controls [5]), and have seen recent adoption in the high-performance games market, they are computationally expensive to implement, especially across low-mid range desktop PC's in real-time. Therefore, we have adopted a skeletal system for our facial segmentation.

2.2 The MPEG-4 Facial Animation Process

As mentioned previously the specifications cater for both Facial and Body animations, for the sake of brevity this section focuses solely on Facial Animation. Figure 1 illustrates the object-based representation of the standards [3].

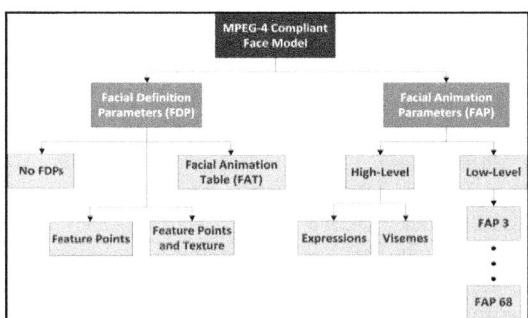

Fig. 1. High-Level Overview of the FA Specification for the MPEG-4 Standards

The structure of the face model within the standard is defined in terms of *Facial Definition Parameters* (FDP). FDP comprises a set of *feature points*. A feature point represents a key structural location on a human face (i.e. the corner of the mouth). MPEG-4 has defined a set of 84 feature points, used both for the calibration and the animation of a synthetic face. However, only 46 of these feature points are actually used for the animation. Feature points are subdivided in groups according to the region of the face they belong to, and numbered accordingly (e.g. Group 3 relates to eye feature points, where point 1 is the positioned at the top of the left eye-lid: *feature point 3.1*). Facial Animation using MPEG-4 is a parameterised system. The various areas of the face are controlled using *Facial Animation Parameters* (FAP). There are two *high-level* FAPs for controlling visemes and expressions, and 66 *low-level* FAPs. High-level FAPs are used to represent, with a single parameter, the most common facial expressions (joy, sadness, anger etc) and the visemes (a viseme is a mouth posture correlated to a phoneme). The majority of the low-level FAP indicate the translation of the corresponding feature point, with respect to its position in the *Neutral Face* (the default posture which the face model takes with no FAPs applied), along one of the coordinate axes (e.g. FAP 19 translates feature point 3.1 downwards along the vertical axis: closing the left eyelid). There are exceptions, with some FAPS controlling rotations such as the rotation of the head (yaw, pitch, and roll) and each of the eyeballs. The amount of the displacement described by a FAP is expressed in specific measurement units, called *Facial Animation Parameter Units* (FAPU), which represent fractions of key facial distances. There are six FAPUs in total, based on the following measurements: *IRIS Diameter* (IRISD): the distance between upper and lower eyelid; *Eye Separation* (ES): the distance between the left and right eye pupil centres; *Eye to Nose Separation* (ENS): the distance between the tip of the nose and the centre point between the eyes; *Mouth to Nose Separation* (MNS): the distance between the mid top-lip and nose tip; *Mouth Width Separation* (MW): the distance between the two corners of the mouth; and finally, *Angular Unit* (AU): The rotation angle as fractions of a radian. A single unit is typically expressed as $1/1024^{th}$ of these distances. Thus, a face is animated by a series of frames containing the displacements of the feature points by their associated FAP values. The *Facial Animation Table* (FAT) is used to express a set of vertices which are also transformed when a given feature point is manipulated by a FAP. This creates a smoother, higher fidelity animation. In our technique, these values are redundant, as explained in Section 4.

2.3 MPEG-4 Facial Animation: Existing Applications and Usage

The MPEG-4 Standards approach has been adopted by a number of applications and frameworks. A significant number of papers have been concerned with the performance-driven aspects (such as [11], [12], [13], [14]) of the specification, focusing on areas such as facial recognition and expressivity testing from MPEG-4 streams [12] [13] and the calibration of the MPEG-4 facial models from video streams [11] [14]. Our primary focus is on the low-level animation and rendering of the MPEG-4 compliant streams. There are several existing applications which implement such functionality, these include: Greta, an open source ECA application, which uses the MPEG-4 FAP and BAP systems to animate their character's within an OpenGL based rendering system [1] [9] [15]. Xface [8] [16] is a toolkit for the construction MPEG-4 compliant virtual animated characters, written in C++ and OpenGL. The toolkit provides four applications, the core library representing the animation framework, an editor for mapping the feature points and vertex influence onto the model, as well as player and remote client applications. FAE [14] is a proprietary system which implements the calibration and FAP-controlled animation and rendering systems. This work has been replaced by the FAE-derived EptaPlayer and has also spawned a web-based Java Applet variant called MpegWeb [17] [13]. However, there are several problems with these existing solutions. In all cases, the animation system is intrinsically linked with the rendering system, allowing minimal customization of scenes, or integration with more powerful rendering toolkits and engines. Whilst XFace is modular in design, its custom math classes mean integration with separate subsystems is a difficult task. The MpegWeb version suffers from poor performance (with 1.4 FPS achieved for a single 16917 polygon model on a Pentium 3 1000MHZ machine [17]). At the time of writing, Greta has two different rendering variants available, a custom OpenGL system (Psyclone and Offline versions) and an Ogre-based version (ActiveMQ). The custom OpenGL version uses the inefficient immediate-mode GL rendering mode and does not make use of display lists, or Vertex Buffer Objects (VBO) to improve rendering performance. Greta also utilizes stack-based transforms, making independent model manipulations a difficult task; The framework is platform-dependent and does not utilize the GPU processing capabilities of modern machines. Therefore, we propose the creation of a platform-independent MPEG-4 compliant animation toolkit, which is completely separated from rendering implementation and is capable of integration across multiple targets (browsers, desktop, mobile etc). The system will be scalable across the gamut of hardware available, whilst making use of GPU processing. The animation toolkit must also be capable of integrating with other animation systems (morph-target, IK etc) as well other common game engine components such as Rigid-body dynamics (for ragdoll control) and support instancing of multiple characters as well as several independent characters within the same scene. The system must also be able to utilize the industry-standard tools for model creation, both proprietary (3D Studio Max, Maya, Mudbox, XSI) and open source (Blender). The system will utilize the segmented skeletal approach to representing the animations instead of the morph target-based approach which the aforementioned applications all use. This system is to be known as Charisma and is described in the subsequent sections of this paper.

3 Implementation of the Charisma API

In order to simplify the MPEG-4 modelling and animation process, as described in section 2.2, we have constructed a rendering-agnostic API for the creation of skeletal-based MPEG-4 compliant animations. Figure 2 illustrates the various components and their hierarchy within the API. There are four key components which comprise the API:

The *model* component represents the definition of the MPEG-4 FA standards in Java object form. The *IFAPSData* interface represents a single FAP. This interface specifies that a FAP must store its transform type (translation, rotation, scaling etc); its position of positive movement (e.g. Feature Point 3.1 is down, as described in section 2.2); its category (FAP Group e.g. Group 3); its feature point number (e.g. 1)

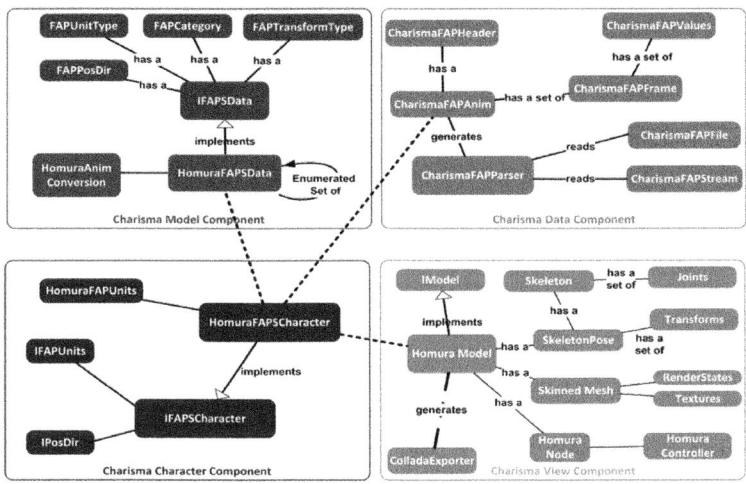

Fig. 2. Modular Architecture of the Charisma Animation API which is comprised of four distinct class separations, based on functional usage

along with descriptive data relating to standard name for the FAP and its description. Each supported renderer implements their own concrete version of this interface (e.g. *HomuraFAPSData*).

Each of the FAPs is represented as an enumerated class instance inside a Java enumeration class, holding the specification of each of the 68 FAPs. This allows the data to be encapsulated within the class at compile time, whilst providing constant time access to the data. The renderer-specific class also uses a conversion class to map the information such as the transform type and direction to the renderer's mathematical representation (e.g. Vector and Quaternion classes), in addition to keeping the base animation library independent of coordinate system.

The *data* component is responsible for representing a FAPS-based animation stream and handling the I/O for source data from streams, sockets, files etc. The data is encapsulated in a *CharismaFAPAnim* object, which is comprised of a list of FAP frames (Charisma*FAPFrame)*, whose purpose is to encapsulate the animation data in a succinct fashion; and an animation header (*CharismaFAPHeader*). The latter holds high-level information relating to the source of the animation data, the number of frames and frame rate of the animation. Each of the FAP frames is composed of a frame index and a List of data representing the state and transformation value of each of the 68 FAPs. This data structure is designed to minimize memory consumption for the animations.

The *view* component, encapsulates the engine-specific view of the 3D face model and animation classes. The component provides a set of Java interfaces and abstract classes which define the contract that all concrete implementations of the view must implement (*IModel*). Currently, we have implemented one concrete implementation: a version for the Homura games framework [18]: an open source, Java/OpenGL-based rendering system and games application framework. The model is represented as a Skinned Mesh, storing the vertex, colour and index buffers in a VBO (where supported). The mesh also stores references to the model's textures and render states (which represent effects placed in the model such as lighting, alpha-blending, GPU shader programs etc). The skeletal information is stored independently in a Skeleton-Pose which stores both a Skeleton (the list of bones comprising the skeleton and their vertex weights) and the local and global transforms. This allows the MPEG-4 system to be used in conjunction with standard key frame and IK-based animation. The skeleton is moved via manipulation of these transforms. The model is also stored as a Node object, which allows it to be placed in the scene graph for interaction with subsystems such as AI and Physics, in addition to ability of being transformed independently of its animations.

The *character* component provides a unified, object representation of a virtual character. The highest level member of the API, its roles vary from maintaining the high-level FAPS animation data; referencing the engine-specific mapping of the MPEG-4 FA standard to the renderer's co-ordinate and mathematical view; as well as a renderer-specific representation of the 3D model, skeletal structure. This class also performs several important tasks such as validating that the skeletal structure and mesh of the character possess the minimum data required for both calibration and animation. The character is also accountable for the animation control scheme, which espouses one of two modes of operation: *interpreted* or *compiled*. In the *interpreted* mode, the model's skeletal structure is manipulated directly from the *data* objects using the concrete *model* implementation (in this case, *HomuraFAPSData*) to translate this data into the direct transforms which manipulate the skeletal structure of the face mesh. This mode is designed for real-time procedural manipulation of the FAP animation values. Code injection points are structured within the algorithm, allowing additional animation processing such as Gaussian noise or interpolation to be performed in conjunction with the FAP-based animation. Whilst the process is efficient enough for real-time execution, it is the most CPU/GPU intensive, since it is translating the high-level FAP data into joint transforms on the fly. Therefore, the *compiled* mode is intended for use when the FAP data is fixed (e.g. input from files or streamed data). In this case, the high-level data is converted into the joint transforms and stored as key-frame animation data in the given rendering system's native animation format. This allows the animations to be executed in a manner analogous to animation exported directly from higher-level art pipeline tools (such as those described in Section 4).

4 The Rigging Process and Artist Pipeline

In order to simplify the animation process and reduce the amount of code required to integrate a smooth MPEG-4 compliant animation system, we have developed a system where the artist is tasked with defining the structure of feature points and the corresponding vertices to be manipulated. Due to the power and flexibility of the current set of 3D modelling tools available, and the WYSIWYG approach to positioning of feature points, moving the calibration step into the art pipeline provides a more intuitive and productive approach to constructing the virtual character, facilitating the iterative development of face models to meet the demands of realism and expressiveness. In order to construct a head model which is compliant with the Charisma API and with sufficient detail to display the required amount of emotional response, we chose to use Autodesk 3D Studio Max 2010 as our modelling tool, due to its industry-standard status (although the technique is equally transferrable to alternative tools such as Maya and Blender). The process is as follows:

Using a high-quality reference stock photo from the front and side perspective, the shape of the head is produced using a poly modelling *edge-cloning* technique. The mesh is then unwrapped and textured using the reference photographs as a base. To rig the head for animation, a framework of bones must be created. The position of the bones is determined from the feature points taken from the MPEG-4 specification 3]. Once the bones are positioned on the face, they are labelled in accordance to the FDP feature point they are representing (e.g. 3.1 becomes bone _3_1). In order to eliminate the need for FATs, we have substituted them with their skeletal equivalent: blended bone weights. Figure 3 illustrates a modelled, textured, bone positioned and weighted face model.

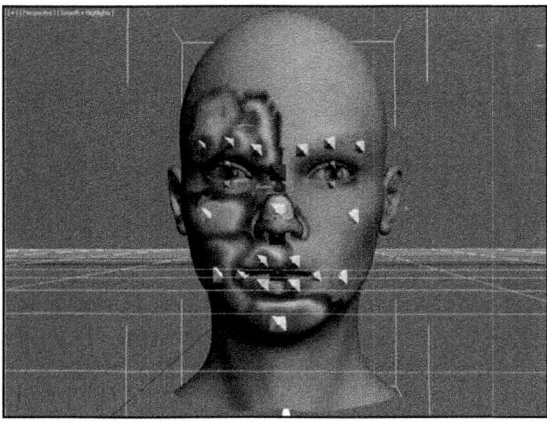

Fig. 3. An example of a Charisma compliant virtual character model; textured and rigged with the skeletal structure which corresponds to the FDP specification. The weight influences of the bones are highlighted, increasing in strength from blue through yellow, to red.

The adoption of blended weights allows for fine-grained control over the vertex manipulation, with every vertex on the model being influenced by one or more bones but with a total weighting of 1.0 (an example of the bone positions and influence weights is shown in Figure 3). This also gives the skin a natural fold and deformation when the facial features are stretched or moved, and allows for smooth interactions when a given vertex is manipulated by more than one bone (e.g. the corner mouth may be affected by jaw movement as well). The model is exported using the ubiquitous standard model interchange format, *collada*. This preserves all the data, whilst providing interoperability with a wide selection of additional modelling and texturing tools. This approach also allows all steps for producing a Charisma-compliant model to be completed inside a single application. In comparison, previous work such as XFace has relied on an independent editor for construction of FATs. This means that the model must be exported to a less-expressive data format from source, then feature points are mapped again and re-exported. Should the model change at source level, the entire process must be repeated. Using our approach, changes to the vertices of the model will instantly update the weighting. Our pipeline also allows the bones to be manipulated inside the editor, in order to rapidly test and incrementally improve the fidelity of the model.

The MPEG-4 calibration step (as described in 2.2) is performed directly from the model's data. We utilise the skeletal structure of the model to determine the FAPUs. In order to control the process entirely from the artist's tool chain, additional marker bones in between the eyes and the fulcrum were added, whilst the other reference points are taken from the feature point bones. Figure 4 illustrates this concept.

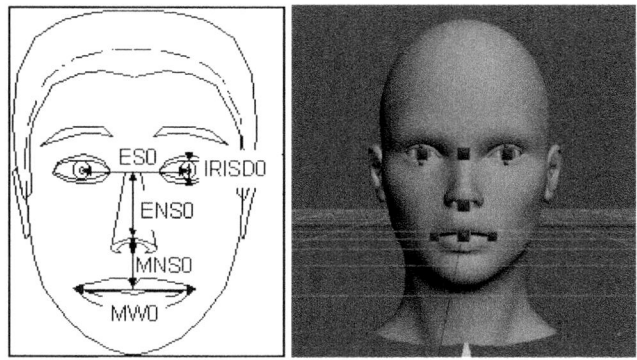

Fig. 4. The FAPU definition areas [3], used to calibrate the model and define the unit transformation effect which each FAP has on its associated feature point (bone). As seen in the right hand image, in our technique, these are determined intrinsically from the model itself, requiring no additional meta-data or data processing.

This approach means that models can be made with varying levels of expressiveness, whilst also automatically supporting cartoon and caricature model, where non-realism is preferred. The model is coloured using a multi-texturing approach. When the model is loaded in to the renderer, standard OpenGL lighting can make it look lifeless. To combat this we use a GLSL-based hardware lighting Shader, in combination with *normal* and *specular* maps. The GLSL shader is executed directly on GPU, performing all transforms and lighting in a parallel manner. Lighting is dynamically changeable at runtime through manipulation the renderer's lighting system.

5 Testing Charisma Using Homura: A Use-Case Analysis

This section describes the applications which were developed to test the API. Our use case for this system was implemented using our own Homura games framework. The framework is designed to construct hardware-accelerated, cross-platform games applications which can be run online as either Java Applets or Java Web Start applications, in a cross-browser manner [19]. The use-case involved the development of two applications – the *Charisma Player* and *Charisma Editor*. Figure 5 illustrates the Java Web Start variants of the test applications.

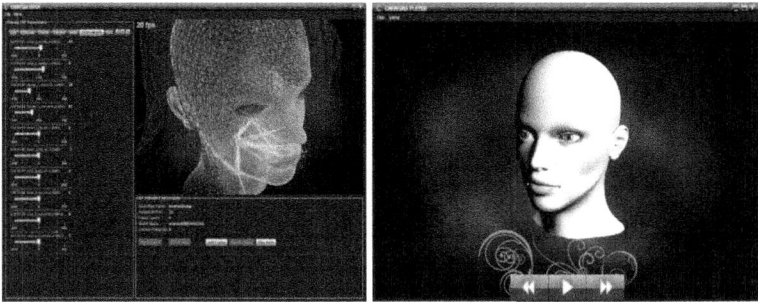

Fig. 5. The prototype applications (left to right): *Charisma Editor* and *Charisma Player*. The applications run as either standalone Java applications or JWS/Applets for web integration.

The web integration provided by Homura allows the applications to be directly embedded or launched from a web context, with access to all the power and performance that OpenGL and GLSL provide, providing a novel solution for web-based MPEG-4 animations.

The editor provides a tool which can load a Charisma-compliant model and provides a Swing-based editing framework for the construction of new/modification of existing FAP-driven animations. The editing framework utilizes the interpreted animation mode in combination with slider controls for each of the 68 FAPs, to directly control the face pose. As shown in Figure 5, the tool provides the ability for wireframe and views of the skeletal structure. Similarly, the player tool also loads a charisma compliant model allowing the player to execute an animation from the variety of sources described in section 3. Currently the player supports voice sync with the animations via WAV and Vorbis files, with Text to Speech (TTS) being added in subsequent development iterations.

Figure 6 illustrates the execution of the animation process in interpreted mode. The compiled mode executes in a similar manner, with the pose transforms cached as animation key-frames, once they have been computed for the entire animation.

In order to validate our system, we decided to use data directly from another MPEG-4 compliant solution: Greta. Greta uses FAP and BAP animations to drive its animation player. The test data used was the *.FAP* format output files generated from Greta's toolset. These were used as the input for both the Charisma player and editor. In order to test the performance of Charisma framework, we used our *Gabrielle* model in combination with the player. The character is a high-polygon (62512 triangles) model, with three light sources placed within the scene on elliptic orbits around the model, tested with a 540 frame FAP animation, with a desired execution rate of 25FPS. The results are averaged over 20 executions of the tests and were performed on the Java Web Start variant of the applications. Typically, we expect an 8-10% performance drop when executed as an Applet embedded inside a web page [19]. In order to test the effect of the animation on the processing, we tested the application with full rendering of the model, but with no animation (animation loaded with updates disabled), the interpreted mode version and the compiled version. Three test types were performed: CPU-based skinning, where all transforms, lighting, texturing and bone manipulation are done on the CPU (typically used in low-end machines, without

```
public void updateAnim(Timer timer,Model model,CharismaFAPAnim anim,int faceroot)
{
    while(anim.incFrame())
    {
        SkeletonPose        pose         = model.getSkins(faceroot).getPose();
        HomuraFAPFrame      frame        = anim.getCurrentFrame();
        Transform[]         local_trans  = pose.getLocalJointTransforms();
        List<CharismaFAP>   faps         = frame.getFAPSData();

        for(HomuraFAP fap : faps)
        {
            HomuraFAPSData lookupdata    = model.lookupFAP(fap.getFAPIndex());
            int feature_point_index      = pose.getSkeleton().findJointByName(lookupdata.getFeaturePoint());
            Transform fap_bone_transform = lookupdata.performTransform(fap.getValue(),model.getFAPUUnits());
            local_trans[feature_point_index].setTransform(fap_bone_transform);
        }
        pose.updateTransforms();
    }
}
```

Fig. 6. Code snippet for animation update loop using Charisma and Homura. For each frame of animation, the default pose positions of the skeleton are retrieved from the model. For each FAP value in the current animation frame, the feature point corresponding to the FAP is fetched via lookup from the *HomuraFAPSData* enumeration. The bone corresponding to the feature point is then found, and the transform is determined using the FAP value from the animation and the model's FAPU collection, again using the enumeration to determine the transform type and direction/axis of transformation. This transform is then applied to the bone. Once all bones have been updated, the model's pose is updated.

Shader Model 2.0 supporting GPUs); GPU-based skinning, where the aforementioned processes are carried out on the GPU and finally, flat shading, dynamic texturing and lighting is omitted – this is also a CPU bound process. Table 1 shows the results of the performance testing.

Table 1. Results of the Charisma performance test using the Homura-based Charisma Player on a Window 7-based PC with an Intel Core 2 Duo E6600, Nvidia 8500GT 256MB GPU, and 2GB DDR-800 System RAM. The testing was performed inside the Eclipse 3.5 IDE, running Java SDK 6u20 with standard VM options.

	Testing Mode (Results expressed as Frames Per Second)		
Test Type:	Static Model – No Anim	Intepreted Anim Mode	Compiled Anim Mode
CPU Skinned:	146.6	38.5	134.6
GPU Skinned:	85.1	36.7	74.5
Flat Shading:	268.3	40.4	222.5

The results in regards to the test types were as expected; Flat shading was the fastest technique, due to the lack of lighting calculations, with CPU skinning also running faster than GPU-based skinning, due to the per-pixel lighting calculations in comparison to the per-vertex lighting used by the CPU process. The results across the animation techniques indicate that interpreted mode has a dramatic affect on the frame-rate with a reduction by a factor of between 2x-6x. This also produced in a levelling effect upon the frame rate across the test types, indicating that the interpreted update consumes the majority of computational time per frame. In comparison, the compiled mode indicated an 8-17% reduction in frame-rate, maintaining high frames per second output. On further inspection of the interpreted animation mode, the application of the FAPs values to the character accounted for, on average, just 3.6ms of processing time.

However, recalculation of the global pose transforms for each of the bones within the character used 16ms of processing time, a significant proportion of the computational time per frame. In each the test cases, the actual frame rate achieved by the application exceeded the animation frame rate, resulting in smooth animations.

6 Conclusions and Future Work

In this paper, we have presented an MPEG-4 compliant animation system, based on skeletal skinned animation instead of the morph-target based approach typically implemented. We have described the animation API's functionality and integration with the asset pipeline and subsequently demonstrated the integration of the framework with a real-time graphics engine. We have illustrated the real-time performance and the ability to utilise high polygon models in combination with advanced texturing and lighting effects, whilst still achieving high frame rates, particularly when the skeletal poses can be calculated prior to the execution of the animation. This allows for the production of smooth, detailed and expressive high fidelity facial animation, interoperable with alternative animation schemes such as IK and within an industry-standard art pipeline. The use of Homura allows us to provide a web-based animation player system which can be applied as a "Talking Head", web applications and in web-based graphics and games applications. Future work will involve the application of this animation system to digital interactive storytelling and integration of MPEG-4 compliant body animation. This research was partially funded by the British Broadcasting Corporation.

References

[1] Poggi, I., Pelachaud, C., de Rosis, F., Carofiglio, V., De Carolis, B.: Greta: A Believable Embodied Conversational Agent. In: Multimodal Intelligent Information Presentation, pp. 3–25. Springer, Heidelberg (2005)
[2] Grammalidis, N., Sarris, N., Deligianni, F., Strintzis, M.G.: Three-Dimensional Facial Adaptation for MPEG-4 Talking Heads. EURASIP Journal on Applied Signal Processing (1), 1005–1020 (2002)
[3] Forchheimer, R., Pandzic, I., Pakstas, A.: MPEG-4 Facial Animation: The Standard, Implementation and Applications. Wiley and Sons Inc., New York (2002)
[4] Blanz, V., Vetter, T.: A morphable model for the synthesis of 3D faces. In: Proceedings of the 26th Annual Conference on Computer Graphics and Interactive Techniques, pp. 187–194 (July 1999)
[5] Deng, Z., Chiang, P.Y., Fox, P., Neumann, U.: Animating blendshape faces by cross-mapping motion capture data. In: Proceedings of the 2006 Symposium on Interactive 3D Graphics and Games, pp. 43–48 (2006)
[6] Tolani, D., Goswami, A., Badler, N.I.: Real-Time Inverse Kinematics Techniques for Anthropomorphic Limbs. Graphical Models 62(5), 353–388 (2000)
[7] Ersotelos, N., Dong, F.: Building highly realistic facial modeling and animation: a survey. The Visual Computer: International Journal of Computer Graphics 24(1), 13–30 (2007)
[8] Balci, K.: Xface: MPEG4 Open Source Toolkit for 3D Facial Animation. In: Proceedings of the Working Conference on Advanced Visual interfaces, Gallipoli, Italy, pp. 399–402 (2004)

[9] de Rosis, F., Pelachaud, C., Poggi, I., Carofiglio, V., De Carolis, B.: From Greta's Mind to her Face: Modeling the Dynamics of Affective States in a Conversational Embodied Agent. International Journal of Human-Computer Studies 59(1-2), 81–118 (2003)
[10] Image Metrics - Official Site (July 2010), http://www.image-metrics.com/
[11] Zhang, Y., Ji, Q., Zhu, Z., Yi, B.: Dynamic Facial Expression Analysis and Synthesis With MPEG-4 Facial Animation Parameters. IEEE Transactions on Circuits and Systems for Video Technology 18(10), 1383–1396 (2008)
[12] Pardas, M., Bonafonte, A., Landabaso, J.L.: Emotion Recognition based on MPEG4 Facial Animation Parameters. In: Proceedings of IEEE Acoustics, Speech, and Signal Processing, pp. 3624–3627 (2002)
[13] Ahlberg, J., Pandzic, I., You, L.: Evaluating Face Models Animated by MPEG-4 FAPs. In: OZCHI Workshop on Talking Head Technology, Fremantle, Western Australia (November 2001)
[14] Lavagetto, F., Pockaj, R.: The Facial Animation Engine: Toward a High-Level Interface for the Design of MPEG-4 Compliant Animated Faces. IEEE Transactions on Circuits and Systems for Video Technology 9(2), 277–289 (1999)
[15] Pasquariello, S., Pelachaud, C.: Greta: A Simple Facial Animation Engine. In: Proceedings of the 6th Online World Conference on Soft Computing in Industrial Applications (2001)
[16] Balci, K., Not, E., Zancanaro, M., Pianesi, F.: Xface open source project and smil-agent scripting language for creating and animating embodied conversational agents. In: Proceedings of the 15th International Conference on Multimedia, Augsburg, Germany, pp. 1013–1016 (2007)
[17] Pandzic, I.S.: Facial Animation Framework for the Web. In: Proceedings of the Seventh International Conference on 3D Web Technology, Arizona, USA, pp. 27–34 (2002)
[18] Carter, C.: Networking Middleware and Online-Deployment Mechanisms for Java-Based Games. In: 6th International Conference in Computer Game Design and Technology (GDTW), Holiday Inn., Liverpool, UK, November 12-13 (2008)
[19] Carter, C., El-Rhalibi, A., Merabti, M., Price, M.: Homura and Net-Homura: The Creation and Web-based Deployment of Cross-Platform 3D Games. In: Proceedings of Ubiquitous Multimedia Systems and Applications (UMSA), St. Petersburg (2009)

Knowledge-Based Probability Maps for Covert Pathfinding

Anja Johansson and Pierangelo Dell'Acqua

Dept. of Science and Technology
Linköping University
Sweden
{anja.johansson,pierangelo.dellacqua}@liu.se

Abstract. Virtual characters in computer games sometimes need to find a path from point A to point B while minimizing the risk of being spotted by an enemy. Visibility calculations of the environment are needed to accomplish this. While previous methods have focused on either general visibility calculations or calculations based only on current enemy positions, we suggest a method to incorporate the agent's knowledge of previous enemy positions. By creating a probability distribution of the positions of the enemies and using this in the visibility calculation, we create a more accurate visibility map.

Keywords: Artificial intelligence, pathfinding, visibility maps, knowledge-based agents, probability maps.

1 Introduction

Pathfinding is one of the most noticeable AI features of virtual characters in computer games. While various solutions to the pathfinding problem have been around for a long time, there are still many problems concerning the realism of character movements. Within the field of robotics pathfinding tends to focus on finding the shortest collision-free path. Although this works fine for robots it is hardly human-like, something that is usually a goal for game characters. Humans do not necessarily look for the shortest path from point A to point B. Neither do they keep a constant speed along the entire path. *Covert pathfinding*[1] focuses on finding a path where the agent will be the least visible. This is for example useful when the agent tries to avoid being detected by enemies.

Our work focuses on virtual characters in computer games and interactive applications. The agent architecture that we have developed contains several modules but those of importance to this paper are the knowledge base (containing the agent's memories of previous events) and the decision making module. The latter chooses high-level actions that best fit the context the agent is currently in (for more information, see [8,9]). One such action is covert pathfinding.

---

[1] While pathfinding using visibility maps have existed for some time, there seems to be no consensus on what name to use for it. For example, some use *covert pathfinding* others *stealthy pathfinding*.

Assume the agent is located in an uncertain, hostile environment. Furthermore, assume it has partial knowledge of where enemies were previously situated. In such a context, the goal of covert pathfinding is to make the agent move from A to be B while minimizing the risk of exposure.

This paper demonstrates the use of knowledge-based probability maps when calculating visibility information in the environment. We let the agent's former knowledge of enemy positions affect the visibility calculations. This enables the creation of a more accurate map for covert pathfinding. We also propose a scheme for speeding up the calculation by using only a subset of the probability map for the visibility calculation. Furthermore, we suggest that the speed of the character at each point along the final path should be affected by the visibility at that point.

The structure of this paper is as follows. First, we discuss different pathfinding techniques. Second, we describe previous visibility calculations. In Section 3 and 4 we describe what pathfinding techniques we use and how we define the cost function in terms of visibility information. Finally, we will round up with some conclusions and suggestions to future work.

2 Background

2.1 Pathfinding

There are many different types of pathfinding algorithms. Below, we will outline a few of the more known methods.

The A*-algorithm is one of the most widely used graph-search algorithm for pathfinding purposes. It is guaranteed to find the least-cost path given a graph with links of different costs. The *cost function* determines the costs for each link in the graph. The A*-algorithm uses heuristics to expand the graph search in more probable directions.

The D*-algorithm [16] is closely related to the A*-algorithm. In contrast, it attempts to efficiently replan paths in a partially-known environment. It can handle a wide spectrum of map information ranging from full information to no information. When changes to the map information are made, the algorithm efficiently updates the search without having to replan the entire path.

Visibility graphs [11] are used to do pathfinding when the obstacles in the scene are represented by polygons. A visibility graph consists of links. The links are made between corners of polygons that are visible to each other. A drawback of visibility graphs is that they do not allow the search space to be divided into anything but obstacles and free space. Also, they do not allow uncertainties or half-visible obstacles. By uncertainties we mean areas that are completely or partly unknown to the agent.

Another common approach is to use navigation meshes [17]. The open area is divided into convex polygons that the character may move freely in. Each area may be assigned different movement costs and additional information. For a navigation mesh to be useful it should be rather sparse. This in turn makes it inadequate when doing visibility calculations. It also suffers from the same drawbacks as visibility graphs.

Overmars, Gerarts and Karamouzas [6,15] have done extensive work on what they call the corridor map method (CMM). The method creates a collision-free set of corridors in which the character can move. Inside the corridors, the agent is guided by forces from the corridor walls towards the goal, along a backbone path. The CMM consists of an off-line phase and an on-line query-phase. The off-line phase takes a considerable amount of time and has to be done prior to the real-time simulation. While the results are very good-looking, the off-line phase presents certain restrictions. One of the bigger drawbacks is that it limits the world to a static scene, hence the robot or agent must have full knowledge of the scene to begin with. This method is also limited to space being occupied only by free space or obstacles. In 2008 the authors published a paper [10] describing how to add variation to paths using Perlin noise. They also varied the speed of the character along the path according to its emotional state.

Hierarchical structures have been used successfully to minimize the search space of the pathfinding. Behnke [2] uses a multi-resolution approach where the resolution is the highest closest to the robot. By using A* heuristics he limits the search time even further. Rapid changes in direction are prevented by using a fake obstacle field right behind the agent. Obstacles further away from the robot are represented with increasingly larger and decreasingly less costly disks. Replanning is fast because of the hierarchical structure, hence the method is suited for dynamic scenes.

Our pathfinding structure is inspired by the hierarchical structure suggested by Behnke [2]. Since our environment is dynamic and uncertain, the agent must be able to replan the path often. Using a multi-resolution approach enables fast searches of the environment. By setting the resolution higher in areas closer to the agent's current position, the accuracy of the immediate path is much higher than the path which is further away. Resolution is also set higher nearby obstacles, in contrast to the work by Behnke [2]. This allows us to speed up the pathfinding.

2.2 Visibility Maps

In general, a *visibility map* is used to describe how much of one position is visible to all the other positions in the environment. Below, we will describe a number of algorithms used to compute such a map.

Geraerts *et al.* propose a stealth-based path-finding algorithm [7] that is based on their work on the corridor map method [6,15]. They calculate visibility polygons, given some observer positions. The visibility is also influenced by the field-of-view of the observer as well as the distance from the observer to the observed point. A final visibility map is created which is incorporated into their corridor map method.

Marzouqi *et al.* [12,13] suggest a way to calculate a global visibility map using distance transforms. They suggest methods to do covert pathfinding on both known and unknown environments. Obstacles are represented discretely on a regular grid. Distance transform techniques are used to do calculate a general visibility map.

In 2000 Cohen-Or et al. [5] made a survey of visibility algorithms for walkthrough applications. Their main focus is on different occlusion culling algorithms. While the focus of the survey is to speedup the rendering process, they describe many algorithms of interest concerning general visibility calculation. However, most algorithms are polygon-based instead of discrete (grid-based). Likewise, a similar survey by Bittner et al. [4] also address polygonal representation of obstacles rather than a discrete representation.

Isovist calculation [3] determines the visible space given the observer point and the obstacle geometry. Isovist fields are calculated by calculating the area (or other measures, Michael Batty has explored different isovist field measures [1]) of the isovist polygon for each point in the scene. While often calculated in two dimensions for simplicity, visibility calculations using isovists in three dimensions have been proposed by Van Bilsen [18,19]. Isovists have generally been used as spatial analysis tools in urban/architectural design. There exist open-source libraries [14] for isovist calculations in two dimensions. However, for real-time purposes they suffer from speed problems when the complexity of the geometry increases.

Error occurs when using a discrete representation of obstacles instead of a continuous representation. However, Van Bilsen et al. show [20] that the average error for using a discrete representation rather than an exact polygonal representation when calculating visibility using the average distance measure [1,19] is roughly 11%. While this presents a problem in urban design where accuracy is important, for a virtual character this error is acceptable.

3 Hierarchical Pathfinding

This paper concentrates on the cost function related to the pathfinding, since this is where we use visibility. However, we will give a brief description of the pathfinding model we use.

The environment is divided into a quad-tree structure. The quad-tree has a fixed number of levels everywhere in the environment, but a level-of-detail function is used to decide if a node should be considered a leaf or not. The connectivity in the quad-tree is horizontal and vertical. Since the level-of-detail in the quad-tree may be different from one node to its neighbors, the connectivity may have an irregular form. An example of such connectivity can be seen in Figure 1.

The search through the graph is done using the common A*-search algorithm with euclidean distance for heuristics. Afterwards, the path is smoothed using a simple smoothing function to obtain a more natural-looking path for the character to walk along. The focus of this work is on how we define the link cost function for the A*-algorithm.

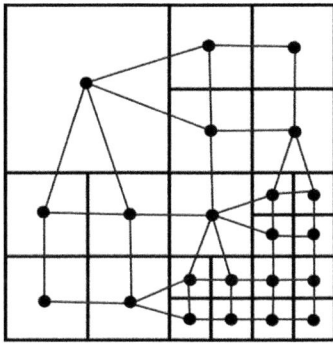

Fig. 1. Connectivity in a hierarchical grid

4 Link Cost Function and Influence Maps

The interesting part of pathfinding is defining the link cost function that the search algorithm uses. While most applications focus on finding the shortest path, the link cost function can be used to model many different things. For instance, it can easily be used to make the character avoid certain areas or to find the safest path in terms of closeness to obstacles. The latter has often been used in robotics to find safe passages where the robot is less likely to collide with obstacles.

The cost function of the A*-algorithm is defined as:

$$f(n) = g(n) + h(n). \qquad (1)$$

where $g(n)$ is the cost of the calculated path from the start node to n and $h(n)$ is a heuristic estimate of the cost of the remaining path to the goal node.

We need several types of information to perform our pathfinding. This information we represent as influence maps. Influence maps [17] are a concept commonly used in strategy games to assist tactical decision making. The term influence map is loosely defined, but we choose to define it as any information represented as a 2D environmental map that is used to assist the agent's decisions, or in our case, the pathfinding process.

An influence map in our system is a two-dimensional uniform rectangular grid, where each point in the grid contains information concerning the environment at that particular point. The influence maps needed for our pathfinding are described in more detail below.

4.1 Obstacle Map

An obstacle map is a simple binary grid where a node contains a 1 if the node contains an obstacle, 0 otherwise. This information is necessary for pathfinding as the character should not create a path that leads through an obstacle.

We implement the obstacle map by extracting geometry information from the agent's memory in the form of bounding boxes. These bounding boxes are projected onto the ground plane and by using various geometric calculations it is possible to determine if a node contains an obstacle or not.

The obstacle map of an example scene is depicted in Figure 2 a). This example scene is used for the other influence maps as well.

4.2 Probability Map

To estimate the enemy positions, we create a probability map. All the positions of enemies that the agent has collected previously are used to create a probability distribution of where the enemies have been. A small falloff radius[2] is used to map the positions onto neighboring points. The reason behind this is that in real life one would not assume that the enemies would turn up in exactly the same position, but rather in the general area around that position.

To calculate the probability map we use the following algorithm:

- For each point p in the probability map:
  - Set $sum(p)$ to 0 and for each known enemy position x:
    * Calculate the distance d between x and p.
    * Filter the value using a sigmoid falloff $f = 1 - \frac{1}{1 + e^{-(d-0.5)/0.2}}$.
    * Add f to $sum(p)$
  - Save $sum(p)$ at position p on the visibility map.
- Normalize all the values in the probability map to lie between 0 and 1.[3]

A probability map computed from 410 known enemy positions generated from 4 separate enemies can be seen in Figure 2 b). The probability map took approximately 0.4 seconds to compute. The algorithm scales linearly with the number of known enemy position. However, since it is possible to incrementally add new information to the probability map whenever an agent sees enemies, the runtime computations are negligible. A possible drawback of using this speedup technique is that it might, depending on the memory architecture used, disable the notion of forgetfulness in the agent's memory if such exists.[4]

4.3 Visibility Map

Most algorithms mentioned in Section 2.2 have no need for real-time speed as they are most often used for analysis of static geometry data. However, when the target is to analyze a dynamic, partly unknown world in real-time, speed becomes an issue.

---

[2] We use a falloff radius of one unit in all of our examples in this paper.
[3] Note that this will not result in a traditional probability distribution. To obtain such, one must instead divide with the total number of enemy positions. We do it this way because it makes it easier to define h in the visibility map.
[4] It is possible to remove forgotten information from the probability map. However, the precise information that has been forgotten needs to be known.

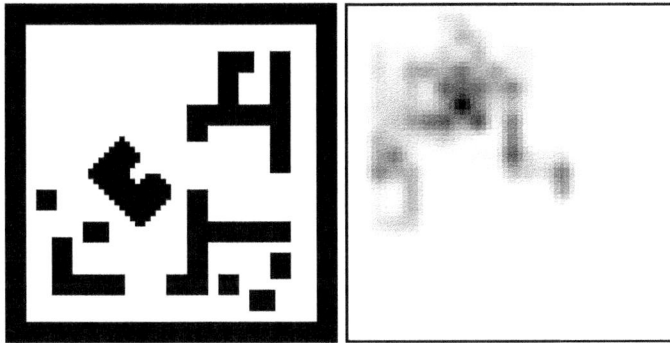

Fig. 2. a) The obstacle map of the environment. b) A 64 by 64 probability map calculated from 410 enemy positions.

None of the methods take into consideration any previous knowledge the agent may have concerning the whereabouts of enemies. Enemies' current positions are used in some cases, but this is inaccurate as the agent should not be aware of the current positions of enemies that it cannot currently see. Using the last known position of an enemy is also incorrect as this will only work if enemies do not move around much.

Calculating visibility at every single grid point can be computationally heavy no matter what method is used. Therefore, we want to create a useful visibility map using as few enemy nodes as possible.

We use the obstacle map described in Section 4.1. The algorithm comprises the following steps:

- For each non-obstacle position x in the grid:
  - For each point p in the probability map that exceeds a given threshold h, move from x towards p until p is reached or there is a collision with an obstacle.
  - Sum up all the probability values for each visible point in the probability map.
  - Save the sum at position x on the visibility map.
- Normalize all the values to lie between 0 and 1.

Note that the threshold h will determine the number of nodes in the grid that are taken into consideration when calculating the visibility. In Figure 3, a comparison is shown of visibility maps where different numbers of nodes are used. In Figure 3 a) only one node is used, the node with the strongest probability value. As the number of nodes increases the visibility map becomes more complex. The visual similarity between using 300 nodes and using all 4096 nodes is great. Table 1 shows the *root mean square error* (RMSE) for different number of nodes used. The computation times for each case are also listed. Since the algorithm is linearly dependent on the number of nodes used, the computation speedup is theoretically 13 times, if only 300 nodes are used. In practice we see a speedup of 11 times.[5]

Note that for different probability maps, a certain h may give a different number of nodes. It may therefore be necessary to adjust the value of h. The RMSEs for the given number of nodes are also likely to change from case to case.

Table 1. RMSE values and computational times (seconds) for visibility maps calculated with different number of nodes

Nr of nodes	1	10	100	300	500	4096(all)
RMSE	0.16243	0.14391	0.07883	0.02925	0.01303	0.0
Time	0.13	0.14	0.29	0.60	0.91	6.6

Fig. 3. Visibility maps (64 by 64) calculated with different values for the threshold h. Left to right, top to bottom: a) 1 node and $h \approx 1.0$, b) 10 nodes and $h \approx 0.7$, c) 100 nodes and $h \approx 0.5$, d) 300 nodes and $h \approx 0.3$, e) 500 nodes and $h \approx 0.2$, f) all 4086 nodes.

Despite the speedup of ~11 times, the visibility calculation takes approximately 0.6 seconds to perform. By using the distance falloff suggested in [7], we can reduce the computational cost because we do not have to consider nodes that lie far from the enemy node. This speedup is particularly important when

---

[5] For these tests we used an Intel Pentium Dual Core 2.80GHz processor.

Fig. 4. a) Visibility maps (64 by 64) calculated with respect to distance. Left to right: a) 1 node, b) 300 nodes.

the scale of the environment is large and the maximum perceivable distance is less than the size of the whole map. It also improves the result, as it is more proper for the detectability to decrease with distance, since the agent would take up a much less portion of the enemy's field of view. The result of using the distance can be seen in Figure 4 a) and b). In Figure 4 a) only one node is used (equivalent to Figure 3 a)). In Figure 4 b) 300 nodes are used.

4.4 Link Cost Function and Character Speed

Our final link cost function C is defined as:

$$C = 1 + a * visibility. \qquad (2)$$

where a is a constant that affects how much the visibility should be taken into consideration. a can be changed dynamically to suit the needs of the character. Note that C has a minimum value of 1 which ensures that using the euclidean distance as the heuristics function is optimistic. This in turn satisfies the demand by the A*-agorithm. In the tests below we have used $a = 5$.

Finally, we let the character's speed along the path be affected by the visibility. For each point p along the path we calculate the speed $s(p)$:

$$s(p) = v + b * visibility(p). \qquad (3)$$

where v is the basic speed of the agent and b is a parameter that determines how much the visibility should affect the speed. The speed of the character can depend on other things, such as mood or fatigue, as well. However, in our examples in this paper we have chosen to use only visibility to better illustrate our idea.

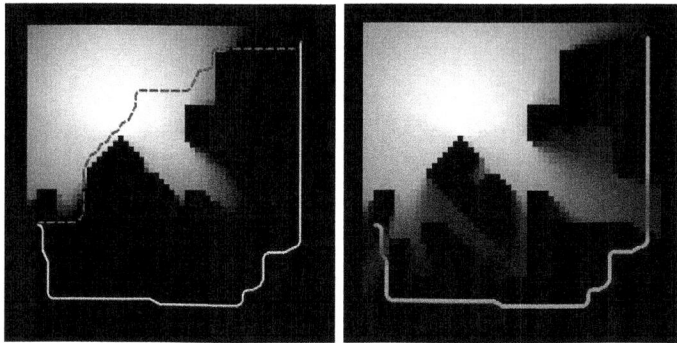

Fig. 5. a) A comparison between using the Euclidean distance as cost function (red dotted path) and using our visibility cost function (cyan path), b) Character speed (red color means faster speed) varies with visibility value

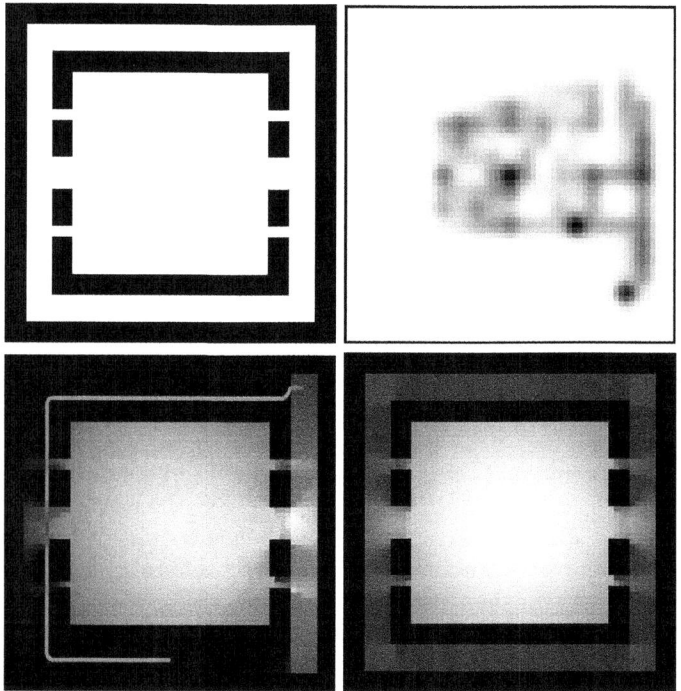

Fig. 6. Maps for the second test. All maps are 64 by 64 in size. a) The obstacle map for our second test. b) The corresponding probability map of 310 enemy positions. c) The visibility map calculated using 300 nodes and the final path with varying speed. d) A general visibility map.

5 Pathfinding Results

In Figure 5 a) one can see the difference between using standard Euclidean distance as cost function and using our visibility function. The difference is evident. Should the agent choose to take the shortest route, it would walk right into an area which it knows is popular among enemies. Our path offers a much safer and realistic way. Figure 5 b) displays how we vary the character's speed according to the visibility information. As the character moves into the area where is might be more visible it moves faster.

In Figure 6 we have used a different environment. We wish to demonstrate the usefulness of our approach by showing how the agent chooses the corridor/room which enemies do no frequent. This can be seen in Figure 6 c). In Figure 6 d) one can see a general visibility map calculated without using our approach.

6 Discussion

We have presented a method for covert pathfinding that uses the agent's knowledge of past enemy positions to compute a probability map. This probability map is then used when calculating the visibility map.

The benefit of using a discrete obstacle map in the visibility calculations instead of using the polygonal representations of the geometry directly is great. It enables the visibility calculation to be independent of the number of obstacles in the scene. Naturally, the obstacle map is dependent on the number of obstacles, but since the obstacle map is needed for our pathfinding in any case, the overall computational cost is reduced. Using a discrete representation of obstacles has its drawbacks both in the pathfinding and in the visibility calculation because it is impossible to represent small obstacles and small spaces between obstacles. This is particularly a problem when calculating visibility in for instance forests, where the sizes of tree trunks and the space between trees are less than the grid size. This is a common problem in any discretization process and something we wish to address in the future. However, if the geometry information is large, using the geometry information directly can be computationally too expensive and the discrete version is hence more suitable.

Our use of the probability map in the visibility algorithm is both an advantage and a drawback. While our algorithm takes into consideration the knowledge of the agent concerning previous enemy positions, it is not currently possible to include more specific information. For instance, the agent may know that an enemy is currently in a given area. It may also know that there is only enemy in the environment. If the agent knows this, it is quite irrelevant where the enemy has been previously. Our algorithm is therefore useful only when one deals with general distributions of enemy positions. In the future we would like to extend this algorithm to include more complex types of information, such as motion prediction for recent enemy positions.

References

1. Batty, M.: Exploring isovist fields: space and shape in architectural and urban morphology. Environment and Planning B: Planning and Design 28, 123–150 (2001)
2. Behnke, S.: Local multiresolution path planning. In: Polani, D., Browning, B., Bonarini, A., Yoshida, K. (eds.) RoboCup 2003. LNCS (LNAI), vol. 3020, pp. 332–343. Springer, Heidelberg (2003)
3. Benedikt, M.L.: To take hold of space: isovist fields. Environment and Planning B: Planning and Design 6(1), 47–65 (1979)
4. Bittner, J., Wonka, P.: Visibility in computer graphics. Journal of Environmental Planning 30, 729–756 (2003)
5. Cohen-or, D., Chrysanthou, Y., Silva, C.T., Durand, F.: A survey of visibility for walkthrough applications, tech. report (2000)
6. Geraerts, R., Overmars, M.: The corridor map method: A general framework for real-time high-quality path planning. Computer Animation and Virtual Worlds 18(2), 107–119 (2007)
7. Geraerts, R., Schager, E.: Stealth-based path planning using corridor maps. In: Proc. Computer Animation and Social Agents, CASA 2010 (2010)
8. Johansson, A., Dell'Acqua, P.: Introducing time in behavior networks. In: Proceedings of 2010 IEEE Conference on Computational Intelligence and Games (CIG), pp. 297–304 (2010)
9. Johansson, A., Dell'Acqua, P.: Affective states in behavior networks. In: Plemenos, D., Miaoulis, G. (eds.) Intelligent Computer Graphics 2009. Studies in Computational Intelligence, vol. 240, pp. 19–39. Springer, Heidelberg (2009)
10. Karamouzas, I., Overmars, M.H.: Adding variation to path planning. Computer Animation and Virtual Worlds 19(3-4), 283–293 (2008)
11. Lozano-Pérez, T., Wesley, M.A.: An algorithm for planning collision-free paths among polyhedral obstacles. Communications of the ACM 22(10), 560–570 (1979)
12. Marzouqi, M., Jarvis, R.: Covert path planning for autonomous robot navigation in known environments. In: Proc. of the Australian Conference on Robotics and Automation (2003)
13. Marzouqi, M., Jarvis, R.: Covert robotics: Covert path planning in unknown environments. In: Proc. of the Australian Conference on Robotics and Automation (2003)
14. Obermeyer, K.J.: The VisiLibity library, r-1 (2008), http://www.VisiLibity.org
15. Overmars, M., Karamouzas, I., Geraerts, R.: Flexible path planning using corridor maps. In: Halperin, D., Mehlhorn, K. (eds.) ESA 2008. LNCS, vol. 5193, pp. 1–12. Springer, Heidelberg (2008)
16. Stenz, A.: Optimal and efficient path planning for partially-known environments. In: Proc. of the IEEE International Conference on Robotics and Automation, pp. 3310–3317 (1994)
17. Tozour, P.: Game programming gems 2, chap. 3.6. Cengage Learning (2001)
18. Van Bilsen, A.: How can serious games benefit from 3d visibility analysis? In: Proc. of International Simulation and Gaming Association 2009 (2009)
19. Van Bilsen, A.: Mathematical Explorations in Urban and Regional Design. Ph.D. thesis, Delft University of Technology, Netherlands (2008)
20. Van Bilsen, A., Stolk, E.: Solving error problems in visibility analysis for urban environments by shifting from a discrete to a continuous approach. In: ICCSA 2008: Proceedings of the 2008 International Conference on Computational Sciences and Its Applications, pp. 523–528. IEEE Computer Society Press, Washington (2008)

Modification of Crowd Behaviour Modelling under Microscopic Level in Panic Situation

Hamizan Sharbini[1] and Abdullah Bade[2]

[1] Department of Multimedia Computing,
Faculty of Computer Science and Information Technology, Universiti Malaysia Sarawak,
94300 Kota Samarahan, Malaysia
`deehanieza@gmail.com`
[2] School of Engineering and Information Technology,
Universiti Malaysia Sabah, 88999, Kota Kinabalu, Sabah, Malaysia
`abade08@yahoo.com`

Abstract. Applications of crowd simulation are numerous such as in computer science that includes study of crowd behaviour under different environmental conditions like panic situation. Therefore, this project will focus on analysis via available literature and to produce more realistic simulation based the microscopic models, namely Social Force Model (SF) that can reproduce individual and collective behaviours observed in real emergency evacuation situation. This paper intend to modify the model that will only pertain to crowd movement behaviour modelling in order to find an exit from a room in panic situation that integrate additional component named Anti-Arching. This also incorporates simple visualization such that it has only a single point per object, whereby the people in crowd will be modelled using particle The modification is also proven aligned with the 'faster-is-slower' effect proposed by Parisi and Dorso [1].

Keywords: Crowd behaviour, crowd modelling, realism, microscopic model, panic situation, 'faster-is-slower' theory.

1 Introduction

In this modern era, human population had evolved and there are even questions: are we too many in this planet? Do we need to plan all the movement in computer graphics when it comes to crowd simulation? Do we need to have a huge database just to remember the movements of each individual? Could we just make a module to decide the movement intelligently? Nowadays, crowd simulation is very essential and become useful in many areas such as in education, entertainment, architecture, urban engineering, training, and virtual heritage. There are various crowd simulation techniques which generally can be divided into two main categories, namely crowd visualization and crowd realism.

Crowd simulation under panic evacuation situation is fairly important as we need to acknowledge several approaches done in this area. The simulation plays an important role as a platform to demonstrate the models in more visualize way and to achieve realism of crowd movement behaviour in escape panic situation. During an escape panic, few of characteristics [2] will be shown such as listed below:

- Individuals trying to move faster than normal.
- Individuals push or interactions become physical.
- Arching and clogging observed at exits.
- Escape slowed by fallen or dead individuals serving as obstacles.
- Tendency toward mass or copied behaviour.
- Alternative or less used exits are overlooked.

Generally, there are still three major problems as mentioned by [3] in the research area pertaining to crowd simulation: the realism of the behaviours during the simulation, high-quality visualization, and computational cost.

Why crowd modelling? Crowd modelling is very useful in handling crowds, so here it is emphasize more on panic situation. The model has to be constructed such that it can depict the human behaviour in 'virtual' scene like the one in evacuation scenario. This could be further enhanced with additional crowd behaviour such that to add more realism in the simulation. A recent research into crowd simulation has been inspired by the work of Reynolds [4]. Therefore, numbers of modelling and simulations have been created to aid or visualize regarding crowd situation. Computer models for emergency and evacuation situations have been developed and most research into panics have been of empirical nature and carried out by social psychologists and others [5]. However, this area still need an in-depth exploration regarding the fundamental issues that must be resolved based on humans character in panic situation.

2 Literature Review

Crowd in an urban environment, the spaces for individual can be limited by obstacle like walls and fences. [6] have explained in their work that what can attract individual in a crowd during panic situation is the exit point. Again, here it is clearly emphasized that crowds can be seen as a set of groups composed by individuals with more than just singular behaviour.

- First, the environment with normal situation has been setup. At this moment, it is only normal behaviour and crowd movement until panic erupted.
- Secondly, the panic event occurred and user (crowd) will choose either strategies that are available and the evacuation time will be calculated.

2.1 Crowd Model under Macroscopic and Microscopic Model in Panic Situation

Davoda and Ketterl [7] as cited in their work regarding crowds behavioural control can be divided into two namely microscopic and macroscopic levels. Same as cited in [5], simulation of crowd can also be categorized into these two techniques. Table 2.1,

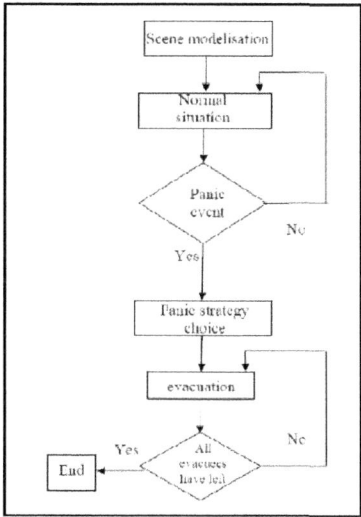

Fig. 1. System Architecture for Evacuation

it described the differences in both microscopic and macroscopic used in panic and emergency situation.

2.2 Social Force Model (SF)

In modelling crowd behaviour, the most said models are such as Cellular Automata, Agent-Based model and Social Force Model. [5] has developed other social force model to simulate crowd in panic situations. According to his idea behavioural changes are guided by so-called social field or social forces. The social force model is useful to depict the motion of pedestrian. The motion is determined by main effects of either the person wants to reach certain destination, to keep certain distance from other pedestrians, to keep certain distance from borders of obstacles such as walls, the person can be attracted by other persons or objects.

Most of the work has been done using social force model in modelling crowd simulation in panic situation. [1] have applied the social force model [5] to revise the evacuation from the room with an exit, and then later they modified slightly the social force model to simulate the evacuation of pedestrians from a room with one exit under panic situation [1].

[8] have presented an agent-based system for crowd evacuation in an emergency situation based on the social force mode [5]. The model can detain the basic characteristics of pedestrian evacuation such as arching and clogging behaviour.

In Helbing's work [5], the workflow of the program can be depicted such in Figure 2 which shows the simulation that is run based on the program. The Arching Module is the program that sends data of current situation for the simulation module to provide a plan or script to the program on what to do next. The simulation module here is the particle that resembles the crowd; meanwhile the script is the code that contains the instructions.

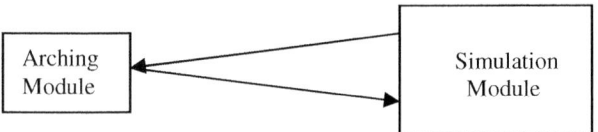

Fig. 2. The basis for the simulation structure

The crowd simulation sometimes used particle to resemble the crowd. The particle movement and description are mostly in mathematical basis; hence it is widely used for physical forces, like motion or fluids [9]. Another work has been done by [10] to describe motion of pedestrian using particle system.

The Social Force model has been the preferred model. Figure 3 concludes the differences between Social Force model and Cellular Automata model. For illustration purpose, the particle in grid (4, 4) tries to move to grid (1, 1). As such, there are two possible ways to move to the said position namely Cellular Automata and Social Force. This means that the first will follow grid based or discrete based whilst the second in continuous form. When it comes to computational fast factor, Cellular Automata yields the best result because it is faster than Social Force model.

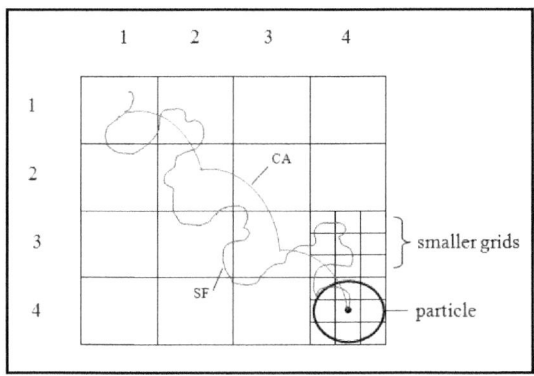

Fig. 3. Differences between the movement of CA and SF model

The SF model resembles more realism in crowd movement despite of its disadvantages of computational difficulties. Although the CA model seems to be computationally fast and scalable, but the realism may not suffice since it does not support integration of realistic visualization of the crowd.

As a conclusion from the literature, there is a necessity for improvement in certain area such as the better the behavioural description specified, more realistic and plausible simulations of panic situations can be provided [11], and more natural looking and realistic in generating crowd behaviour [12].

According to [1], the Social Force model [5] is the most referred since it portrays a continuous model to describe a pedestrian dynamics in panic situation which qualitatively reproduces many self-organizing situation, for example the phenomenon of lane formation, clogging and 'faster-is-slower' effect. The model can be the appropriate one when it is used to reproduce the basic physical movements in quite simple geometries.

When it comes to simulation, the theory comes in handy to observe any possibility of individual in a crowd can face fatality during the panic situation such as stampede, rushing and pushing behaviour. The theory is useful since it always and has been prove to be happened in panic situation and thus will be compared with the new proposed anti-arching component to see that it is adhere to the said theory.

Lastly, it is crucial to integrate responsible module to produce movement realistic like walk, to run it and other movement related to the crowd in emergency or panic situation [6].

3 Modification on Social Force Model

Theoretically, human behaviour during panic situation will form arching and clogging manner. This is what has been done by [5] in order to simulate the crowd in panic situation usually resulting in arching and clogging phenomenon like shown below in Figure 4.

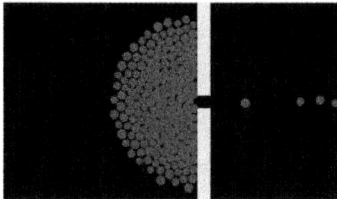

Fig. 4. Simulation of arching and clogging phenomenon [5]

So far, crowd simulation is attempting to simulate the actual human behaviour. In this work, it is clearly shown that the crowd movement simulation under panic situation is exhibiting arching and clogging phenomena. To add more realism, the normal situation has been modified such that it will become anti-arching during panic situation. Figure 5, it shows the schematic diagram that involved the modified component which reflects how the anti-arching phenomenon can occur.

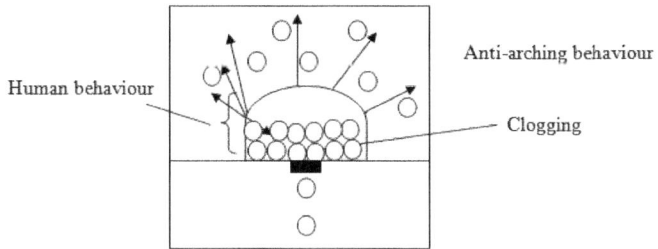

Fig. 5. Modified schematic of arching and clogging phenomenon during panic situation

Based on the equation of motion,

$$\frac{d\vec{x}_i(t)}{dt} = \vec{v}_i(t) \qquad (1)$$

where $\vec{x}_i$ = Place or location; t = Time; $\vec{v}_i$ = Speed.

The acceleration equation for pedestrian can be derived from (1).

$$\underbrace{m_i \frac{d\vec{v}_i(t)}{dt}}_{Acceleration} = \underbrace{\frac{m_i}{\tau_i}(v_i^0 \vec{e}_i(t) - \vec{v}_i(t))}_{Driving\ Force} + \underbrace{\sum_{j(\neq i)} \vec{F}_{ij}^{ww}(t)}_{Interactions} + \underbrace{\vec{F}_i^b(t)}_{\substack{Borders,\\Fire}} + \underbrace{\sum_k \vec{F}_{ik}^{att}(t)}_{Attractions} + \underbrace{\vec{\xi}_i(t)}_{Fluctuations} \qquad (2)$$

$$\vec{F}_{ij}^{ww}(t) = \underbrace{\vec{F}_{ij}^{psy}(t)}_{\substack{Psychological\\Repulsion}} + \underbrace{\vec{F}_{ij}^{ph}(t)}_{\substack{Physical\\Interactions}} + \underbrace{\vec{F}_{ij}^{att}(t)}_{\substack{Interaction\\Between\\People}}$$

where $\vec{x}_i$ = Place or location; t = Time; $\vec{v}_i$ = Speed; m_i = Mass, τ_i = Acceleration on Time; v_i^0 = Desired Velocity; $\vec{e}_i$ = Desired Direction

The modification takes place as one of the social force module as presented in the model, which is the 'Interactions' module that can affect the crowd movement. This is the chosen module because it consists of the interaction among each individual in crowd, in this case the particle. The 'Interactions' is a module that cause the movement to be in such a way that crowd can form arching based on individual forces. The meaning of this Interaction module or component can be the sum of all forces, such that depicted in (3).

$$\sum_{j(\neq i)} \vec{F}_{ij}^{ww}(t) = F_{ij}(t) + F_{ij}(t) + F_{ij}(t) \qquad (3)$$

From (3), the sum of all forces is represented by $F_{ij}(t) + F_{ij}(t) + F_{ij}(t)$ components such that i represent the i-th node and j represents the j-th node. The node here is the point or particle. Meanwhile, ww is having the set of $ww = \{psy, ph, att\}$.

The equation can be expressed in other way such that it also equivalent with

$$\sum_{j(\neq i)} F_{ij}^{ww} = F_{ij}^{psy} + F_{ij}^{ph} + F_{ij}^{att}$$

The modification is the addition of the fourth component,

$$\vec{F}_{ij}^{ww}(t) = \underbrace{\vec{F}_{ij}^{psy}(t)}_{\substack{Psychological \\ Repulsion}} + \underbrace{\vec{F}_{ij}^{ph}(t)}_{\substack{Physical \\ Interactions}} + \underbrace{\vec{F}_{ij}^{att}(t)}_{\substack{Interaction \\ Between \\ People}} + \underbrace{\vec{F}_{ij}^{r}(t)}_{Anti-arching} \qquad (4)$$

(Fourth component)

where $\vec{F}_{ij}^{r}$ denote the movement of the crowd for anti-arching in the simulation in this social force model under Interaction module.

In any panic situation, the attraction force which is the escape door is not always unidirectional. When clogging occurred, there is a group of people tried to look for different escape route. This is what the modification is trying to achieve in terms of the realism of the crowd behaviour.

3.1 Anti-arching Algorithm and "Faster-is-Slower" Effect

The modified element from the Social Force model under Interaction module has been carried out. Under this phase, the parameter in the program which is the number of the crowd has to be increased, with fixed parameter of room size. Each of the changes in crowd number will be observed together with the timing of evacuation. The theory says that the increasing level of density of crowd, the more it takes for time to leave the room. Second part is to observe the modified panic situation algorithm such that to make sure it is according to the theory, the more density with changes in behaviour of crowd also affect the evacuation time.

The anti-arching behaviour would definitely increase the activity of each particle in finding alternative escape route. The increase of activities is proportional with the increase of speed or in other words, the anti-arching would increase the overall speed of the particle. This is in line with the reality of the anti-arching phenomenon where crowd would start to move randomly with higher speed if arching occurs.

One way to prove the proposed anti-arching component realism is by measuring the escape time taken. According to [1] and [5] as mentioned in the literature, the "faster-is-slower" theory is used as a basis in proving the realism of the proposed component. The speed and evacuation time are counter-proportional with each other. Hence, the time taken with the anti-arching component and time taken without anti-arching is proportional with a multiplier p that is greater than 1. It could be simplified with the following equation

$$T = pT_a \qquad (5)$$

where T= time taken with the anti-arching component,
 T_a= escape time taken with the arching phenomenon,
 p= multiplier.

The testing will be done by comparing the evacuation time of the original panic situation with the modified panic situation. Thus, in order to get the value p, the sum of the modified component panic situation evacuation time will be divided by the sum of the original panic situation evacuation time as shown in equation below.

$$\text{Multiplier (p)} = \frac{\sum Evacuation\ Time\ (M)/n}{\sum Evacuation\ Time\ (O)/n}$$

The anti-arching component shall expect p to be larger than one to prove that the component is realistic as describe in below notation:

$$Realism = \begin{cases} realistic, & p > 1 \\ not\ realistic, & p \leq 1 \end{cases}$$

As for validation matters, the result of Anti-Arching component is validated using value of velocity by crowd density over time as what has been proposed by [5]. This means, formation of arching and anti-arching in panic situation are compared with normal (no panic) situation based on high and low value velocity; velocity is increased for panic situation and velocity is decreased for no panic situation.

4 Results of Additional Anti-arching Component and 'Faster-is-Slower' Effect

In this study, the first part is to study regarding the differences in crowd's movement characteristic in panic situation with and without Anti-Arching component. Secondly, the experiment of the modification of the model has been tested and the results are visually compared to the original one in order to produce crowd realism behaviour in terms of movement during panic situation. Finally, the result of the 'faster-is-slower' effect will be described and explained on how it can converge to the theory based on the experiment that has been done. The environment of the panic situation is being setup based on shown in Table 1.

Table 1. Setting of main parameter list

Parameter	Value
Crowd number(particles)	100 particles
Room size	10m x 10m
Door width	1.2m
Wall width	1.0m

The component has been tested with different velocity. The graphical structure could be seen in Table 2 below.

Table 2. Differences between the crowd movement in panic situation with and without Anti-Arching component

Time t (s)	Situations		
	No Panic (v= 4m/s)	Panic (v=20m/s)	Panic + Anti-arching component (v=20m/s)
10s			
20s			
30s			
>40s			

In this experiment, the panic parameter which is the velocity has been decreased to 4m/s to simulate no panic situation. Meanwhile in panic situation, it remains at the same velocity of 20 m/s.

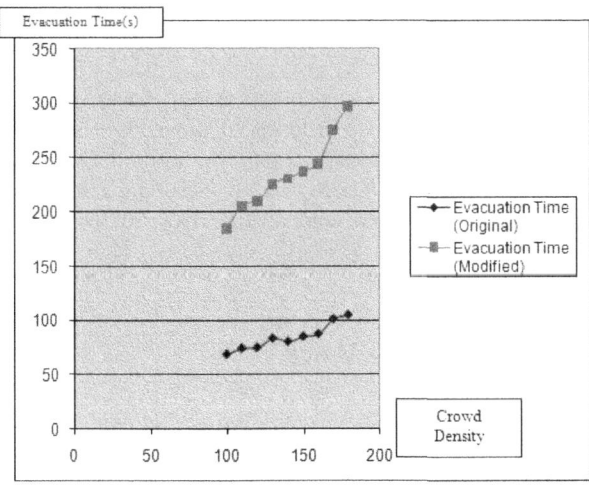

Fig. 6. Graph showing the 'trend' of the original and modified of the model

Table 3 shows the crowd number has been altered by increasing the number to see the result of the effect towards the evacuation time. Figure 6 shows the graph trend based on the data plotting from Table 3, which is to denote the evacuation time is proportional to the crowd density.

Table 3. Validation of Crowd Density proportional to Evacuation Time

Crowd Density	Evacuation Time (Original)	Evacuation Time(Modified)
100	69	184.7
110	74.4	205.8
120	75.1	209.8
130	83.4	225.5
140	80.5	230.7
150	85	237.6
160	87.5	244.4
170	101.3	275.3
180	105.1	297
Total:	761.3	2110.8
p:	2.77	

From Table 3, the value of p has been calculated from the formula in (5). The result for p must be larger than 1 in order to prove that it is realistic based on what has been mentioned in section 3.2.2.3 in Chapter 3. It yields that $p = 2.77$, so significant with theory of 'faster-is-slower' effect as mentioned previously. Hence, the propose component is realistic.

The modification is based on Social Force model and the additional component is very much related to crowd realism in terms of movement. The differences between the panic, no panic and with the modified panic situation have been done. The theory of the 'faster-is-slower' effect has been proven such that it adheres to the trend whereby the evacuation time is proportional to the crowd density. Apart from that, the calculation has been taken based on the formula to emphasize that the theory adhere to crowd realism.

5 Conclusions and Future Work

The additional component is the Anti-Arching component that gives the effect of realism in terms of crowd movement, apart from the other components. The proposed Anti-Arching component is also been analyzed in order to prove that it is consistence with the 'faster-is-slower' theory. Based on the experiment that has been done on this study, some conclusions can be summarized.

The crowd movement also been compared to the original model in panic situation such that modified version shows more behaviour in terms of reducing the arching phenomenon towards other direction. During no panic or less panic situation, the crowd might still have arching behaviour but the flow is in regular basis; no obvious clogging can be observed at the exit point. This is what adheres to crowd in normal situation that have calm behaviour. During panic situation, the arching and clogging is obvious at certain time during the simulation and thus continue to act as such until last person going out.

With the modification made to the original panic situation algorithm, it is intended to give a plausible in crowd behaviour during panic situation, which is less arching and not waiting for others in the exit point, instead it roams around at certain time as if to have another effort to find for other exit, but in this case the other exit is not included as the main intention is for analyzing the movement of the crowd.

There are several suggestions for the future works to improve the work of this project. The modification can still further be experimented using different values in order to make it suitable with different size of room and crowd number in order to achieve more plausible simulation of crowd behaviour. Additionally, an optimize way is required in order to generate more natural and plausible behaviour during panic situation. There is also important to highlight the issues such as the computational cost and memory efficiency to make the model more accurate if it's incurred a lot of crowd numbers in one particular confinement, but still consistent with the crowd realism. The current validation method also can be upgraded by using validation software to improve the validation's result.

References

1. Parisi, D.R., Dorso, C.O.: Microscopic dynamics of pedestrian evacuation. Physica A 354, 606–618 (2005)
2. Altshuler, E., et al.: Symmetry Breaking in Escaping Ants. The American Naturalist 166, 6 (2005)
3. Murakami, Y., Minami, K., Kawasoe, T., Ishida, T.: Multi-Agent Simulation for Crisis Management. Department of Social Informatics, Kyoto University, JST CREST Digital City Project (2002),
 http://www.lab7.kuis.kyoto-u.ac.jp/publications/02/yohei-kmn2002.pdf
4. Reynolds, C.: Flocks, Herds and Schools: A Distributed Behavioural Model. Computer Graphics 21(4), 25–34 (1987)
5. Helbing, D., Farkas, I., Vicsek, T.: Simulating Dynamical Features of Escape Panic. Nature 407, 487–490 (2000)
6. Cherif, F., Djedi, N.: A computer simulation model of emergency egress for space planners. Georgian Electronic Scientific Journal 8(1), 17–27 (2006)
7. Davoda, J., Ketterl, N.: Crowd Simulation. In: Research Seminar on Computer Graphics and Image Processing, SS 2009, Institut für Computer Graphik und Algorithmen, TU Wien, Austria (2009)
8. Lin, Y., Fedchenia1, I., LaBarre, B., Tomastik, R.: Agent-Based Simulation of Evacuation: An Office Building Case Study. In: Proceedings of PED 2008, Wuppertal, Germany (2008)

9. Reeves, W.T.: Particle systems - a technique for modeling a class of fuzzy objects. ACM Transactions on Graphics 2, 359–376 (1983)
10. Foudil, C., Rabiaa, C.: Crowd Simulation Influenced by Agent's Sociopsychological State. Lesia Laboratory, Biskra University (2009), `http://lesia.univbiskra.net/IMAGE2009/Documents/Communications/08-Chighoub.pdf`
11. Davoda, J., Ketterl, N.: Crowd Simulation. In: Research Seminar on Computer Graphics and Image Processing, SS 2009, Institut für Computer Graphik und Algorithmen, TU Wien, Austria (2009)
12. Sun, L., Liu, Y., Sun, J., Bian, L.: The hierarchical behavior model for crowd simulation. In: Spencer, S.N. (ed.) Proceedings of the 8th International Conference on Virtual Reality Continuum and Its Applications in Industry, VRCAI 2009, Yokohama, Japan, December 14 - 15, pp. 279–284. ACM, New York (2009)

Expressive Gait Synthesis Using PCA and Gaussian Modeling

Joëlle Tilmanne and Thierry Dutoit

TCTS Lab, University of Mons, Mons, Belgium
joelle.tilmanne@umons.ac.be

Abstract. In this paper we analyze walking sequences of an actor performing walk under eleven different states of mind. These walk sequences captured with an inertial motion capture system are used as training data to model walk in a reduced dimension space through principal component analysis (PCA). In that reduced PC space, the variability of walk cycles for each emotion and the length of each cycle are modeled using Gaussian distributions. Using this modeling, new sequences of walk can be synthesized for each expression, taking into account the variability of walk cycles over time in a continuous sequence.

Keywords: Motion synthesis, expressivity, PCA, variability.

1 Introduction

Modeling of all kind of human behaviors is a very challenging field of study, as those behaviors which are so natural for the human eye are very often extremely difficult to model, and even more difficult to mimic. It is also the case for human motion, which is a complex phenomenon involving our physiological structure as well as our capacity to adapt to external constraints and to feedbacks from our body.

In the field of virtual human animation, various approaches can be taken to synthesize realistic human motion. In particular, there has been a lot of interest in the ways of using and re-using motion capture data [5], a technology that brings the movements of real humans into the virtual world. The main problems encountered with motion data are its variability and its high dimensionality; which make it hard to retrieve, analyze, adapt and modify motion patterns either made "on demand" or coming from an existing motion database.

Two main approaches are encountered regarding the use of motion capture data for animation. The first one consists in building a database, developing techniques to retrieve motion parts in this database, editing these motion parts if needed, and blending them together [6].

The second one uses various machine learning techniques in order to build models based on training motion capture data. The models can later be used to synthesize new motion sequences without resorting to the database initially used for training [2,7,1].

In this paper, we focus on the second approach. We model walk cycles performed with eleven different styles as well as the cycle variation over time during each walk sequence, using a finite number of parameters. Contrarily to most studies addressing this topic, we do not only model style variations but also the variability of motion cycles over time, as these variations are an intrinsic part of the plausibility of the synthesized motions.

Our method, based on the method by Glardon et al. [7,8] uses principal component analysis (PCA) to reduce the dimensionality of each walk cycle and to model the different style components. A Gaussian modeling of the data represented in the PC space is then conducted and enables us to model the variability of the walk cycles over time and thus to introduce some randomness in the synthesized sequences.

This paper is organized as follows. Section 2 makes a brief review of related work. The recording of the database is then presented in Section 3 and is followed by the preprocessing of the data prior PCA in Section 4. Section 5 presents the PCA of the original data and in Section 6 we explain how the PC subspace was modeled. Section 7 presents how this modeling enabled us to extrapolate and synthesize new motion sequences and the results are discussed in Section 8. Section 9 will conclude this paper by presenting future work.

2 Related Work

2.1 PCA for Dimensionality Reduction

PCA [13] is a widespread technique, consisting in finding, by rotation of the original set of axis, the best set of orthogonal axis (corresponding to principal components) to represent the data. Variables in the PC space are thus uncorrelated. The principal components are ordered so that the first ones retain most of the variation present in all of the original data.

This approach is widely used as a first step in motion data analysis and synthesis, mainly to reduce the dimensionality of the data vector needed to describe the pose of the character at each frame (see for instance [17,2,4,10]). This is based on the assumption that despite the high dimensionality of the original motion description space, most human movements have an intrinsic representation in a low dimensional space [3].

Only a few studies use PCA not for reducing the dimensionality of the angle data, but as a way of modeling motion units composed of a sequence of frames. Thanks to their periodicity, walk cycles are especially well suited for such an algorithm. This approach as been taken for instance by Glardon et al. [7,8] and Troje [18].

2.2 Statistical Motion Data Modeling

Walk synthesis techniques taking into account the variability of the walk cycle over time also exist. They use statistical learning techniques to automatically extract the underlying rules of human motion, without any prior knowledge, directly from training on 3D motion capture data. Starting from the statistical models trained that way, new motion sequences can be generated automatically, using only some high-level commands from the user. Two movements generated by the same command (for example executing two walk cycles) will never be exactly identical. The result presents a random aspect as can be found in the human execution of each motion, and becomes potentially more realistic than the repetition of the same motion capture sequence over and over. The motions produced that way are thus visually different, but are all stochastically similar to the training motions.

In order to take into account the high dynamic complexity of human motion, most of the researches in this path base their training on variations of hidden Markov models, Markov chains or other kind of probabilistic transitions between motions [17,20,14].

3 Database Recording

The performance of models trained on data will highly depend on the quality of the data and its accuracy of description of the phenomenon that has to be modeled. As motion capture is the only way to obtain realistic 3D human motion data [15], it is the only way to obtain representative training data for statistical modeling of human motion.

Most motion capture recordings are performed using optical motion capture devices. This technology very often implies space constraints and treadmills have to be used to record walk databases, which impairs the naturalness of walking. Our database was created using the IGS-190 [11], a commercial motion capture suit that contains 18 inertial sensors consisting of a three axis accelerometer, a three axis gyroscope and a three axis magnetometer. This kind of motion capture suit has no space limitation and walk can thus be recorded in a more natural way. This is especially interesting for expressive gait where the subject does not always follow a perfectly straight trajectory and is thus given more freedom when he is not constrained to a given speed and trajectory like he would be with a treadmill.

The inertial motion capture suit captures directly angles between the body segments hence no mapping is necessary between tracked 3D positions of markers and joint angles.

Each motion file contains two parts: the skeleton definition and the motion data. The first part consists in defining the hierarchy of the skeleton, an approximation of the human body structure used in all motion capture systems which consists in a kinematic tree of joints modeled as points separated by segments of known constant lengths.

In the motion data part, the first three values of each frame give the 3D position of the root of the skeleton. They were discarded, as they depend on the displacement and orientation of the walk and can be recalculated given the foot contact with the ground and the leg segments lengths. The pose of the skeleton

Table 1. Database walk styles and corresponding number of cycles recorded

Walk Nbr	Style	Number of Cycles
1	Proud	21
2	Decided	15
3	Sad	31
4	Cat-walk	25
5	Drunk	38
6	Cool	23
7	Afraid	16
8	Tiptoeing	18
9	Heavy	23
10	In a hurry	19
11	Manly	18
Total		247

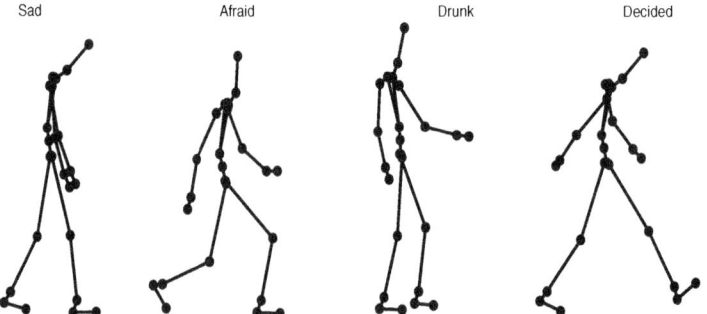

Fig. 1. Four example postures taken from the motion capture database (sad, afraid, drunk and decided walks)

at each frame is then described by 18 tridimensional joint angles, which gives 54 values per frame to describe the motion. In our database, the data was recorded at a frame rate of 30 fps.

For the recording session, an actor wore the motion capture suit and walked back and forth on a scene. Before each capture sequence, he was given instructions about the "style" of walk that he had to act.

The walk sequences where then manually segmented into walk cycles (one cycle including two steps, one with each leg). We defined arbitrarily the boundary of our walk cycles as the first contact of the right foot with the ground (heel or toe depending on the walk style). As the actor walked back and forth, a turn was captured after each straight walk trajectory. Only the perfectly straight walk cycles were kept in this database, removing the turn steps and the transitions between turn and straight walk. Depending on the style of walk performed and its corresponding step length, a different number of walk cycles was recorded for each style. The eleven styles and their corresponding number of cycles are presented in Table 1. These eleven styles were arbitrarily chosen as they all have a recognizable influence on walk, as illustrated in Fig. 1.

4 Data Preprocessing

First of all, in order to avoid discontinuity problems associated with Euler angles, our original joint angle format, the motion data is converted in its quaternion form. In addition, this conversion enables us to interpolate between two motion poses using the SLERP algorithm [16].

The actor moved across a scene, walking back and forth. The global orientation of the actor, encapsulated in the joint angle values of the root of the skeleton, were rotated so that the walk sequences in the database always face the same direction.

As the number of frames for each walk cycles varies broadly across styles and over time for a single walk sequence, the time length of the walk cycles was normalized in order to give the same weight to each cycle in the subsequent PCA. This data resampling was performed using the SLERP algorithm for interpolation and all cycles were resampled to 40 frames. Furthermore, the same number of cycles had to be kept for each walk style when building the PC space. As 15 is the maximum common number of cycles (walk number 2 (*decided* style) has only 15 examples (see Tab. 1)), the first 15 cycles of each walk were kept for the PCA step, for a total of 165 cycles out of 247.

Unfortunately, PCA is a strictly linear algorithm and cannot be applied on quaternions as they do not form a linear space. The non linear quaternion rotations have thus to be converted into a linear parameterization. Our quaternion representation of joint rotations was reparameterized into exponential maps [9,12] that are locally linear and where singularities can be avoided. In addition to that, this transformation maps the four values of quaternion angles to three values for exponential map representation and reduces the dimensionality of our data before PCA. Each walk cycle is thus represented after data preprocessing by a vector with a fixed number of variables:

$40\ frames * 18\ joints * 3\ dimensions\ for\ exponential\ maps = 2160\ values.$

5 Principal Component Analysis

Given a set of numerical variables, the aim of PCA is to describe that original set of data by a reduced number of uncorrelated new variables. Those new variables are linear combinations of the original variables. Reducing the number of variables causes a loss of information, but PCA ensures that loss of information to be as small as possible. This is done by ordering the new variables by the amount of variance of the original data set that they capture.

In our case, the variables of the original data matrix on which the PCA will be performed are the 2160 values representing each walk cycle. The observations of these variables are the:

15 *cycles per style* ∗ 11 *walk styles* = 165 *observations of the walk cycle*.

When performing PCA, a mean centering is first necessary for the first principal component to describe the direction of maximum variance and not the mean of the data. We thus compute the mean vector out of our 165 walk cycles and remove it from our data matrix before PCA. The PCA can then be carried and the subspace of the principal components is calculated. Given that in our case the number of variables (P=2160) is higher than the number of observations (N=165), the number of principal components is reduced to the number of observations minus one (N-1=164). The new variables in the PC space can then be expressed as follows:

$$Z = XA \qquad (1)$$

with X the original data matrix minus the mean of size NxP , Z the matrix of principal component scores (or the original data in the PC space) of size NxN-1 and A the loadings matrix (or the weights for each original variable when calculating the principal components) of size PxN-1.

As was stated before, PCA orders the principal components according to how much information (or variance from the original data) they represent. The contribution of each component to the whole original information is represented as a cumulative percentage in Fig. 2.

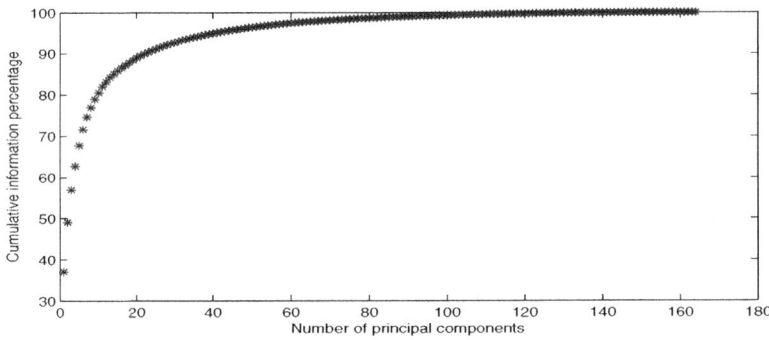

Fig. 2. Cumulated percentage of information contained in the 164 principal components

As PCA is performed in the first place to reduce the dimensionality of the original data, one has to decide how many principal components are to be kept. A very usual way of doing this is to keep the first k principal components that represent 80% of the cumulative percentage of information. Unfortunately, this is an empirical criterion as, depending on the variation present in the original data, 80% of these variations will represent very different levels of detail in the original motion. In our case, 80% of cumulated percentage was reached with 10 PCs but

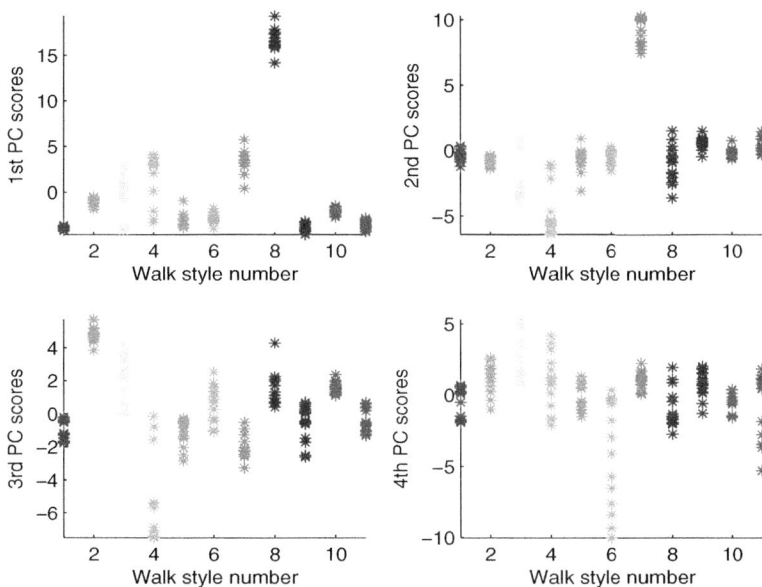

Fig. 3. Scores of the first four principal components, for the 15 occurrences of each of the eleven different styles (see Tab. 1)

the data reconstruction using only 10 PCs was visually significantly impoverished compared to the original data as the style variations were smoothed. So we chose to increase the cumulated percentage of information in our PC subspace by using more principal components. Taking into account 90% of the cumulated percentage of information, which corresponds to 23 principal components, gave data reconstruction that was very difficult to differentiate from original data by the human eye.

As our original 165 walk cycles differ mainly in their style, the first PCs that represent the more variation in the original data will represent mainly the style variations. This assumption is verified and represented in Fig. 3 where the scores of the first four principal components for 15 sequences of each of the 11 different styles are illustrated. We can choose any pair of styles: one or several of the PCs will always enable us to differentiate them. For instance even if the first style (*decided*) is very similar to the second one (*sad*) if we look at the 2nd or 4th PCs, the 1st and 3rd PCs enable to differentiate those two styles very well.

6 Principal Component Space Modeling

Once principal component analysis is performed, both Z, the matrix of variables expressed in the new PC space (scores matrix), and A, the transform matrix from the original space to the PCA space (loadings), are available. In Section 5,

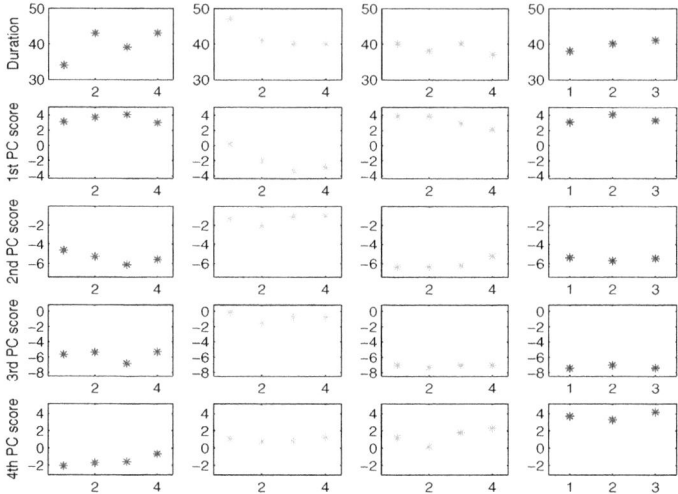

Fig. 4. Variations of the durations and of the scores of the first four PCs for 15 walk cycles with a *decided* style. The different windows in one row show the cycles that follow each other in one straight walk trajectory between two turns (i.e. the continuous walk sequences).

only 15 examples of walk cycle were kept for each motion style so that each walk style was represented by the same number of cycles. Once the PC space is determined giving the same importance to each style, the transformation matrix A can be used to transform the remaining walk cycles, that were not used for principal component analysis, into the PC space. This enables us to take fully advantage of our database for the analysis of the inter-cycles variability of each motion style.

As we have seen in Section 5, PCA enabled us to capture the style variation of walk cycles in a reduced number of principal components. But as could already be seen in Fig. 3, the principal components do not only vary according to style but they also vary along several occurrences of walk cycles for a given style. This phenomenon can already be noticed in straight natural walk, but is heavily amplified in these acted style walk sequences. In this kind of walks, the variability of walk cycles over time is an intrinsic part of the plausibility of the whole sequence. A single walk cycle repeated again and again will rapidly loose all believability in the eye of the spectator. This is why we decided to model the time variability of each style expressed in the PC space rather than taking only one sample of each style for building the PC space or taking only the mean of the principal component scores for each style.

Following the same reasoning, the time variability of the duration of each walk cycle before time normalization to 40 frames per cycle was also modeled for each style independently.

Figure 4 shows the variability of the scores of the first four principal components and of the cycle duration over the first 15 examples of walk cycles for style 4 (*drunk walk*).

We analyzed how principal component scores varied over time for walk cycles that follow each other but there appeared to be no obvious law directing the variations, except that they always stayed in the same range and that the variations between two adjacent cycles did not exceed a given threshold. We decided to model these variations using a very basic description. Each principal component score is modeled as a Gaussian distribution, whose mean $Mean_i$ and standard deviation $StdDev_i$ are computed using all the available walk cycles for the style being modeled. The maximum variation between two adjacent walk cycles is also calculated and used later in the motion synthesis step. The same process is repeated with the time duration for each walk style.

Once this is done, our whole database is modeled using a finite number of parameters, that can then be used to produce walk sequences that vary accordingly to the original data at each cycle.

7 Walk Synthesis

7.1 New PC Space Variable Production

Once all our models are trained, they can be used to produce new sequences of motion that have the same characteristics as the original data and make thus new plausible motions. In this work, we did not study the possibility of morphing one walk style into another one in the PC space, so we only produce walk sequences for one given walk style at a time. Once the style is chosen, the first step of this synthesis is to produce new values of the scores in the PC space for each cycle of the walk sequence to be synthesized. Given the Gaussian distributions calculated in Section 6 and a *RandGauss* function that outputs Gaussian distributed random values with mean zero and standard deviation one, the new score corresponding to the ith PC of the tth walk cycle is calculated as follows:

$$Z^t_{synth}(i) = Mean_i + StdDev_i * RandGauss^t_i \qquad (2)$$

To help synthesizing plausible walk sequences and cycles that can be smoothly concatenated, a post-processing step is then performed to ensure that the variation between the scores of two subsequent cycles does not exceed the threshold of the original data, and reduce the gap by recalculating the concerned score if it was to be the case.

7.2 From PC Space to Original Data Format

Once we have our synthesized scores in the PC space for each one of the walk cycles, the transformation from the PC space to the original motion space can easily be performed using the following equation:

$$X^t_{synth} = Mean_{OrigData} + Z^t_{synth} * A^T \qquad (3)$$

The data X_{synth}^t can then be brought from its exponential map form to the quaternion representation of joint angles. In the quaternion space, a resampling of the walk cycles can be performed using the SLERP algorithm, according to the synthesized durations obtained in the same manner as the PC scores:

$$Duration_{Synth}^t = Mean_{Dur} + StdDev_{Dur} * RandGauss^t \qquad (4)$$

The cartesian coordinates of the root of the skeleton can then be computed. Using our knowledge of the boundaries of the synthesized walk cycles and calculating the height of each foot thanks to the known leg segment lengths, we determine which foot is in contact with the ground. From that fixed 3D position, we calculate the position of the whole body until the other foot becomes the reference, and so on for the whole sequence.

This method enables us to ensure that no foot sliding effect can occur, as the displacement of the whole body is driven by the foot contact point with the ground.

7.3 Cycles Concatenation

Given that the $Mean_{OrigData}$ of the PC space to original space recomposition is the same for all sequences, and that the variations between PC scores from adjacent motion cycles were kept under values encountered in the training database, no huge differences appear between the end of one cycle and the beginning of the following one. A very simple smoothing was thus sufficient to ensure that the cycle transitions were not disturbing for the human eye.

8 Results

Thanks to the method presented in Sections 5 and 6 we modeled our original data in a PC space that makes obvious the style differences in our walk cycles. In Section 7, these different styles were then modeled in that PC space and the use of random Gaussian values enabled us to introduce variability over time into the synthesizing process in a very simple way, while keeping the new motion data plausible. With a finite and reduced number of parameters we are now able to produce an infinite number of new motion sequences, as one cycle is not looped over and over for each style but a new cycle is produced each time. Some examples of synthesized motion sequences can be found at *http://tcts.fpms.ac.be/~tilmanne/*.

The method presented in this paper uses very simple algorithms and modeling techniques but still outputs very interesting results even convincing to the human eye, which is very sensitive to motion naturalness. In this study, we analyzed motions presenting very different style characteristics, which is quite unusual for such a study but still very interesting as characters in the virtual world very often present exaggerated or over-acted behaviors. With as low as 23 components, eleven completely different walk styles were represented, some of them like the *drunk walk* presenting a very high intra style variability.

9 Future Work

One recurrent problem with motion data analysis and synthesis is the difficulty to evaluate the produced motion sequences. The next step for this study will thus be to perform an user evaluation to assess the naturalness of the produced motion and whether the loss of information when reducing the dimensionality of the data using PCA is perceived by the user. Several factors influencing the final results have to be studied, like how different from each other the original walk styles appear to the subject, how the subject perceives the difference between original motions and synthesized motions, how time variability influences the naturalness of the motion compared to a single walk cycle looped, and how the number of principal components influences the reconstructed motion.

With very simple algorithms we were able to build a perpetual walking synthesizer. As we performed our principal component analysis on the whole database, all walk styles are represented in the same PC space. Even if we did not use this property here, the aim is now to be able to produce smooth style transitions into the PC space, so that our perpetual walker could not only walk with time variability but also move its expression from one style to any other style. This could include direct trajectories in the PC space or a transitional neutral walk.

Acknowledgements

The authors would like to thank the comedian Sébastien Marchetti for his participation in the motion capture database recording.

J.Tilmanne receives a PhD grant from the Fonds de la Recherche pour l'Industrie et l'Agriculture (F.R.I.A.), Belgium.

References

1. Brand, M., Hertzmann, A.: Style machines. In: Proceedings of the 27th Annual Conference on Computer Graphics and Interactive Techniques, pp. 183–192 (2000)
2. Calinon, S., Guenter, F., Billard, A.: On Learning, Representing, and Generalizing a Task in a Humanoid Robot. IEEE Transactions on Systems, Man and Cybernetics 37(2), 286–298 (2007)
3. Elgammal, A., Lee, C.S.: The Role of Manifold Learning in Human Motion Analysis Human Motion Understanding, Modeling, Capture and Animation, pp. 1–29 (2008)
4. Forbes, K., Fiume, E.: An efficient search algorithm for motion data using weighted PCA. In: Proceedings of the 2005 ACM SIGGRAPH/Eurographics Symposium on Computer Animation, pp. 67–76 (2005)
5. Forsyth, D.A., Arikan, O., Ikemoto, L., O'Brien, J., Ramanan, D.: Computational Studies of Human Motion: Part 1, Tracking and Motion Synthesis. In: Foundations and Trends in Computer Graphics and Vision, vol. 1(2-3). Now Publishers Inc. (2006)
6. Geng, W., Yu, G.: Reuse of Motion Capture Data in Animation: A Review. In: Kumar, V., Gavrilova, M.L., Tan, C.J.K., L'Ecuyer, P. (eds.) ICCSA 2003. LNCS, vol. 2667, pp. 620–629. Springer, Heidelberg (2003)

7. Glardon, P., Boulic, R., Thalmann, D.: PCA-based walking engine using motion capture data. In: Computer Graphics International, pp. 292–298 (2004)
8. Glardon, P., Boulic, R., Thalmann, D.: A Coherent Locomotion Engine Extrapolating Beyond Experimental Data. In: Proceedings of Computer Animation and Social Agent (CASA), Geneva, Switzerland, pp. 73–84 (2004)
9. Grassia, F.S.: Practical parameterization of rotations using the exponential map. Journal of Graphics Tools 3, 29–48 (1998)
10. Grudzinski, T.: Exploiting Quaternion PCA in Virtual Character Motion Analysis. In: Computer Vision and Graphics, pp. 420–429 (2009)
11. IGS-190. Animazoo, http://www.animazoo.com (2010)
12. Johnson, M.P.: Exploiting quaternions to support expressive interactive character motion. PhD Thesis (2002)
13. Jolliffe, I.T.: Principal Component Analysis, 2nd edn. Springer Series in Statistic. Springer, New York ((2002)
14. Li, Y., Wang, T., Shum, H.: Motion texture: a two-level statistical model for character motion synthesis. In: Proc. of SIGGRAPH 2002, New York, NY, USA, pp. 465–472 (2002)
15. Menache, A.: Understanding motion Capture for Computer Animation and Video Games. Morgan Kauffman Publishers Inc., San Francisco (1999)
16. Shoemake, K.: Animating Rotations with Quaternion Curves. In: Proc. of SIGGRAPH 1985, San Francisco, vol. 19(3), pp. 245–254 (1985)
17. Tanco, L.M., Hilton, A.: Realistic synthesis of novel human movements from a database of motion capture examples. In: Proc. of the Workshop on Human Motion (HUMO 2000), Washington DC, USA, pp. 137–142 (2000)
18. Troje, N.F.: Retrieving information from human movement patterns. In: Understanding Events: How Humans See, Represent, and Act on Events, pp. 308–334. Oxford University Press, Oxford (2008)
19. Urtasun, R., Glardon, P., Boulic, R., Thalmann, D., Fua, P.: Style-Based Motion Synthesis. Computer Graphics Forum 23(4), 799–812 (2004)
20. Wang, Y., Liu, Z., Zhou, L.: Automatic 3d motion synthesis with time-striding hidden markov model. In: Yeung, D.S., Liu, Z.-Q., Wang, X.-Z., Yan, H. (eds.) ICMLC 2005. LNCS (LNAI), vol. 3930, pp. 558–567. Springer, Heidelberg (2006)

Autonomous Multi-agents in Flexible Flock Formation

Choon Sing Ho, Quang Huy Nguyen, Yew-Soon Ong, and Xianshun Chen

School of Computer Engineering, Nanyang Technological University,
50 Nanyang Avenue, Singapore 639798
{csho,qhnguyen,asysong}@ntu.edu.sg, chen0469@e.ntu.edu.sg

Abstract. In this paper, we present a flocking model where agents are equipped with navigational and obstacle avoidance capabilities that conform to user defined paths and formation shape requirements. In particular, we adopt an agent-based paradigm to achieve flexible formation handling at both the individual and flock level. The proposed model is studied under three different scenarios where flexible flock formations are produced automatically via algorithmic means to: 1) navigate around dynamically emerging obstacles, 2) navigate through narrow space and 3) navigate along path with sharp curvatures, hence minimizing the manual effort of human animators. Simulation results showed that the proposed model leads to highly realistic, flexible and real-time reactive flock formations.

Keywords: Flocking, reactive formation, collision avoidance, path-following.

1 Introduction

The art of flock modeling and simulation has received increasing interests and applications in the multi-media and entertainment industry. Particularly in advertising and film production, automating the generation of group animations have been widely pursued as a means of reducing overall production cost or time. Due to the high resource requirement, the generation of creative imaginative scenarios is often regarded as an extravagance in reality. Among various forms of flock simulation, formation constraint flocking, widely used to simulate massive battle scenes in games and movies (e.g., Real time strategy game such Total War, and even blockbusters like the "Avatar" and "The lord of the Ring"), remains a challenge to researchers as it requires efficient algorithms to handle large number of characters while still maintaining the formation shape and autonomy among the flock.

Among the first to introduce models that describe the behavior of large groups of birds, herds, and fish with perceptual skills existing in the real world, Reynolds [1] in his work showed that complex group behavior can be decomposed into several simple steering behaviors at individual level, by concentrating on local coordination among the agents. The set of possible steering behaviors was later extended and modeled based on physical parameters including mass, position, velocity and orientation [2].

Recently, Olfati-Saber presents an exposition relating to algorithms and theory in the field of flocking and multi-agent dynamic system [3]. Several fundamental questions that a flock model may need to address are highlighted as 1) Scalability: how does the model handles massive agents at real-time while maintaining a frame rate of 30fps? 2) Formation control: how does flock maintains a specific formation? 3) Obstacle avoidance: how does a flock avoid colliding with static and dynamic obstacles?

With respect to points 2 and 3, Leonard & Fiorelli [4] and Olfati-Saber & Murray [5] have described graph approaches for constructing an effective communication model that guarantees collision-free system between multiple vehicles of desired formation. Alternatively, Anderson et al. [6] considered an iterative sampling method to generate group animations with predefined formation. The method nonetheless was deemed to be too computationally intensive for real-time rendering and they did not consider handling obstacles avoidance during certain phases. Lai et al. [7] on the other hand was able to produce controlled flock behaviors at significantly reduced computational effort by building group motion graphs with each node representing a cluster of agents. In their work, the form of path and formation constrains are however limited to straight paths and simple planar shapes, respectively. More recently, Xu et al. [8] proposed a shape constrained flock algorithm to handle complex 2D and 3D shape constrained flocks that move along predefined paths. It is worth noting that most existing work adopts the common process of first sampling the desired formation shape, followed by a path planning stage for each sample points. Subsequently, agents in the flock follow the corresponding sample points, while at the same time exhibiting flocking behaviors. However, to the best of our knowledge, to date little emphasis has been placed on flexible flock formation control, such as the ability to react and adapt the formation according to the dynamically changing game play scenarios. Examples include typical game scenario whereby a group of game characters encountering an emergent situation (such as being attacked) would react and change its formation accordingly to user's activity as well as the surrounding conditions.

In this paper, we present an agent-based flocking model for flexible formation handling. In particular, we focus on the problems of global navigation along a predefined path. When a flock travels along paths with sharp curvature, the formation shape will bend naturally along the path's curvature and reverts to its original formation upon reaching gentle path curvature. In this manner, the flock thus animates naturally and automatically according to the curvature of the given path. Further, a geometrical approach to real-time obstacle avoidance is first demonstrated developed to steer the flock away from newly emerged obstacles or squeeze through narrow paths in the scene, by allowing each agent to automatically disengage from the user-defined flock formation when necessary, and subsequently regroup into the original formation upon clearing the obstacles.

The rest of the paper is organized as follows. In section 2, the formal problem definition on autonomous multi-agents in flexible flock formation is presented. Section 3 outlines the formulation of the proposed model, followed by some analysis and discussion. Section 4 demonstrates the capabilities of the proposed flocking model on Unity 3D animation engine. Section 5 concludes the paper with some brief highlights for future work.

2 Problem Definition

In computer games or crowd simulations, the agents are often placed in a dynamic environment whereby obstacles could move or introduce into the scene during runtime. Here, the interest of our present work is to achieve natural and flexible flock formation motions. First, we envision that while the flock is travelling along the path, each agent can adaptively react to the environment so as to avoid obstacles or squeeze through narrow paths and subsequently moving back into the pre-specified formation (see Fig. 1d & e) instead of splitting into smaller groups or losing its formation shape upon avoiding the obstacles (see Fig 1a & b). Secondly, we seek for realistic shape formation that bends along with the curvature of a given path in a natural manner without the need for any user intervention to compute the new positions of the agents at each time step, and in real time (see Fig. 1f). Note that a simple path following algorithm that translates the formation shape along the path will result in an unnatural and rigid formation as depicted in Fig 1c.

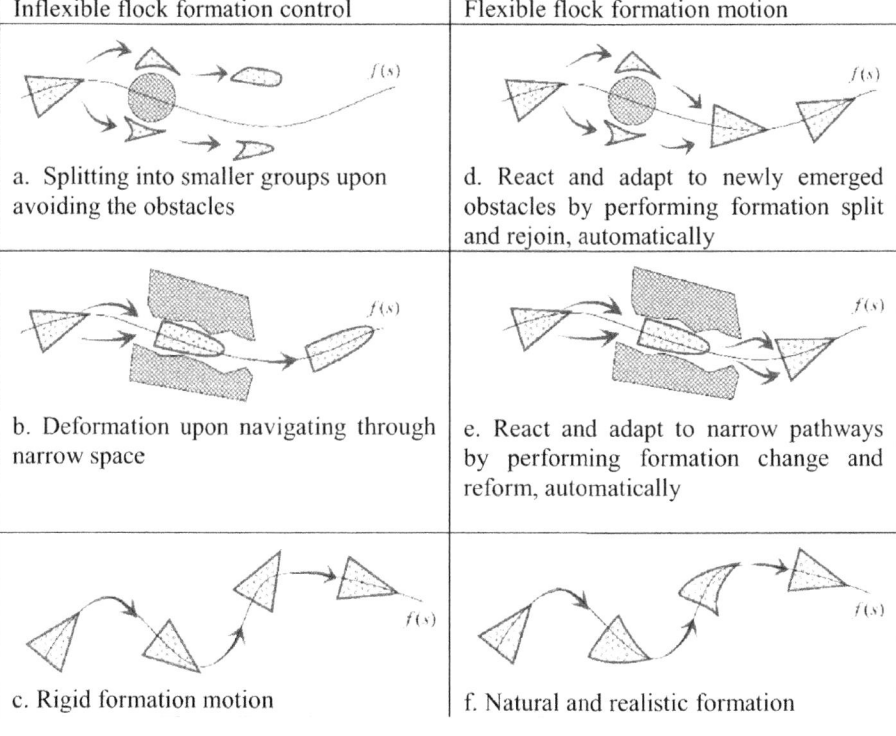

Fig. 1. Challenges towards flexible flock formation

3 Flexible Flock Formation Model

Here we present a flexible flock formation model to tackle the problems of global navigation along a predefined path. At the flock level, the required formation is enforced as a form of soft constraints while at the individual level, flock members are able to react and adapt their own positions when encountering emerged obstacles.

We adopt agent-based approach for modeling the flock behaviors. Agents are modeled by a set of parameters including position, orientation, velocity, maximum speed and mass. The movement of each agent is defined by different behavioral forces including the formation control force that guides the agent along the desired path trajectory while maintaining itself at the appropriate position of the formation and the obstacle avoidance force that steers the agent away from the obstacle within its perception range. These behavioral forces will be accumulated to generate the actual force that acts on the agent.

$$\mathbf{F} = \mathbf{F}_{formation} + \mathbf{F}_{obstacle} \quad (1)$$

In what follows, the set of behavioral forces provided are described.

3.1 Formation Control Force - $\mathbf{F}_{formation}$

The target of this force is to guide the agents along the desire travelling path while maintaining itself at the appropriate position of the formation. Note that the agents do not have information about the formation. Here, we make use of the relative position between the agent and path to control the formation. Let $\mathbf{P}_{(t_i)}$ be the current position of the agent at time step t_i, $\mathbf{Q}_{(t_i)}$ is the projection of $\mathbf{P}_{(t_i)}$ along the path, i.e., $\mathbf{Q}_{(t_i)}\mathbf{P}_{(t_i)}$ is the normal of the path at $\mathbf{Q}_{(t_i)}$ (see Fig. 2).

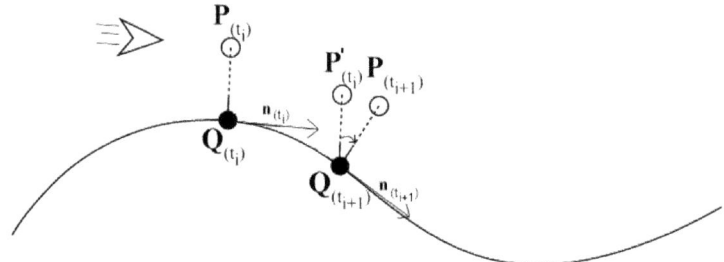

Fig. 2. Formation control along travelling path

Here we assume the path is expressed as a function $f(s)$ where s is the displacement w.r.t the starting point of the path. The updated position of $\mathbf{Q}_{(t_{i+1})}$ at time step t_{i+1} is defined as:

$$\mathbf{Q}_{(t_{i+1})} = f(s_{(t_{i+1})}) = f(s_{(t_i)} + \mathbf{v}*(t_{i+1} - t_i)) \qquad (2)$$

where $\mathbf{v}$ denotes the preferred speed of the flock. Let $\mathbf{n}_{(t_i)}$ and $\mathbf{n}_{(t_{i+1})}$ be the normalized tangent vectors of the curve at the points $\mathbf{Q}_{(t_i)}$ and $\mathbf{Q}_{(t_{i+1})}$ respectively, i.e., $\mathbf{n}_{(t_i)} = f'(s_{(t_i)})$ and $\mathbf{n}_{(t_{i+1})} = f'(s_{(t_{i+1})})$, the new position $\mathbf{P}_{(t_{i+1})}$ is obtained by first translating the vector $\mathbf{P}_{(t_i)}$ along $\mathbf{Q}_{(t_i)}\mathbf{Q}_{(t_{i+1})}$ direction and subsequently rotating it about $\mathbf{Q}_{(t_{i+1})}$ by the angle defined by difference between the slopes of $\mathbf{n}_{(t_i)}$ and $\mathbf{n}_{(t_{i+1})}$. Let

$$\mathbf{P}_{(t_i)} = \begin{bmatrix} x_{\mathbf{P}_{(t_i)}} \\ y_{\mathbf{P}_{(t_i)}} \end{bmatrix}, \quad \mathbf{Q}_{(t_i)} = \begin{bmatrix} x_{\mathbf{Q}_{(t_i)}} \\ y_{\mathbf{Q}_{(t_i)}} \end{bmatrix} \text{ and } \mathbf{P}_{(t_{i+1})} = \begin{bmatrix} x_{\mathbf{P}_{(t_{i+1})}} \\ y_{\mathbf{P}_{(t_{i+1})}} \end{bmatrix}, \quad \mathbf{Q}_{(t_{i+1})} = \begin{bmatrix} x_{\mathbf{Q}_{(t_{i+1})}} \\ y_{\mathbf{Q}_{(t_{i+1})}} \end{bmatrix}, \text{ we have :}$$

$$\mathbf{P}'_{(t_{i+1})} = \begin{bmatrix} x_{\mathbf{P}'_{(t_{i+1})}} \\ y_{\mathbf{P}'_{(t_{i+1})}} \end{bmatrix} = \begin{bmatrix} x_{\mathbf{P}_{(t_i)}} \\ y_{\mathbf{P}_{(t_i)}} \end{bmatrix} + \begin{bmatrix} x_{\mathbf{P}_{(t_{i+1})}} \\ y_{\mathbf{P}_{(t_{i+1})}} \end{bmatrix} - \begin{bmatrix} x_{\mathbf{Q}_{(t_i)}} \\ y_{\mathbf{Q}_{(t_i)}} \end{bmatrix} \qquad (3)$$

The rotation is then performed as:

$$\begin{bmatrix} x_{\mathbf{P}_{(t_{i+1})}} \\ y_{\mathbf{P}_{(t_{i+1})}} \\ 1 \end{bmatrix} = \begin{bmatrix} 1 & 0 & x_{\mathbf{Q}_{(t_{i+1})}} \\ 0 & 1 & y_{\mathbf{Q}_{(t_{i+1})}} \\ 0 & 0 & 1 \end{bmatrix} \begin{bmatrix} \cos\theta & -\sin\theta & 0 \\ \sin\theta & \cos\theta & 0 \\ 0 & 0 & 1 \end{bmatrix} \begin{bmatrix} 1 & 0 & -x_{\mathbf{Q}_{(t_{i+1})}} \\ 0 & 1 & -y_{\mathbf{Q}_{(t_{i+1})}} \\ 0 & 0 & 1 \end{bmatrix} \begin{bmatrix} x_{\mathbf{P}'_{(t_{i+1})}} \\ y_{\mathbf{P}'_{(t_{i+1})}} \\ 1 \end{bmatrix}$$

$$= \begin{bmatrix} \cos\theta & -\sin\theta & x_{\mathbf{Q}_{(t_{i+1})}} \\ \sin\theta & \cos\theta & y_{\mathbf{Q}_{(t_{i+1})}} \\ 0 & 0 & 1 \end{bmatrix} \begin{bmatrix} x_{\mathbf{P}'_{(t_{i+1})}} - x_{\mathbf{Q}_{(t_{i+1})}} \\ y_{\mathbf{P}'_{(t_{i+1})}} - y_{\mathbf{Q}_{(t_{i+1})}} \\ 1 \end{bmatrix} \qquad (4)$$

where θ denotes the angle defined by difference between the slopes of $\mathbf{n}_{(t_i)}$ and $\mathbf{n}_{(t_{i+1})}$. $\cos\theta$ and $\sin\theta$ can be calculated by

$$\cos\theta = \mathbf{n}_{(t_i)} \cdot \mathbf{n}_{(t_{i+1})} \qquad (5)$$

$$\sin\theta = \sqrt{1 - \cos^2\theta} \qquad (6)$$

For the agent to reach $\mathbf{P}_{(t_{i+1})}$, the desired velocity vector is defined in the direction from the current position to the target position as:

$$\mathbf{v}_{desired} = \Delta s / \Delta t = (\mathbf{P}_{(t_{i+1})} - \mathbf{P}_{(t_i)}) / (t_{i+1} - t_i) \qquad (7)$$

Finally, the path following force is arrived based on the desired acceleration

$$\mathbf{F}_{formation} = a_{desired} * m = (\mathbf{v}_{desired} - \mathbf{v}_{agent}) * m / (t_{i+1} - t_i) \qquad (8)$$

where $\mathbf{v}_{agent}$ is the current velocity vector of the flock agent.

3.2 Obstacle Avoidance - $F_{obstacle}$

An obstacle is defined as an object that interferes with the natural action or movement of an agent or flock. Obstacles can be in the form of static or moving objects of a scene. The obstacle avoidance behavior serves to drive the agent away from obstacles. Since obstacle can be in any polygonal forms, we first encapsulate the object with a circle so that every vertices of the polygon are lying on or inside the circle.

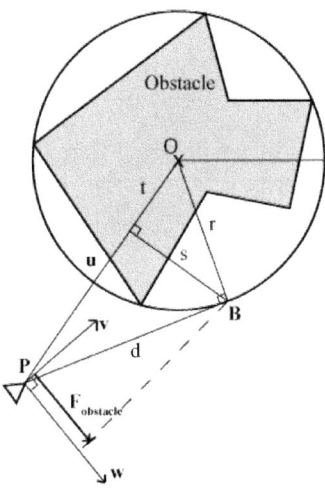

Fig. 3. Obstacle avoidance force

In order to formulate an efficient obstacle avoidance force, we identify the factors that affect the magnitude of the force. Let **v** be the current velocity vector of the agent, **P** be the current position of the agent, **O** be the center of the obstacle, **PB** be the tangent of the obstacle from **P** such that **v** is between **PO** and **PB** (see Fig. 3). When **v** is close to **PO**, the steering force should be larger, one the other hand, when **v** is close to **PB** the steering force is deemed to be smaller. Taking that into account, the steering force is formulated by projecting vector **PB** onto the perpendicular direction of the velocity vector.

Let **u** = **PO**, **d** = **PB**, using simple geometry, vector **d** can be obtained as:

$$t = r^2 / |u|$$

$$s = \sqrt{r^2 - t^2}$$

$$\mathbf{w} = (\mathbf{u} \times \mathbf{v}) \times \mathbf{u}$$

$$d = (|\mathbf{u}|-t)\frac{\mathbf{u}}{|\mathbf{u}|} + s\frac{\mathbf{w}}{|\mathbf{w}|}$$

Now let $\mathbf{n}$ be the perpendicular vector to the current movement direction of agent, i.e., $\mathbf{n} \cdot \mathbf{v} = 0$. The obstacle avoidance force is defined in the direction of the projection of $\mathbf{d}$ onto the direction of $\mathbf{n}$ and the magnitude is scaled inversely proportioned to the distance from the agent to the obstacle.

$$\mathbf{F}_{obstacle} = \frac{\mathbf{n}*(\mathbf{n}\cdot\mathbf{d})}{\mathbf{d}\cdot\mathbf{d}} \qquad (9)$$

3.3 Analysis and Discussion

In this subsection, we analyze the proposed flocking model in terms of formation flexibility, obstacles avoidance and complexity analysis.

a. Formation Flexibility

Here we study how the proposed model maintains a given pre-specified formation. Fig. 4 illustrates how the model work across 3 different time steps, t=0 flock is on a linear path, t=1 at the bend of the given path and then at t=2, the flock proceeds onto a linear path. For the sake of brevity, we consider only 5 agents and the triangle shape formation.

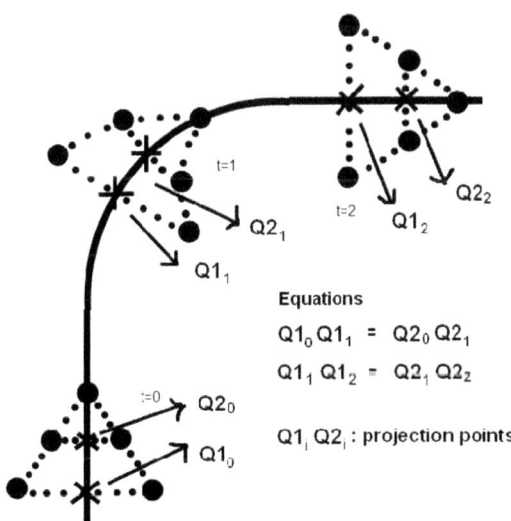

Fig. 4. Illustration on the flexible formation control

It can be observed that by sharing the same knowledge about the path function $f(s)$ and the flock speed $\mathbf{v}$, agents are able to maintain the formation by aligning themselves with the path naturally. Furthermore, at the bend, the agents at the outer corner will automatically move faster (due to further distance to the next points), the (virtual) formation shape will be bended according to the curvature of the path, creating a very realistic and believable animation (similar to a herd of animals running along the bends of the path).

b. Obstacle avoidance

It is noted that the amplitude of the obstacle avoidance force is inversely proportional to the distance from the agent to the obstacle. Therefore, when the obstacle is very near to the agent, obstacle avoidance force will be dominant and the agent will be steered away from the obstacle. On the other hand, when the agent moves away from the obstacle, it will also be away from the path and the expected position in the flock, the obstacle avoidance force will be small and the formation control force will be dominant, which drives the agent back to the correct position within the flock so as to maintain the formation. In Section 4, we will demonstrate that such mechanism allow the flock to naturally split when it encounters an obstacle, and subsequently rejoin to the original formation once the obstacle has been passed.

c. Complexity analysis

Let n be the number of agents in the flock. Since our algorithm uses Quadtree for neighborhood query on possible collisions, the computational complexity incurred in the process of obstacle avoidance at each time step for each agent and entire flock is $O(\log n)$ and $O(n \log n)$, respectively. In formation control, each agent on the other hand incurs a computational effort of $O(1)$ for updating their respective guiding target along the path and new position at each time step. Hence, the total computational complexity incurred for formation control is $O(n)$.

4 Demonstration

The proposed model is implemented and simulated using a PC with Intel Core 2 Duo 2.66 GHz CPU and NVIDIA GeForce 7300SE graphic cards. Unity 3D is employed as the animation engine. We demonstrate the performance of the proposed model with a flock of 400 agents. The initial position of these agents in the formation is generated by performing uniform sampling and K-mean clustering on the formation shape. The screenshots presented comprises of two views, a plan view to show to general scene and a perspective view to provide close up observation.

Fig. 5a shows the initial setup of the simulation where the agents were created and arranged in the formation shape defined. This experiment was designed to test the model on three different scenarios, 1) navigating on path with sharp curvature,

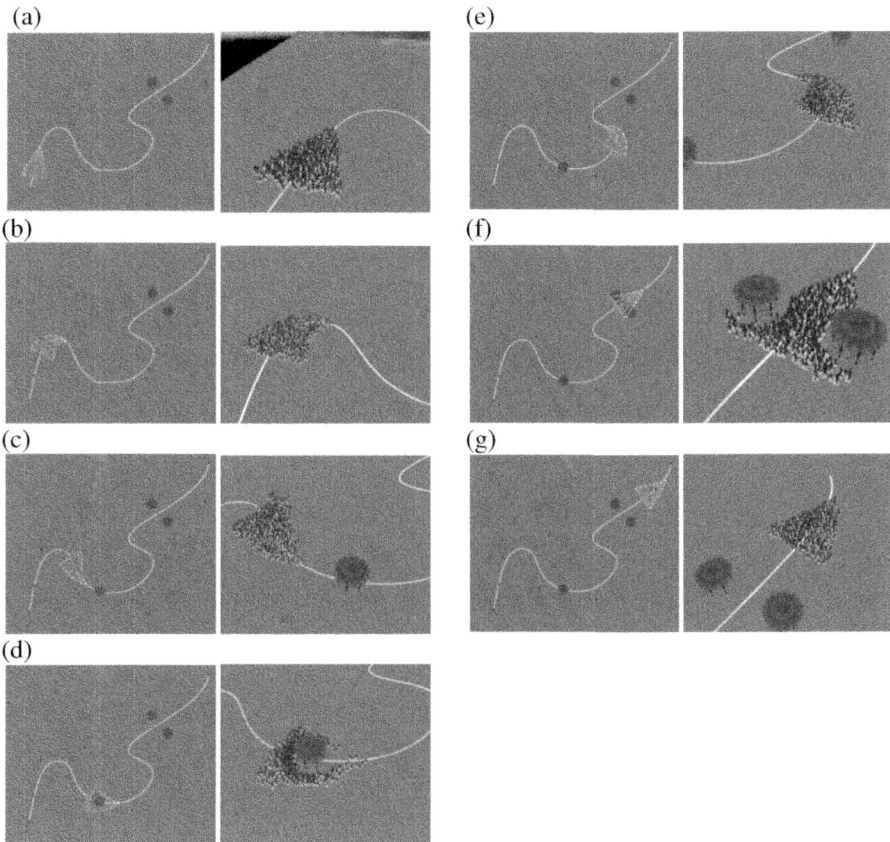

Fig. 5. Flexible flock formations

2) navigating on path with newly emerged obstacles that blocks the flock's path and 3) navigating through narrow space.

As the flock enters the first bend along the path, it is observed that the formation shape is well maintained when the path curvature is gentle. On the other hand, at the path segment with sharp curvature, the formation shape bends naturally, as seen in Fig 5b. Next, when an obstacle is placed dynamically during runtime along the flock's path, it can be observed in Fig. 5c and 5d that the agents will automatically disengaged itself from the formation to steer away from the obstacle. The direction that each agent travels is dependent on its relative position and the obstacle. Subsequently, once the flock passed the obstacle, they automatically regroup to its original formation (Fig. 5e). Similarly for navigating through narrow space, where two obstacles were placed on the sides of the path as shown in Fig 5f, it can be observed that the agents were able to arrange among themselves to squeeze pass the narrow space. The obstacle avoidance mechanism also acts as a separation force to prevent flock members from moving too close, and slowing them down if necessary.

It is worth noting that the members seamlessly rejoin to the original formation after passing the obstacles, as shown in Fig 5g.

Finally, we show the result of the proposed flexible formation model on three different shapes, namely a rectangular column formation, a star-shaped formation, and a V-shape formation in Fig 6, when navigating along a path with a sharp 90 degree turn.

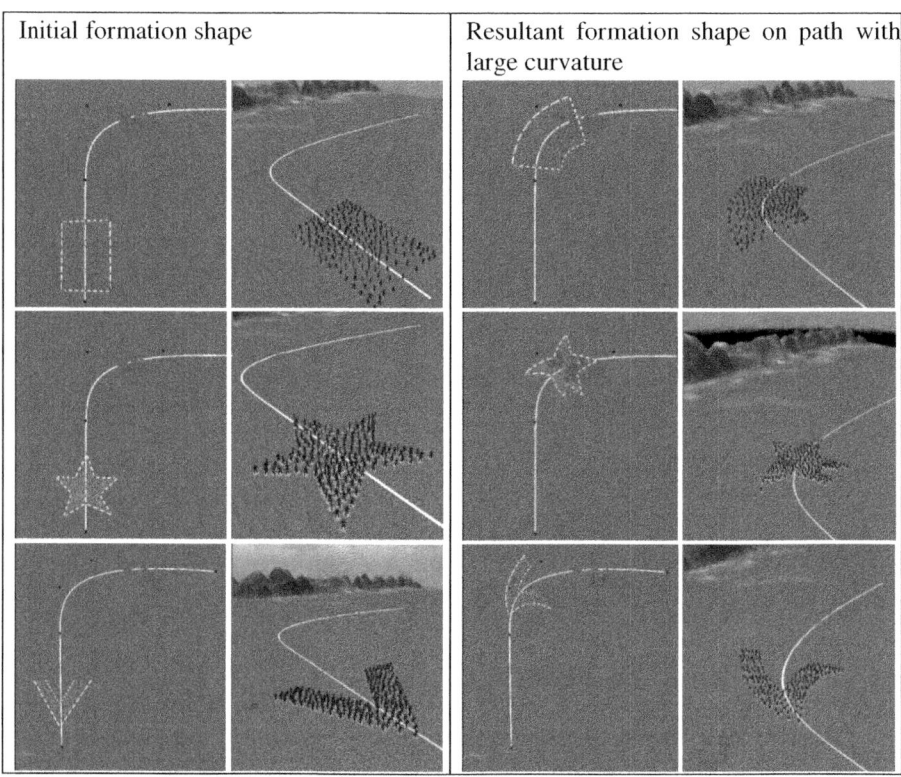

Fig. 6. Flexible formation on different shapes

These screenshots in Fig. 5 and 6 are captured with the demo is running at an interactive frame rate of above 30fps. The computation effort comprises of the animation rendering and agent position update processes. In the present work, our focus is on the latter. Rendering time are typically optimized by means of simple graphical object models, state-of-the-art graphic rendering technologies, i.e., via level-of-details and GPU acceleration [9] [10] [11] [12]. Here, the flock agent position are updated at the rate of 14ms and 23ms for flocks comprising of 400 and 1000 agents, which yields a high computational frame rate of 71 and 43fps, respectively. Nevertheless, by taking an agent-based paradigm, arbitrarily large flocks' computations can be easily scaled up using a distributed architecture.

5 Conclusion

In this paper, we have described a flocking model for attaining natural and flexible flock motions and navigations while maintaining the user specified formation. The flock is also equipped with navigational and obstacle avoidance capabilities that lead to flexible and reactive formations. Demonstration and analysis show that the proposed model gives highly realistic flock formation in real time animations. The model was designed using agent-based approach that required limited knowledge to be shared among the flock members, thus eliminating the degree of communications necessary. Therefore, the model can be easily deployed in a distributed platform where very large scale flock can be simulated.

Acknowledgement

The authors wish to thank Singapore-MIT GAMBIT Game Lab for their support.

Reference

1. Reynolds, C.: Flocks, herds and schools: A distributed behavioral model. In: Proceedings of the 14th Annual Conference on Computer Graphics and Interactive Techniques, pp. 25–34 (1987)
2. Reynolds, C.: Steering behaviors for autonomous characters. In: Game Developers Conference, pp. 763–782 (1999)
3. Olfati-Saber, R.: Flocking for multi-agent dynamic systems: Algorithms and theory. IEEE Transactions on Automatic Control 51(3), 401–420 (2006)
4. Leonard, N., Fiorelli, E.: Virtual leaders, artificial potentials and coordinated control of groups. In: Proceedings of the 40th IEEE Conference on Decision and Control, Orlando, FL, vol. 3, pp. 2968–2973 (2001)
5. Olfati-Saber, R., Murray, R.: Graph Rigidity and Distributed Formation Stabilization of Multi-Vehicle Systems. In: Proceedings of the 42nd Conference on Decision and Control (2002)
6. Anderson, M., McDaniel, E., Chenney, S.: Constrained animation of flocks. In: Proceedings of the 2003 ACM SIGGRAPH/Eurographics Symposium on Computer Animation, pp. 286–297 (2003)
7. Lai, Y., Chenney, S., Fan, S.: Group motion graphs. In: Proceedings of the 2005 ACM SIGGRAPH/Eurographics Symposium on Computer Animation, pp. 281–290 (2005)
8. Xu, J., Jin, X., Yu, Y., Shen, T., Zhou, M.: Shape-constrained flock animation. In: Computer Animation and Virtual Worlds, vol. 19, pp. 313–330 (2008)
9. Trueille, A., Cooper, S., Popović, Z.: Continuum crowds. In: ACM SIGGRAPH 2006, pp. 1160–1168 (2006)
10. van den Berg, J., Lin, M., Manocha, D.: Reciprocal velocity obstacles for realtime multi-agent navigation. In: Proc. IEEE Int. Conf. on Robotics and Automation, Los Alamitos, pp. 1928–1935 (2008)
11. Guy, S., Chhugani, J., Kim, C., Satish, N., Lin, M., Manocha, D., Dubey, P.: ClearPath: Highly Parallel Collision Avoidance for Multi-Agent Simulation. In: ACM SIGGRAPH/Eurographics Symposium on Computer Animation, SCA (2009)
12. Guy, S., Chhugani, J., Curtis, S., Dubey, P., Lin, M., Manocha, D.: PLEdestrians: A Least-Effort Approach to Crowd Simulation. In: Eurographics/ ACM SIGGRAPH Symposium on Computer Animation (2010)

Real-Time Hair Simulation with Segment-Based Head Collision

Eduardo Poyart and Petros Faloutsos

University of California, Los Angeles

Abstract. This work presents a simple, stable and fast hair simulation system for interactive graphics applications whose CPU budget for hair simulation is small. Our main contribution is a hair-head collision method that has very little CPU cost and simulates the volumetric effect of hair resting on top of each other without the need for hair-hair collision. We also present a simulation-based hair styling method for end-users. This easy to use method produces hair styles for the whole head, and it is particularly suitable for abstract and exotic styles. It is applicable for video games in which avatars can be customized by the user.

Keywords: Hair simulation, character animation, physics-based motion.

1 Introduction

Animating hair in real time remains a hard task in computer graphics. An average human head has in the order of 100,000 highly deformable strands of hair that constantly interact with their neighbors and the scalp. Furthermore, real-time applications such as video games often display many characters at the same time, with hair motion being a secondary effect to other more important tasks that also require computational resources. This leaves very small CPU and GPU budgets for hair simulation, which means that a practical hair simulation system must be extremely efficient.

To address this challenge, many real-time applications trade realism for performance. The most realistic method of animating and rendering hair in real time is rarely the most desirable one, due to those performance constraints. Making simplifications that produce a pleasant result, albeit with some physical inaccuracies, can be very useful if the results have good real-time performance.

A good candidate for simplification or elimination is hair-hair collision, which is generally the most costly operation in hair simulation. We can assume the application doesn't require close-up shots and it's difficult or nearly impossible to notice hair crossing each other. However, one side effect remains: hair strands lie flat at the head surface. This causes a loss of hair volume, and also bad interpenetrating artifacts when common texturing and shading methods are used.

The main contribution of this work is a novel hair-head collision technique that solves the problem of hair strands resting on top of each other without the need for hair-hair collision. This technique is very fast and produces believable results, and differently from previous work, it doesn't produce artifacts when the head tilts.

As a secondary result, we present a method to allow an end-user such as a video game player to quickly and interactively create exotic and artistic hairstyles. Hair styling is traditionally a complex and time-consuming process, which involves editing meshes or other parameters in a specialized tool. A system for creating entertaining hair styles quickly and easily can be applied, for example, in video games in which the player creates his/her own characters.

It is important to notice that in this work we focus exclusively on the motion of the hair; we consider rendering a separate problem that we will address in the future.

The remainder of this document is organized as follows. On Section 2, we analyze related work in hair simulation and hairstyle modeling. Section 3 shows how we assign hair to the head. Sections 4 describes the simulation and collision method. Section 5 describes the main aspect of our contribution, *segment-based head collision*. Section 6 discusses how our real-time simulator can be used for hair styling, using a force-field-based virtual combing method. The last section summarize our results and describes future work.

2 Related Work

Since the pioneering work of Terzopoulos et al. [17], a lot of work has been done on deformable models and hair. A very complete survey of hair styling, simulation and rendering can be found in [20]. In the remainder of this section, we only discuss the most relevant prior work.

Chang et al. [4] introduced the concept of *guide hair*, a small number of hair strands which undergo physical simulation and guide the rest of the hair. We use this concept in our work.

Real-time results were achieved by some animation techniques, such as [16], which uses the GPU. Our technique, in contrast, uses only the CPU. In some cases this may be desirable – if the GPU is free from the physical simulation, it can be used to render hair using a more complex algorithm in real time.

Hair strands can be simulated as springs and masses [14] or as more complex models, including rigid bodies connected by soft constraints [6]. Springs are revisited by Plante et al. [13], in which their stretchiness is used to model wisps of curly and almost-straight hair, and by Selle et al. [15], in which the authors model a complex arrangement of springs capable of modeling straight and curly hair, but with non-real-time performance.

To achieve real-time performance, Lee and Ko [9] chose not to model hair-hair collision. They obtain hair volume by colliding the hair with different head shells based on pore latitude on the head. Their method suffers from artifacts if the head tilts too much (e.g. looking down). Our method, in contrast, modulates the head shells based on hair segment, instead of pore latitude. We show that our approach achieves the same volumetric results, doesn't suffer from artifacts when the head tilts, and gives the added benefit of allowing spiky hair modeling.

There are many other possible modeling techniques, of various degrees of complexity. Some examples are: loosely connected particles distributed inside the hair volume [1], simulating hair on a continuum ([7], [8], [10]), or trigonal prisms with limited degrees of freedom based on a pendulum model [5].

Hair styling based on meshes was developed by [21]. In this work, an interactive system is used to allow an artist to model hair styles by performing usual mesh editing operations. This model is very flexible, however it is also time consuming. Our method allows almost instant generation of artistic and exotic hair styles, albeit without a lot of that flexibility and control. Also in the area of hair styling, [19] developed automated hair style generation by example, which is useful for rendering large amounts of unique but similar hair styles.

3 Head Parametrization

We parametrize the head and randomly assign hair roots by using the Lambert azimuthal projection. One of its important properties is area preservation. It provided good results in distributing hair over the hair. The procedure is described as follows. A random point in the plane xy, uniformly distributed in the circle of radius $\sqrt{2}$, is selected. Its coordinates are (X, Y). This 2D point is stored in the data structure for the strand, and it is also projected into the 3D sphere using the Lambert azimuthal projection, as follows:

$$(x, y, z) = \left(\sqrt{1 - \frac{X^2 + Y^2}{4}} X, \sqrt{1 - \frac{X^2 + Y^2}{4}} Y, -1 + \frac{X^2 + Y^2}{2} \right) \quad (1)$$

Since the radius of the originating circle is $\sqrt{2}$, the projection will occupy the lower hemisphere of the head. The vertex position is then transformed to the upper hemisphere, and rotated by 30 degrees towards the back of the head, which approximates real-life hair distribution. Note that the generation process is separated from the animation process. A better, more accurate hair distribution can be envisioned without any penalty for the run-time system.

4 Hair Animation

In the run-time system, the hair is modeled as masses and springs. Each node has a mass, and springs connects the nodes. The root nodes lie at the head surface. One hair strand is a linear connection of nodes and springs in between them, starting from the root node and extending until the last node. All hairs in the head have the same number of nodes.

Two main reasons led us to choose masses and springs as a modeling method. First, we intended to achieve real-time performance, and by using springs and masses combined with forward Euler integration we achieved results that are computationally very fast. Second, it is an easy method to implement, which has a positive impact on its applicability in current real-time systems such as video-games.

The run-time physical simulation system perform collision detection and resolution between hair and head, computes forces applied to the nodes due to the springs and gravity, performs integration, and applies error correction. These steps will be explained on this section. We opted not to simulate hair-hair collision to avoid the associated computational cost; instead we focus on techniques to obtain good results without it.

The following pseudo-code shows the hair simulation algorithm.

```
For every simulation step:
    Collide_with_sphere()
    Assign_spring_forces()
    Euler_integration()
    Apply_ERP()

Procedure Collide_with_sphere():
    For every hair strand:
        For every node (except the root):
            If node is inside sphere:
                Correct position
                Assign force to node

Procedure Assign_spring_forces():
    For every hair strand:
        For every segment:
            Add spring force to both nodes

Procedure Euler_integration():
    For every hair strand:
        For every node (except the root):
            Compute acceleration according to force
            Compute velocity according to acceleration
            Compute position according to velocity

Procedure Apply_ERP():
    For every hair strand:
        For every node (except the root):
            Update node position based on ERP
```

4.1 Head Collision

Collision between the hair and the head is performed by sphere collision. If the distance between a node and the center of the head is smaller than r, this node has collided with the head. The center of the sphere and radius r have to be fine-tuned so as to align the sphere as close as possible with the head model being used for rendering. An ellipse matches human heads more closely. Each node can be transformed by a translation and a non-uniform scaling matrix, such that the collision ellipse becomes a sphere. With that, the usual sphere collision detection can be easily performed.

Once we find a node that penetrates the sphere or ellipse, we need to apply a correction force, which is proportional to the amount of penetration. But before applying this force, we apply a position correction, which will be discussed in Section 4.4. The force is computed using Hooke's law, with a different spring constant ($head\_k_s$) than the one used on the hair segments (k_s). The force is applied to a force accumulator array which has one 3D vector per node per strand. This array is previously set to zero at the beginning of each simulation step.

4.2 Spring Forces

The next step is to add spring forces to the force accumulator array. The forces applied follow Hooke's law:

$$f = -k_s x - k_d v \qquad (2)$$

where k_s and k_d are the spring and damping constants, x is the spring extension from the rest length, and v is the current velocity of the node. All segments have the same rest length.

4.3 Euler Integration

The forward Euler integrator simply takes in the time step value and the array of forces applied to nodes, and computes acceleration, velocity and position for each node, as described in the pseudocode.

Euler integration is subject to instabilities when the springs are too stiff, when the time step is too big, or when the masses are too light. We used a fixed time step of 1/60 second. With this time step, we fine tuned the mass and k_s in order to have stability. This resulted in a low spring constant, and the system suffered from loose springs. Instead of replacing the integration method with a more expensive one, we solved this problem with the technique explained below.

4.4 Error Reduction Parameter (ERP)

With a spring-mass system for hair strands, the length of each hair segment is not guaranteed to be constant, especially when k_s is low. In practice, this means that each spring will extend under the weight of the nodes and the hair will stretch, and worse, it will bounce up and down. This is a major cause of unnatural results for hair modeled this way. If k_s is increased to achieve stiffer springs, the system becomes unstable unless k_d is increased as well.

One solution would be to use a rigid body model for each segment and solving a system of equations for each hair strand. This was not implemented due to the computational cost that would result. The algorithm would no longer be linear on the number of segments per hair, unless a more advanced algorithm such as Featherstone [11] was used.

Table 1. Simulation parameters that provided realistic and stable results

Parameter	Value
head radius	1
number of segments	10
hair length	0.1 to 2.0
k_s	15
k_d	0.0001
mass	0.001
erp	0.6
$head\_erp$	0.5
$head\_k_s$	5
$sphere\_distance\_factor$	0.02

Another solution is to implement an error reduction parameter (ERP), similar to the one implemented in Open Dynamics Engine [12]. ERP is simply a scalar value between 0 and 1 that defines how much of the length error will be corrected at each time step. The correction is performed as a change in position of each node, along the length of the spring, towards the correct length. If ERP=0, there is no correction. If ERP=1, there is full correction: each node will be moved so that the length is fully corrected in one time step.

Why not always use ERP=1, correcting by the full amount every time step? Performing error reduction in this fashion has the practical effect of introducing damping on the system. Tests made with ERP=1 showed that the system became too damped, so smaller values had to be used. Common values for all parameters are presented on Table 1. With these values, k_d introduces virtually no damping while ERP is set to the highest value possible that would not result in too much damping – "too much damping" being an empirically observed behavior. This gave us the best possible constraint on hair length. It is important to notice that this achieves a better constraint/damping ratio than what would be achieved by increasing k_s and k_d alone without the use of ERP.

4.5 Guide Hair

We employ the guide hair technique [4] in order to get as close as possible to 100,000 hair in real time. We physically simulate only a small amount of hair and interpolate the nodes of the remaining hair. We found that physically simulating 2000 guide hair and rendering 20,000 to 40,000 hair typically give adequate, interactive performance.

5 Segment-Based Head Collision

As described so far, the system is real-time and stable, and maintains an almost constant length for simulated hair strand. However, a major drawback is that the hair resting on top of the head is very flat, with no volume being formed due to layers of hair lying on top of each other. In this section we describe previous efforts to overcome this problem, and then our technique to do the same, and we compare it to a more expensive method of hair-hair collision.

Lee and Ko [9] achieve volume by increasing the head collision radius based on pore latitude on the head. The head collision sphere is increased the most when the pores are close to the top of the head. This leads to a result where layers of hair whose pores are higher on the head rest on top of other layers. The biggest drawback of this technique is that, if the head tilts, the order is inverted and the result is unnatural, with hair strands crossing each other. The more the head tilts, the more this crossing over happens.

Let us consider the fact that our hair is a sequence of straight line segments. Between these segments there are nodes, which are indexed, starting with 0, which is the root of the hair strand on the head, and ending with the maximum number of nodes in a strand. In order to maintain layers of hair on the head, the ordering of nodes can be used to slightly increase the sphere radius step by step. That is, the further away from the root a node is, the further away from the head will be its actual collision surface. Fig. 1 depicts this idea. A parameter called sphere distance factor (sdf) represents the increase in radius that is in effect at the last node in the hair strand. Therefore, the per-node increase is sdf/n, n being the number of nodes. Parametrizing sdf this way allows the user to change the number of nodes, e.g. to improve performance, and have consistent visual results, without having to change sdf as well to compensate.

We compared this technique with hair-hair collision. In our hair-hair collision implementation, the main computational primitives used were line-line distance and point-line distance functions. Both penalty forces and position displacement were attempted. We had oscillations in the system, especially at the top of the head. Furthermore, with nothing but guide hair being simulated, there was not enough guide hair to achieve volume. The interpolated hair still appears flat. Chang et al., in [4], attacked this problem by adding triangle strips between guide hair, with an added performance penalty.

Even though hair-hair collision is more general, our approach of index-based head distance comes virtually for free and solves the problem of hair strands resting on top of each other. The flatness of the head is avoided, and hair strands no longer crisscross each other when lying on the head. It works even if only a small number of guide hair are simulated.

As an added bonus, a new hair style can be generated: spiky hair. This can be achieved by making sdf large enough. By making the collision radius grow linearly with the segment index, we had surprisingly interesting results. A high sdf produces great fur simulation. Both spiky hair and fur are depicted on Fig. 2. Other extensions can be devised. The collision radius growth need not be linear. A quadratic increment, for example, makes the hair more spiky. Exotic collision shapes can also be used. Figure 3 shows the result of a collision sphere whose horizontal cross-section is perturbed by a sine wave. This creates vertical "hills and valleys" which shape the hair. This figure also show how volume is retained with segment-based head collision even if long hair is used.

The accompanying video in http://www.erpoyart.com/research/hair shows a comparison between the previous latitude-based head collision and our index-based head collision. It highlights hair layers crossing each other in the latitude-based method, which doesn't happen with index-based collision.

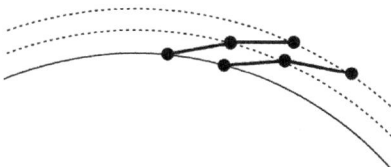

Fig. 1. Segment based collision. The thicker line segments are hair segments; the solid circle represents the head surface where the root nodes lie and the dashed circles are collision spheres for subsequent nodes of the hair.

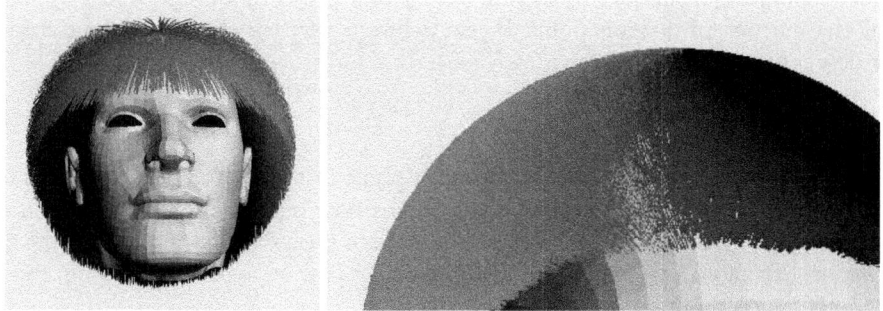

Fig. 2. Spiky hair (left) and fur (right) obtained using segment based collision

6 Hair Styling

Achieving interactive physically-based hair simulation gives rise to interesting possibilities. One of them is the easy modeling of hair styles by pausing the physical simulation at any point in time and interacting with the hair.

This feature was implemented in our system. The user can interactively rotate and move the head, which will cause the hair to move under physics, and he/she can pause and continue the simulation at will. Pausing the simulation causes each node to be attached to its current position in head-space through zero-length springs. The attachment points are called anchors. The hair is still simulated and it responds to head movements and other agents like wind, as if the character applied gel to fix his hairstyle.

We also allow the user to perform "virtual combing". Mouse drag operations are transformed from screen coordinates to world coordinates, lying on a plane parallel to the screen and just in front of the head. This plane intersects with

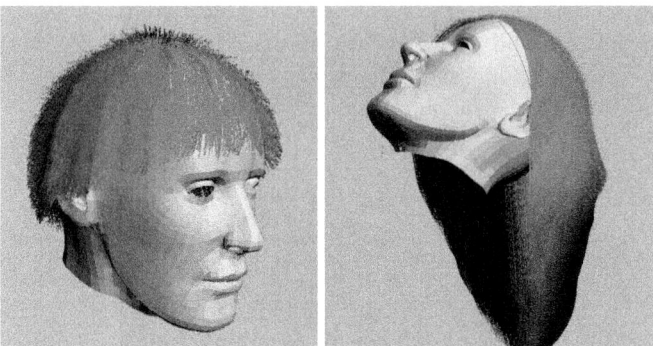

Fig. 3. Left: An exotic collision shape gives "hills and valleys" to the hair. Right: Volume is retained with long hair.

many of the hair strands closest to the user. The movement direction, V, is a vector pointing from the previous to the current projection point. In the vicinity of the current projection point P, we define a vector field whose magnitudes follow a Gaussian fall-off law:

$$m = e^{(-kd^2)} \qquad (3)$$

where d is the distance between P and a point in the vector field, and k is an empirically determined constant that defines how big the area of actuation is. The scalar m is computed for each node in each strand of hair. If it lies above a certain threshold (we used 0.1), it scales the vector V and the resulting vector modifies the node position.

With a little practice, the user can quickly figure out many ways to obtain hair styles with this method. In Fig. 4, a few examples are shown in which only head movements were applied, and in Fig. 5 head movements and virtual combing were applied together.

Due to the simplicity and "fun-factor" of this hair styling technique, and due to the fact that it can generate very artistic and abstract hair styles, we envision its immediate use in video game applications, particularly RPGs or massively multiplayer games in which the user creates and customizes his/her character (avatar) before starting the game. This customization nowadays mainly supports the selection of one out of a number of predefined hair styles that can be customized minimally. Such functionality can be found for exampled in the popular games Oblivion [2] and World of Warcraft [3]. Allowing the user to customize a hair style interactively with a physically based system would be something new and potentially entertaining. Even for systems that can't animate hair in real time due to CPU budget constraints, the physically-based system can run during the character customization phase since the CPU is not heavily required at that point (no game logic is running and no other characters are

Fig. 4. Examples of different hair styles obtained using the system, without combing

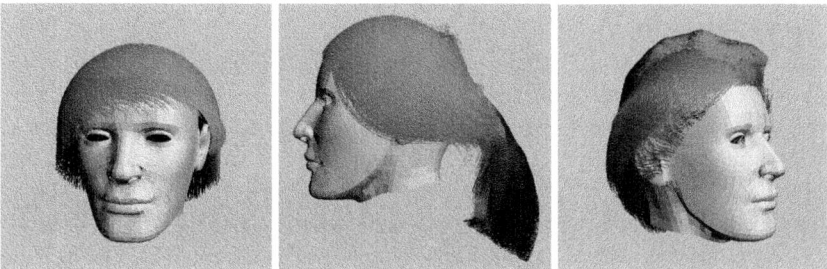

Fig. 5. Hair styles obtained by combing

being rendered), and the character can appear in-game with a static rendering of the styled hair created by the user. This would add an element of fun to the game.

7 Results

The present algorithm is very fast and linear in time complexity, simulating around 20,000 hair strands at 30 frames per second, or around 40,000 at 15 frames per second. It produces good visual results and it is fun to interact with. The performance effect of introducing guide hair is shown on Table 2. An even more important indication of performance is the time taken for the physics simulation, including integration and collision of guide hairs and update of slave hairs: around 4.4 ms – only 1/8th of a frame at 30 fps on a single CPU core. We were able to simulate multiple simultaneous heads with hair. On the accompanying video, we show three heads with hair simulated at 30 frames per second (the frame rate is shown at 25 due to the cost of video capture).

The method has a few drawbacks. Since there is no hair-hair collision, there is hair interpenetration. In a real-world application such as a simulation or videogame, care must be taken so that this effect is hidden from the user as much as possible. We noticed that if the character is not being viewed at close-up, hair-hair interpenetration is not noticeable. The modeling system, although easy to

Table 2. Effect of guide hair on performance. Results obtained on a 2.5 GHz Intel Core 2 Duo, using only one core.

Simulated (Guide) Hair	Slave Hair	Frame Rate	Physics Sim. Time
20,000	0	19 fps	23 ms
1,000	19,000	33 fps	4.4 ms

use, does not allow for detailed control. It is primarily based on trial and error. Our implementation of virtual combing also does not allow precise fine-tuning; it is geared towards major changes in hair positioning.

8 Conclusion

This work introduced a real-time hair simulation system that is very fast and relatively simple to implement. Its time complexity is linear on the number of hair strands and number of nodes per hair strand, which means the number of strands can be greatly increased, up to the same order of magnitude as the human head, and still maintain interactive frame rates. We achieve volume on the hair, as opposed to a flat hair style, by means of segment-based head collision.

A simple modeling method is also introduced. By simply moving the head, freezing the system and applying virtual combing, the user can quickly get interesting hair styles. This is especially useful for abstract and artistic hair styles. This styling method can be applied to situations such as video games where the player is able to customize an avatar. In that kind of situation the requirement is usually that the modeling process shouldn't be lengthy, complex or detailed; it should be simple and fun to use. The fact that our system runs entirely in real-time with attractive results makes this possible.

There are many avenues to extend this work. We would like to experiment with other integration schemes, such as Verlet integration, and articulated body representations of hair.

The styling system can be extended in many different ways so that the user has better control of the results. A combination of user defined vector-fields and physical interaction would be very interesting to pursue. Furthermore, the results of the user-generated hair styles developed here can be fed into a system such as [19], and variations of the hair styles can be automatically generated. Hair rendering is another aspect that will be addressed in the future, since in the current work we have opted to focus exclusively on modeling and animation.

References

1. Bando, Y., Chen, B.-Y., Nishita, T.: Animating Hair with Loosely Connected Particles. Computer Graphics Forum (Eurographics Proc.) 22(3) (2003)
2. Bethesda Softworks: The Elder Scrolls: Oblivion, video game title (2004), http://www.elderscrolls.com/games/oblivion_overview.htm

3. Blizzard: World of Warcraft, video game title (2004), http://www.worldofwarcraft.com/
4. Chang, J.T., Jin, J., Yu, Y.: A Practical Model for Hair Mutual Interactions. ACM Transaction on Graphics (Proc. of SIGGRAPH), 73–80 (2002)
5. Chen, L.H., Saeyon, S., Dohi, H., Hisizuka, M.: A System of 3D hair Style Synthesis based on the Wisp Model. The Visual Computer 15(4), 159–170 (1999)
6. Choe, B., Choi, M., Ko, H.-S.: Simulating complex hair with robust collision handling. In: Proc. of ACM SIGGRAPH/Eurographics Symp. on Comput. Anim., pp. 153–160 (2005)
7. Hadap, S., Magnenat-Thalmann, N.: Interactive hair styler based on fluid flow. In: Proc. of the Eleventh Eurographics Workshop on Computer Animation and Simulation (2000)
8. Hadap, S., Magnenat-Thalmann, N.: Modeling dynamic hair as a continuum. Comp. Graph. Forum (Eurographics Proc.), 329–338 (2001)
9. Lee, D.-W., Ko, H.-S.: Natural Hairstyle Modeling and Animation. Graphical Models 63(2), 67–85 (2001)
10. McAdams, A., Selle, A., Ward, K., Sifakis, E., Teran, J.: Detail Preserving Continuum Simulation of Straight Hair. ACM Transactions on Graphics (Proc. of SIGGRAPH) 28(3) (2009)
11. Mirtich, B.: Impulse-based Dynamic Simulation of Rigid Body Systems. Ph. D. Thesis, University of California at Berkeley (1993)
12. ODE – Open Dynamics Engine, http://www.ode.org/
13. Plante, E., Cani, M.P., Poulin, P.: A Layered Wisp Model for Simulating Interactions Inside Long Hair. In: Proceedings of the Eurographic Workshop on Computer Animation and Simulation, pp. 139–148 (2001)
14. Rosenblum, R.E., Carlson, W.E., Tripp III, E.: Simulating the structure and dynamics of human hair: modelling, rendering and animation. J. Vis. and Comput. Anim. 2(4), 141–148 (1991)
15. Selle, A., Lentine, M., Fedkiw, R.: A mass spring model for hair simulation. ACM Trans. on Graph. 27(3), 1–11 (2008)
16. Tariq, S., Bavoli, L.: Real time hair simulation on the GPU. In: ACM SIGGRAPH Talks, Session: Let's Get Physical (2008)
17. Terzopoulos, D., Platt, J., Barr, A., Fleischer, K.: Elastically deformable models. In: Computer Graphics, pp. 205–214 (1987)
18. Volino, P., Magnenat-Thalmann, N.: Animating Complex Hairstyles in Real-Time. In: Proceedings of the ACM Symposium on Virtual Reality Software and Technology, Hong Kong, pp. 41–48 (2004)
19. Wang, L., Yu, Y., Zhou, K., Guo, B.: Example-based hair geometry synthesis. ACM Transactions on Graphics (Proc. of SIGGRAPH) 28(3), Article 56 (2009)
20. Ward, K., Bertails, F., Kim, T.-Y., Marschner, S.R., Cani, M.-P., Lin, M.C.: A Survey on Hair Modeling, Styling, Simulation, and Rendering. IEEE Transactions on Visualization and Computer Graphics, 213–234 (2007)
21. Yuksel, C., Schaefer, S., Keyser, J.: Hair meshes. In: Proc. of SIGGRAPH, Asia (2009)

Subgraphs Generating Algorithm for Obtaining Set of Node-Disjoint Paths in Terrain-Based Mesh Graphs

Zbigniew Tarapata and Stefan Wroclawski

Military University of Technology, Cybernetics Faculty,
Gen. S. Kaliskiego Str. 2, 00-908 Warsaw, Poland
zbigniew.tarapata@wat.edu.pl, stefan.wroclawski@gmail.com

Abstract. In the article an algorithm (*SGDP*) for solving node-disjoint shortest K paths problem in mesh graphs is presented. The mesh graph can represent e.g. a discrete terrain model in a battlefield simulation. Arcs in the graph geographically link adjacent nodes only. The algorithm is based on an iterative subgraph generating procedure inside the mesh graph (for finding a single path from among K paths single subgraph is generated iteratively) and the usage of different strategies to find (and improve) the solution. Some experimental results with a discussion of the complexity and accuracy of the algorithm are shown in detail.

Keywords: Node-disjoint paths, battlefield simulation games, mesh graph, subgraphs generating, terrain-based graph.

1 Introduction

The disjoint paths problem is a well-known network optimization problem. It may be related to the following problems: maneouver planning of military detachments [17], transport planning of any vehicles, tasks scheduling (sending) in a parallel or a distributed computing system [16], couriers problem, VLSI layout designing [1], routing in telecommunication networks (in particular: optical) [1], [2], [7], [9], [10], etc.

A special application of disjoint paths problem is movement planning and simulation used in military systems (wargame simulators, DSS). The automation of simulated battlefield is a domain of Computer Generated Forces (CGF) systems [11] or semi-automated forces (SAF or SAFOR). CGF or SAF (SAFOR) is the technique which provides the simulated opponent using a computer system that generates and controls multiple simulation entities using software and possibly a human operator. One of the element of the CGF systems is movement planning and simulation of military objects. Any module for redeployment modelling forces is part of such a software [20]. For example the papers [17], [19] present a few problems of movement scheduling for many objects to synchronize their movement and algorithms for solving them with theoretical and experimental analysis in military application. The paper [18] deals with very important problems of multiresolution paths planning and simulation. In these problems algorithms for disjoint path planning are required.

It is known ([5], [10]) that the optimization problem for finding $K>1$ shortest disjoint paths between K pairs of distinct nodes is NP-hard (even for $K=2$). The problem of finding two or more of the disjoint paths between specified pairs of terminals (network nodes) has been well studied. The first significant result in this subject was done by Suurballe [14]. Presented in this paper is the algorithm for the single source – single destination case having a complexity of $O(A\log_{(1+A/N)}N)$, where N – number of network nodes, A – number of network arcs. This procedure solved the problem as a special case of a minimum-cost network flow problem using two efficient implementations of Dijkstra's single–source shortest path algorithm. An efficient algorithm to solve the problem for the single–source all destinations node–disjoint paths was given in [15]. Here, the disjoint pairs of paths from the source node to all the other nodes in the network are determined using a single Dijkstra–like calculation to derive an algorithm having a time complexity of $O(A\log_{(1+A/N)}N)$. Perl and Shiloach [10] studied the complexity of finding two disjoint paths between two different sources and two different destinations in directed acyclic graphs (DAGs). They proposed an algorithm, which is easily generalized in finding the shortest pair of paths (measured by total path length) or finding tuples of d disjoint paths between distinct specified terminals; in the latter case the running time would become $O(AN^{d-1})$. Eppstein in [3] considered the problem of finding pairs of node-disjoint paths in DAGs, either connecting two given nodes to a common ancestor, or connecting two given pair of terminals. He showed how to find the K pairs with the shortest combined length, in time $O(AN+K)$. He also showed how to count all such pairs of paths in $O(AN)$ arithmetic operations. These results can be extended to finding or counting tuples of d disjoint paths, in time $O(AN^{d-1}+K)$ or $O(AN^{d-1})$. Li et al. [8] gives a pseudo-polynomial algorithm for an optimization version of the two-path problem, in which the length of the longer path must be minimized. In the other paper of these authors [9] the difficult biffurcated routing problem was described. They solved the problem when each path corresponds to the routing of a different commodity so that each arc is endowed with a cost depending on the path to which it belongs. In paper [7] the problem of finding two node disjoint paths with minimum total cost in the network was studied, in which a cost is associated with each arc or edge and a transition cost is associated with each node. This last cost is related to the presence of two technologies on the network and is incurred only when a flow enters and leaves the corresponding node or arcs of different types. A good study for a very important problem of finding disjoint paths in planar graphs was presented in paper [13]. A very interesting approach to the time-dependent shortest pair of disjoint paths problem was discussed in [12]. In [16] a new approach to the K disjoint path problem was proposed: it is based on the building, starting from the initial network, so-called K-nodes (K-dimensional vectors of network nodes), K-arcs and "virtual" K-network, and finding in such a K-network the shortest K-path (K-dimensional vector of simple paths) using the original Dijkstra-like pseudo-polynomial algorithm. The specific problem has been considered in the paper. It deals with the parallel or distributed computing system, in which we want to send (or process), in generality, K ($K>1$) tasks from the K^s ($K^s=1$ or $K^s=K$) computers-nodes (local servers) to the K^d ($K^d=1$ or $K^d=K$) destination ones through disjoint paths to minimize sending (or processing) time of all tasks and simultaneously to ensure task sending (or processing) on the most reliable paths (when the elements of the network structure are unreliable). One of the methods being proposed to solve the problem is finding the best paths for K objects iteratively using methods for finding the m-th (1st, 2nd, etc.) best path for each of the K objects [4] and visiting specified nodes [6].

In this paper the subgraphs generating-based disjoint paths (*SGDP*) approximation algorithm is presented. The algorithm is based on the assumption that the topology of the network is represented by a mesh graph, which is a typical environment (terrain) model in a battlefield simulation. In section 2 we define the problem in detail, section 3 contains a description of the *SGDP* algorithm and section 4 presents selected experimental results for the algorithm.

2 Definition of the Problem

Mesh graph, which is the basic data for the problem defined later on can model, for example, regular grid of terrain squares used to plan off-road (cross-country) movement (see Fig.1a). This grid divides the terrain space into squares of equal size. Each square is homogeneous from the point of view of terrain characteristics (dimensions, degree of slowing down velocity, ability to camouflage, degree of visibility, etc.). Structure of such a terrain can be represented by a "mesh" digraph $G = \langle V_G, A_G \rangle$, V_G – set of graph nodes (V_G describes the centre of terrain squares), A_G – set of graph arcs, $A_G \subset V_G \times V_G$, $A = |A_G|$. Arcs are allowed between geographically adjacent squares only (see Fig.1a).

To define the considered problem let us accept the following descriptions: $\mathbf{A} = [a_{ink}]_{V \times M \times K}$ - matrix of source and destination nodes via indirect nodes for each object (a path for each object is divided into $M=N+1$ parts (segments) from one to other indirect nodes, N – number of indirect nodes): $a_{ink}=1$ if the i-th node is the n-th source node for the k-th object, $a_{ink}=-1$ if the i-th node is the n-th destination node for the k-th object; $a_{ink}=0$ otherwise; additionally, the following conditions must be satisfied: $a_{i1k}=1 \Leftrightarrow i=s_k$ (it means that node s_k must be the source node of the first segment of the path for the k-th object), $a_{i1k}=-1 \Leftrightarrow i=i_1(k)$ (the first indirect node $i_1(k)$ for the k-th object is the destination node for the first segment of the path for this object), $a_{iMk}=1 \Leftrightarrow i=i_N(k)$ (the last indirect node $i_N(k)$ for the k-th object is the source node for the last segment of the path for this object), $a_{iMk}=-1 \Leftrightarrow i=t_k$, (node t_k is the destination node of the last segment of the path for the k-th object), $\underset{n\in\{1,...,N\}}{\forall} a_{ink}=-1 \Rightarrow a_{i(n+1)k}=1$ (the destination node of the n-th path segment for the k-th object is, simultaneously, the source node of $(n+1)$-st segment for this path); $\mathbf{H} = [h_{ik}]_{V \times K}$ matrix of nodes (generating subgraphs of G), which are allowed to be taken into account during paths determination for each object: $h_{ik}=1$ if the i-th node can be taken into account during paths determining for the k-th object, $h_{ik}=0$ - otherwise (in particular: $i=s_k \Rightarrow h_{ik}=1$, $i=t_k \Rightarrow h_{ik}=1$); $\mathbf{OUT} = [out_{ij}]_{V \times A}$ - binary crossing matrix of arcs starting in nodes of G: $out_{ij}=1$ if the j-th arc starts in the i-th node, $out_{ij}=0$ - otherwise; $\mathbf{IN} = [in_{ij}]_{V \times A}$ binary crossing matrix of arcs ending in nodes of G: $in_{ij}=1$ if the j-th arc ends in the i-th node, $in_{ij}=0$ – otherwise; $\mathbf{D} = [d_j]_{1 \times A}$ vector of arcs' cost; $\mathbf{X} = [x_{jnk}]_{A \times M \times K}$ - decision variables matrix, $x_{jnk}=1$ if the j-th arc of the G belongs to the n-th segment of the path for the k-th object, $x_{jnk}=0$ – otherwise.

The optimization problem of determining K shortest node-disjoint paths via some indirect nodes in the restricted area can be defined as follows:

$$\sum_{j=1}^{A}\sum_{n=1}^{M}\sum_{k=1}^{K} d_j x_{jnk} \to \min \quad (1)$$

with constraints:

$$\sum_{j=1}^{A}\left(out_{ij}-in_{ij}\right)x_{jnk} = a_{ink}, \quad i=\overline{1,V},\ n=\overline{1,M},\ k=\overline{1,K} \quad (2)$$

$$\sum_{j=1}^{A}\sum_{n=1}^{M}\sum_{k=1}^{K} out_{ij} x_{jnk} \leq 1, \quad i=\overline{1,V} \quad (3)$$

$$\sum_{j=1}^{A}\sum_{n=1}^{M}\sum_{k=1}^{K} in_{ij} x_{jnk} \leq 1, \quad i=\overline{1,V} \quad (4)$$

$$\sum_{j=1}^{A}\sum_{n=1}^{M} out_{ij} x_{jnk} \leq h_{ik}, \quad i=\overline{1,V},\ k=\overline{1,K} \quad (5)$$

$$\sum_{j=1}^{A}\sum_{n=1}^{M} in_{ij} x_{jnk} \leq h_{ik}, \quad i=\overline{1,V},\ k=\overline{1,K} \quad (6)$$

$$x_{jnk} \geq 0, \quad j=\overline{1,A},\ n=\overline{1,M},\ k=\overline{1,K} \quad (7)$$

The first constraint (2) assures that for each node (excluding source and destination nodes), for each object and for each path segment, the sum of arcs starting from the node and the sum of arcs ending at the node, which are selected to the path is the same (further constraints assure that this value is ≤ 1). For source node this difference is equal 1 (only the single path segment can start at the source node) and for destination node -1 (only the single path segment can end at the destination node). Constraints (3) and (4) supplement constraint (2) to assure that for each node only, at least one arc starting and ending at that point can belong to any path. Constraints (5) and (6) guarantee that only allowed nodes are on the path for the k-th object (definition of the restricted area). Additionally, it can be observed that the matrix of constraint coefficients (built on the basis of the left sides of the constraints (2)÷(6)) is totally unimodular and a_{ink}, h_{ik} (right sides) are integer, hence we obtain the continuous linear programming problem (instead of the binary linear programming) and constraint (7) (instead $x_{jnk} \in \{0,1\}$). In the presented optimization problem we have the AMK decision variables and $V(MK+K+2)$ constraints (excluding (7)).

3 Description of the SGDP Algorithm

For solving the (1)-(7) problem we propose the subgraphs generating-based algorithm (*SGDP*). The algorithm is searching for a bundle of node-disjoint paths for K objects, each path consists of 2 or more indirect nodes (including source and destination). The idea of the algorithm is to generate subgraphs (see Fig.1b) in the network of terrain squares (for each moved object we generate a separate subgraph) and afterwards, in each of the subgraphs the Dijkstra's shortest path algorithm is run. Each of these subgraphs is created as follows. We link nodes: source and destination for the given object (if we have e.g. 4 indirect nodes we set following pairs source-destination: 1-2, 2-3, 3-4) and next we "mark" right and left from the line linking these node stripes with a width of $0{,}5sw_k$, where sw_k describes the width of the stripe, in which the object should move. Nodes of graph $G = \langle V_G, A_G \rangle$, which centre coordinates are located at this stripe generate the subgraph. It means that the PG_k subgraph for the k-th object is defined as follows:

$$PG_k = \langle V_{Gk}, A_{Gk} \rangle \qquad (8)$$

where V_{Gk} – set of subgraph nodes for the k-th object,

$$V_{Gk} = \left\{ \begin{array}{l} n \in V_G : x(s_k) + \tan g_k \cdot (y(n) - y(s_k)) - \dfrac{sw_k}{\cos g_k} \leq x(n) \leq \\ \leq x(s_k) + \tan g_k \cdot (y(n) - y(s_k)) + \dfrac{sw_k}{\cos g_k} \end{array} \right\} \qquad (9)$$

where s_k denotes source node for the k-th object, and $x(n), y(n)$ - coordinates of the n-th node; A_{Gk} – set of subgraph's arcs, $A_{Gk} = \{(n,n') \in V_{Gk} \times V_{Gk} : (n,n') \in A_G\}$. It is possible to exclude some arcs during paths searching by using the passability threshold parameter, which is introduced to reject each arc with cost greater than the given parameter value. Having the subgraph generated for each object we can determine the shortest path for each one in the network based on this subgraph using a few searching strategies.

Three strategies are being used to generate order of objects, for which we find paths: *stripeOrderStrategies*={*Ascending, Descending, Random*}. The first two strategies are basing on order of requests: *Ascending* – order is the same as in the given paths to find; *Descending* – order is reversed. In *Random* the strategy generated order is randomized with a uniform distribution. Searching with nondeterministic strategy *Random* allows the algorithm to try each $K!$ possible orders, where K is the number of objects. Number of examined orders is restricted by stop conditions defined in *stopStrategiesSets* (see further).

There are two modes of path searching: *SameWidth* – all stripes must have the same width, *VariousWidth* – each stripe may have a different width.

Two different strategies for generating the width of stripes are implemented: *widthOfStripeGenerationStrategy*={*Constant, Random*}, where *Constant* – width of the stripe is given and never changed; *Random* – width of stripe is randomized with a uniform distribution. Random strategy implementation generates a new width for the stripe with respect to the previous generated width, so that only greater values are allowed. Minimal width increasing is 0.5 unit.

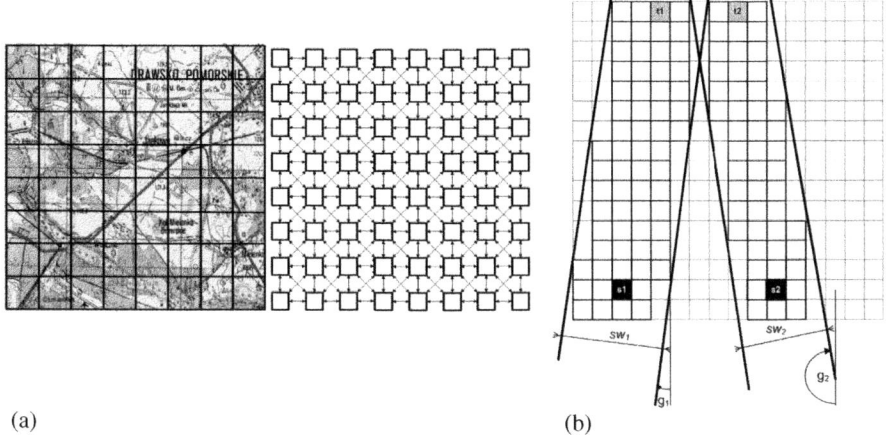

Fig. 1. (a) Fragment of the terrain divided into 8×8 squares (left-hand side) and mesh graph for this terrain (right-hand side). (b) The idea of "cutting" the subgraphs (in mesh graph) into stripes with a width of sw_k for two moved objects with none indirect nodes.

Additionally, we used four different stop strategies, which could be used in any combination:

stopStrategies={MaxIterationsNumber, MaxFeasibleSolutionsFound,
 NextSolutionIsBetter, TimeLimit}.

In *MaxIterationNumber* strategy algorithm ends when the maximum iteration number is reached, where a single iteration is the one searched with a fixed order and the width of stripes. With the *MaxFeasibleSolutionsFound* strategy the algorithm ends, when a specific number of the feasible (acceptable) solutions is found; *NextSolutionIsBetter* stop strategy ends, when a next feasible solution is no better than the previous one plus specific epsilon value. The last strategy, *TimeLimit*, stops the algorithm when the execution time reaches a specified time limit.

We can remember or not paths being found during previous iterations for objects (*PathMemory={true, false}*): if next iteration use the same stripe width like for previously found path we can use it to decrease computational time of the iteration.

Let us analyse the computation complexity of the *SGDP* algorithm. Generating K subgraphs for each pairs source-destination in each path segment is an operation, which complexity is $O(MKV)$. Determining shortest path in each subgraph has complexity $O(\Lambda \log V)$ using Dijkstra's algorithm with binary heaps, because we do it MK times (M path segments for each of K objects) so we have $O(MKA \log V)$. The number of possible combination of paths determining the order is equal to the number of permutation among K elements, that is $K!$. If we check it for each possible action stripe width (let the number of possible action stripe width for each object be equal L) then we can do it, in the worst case, $L^K K!$ times. Since the complexity of *SGDP* algorithm is $O(L^K K! MKA \log V)$. Average case complexity of the algorithm is better than the worst case, because we use some techniques to avoid checking all possible combinations

($L^K K!$): randomization, different stop conditions, remembering paths being found earlier, etc.

Let us notice that the considered algorithm superbly fits the parallel computations by using, for example, K processors (each of the processors generates a subgraph and determines the shortest path in this subgraph).

The pseudo code of the *SGDP* algorithm is as follows:

Input data (see section 2): K, $G = \langle V_G, A_G \rangle$, $\mathbf{A} = [a_{ink}]_{V \times M \times K}$ - matrix of source and destination nodes via indirect nodes for each object (generating set of path segments $PS(k)$ for the k-th object, number of path segments for each object is equal M)

```
0) Save initial graph state
1) While (none of the stop conditions is fulfilled) {
   2) Generate stripe order using stripeOrderStrategy;
   3) If no unchecked stripe order remains -> END;
   4a) If searching mode equals SameWidth{
      5a)While (none of the stop conditions is fulfilled){
         6a) Generate width of stripes using
         widthOfStripeGenerationStrategy;
         7a) If no unchecked width remains -> Restore
         initial graph state and go to 5a);
         8a) For each path k among K objects to find{
            9a) For each segment in PS(k) {
               10a) Search path for segment in G;
            }
            11a) If path was found -> save path for object
            and remove used nodes and arcs from graph
         }
         12a) If all objects have found paths -> save
         feasible solution
   }}
   4b) If searching mode equals VariousWidth{
      5b)While (none of the stop conditions is fulfilled){
         6b) For each path k among K objects to find {
            7b) Generate width of stripes using
            widthOfStripeGenerationStrategy;
            8b) If no unchecked width remains -> Restore
            initial graph state and go to 5b);
            9b) For each segment in PS(k) {
               Search path for segment in G;
            }
            10b) If path was found -> save path for object
            and remove used nodes and arcs from graph
            Else -> restore initial graph state

         }
         11b) If all objects have found paths -> save
         feasible solution and restore initial graph state
   }}}
```

4 Experimental Analysis of the SGDP Algorithm

We have conducted computations for real terrain areas with different number of nodes: 5000, 7540 (Fig.2a), 10000, 20500, 25000 (Fig.2b), 35000 (two of them are presented in Fig.2). We have used random pairs of source-destination nodes (single segments only (M=1)) for K objects ($K \in \{2, 3, 4, 5, 6\}$). We have performed research for almost every possible combination of parameters defined in section 3.

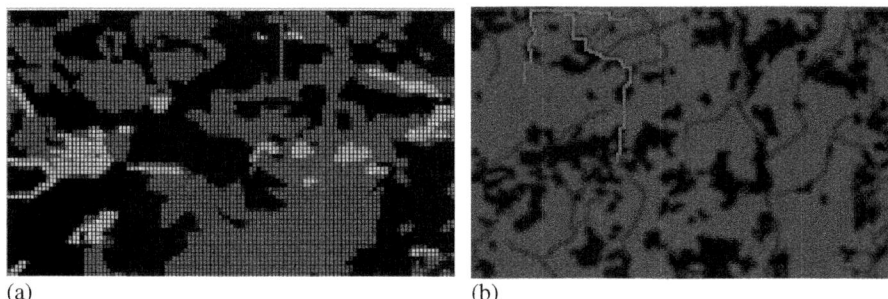

(a)   (b)

Fig. 2. Typical mesh graphs representing fragment of the terrain. Colour represents cost of nodes: the light colour of the nodes (square) describes open (well passable) terrain, the dark colour describes obstacles (forests, lakes, rivers, buildings), the lighter colour the smaller cost value. (a) Graph with 7540=65×116 nodes representing terrain near Drawsko (Poland). (b) Graph with 25000=125×200 nodes representing terrain near Radom (Poland) with two node-disjoint paths found by *SGDP* algorithm (lighter colour).

The following stopStrategiesSets have been used: {{TimeLimit, MaxIterationNumber}, {TimeLimit, MaxIterationNumber, NextSolutionIsBetter}, {*TimeLimit, MaxIterationNumberStrategy, NFeasibleSolutionsFound*}} with the following values: *MaxIterationNumber*=10, *NFeasibleSolutionsFound*=4, in *NextSolutionIsBetter* we set the minimum decrement of cost to 5.0, in *TimeLimit* strategy we have restricted the execution time of each iteration to 5000ms.

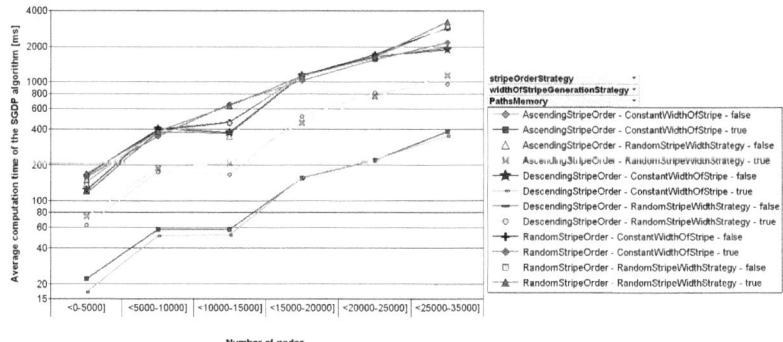

Fig. 3. Average computation time (milliseconds, logarithmic scale) of the *SGDP* algorithm in relation to *stripeOrderStrategy*, *widthOfStripeGenerationStrategy* and *PathsMemory*

All computations have been done using computer with Core 2 Duo 2.2 GHz Intel processor and 3GB RAM.

In Fig.3 we present average computation time of the *SGDP* algorithm in relation to *stripeOrderStrategy*, *widthOfStripeGenerationStrategy* and *PathsMemory* for different number of nodes. It is easy to observe that we have obtained the shortest computation times for *stripeOrderStrategy*∈{*Ascending, Descending*}, *widthOfStripeGenerationStrategy=Constant*. Moreover we can observe that for each pair *stripeOrderStrategy-widthOfStripeGenerationStrategy* computation time for *PathsMemory=true* is significantly shorter than otherwise (from about 3 to 10 times). Analysis of results in Fig.4 supplements results presented above for different values of *stopStrategy*. We obtain the shortest computation time for {*TimeLimit, MaxIterationNumber, NextSolutionIsBetter*} set of stop strategies.

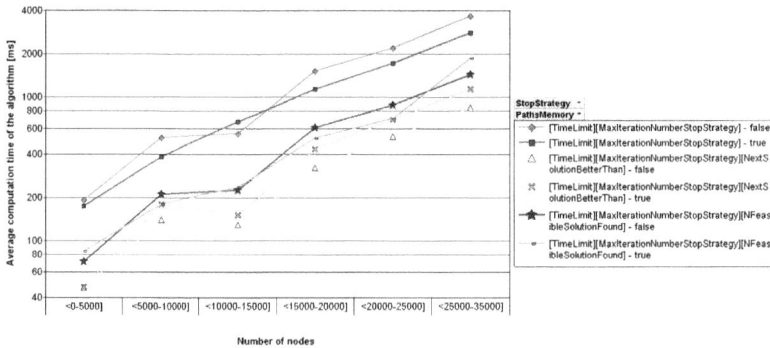

Fig. 4. Average computation time (milliseconds, logarithmic scale) of the *SGDP* algorithm in relation to *stopStrategy* and *PathsMemory*

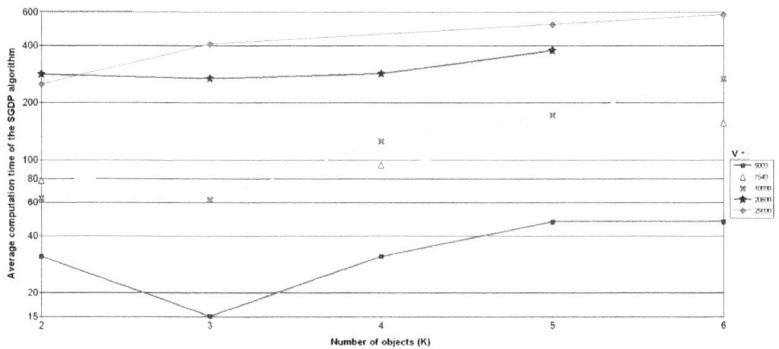

Fig. 5. Average computation time (milliseconds, logarithmic scale) of the *SGDP* algorithm in relation to number of graphs nodes (V) and number of objects (K)

In Fig.5 we show average computation time (milliseconds, logarithmic scale) of the *SGDP* algorithm in relation to number of graphs nodes (V) and number of objects (K).

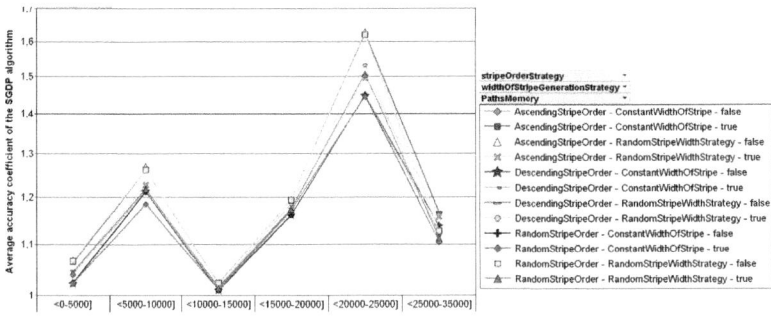

Fig. 6. Average accuracy coefficient (*avgAC*) of the *SGDP* algorithm in relation to *stripeOrderStrategy*, *widthOfStripeGenerationStrategy* and *PathsMemory* (accurate solution obtained using CPLEX 7.0 solver)

Table 1. Comparison of the computation time and accuracy of the *SGDP* algorithm with characteristics of optimal solution obtained by solving problem (1)-(7) using CPLEX 7.0 solver for K=2. *OOF* – optimal value of objective function (1); *minOF_SGDP*, *maxOF_SGDP*, *avgOF_SGDP* – minimal, maximal and average values of objective function, respectively, from *SGDP* algorithm; *AC* – percentage approximation coefficient of *SGDP* algorithm=value of objective function from *SGDP* algorithm/optimal value of objective function in percents; *minAC*, *maxAC*, *avgAC* – minimal, maximal and average values of *AC*; *minCT_SGDP*, *maxCT_SGDP*, *avgCT_SGDP* – minimal, maximal and average values of computation time (in msec) using *SGDP* algorithm; *CT_CPLEX* – computation time (in msec) for finding optimal solution using CPLEX 7.0 solver; *minCTAC*, *maxCTAC*, *avgCTAC* – computation time acceleration coefficient *CTAC* values (respectively: minimal, maximal, average), *CTAC*=computation time using CPLEX solver/computation time using *SGDP* algorithm.

V	5000	7540	10000	20500	25000	35000
minOF_SGDP	162	1247	355	203	2729	23775
maxOF_SGDP	191	1464	362	208	3334	25488
avgOF_SGDP	167	1333	356	204	2950	24364
OOF	158	956	353	190	1955	21038
minAC	2,2%	30,4%	0,7%	6,7%	39,6%	13,0%
maxAC	20,4%	53,1%	2,5%	8,9%	70,5%	21,2%
avgAC	5,5%	39,4%	1,0%	7,3%	50,9%	15,8%
minCT_SGDP	10	31	15	140	109	250
maxCT_SGDP	156	375	282	1375	1250	3079
avgCT_SGDP	98,2	224,4	180,2	852,4	789,2	1786,7
CT_CPLEX	5200	8910	10690	73140	26030	114410
minCTAC	33	24	38	53	21	37
maxCTAC	520	287	713	522	239	458
avgCTAC	52,9	39,7	59,3	85,8	33,0	64,0

In Fig.6 we show average accuracy coefficient *avgAC* (*AC*=value of objective function from *SGDP* algorithm/optimal value of objective function) of the *SGDP* algorithm in relation to *stripeOrderStrategy*, *widthOfStripeGenerationStrategy* and *PathsMemory* (an accurate solution of the problem (1)-(7) obtained using CPLEX 7.0 solver). Value of *avgAC* fluctuates from ~1.02 to ~1.6. It means that value of objective function (sum of the cost of the *K* paths) obtained from *SGDP* algorithm was worse from ~2% to ~60% in relation to optimal one (see also Table 1). The *SGDP* algorithm gives the best values of the *avgAC* for *PathsMemory=true*, *stripeOrderStrategy=Random*, and *widthOfStripeGenerationStrategy=Constant*. In Table 1 we present comparison of the computation time and accuracy of the *SGDP* algorithm with characteristics of optimal solution obtained by solving problem (1)-(7) using CPLEX 7.0 solver for *K*=2. Values of *minAC*, *maxAC* and *avgAC* indicate that value of objective function (sum of the cost of the *K* paths) obtained from *SGDP* algorithm was worse from ~1% to ~50% (average) in relation to optimal one but computation time for *SGDP* algorithm was shorter from ~30 to ~85 times in relation to GAMS/CPLEX solver (in selected cases even >700 times faster, *maxCTAC*(10000)=713).

5 Summary

In the article an algorithm (*SGDP*) for solving node-disjoint shortest *K* paths problem in mesh graphs has been presented. The algorithm can be used for transportation, e.g. maneouver planning of military detachments [17]. It has been shown that it is fast (in comparison with GAMS/CPLEX solver) and gives satisfying solution of the problem (experimental average approximation coefficient of the algorithm is equal from 1% to 50%). Since the algorithm is approximation it seems to be essential to provide necessary and sufficient conditions for obtaining optimal solutions and estimate theoretical approximation coefficient. Moreover, it seems to be essential to examine sensitivity of the algorithm changing number of indirect nodes in paths for each object and values of the parameters: *MaxFeasibleSolutionsFound*, *NextSolutionIsBetter*, *NFeasibleSolutionsFound*. It is possible to extend considered problem using more criteria (e.g. minimization of maximal path cost for any object) and obtaining multicriteria disjoint shortest paths problem. Moreover, the *SGDP* algorithm superbly fits the parallel computations by using, for example, *K* processors (each of the processors generates a subgraph and determines the shortest path in this subgraph). Presented suggestions may contribute to further works.

Acknowledgements

This work was partially supported by projects: MNiSW OR00005506 titled "Simulation of military actions in heterogenic environment of multiresolution and distributed simulation" and MNiSW OR00005006 titled "Integration of command and control systems".

References

1. Aggarwal, A., Kleinberg, J., Williamson, D.: Node-Disjoint Paths on the Mesh and a New Trade-Off in VLSI Layout. SIAM Journal on Computing 29(4), 1321–1333 (2000)
2. Andersen, R., Chung, F., Sen, A., Xue, G.: On Disjoint Path Pairs with Wavelength Continuity Constraint in WDM Networks. In: Proceedings of the IEEE INFOCOM 2004, Hong Kong (China), March 7-11, pp. 524–535 (2004)

3. Eppstein, D.: Finding common ancestors and disjoint paths in DAGs, Technical Report 95-52, Department of Information and Comp. Science, Univ. of California, Irvine (1995)
4. Eppstein, D.: Finding the K shortest Paths. SIAM J. Computing 28(2), 652–673 (1999)
5. Even, S., Itai, A., Shamir, A.: On the complexity of time-table and multicommodity flow problems. SIAM Journal on Computing 5, 691–703 (1976)
6. Ibaraki, T.: Algorithms for obtaining shortest paths visiting specified nodes. SIAM Review 15(2), Part 1, 309–317 (1973)
7. Jongh, A., Gendreau, M., Labbe, M.: Finding disjoint routes in telecommunications networks with two technologies. Operations Research 47, 81–92 (1999)
8. Li, C.L., McCormick, S.T., Simchi-Levi, D.: The complexity of finding two disjoint paths with min-max objective function. Discrete Applied Math. 26, 105–115 (1990)
9. Li, C.L., McCormick, S.T., Simchi-Levi, D.: Findind disjoint paths with different path-costs: Complexity and algorithms. Networks 22, 653–667 (1992)
10. Perl, Y., Shiloach, Y.: Finding two disjoint paths between two pairs of vertices in a graph. Journal of the ACM 25, 1–9 (1978)
11. Petty, M.D.: Computer generated forces in Distributed Interactive Simulation. In: Proceedings of the Conference on Distributed Interactive Simulation Systems for Simulation and Training in the Aerospace Environment, The International Society for Optical Engineering, Orlando, pp. 251–280 (1995)
12. Sherali, H., Ozbay, K., Subramanian, S.: The time-dependent shortest pair of disjoint paths problem: complexity, models and algorithms. Networks 31, 259–272 (1998)
13. Schrijver, A., Seymour, P.: Disjoint paths in a planar graph – a general theorem. SIAM Journal of Discrete Mathematics 5(1), 112–116 (1992)
14. Suurballe, J. W.: Disjoint paths in a network. Networks 4, 125–145 (1974)
15. Suurballe, J.W., Tarjan, R.E.: A quick method for finding shortest pairs of disjoint paths. Networks 14, 325–336 (1984)
16. Tarapata, Z.: Multi-paths optimization in unreliable time-dependent networks. In: Proceedings of the 2nd NATO Regional Conference on Military Communication and Information Systems, Zegrze (Poland), October 04-06, vol. I, pp. 181–189 (2000)
17. Tarapata, Z.: Approximation Scheduling Algorithms for Solving Multi-objects Movement Synchronization Problem. In: ICANNGA 2009. LNCS, vol. 5495, pp. 577–589. Springer, Heidelberg (2009)
18. Tarapata, Z.: Multiresolution models and algorithms of movement planning and their application for multiresolution battlefield simulation. In: Nguyen, N.T., Le, M.T., Świątek, J. (eds.) ACIIDS 2010. LNCS, vol. 5991, pp. 378–389. Springer, Heidelberg (2010)
19. Tarapata, Z.: Movement Simulation and Management of Cooperating Objects in CGF Systems: a Case Study. In: Jędrzejowicz, P., Nguyen, N.T., Howlet, R.J., Jain, L.C. (eds.) KES-AMSTA 2010. LNCS, vol. 6070, pp. 293–304. Springer, Heidelberg (2010)
20. Tuft, D., Gayle, R., Salomon, B., Govindaraju, N., Lin, M., Manocha, D.: Accelerating Route Planning And Collision Detection for Computer Generated Forces Using GPUS. In: Proc. of Army Science Conference, Orlando (2006)

Path-Planning for RTS Games Based on Potential Fields

Renato Silveira, Leonardo Fischer, José Antônio Salini Ferreira,
Edson Prestes, and Luciana Nedel

Universidade Federal do Rio Grande do Sul, Brazil
{rsilveira,lgfischer,jasferreira,prestes,nedel}@inf.ufrgs.br
http://www.inf.ufrgs.br

Abstract. Many games, in particular RTS games, are populated by synthetic humanoid actors that act as autonomous agents. The navigation of these agents is yet a challenge if the problem involves finding a precise route in a virtual world (path-planning), and moving realistically according to its own personality, intentions and mood (motion planning). In this paper we present several complementary approaches recently developed by our group to produce quality paths, and to guide and interact with the navigation of autonomous agents. Our approach is based on a BVP Path Planner that generates potential fields through a differential equation whose gradient descent represents navigation routes. Resulting paths can deal with moving obstacles, are smooth, and free from local minima. In order to evaluate the algorithms, we implemented our path planner in a RTS game engine.

Keywords: Path-planning, navigation, autonomous agent.

1 Introduction

Recent advances in computer graphics algorithms, especially on realistic rendering, allow the use of synthetic actors visually indistinguishable from real actors. These improvements benefit both the movie and game industry which make extensive use of virtual characters that should act as autonomous agents with the ability of playing a role into the environment with life-like and improvisational behavior. Despite a realistic appearance, they should present convincing individual behaviors based on their personality, moods and desires. To behave in such way, agents must act in the virtual world, perceive, react and remember their perceptions about this world, think about the effects of possible actions, and finally, learn from their experience [7].

In this complex and suitable context, navigation plays an important role [12]. To move agents in a synthetic world, a semantic representation of the environment is needed, as well as the definition of the agent initial and goal position. Once these parameters were set, a path-planning algorithm must be used to find a trajectory to be followed.

However, in the real world, if we consider different persons – all of them in the same initial position – looking for achieving the same goal position, each path followed will be unique. Even for the same task, the strategy used for each person to reach his/her goal depends on his/her physical constitution, personality, mood, reasoning, urgency, and so on. From this point of view, a high quality algorithm to move characters across virtual environments should generate expressive, natural, and occasionally unexpected steering behaviors. In contrast, especially in the game industry, the high performance required for real-time graphics applications compels developers to look for most efficient and less expensive methods that produce good and almost natural movements.

In this paper, we present several complementary approaches recently developed by our group [4, 15, 14, 6, 18] to produce high quality paths and to control the navigation of autonomous agents. Our approach is based on the BVP Path Planner [14], that is a method based on the numeric solution of the boundary value problem (BVP) to control the movement of agents, while allowing the individuality of each one.

The topics presented in this article are: (i) a robust and elegant algorithm to control the steering behavior of agents in dynamic environments; (ii) the production of interesting and complex human-like behaviors while building a navigational route; (iii) a strategy to handle the path-planning for group of agents and the group formation-keeping problem, enabling the user to sketch any desirable formation shape; (iv) a sketch based global planner to control the agent navigation; (v) a strategy to deal with the BVP Path Planner in GPU; (vi) a RTS game implementation using our technique.

The remaining of this paper is structured as follows. Section 2 reviews some related works on path-planning techniques that generate quality paths and behaviors. Section 3 explains the concepts of our planner and how we deal with agents and group of agents. Section 4 proposes an alternative to our global path planner, enabling the user interaction. Improvements in the performance are discussed in Section 5. Some results, including a RTS game implementation using our technique, is shown in Section 6. Section 7 presents our conclusions and some proposals for future works.

2 Related Work

The research on path-planning has been extensively explored on the games domain where the programmer should frequently deal with many autonomous characters, ideally presenting convincing behavior. It is very difficult to produce natural behavior by using a strategy focusing on the global control of characters. On the other hand, taking into account the individuality of each character may be a costly task. As a consequence, most of the approaches proposed in computer graphics literature do not take into account the individual behavior of each agent, compromising the planner quality.

Kuffner [8] proposed a technique where the scenario is mapped onto a 2D mesh and the path is computed using a dynamic programming algorithm like Dijkstra.

He argues that his technique is fast enough to be used in dynamic environments. Another example is the work developed by Metoyer and Hodgings [11], that proposes a technique where the user defines the path that should be followed by each agent. During the motion, the path is smoothed and slightly changed to avoid collisions using force fields that act on the agent.

The development of randomized path-finding algorithms – specially the PRM (Probabilistic Roadmaps) [10] – allowed the use of large and more complex configuration spaces to efficiently generate paths. There are several works in this direction [3,13]. In most of these techniques, the challenge becomes more the generation of realistic movements than finding a valid path. Differently, Burgess and Darken [2] proposed a method based on the *principle of least action* which describes the tendency of elements in nature to seek the minimal effort solution. The method produces human-like movements through realistic paths using properties of fluid dynamics.

Tecchia et al. [16] proposed a platform that aims to accelerate the development of behaviors for agents through local rules that control these behaviors. These rules are governed by four different control levels, each one reflecting a different aspect of the behavior of the agent. Results show that, for a fairly simple behavioral model, the system performance can achieve interactive time. Treuille et al. [17] proposed a crowd simulator driven by dynamic potential fields which integrates global navigation and local collision avoidance. Basically, this technique uses the crowd as a density field and constructs, for each group, a unit cost field which is used to control people displacement. The method produces smooth behavior for a large amount of agents at interactive frame rates. Based on local control, van den Berg [1] proposed a technique that handles the navigation of multiple agents in the presence of dynamic obstacles. He uses a concept, called *velocity obstacles*, to locally control the agents with few oscillation.

As mentioned above, most of the approaches do not take into account the individual behavior of each agent, his internal state or mood. Assuming that realistic paths derive from human personal characteristics and internal state, thus varying from one person to another, we [14] recently proposed a technique that generates individual paths. These paths are smooth and dynamically generated while the agent walks. In the following sections, we will explain the concepts of our technique and the extensions implemented to handle several problems found in RTS games.

3 BVP Path Planner

The BVP Path Planner, originally proposed by Trevisan et al. [18], generates paths using the potential information computed from the numeric solution of

$$\nabla^2 p(\mathbf{r}) = \epsilon \mathbf{v} \cdot \nabla p(\mathbf{r}), \tag{1}$$

with Dirichlet boundary conditions. In Equation 1, $\mathbf{v} \in \Re^2$ and $|\mathbf{v}| = \mathbf{1}$ corresponds to a vector that inserts a perturbation in the potential field; $\epsilon \in \Re$ corresponds to the intensity of the perturbation produced by $\mathbf{v}$; and $p(\mathbf{r})$ is the potential at position $\mathbf{r} \in \Re^2$. Both $\mathbf{v}$ and ϵ must be defined before computing this equation. The gradient descent on these potentials represents navigational routes from any point of the environment to the goal position. Trevisan et al. [18] showed that this equation does not produce local minima and generates smooth paths.

To solve numerically a BVP, we can consider that the solution space is discretized in a regular grid. Each cell (i,j) is associated to a squared region of the environment and stores a potential value $p(i,j)$. Using Dirichlet boundary conditions, the cells associated to obstacles in the real environment store a potential value of 1 (*high potential*) whereas cells containing the goal store a potential value of 0 (*low potential*).

A high potential value prevents the agent from running into obstacles whereas a low value generates an attraction basin that pulls the agent. The relaxation method usually employed to compute the potentials of free space cells is the Gauss-Seidel (GS) method. The GS method updates the potential of a cell c through:

$$p_c = \frac{p_b + p_t + p_r + p_l}{4} + \frac{\epsilon((p_r - p_l)v_x + (p_b - p_t)v_y)}{8} \qquad (2)$$

where $\mathbf{v} = (v_x, v_y)$, and $p_c = p_{(i,j)}$, $p_b = p_{(i,j-1)}$, $p_t = p_{(i,j+1)}$, $p_r = p_{(i+1,j)}$ and $p_l = p_{(i-1,j)}$. The potential at each cell is updated until the convergence sets in.

After the potential computation, the agent moves following the direction of the gradient descent of this potential at its current position (i,j),

$$(\nabla \mathbf{p})_c = \left(\frac{p_l - p_r}{2}, \frac{p_b - p_t}{2}\right).$$

In order to implement the BVP Path Planner, we used global environment maps (one for each goal) and local maps (one for each agent). The global map is the Global Path Planner which ensures a path free of local minima, while the local map is used to control the steering behavior of each agent, also handling dynamic obstacles.

The entire environment is represented by a set of homogeneous meshes $\{\mathcal{M}_k\}$, where each mesh $\mathcal{M}_k$ has $L_x \times L_y$ cells, denoted by $\{\mathcal{C}_{i,j}^k\}$. Each cell $\mathcal{C}_{i,j}^k$ corresponds to a squared region centered in environment coordinates $r = (r_i, r_j)$ and stores a particular potential value $\mathcal{P}_{i,j}^k$. The potential associated to each cell is computed by GS iterations (Equation 2) and then used by the agents to reach the goal. In order to delimit the navigation space, we consider that the environment is surrounded by static obstacles.

3.1 Dealing with Individuals

Each agent a_k has one local map m_k that stores the current local information about the environment obtained by its own sensors. This map is centered in the current agent position and represents a small fraction of the global map. The size of the local map influences the agent behavior. A detailed analysis of the influence of the size of the local map on the behavior of the agent can be found in [14].

The local map is divided in three regions: the update zone (*u-zone*); the free zone (*f-zone*); and the border zone (*b-zone*), as shown in Figure 1. Each cell corresponds to a squared region, similar to the global map. For each agent, a goal, a particular vector $\mathbf{v}_k$ that controls its behavior, and a ϵ_k should be stated. If $\mathbf{v}_k$ or ϵ_k is dynamic, then the function that controls it must also be specified.

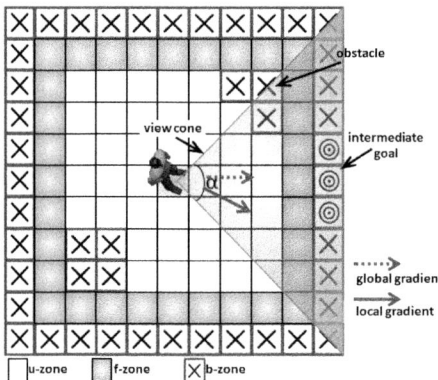

Fig. 1. Agent Local Map. White, green and red cells comprise the *update, free* and *border zones*, respectively. Blue, red and light blue cells correspond to the intermediate goal, obstacles and the agent position, respectively.

To navigate into the environment, an agent a_k uses its sensors to perceive the world and to update its local map with information about obstacles and other agents. The agent sensor sets a view cone with aperture α. Figure 1 sketches a particular instance of the agent local map. The *u-zone* cells that are inside the view cone and correspond to obstacles or other agents have their potential value set to 1. Dynamic or static obstacles behind the agent do not interfere in its future motion.

For each agent a_k, we calculate the global descent gradient on the cell of the global map containing its current position. The gradient direction is then used to generate an intermediate goal in the border of the local map, setting the potential values to 0 of a couple of *b-zone* cells, while the other *b-zone* cells are considered as obstacles, with their potential values set to 1. The intermediate goal helps the agent a_k to reach its goal while allowing it to produce a particular motion.

F-zone cells are always considered free of obstacles, even when there are obstacles inside. The absence of this zone may close the connection between the current agent cell and the intermediate goal due to the possible mapping of obstacles in front of the intermediate goal. When this occurs, the agent gets lost because there is no information coming from the intermediate goal to produce a path to reach. *F-zone* cells handle the situation, always allowing the propagation of the information about the goal to the cells associated to the agent position.

After the sensing and mapping steps, the agent k updates the potential value of its map cells using Equation 2 with its pair $\mathbf{v}^k$ and ϵ^k. Hereinafter, it updates its position according to:

$$\Delta \mathbf{d} = v\, (\cos(\varphi^t), \sin(\varphi^t))\, \Psi(|\varphi^{t-1} - \zeta^t|) \tag{3}$$

with

$$\varphi^t = \eta\, \varphi^{t-1} + (1 - \eta)\, \zeta^t \tag{4}$$

where v defines the maximum agent speed, $\eta \in [0, 1)$, φ is the agent orientation and ζ is the orientation of the gradient descent computed from the potential field stored on its local map in the central position. Function $\Psi : \Re \to \Re$ is

$$\Psi(x) = \begin{cases} 0 & \text{if } x > \pi/2 \\ \cos(x) & \text{, otherwise} \end{cases}$$

If $|\varphi^{t-1} - \zeta^t|$ is higher than $\frac{\pi}{2}$, then there is a high hitting probability and this function returns the value 0, making the agent stops. Otherwise, the agent speed will change proportionally to the collision risk.

In order to demonstrate that the proposed technique produces realistic behaviors for humanoid agents, we compared the paths generated by our method with real human paths [14] . Associating specific values to the parameters ϵ and $\mathbf{v}$ in the agent local map, the path produced by the BVP Path Planner almost mimics the human path. Figure 2 shows a path generated by dynamically changing ϵ and $\mathbf{v}$. Up to the half of the path, the parameters $\epsilon = 0.1$ and $\mathbf{v}= (0, -1)$ were used. Half the path forward, the parameters were changed to $\epsilon = 0.7$ and $\mathbf{v}= (1, 0)$. These values were empirically obtained. We can visually compare and observe that the calculated path is very close to the real one.

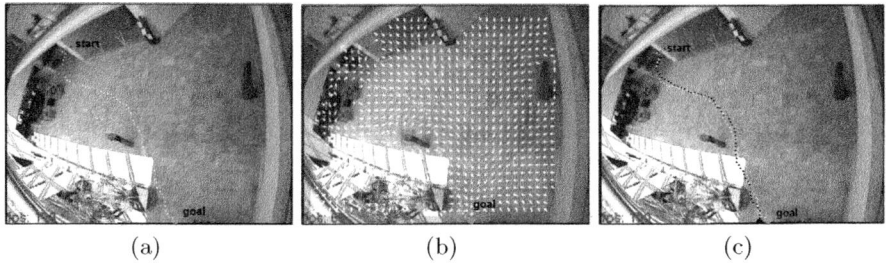

(a)   (b)   (c)

Fig. 2. Natural path generated by the BVP Path Planner: (a) the environment with the real person path (yellow); (b) the environment representation discretized by our planner; (c) the agent path (in black) calculated by our planner after adjusting the parameters

3.2 Dealing with Groups

The organization of agents in groups has two main goals: to allow the control of groups or armies by the user, and to decrease the computational cost through the management of groups instead of handling individuals. In a previous work [15], an approach to integrate group control in the BVP Path Planner was proposed. This approach also support the group formation-keeping while agents move following a given path. Kallmann [9] recently proposed a path planner to efficiently produce quality paths that could also be used for moving groups of agents. Our technique, illustrated in Figure 3, focus on the group cohesion and behavior while enabling the user to interact with the group.

In this approach, each group is associated to a specific formation and a map, called *group map*. The user should provide – or dynamically sketch – any desired shape for the group formation. This formation is then discretized into a set of points that can be seen as attraction points associated to agents – one point attracting each agent towards him. Analogous to an agent local map, the group map is then projected into the environment and its center is aligned with the center of the group of agents in the environment. The center of the formation shape is also aligned with the center of the group map.

Obstacles and goals are mapped to this map in the same way that we have done for the local maps. However, in the group map there is no view cone. Each agent in a group is mapped to its respective position on the group map as an obstacle. In order to obtain the information about the proximity of an agent in relation to the obstacles, we divide the group map into several regions surrounding the obstacles. Each cell in a region has a scalar value that is used to weight the distance between an agent and an obstacle. When the agent is in a cell associated to any of these regions, it will be influenced not only by the force exerted by the formation, but also by the gradient descent computed at its position in the group map. After that, the vector field is extracted and the agent motion is then influenced by two forces: formation force, and the *group map* vector field. As both forces influence in the path definition, they should be properly established, in order to avoid undesired or non realistic behaviors (for more details on combining these forces, refers to [15].

The motion of the agents which are inside the group map is established using these forces while the motion of the entire group is produced by moving the group map along the global map. For this, we consider the group map center point as a single agent and any path planner algorithm can be used to obtain a path free of collisions.

Fig. 3. Group control: agents can keep a formation or move freely; the user can interact and sketch trajectories to be followed by the groups

4 User Interaction with RTS Games

As seen in Section 3, our technique uses a global map to make a global path-planning, and local or group maps to avoid the collision with dynamic elements of a game (e.g. units, enemies, moving obstacles). This fits very well to most RTS games, where the player selects some units and click on a desired location in order to give a "go there" command.

This kind of game-play commonly requires several mouse clicks to give a specific behavior. RTS players commonly want that a group of units overcome an obstacle by following a specific path that conforms to his/her own strategy. Define a strategy in a high level of abstraction is generally hard to be done with only a few mouse clicks.

Based on the ideas of a previous work from our group [5], we suggested an interaction technique where the units are controlled by a sketch based navigation scheme, as an alternative to the global map . The player clicks with the mouse on the screen and draws the desired path for the currently selected units. The common technique of just clicking on the goal location is also supported. This way of sketching the path to be followed is the same one used in paint-like applications, and can be easily adapted to touch screen displays, like Microsoft Surface® and Apple iPad®, for instance.

4.1 Basic Implementation

The technique is divided in two steps: an *input capture* step, where the user draws the path to be followed; and a *path following* step, where the army units run to their goals. In the first step, the path can be drawn by dragging the mouse with a button pressed, or by dragging his/her finger over a touch screen surface. When the user releases the mouse button or removes his/her finger from the screen surface, a list of 2D points is taken and projected against the battlefield, resulting in a list of 3D points.

In the second step, the points generated on the first step are put in a list and used as intermediary goals for the units. Following this list, each point is used as the position goal for all selected units. When the first unit reaches this goal, it is discarded and the next one in the list is used. This continues until the first unit of the group reaches the last goal.

4.2 Splitting the Army

Imagine a situation where the player has walking soldiers and tanks available to attack, and the enemy headquarters is protected by one of these sides by a swamp. Only walking soldiers can walk through it. Then, the player may choose to attack the enemy by one side with the tanks, and the protected side with the walking soldiers.

An extension of the sketching technique may let the player use this strategy by simply drawing one path to the enemy headquarters, where in a certain division point the path divides into two parts: one goes through the swamp and the other avoids it. Then, all units (walking soldiers and tanks) follow the path, and in the division point, the army divides into two groups, one that can trespass the swamp and another that goes through mainland.

In order to allow this maneuver, we suggested a structure where the user sketches are stored in a tree. In this tree, each node represents a division point, start point or end point of the motion, and the links between nodes store one section of the path drawn by the player. As in the Section 4.1, each path section is stored as a list of 3D points.

When the player draws the first path, a node a is created for the 3D point where the path starts, and a node b where the path ends. A link storing all the points between a and b is created, with b being child of a. Then, the user draws a second path, starting nearly the point of division chosen by the player along the link. Considering a tree composed by the previously drawn paths, we search for the link l with the path section that contains the closest point to the start point of the newly drawn path. This link l is broken in two parts, l_1 and l_2, and a new end node p is created between both. Finally, we create a new end node c, a link between p and c, and attach the recently drawn path section to the tree.

5 Improvements and Speedup Using GPU

Solving the Laplace equation is the most expensive part of the BVP Path Planner algorithm. The iterative convergence process of an initial solution to an acceptable solution demands several iterations. So, if we want to improve the convergence speed, we must focus our attention in the relaxation process. We present an approach to improve the convergence on the local maps based on a GPU implementation.

The technique presented here is highly parallelizable, mainly the update of local maps and the computation of Laplace's equation. Each of these steps has several computations to be done, and it can be accomplished in parallel for each agent. We proposed an approach to implement it using nVidia CUDA® and will present it here assuming that the reader knows the CUDA architecture (a detailed explanation can be found in [6]).

Intuitively, each agent detects its observable obstacles in parallel, and each one has its own objective. So, the update of each local map can be also done in parallel. In our algorithm, each agent must seek for obstacles in its own view cone, setting the corresponding cells in the local map as "blocked". Note that, for a given agent, all other agents are seen as obstacles too. We assume that, in the beginning of this step, all the space occupied by the local map is free of obstacles. Then, for each agent, we launch one thread for each obstacle that the agent can potentially see. In each thread, we check if the obstacle is inside the view cone and inside the local map. If it is, then the thread sets each one of the cells that contains part of the obstacle with a tag "blocked".

This scheme of launching one thread for each obstacle of each agent fits very well in the nVidia CUDA architecture, where the processing must be split into blocks of threads. Assigning one block for each agent, each thread of the block

Fig. 4. Two autonomous army fighting (a). The unit local map and the path produced by the BVP Path Planner, illustrated by green dots (b).

can look for one obstacle. Also, each thread can update the local map without synchronization, because all the threads will, if needed, update one cell from a "free" state to a "blocked" state.

After the update of the local maps with the obstacles position, we need to set the intermediate goal. For this, we simply need one thread per agent, each one updating the cells with a "goal" tag. This must be done sequentially to the previous step, to avoid race conditions and the need to synchronize all threads of each block of threads. When each local map has up-to-date information about what cells are occupied, free, or goals, we must relax it to get a smooth potential field. To do this, we assigned one local map to one block of threads in CUDA. In each block of threads, each thread is responsible for updating a value of potential to a single cell. Each thread stays in loop, with one synchronization point between the cells at the end of each iteration. Each thread updates the potential value of the cell using the Jacoby relaxation method.

Finally, to avoid unnecessary memory copies between the GPU and the main memory, we store each attribute of all agents in one single structure of contiguous memory. With this, some parameters that do not change so frequently (e.g. the local map sizes, and the current goal) can be sent only once to the GPU. When the new agent position must be computed, we fetch from the GPU only the current gradient descent on the agent positions.

With our GPU-based strategy we achieved a speed up to 56 times the previous CPU implementation. For a detailed evaluation of our results, refers to [6].

6 A RTS Game Implementation Using the BVP Path Planner

In order to demonstrate the applicability of the proposed technique, we implemented the BVP Path Planner in the Spring®Engine for RTS games. Spring is an open source and multi-platform game engine available under the GPLv2 license to develop RTS games. We have chosen this engine since it is well known in the RTS community and there are several commercial games made with it and available for use. With our planner, we can populate a RTS game with hundred of agents at interactive frame rates [6]. Figure 4(a) shows a screenshot of the game where two army are fighting using the BVP Path Planner. Figure 4(b) shows one unit and its local map with 33×33 cells. Red dots represent blocked cells, while the yellow dots represent free cells. The path followed by the unit is illustrated by green dots. An executable demo using the BVP Path Planner implemented with the Spring Engine can be found at http://www.inf.ufrgs.br/~lgfischer/mig2010, as well as a video including examples that demonstrate all the techniques presented in this paper. All animations in the video were generated and captured in real time.

7 Conclusion

This paper presented several complementary approaches recently developed by us to produce natural steering behaviors for virtual characters and to allow interaction with the agent navigation. The core of these techniques is the BVP Path Planner [14], that generates potential fields through a differential equation whose gradient descent represents navigation routes. Resulting paths are smooth, free from local minima, and very suitable to be used in RTS games.

Our technique uses a global and a local path planner. The global path planner ensures a path free of local minima, while the local planner is used to control the steering behavior of each agent, handling dynamic obstacles. We demonstrated that our technique can produce natural paths with interesting and complex human-like behaviors to achieve a navigational task, when compared with real paths produced by humans. Dealing with groups of agents, we shown a strategy to handle the path-planning problem while keeping the agent formations. This strategy enables the user to sketch any desirable formation shape. The user can also sketch a path to be followed by agents replacing the global path planner.

We also described a parallel version of this algorithm using the GPU to solve the Laplace's Equation. Finally, to demonstrate the applicability of the proposed technique we implemented the BVP Path Planner in a RTS game engine and released an executable demo available on the Internet.

We are now developing a hierarchical version of the BVP Path Planner that spends less than 1% of the time needed to generate the potential field using our original planner in several environments. We are also exploring strategies to use this planner in very large environments. Finally, we are also working on the generation of benchmarks for our algorithms.

References

1. van den Berg, J., Patil, S., Sewall, J., Manocha, D., Lin, M.: Interactive navigation of multiple agents in crowded environments. In: Proc. of the Symposium on Interactive 3D Graphics and Games, pp. 139–147. ACM Press, New York (2008)

2. Burgess, R.G., Darken, C.J.: Realistic human path planning using fluid simulation. In: Proc. of Behavior Representation in Modeling and Simulation, BRIMS (2004)
3. Choi, M.G., Lee, J., Shin, S.Y.: Planning biped locomotion using motion capture data and probabilistic roadmaps. ACM Trans. Graph. 22(2), 182–203 (2003)
4. Dapper, F., Prestes, E., Nedel, L.P.: Generating steering behaviors for virtual humanoids using BVP control. In: Proc. of CGI, vol. 1, pp. 105–114 (2007)
5. Dietrich, C.A., Nedel, L.P., Comba, J.L.D.: A sketch-based interface to real-time strategy games based on a cellular automation. In: Game Programming Gems, vol. 7, pp. 59–67. Charles River Media, Boston (2008)
6. Fischer, L.G., Silveira, R., Nedel, L.: Gpu accelerated path-planning for multi-agents in virtual environments. SB Games, 101–110 (2009)
7. Funge, J.D.: Artificial Intelligence For Computer Games: An Introduction. A. K. Peters, Ltd., Natick (2004)
8. James, J., Kuffner, J.: Goal-directed navigation for animated characters using real-time path planning and control. In: Magnenat-Thalmann, N., Thalmann, D. (eds.) CAPTECH 1998. LNCS (LNAI), vol. 1537, pp. 171–186. Springer, Heidelberg (1998)
9. Kallmann, M.: Shortest Paths with Arbitrary Clearance from Navigation Meshes. In: Symposium on Computer Animation, SCA (2010)
10. Kavraki, L., Svestka, P., Latombe, J.C., Overmars, M.: Probabilistic roadmaps for path planning in high-dimensional configuration space. IEEE Trans. on Robotics and Automation 12(4), 566–580 (1996)
11. Metoyer, R.A., Hodgins, J.K.: Reactive pedestrian path following from examples. Visual Comput. 20(10), 635–649 (2004)
12. Nieuwenhuisen, D., Kamphuis, A., Overmars, M.H.: High quality navigation in computer games. Sci. Comput. Program. 67(1), 91–104 (2007)
13. Pettre, J., Simeon, T., Laumond, J.: Planning human walk in virtual environments. In: Int. Conf. on Intelligent Robots and System, vol. 3, pp. 3048–3053 (2002)
14. Silveira, R., Dapper, F., Prestes, E., Nedel, L.: Natural steering behaviors for virtual pedestrians. Visual Comput. (2009)
15. Silveira, R., Prestes, E., Nedel, L.P.: Managing coherent groups. Comput. Animat. Virtual Worlds 19(3-4), 295–305 (2008)
16. Tecchia, F., Loscos, C., Conroy, R., Chrysanthou, Y.: Agent behaviour simulator (abs): A platform for urban behaviour development. In: Proc. Game Technology, 2001 (2001)
17. Treuille, A., Cooper, S., Popović, Z.: Continuum crowds. In: ACM SIGGRAPH, pp. 1160–1168. ACM Press, New York (2006)
18. Trevisan, M., Idiart, M.A.P., Prestes, E., Engel, P.M.: Exploratory navigation based on dynamic boundary value problems. J. Intell. Robot. Syst. 45(2), 101–114 (2006)

Learning Human Action Sequence Style from Video for Transfer to 3D Game Characters

XiaoLong Chen[1], Kaustubha Mendhurwar[1], Sudhir Mudur[1],
Thiruvengadam Radhakrishnan[1], and Prabir Bhattacharya[2]

[1] Department of Computer Science and Software Engineering, Concordia University,
1455 de Maisonneuve W., Montreal, H3G 1M8, Quebec, Canada
as.allen1128@gmail.com, k_mendhu@cs.concordia.ca,
mudur@cs.concordia.ca, krishnan@cs.concordia.ca
[2] Department of Computer Science, College of Engineering and Applied Sciences
814B Rhodes Hall, University of Cincinnati, Cincinnati, OH 45221-0030, USA
bhattapr@ucmail.uc.edu

Abstract. In this paper, we present an innovative framework for a 3D game character to adopt human action sequence style by learning from videos. The framework is demonstrated for kickboxing, and can be applied to other activities in which individual style includes improvisation of the sequence in which a set of basic actions are performed. A video database of a number of actors performing the basic kickboxing actions is used for feature word vocabulary creation using 3D SIFT descriptors computed for salient points on the silhouette. Next an SVM classifier is trained to recognize actions at frame level. Then an individual actor's action sequence is gathered automatically from the actor's kickboxing videos and an HMM structure is trained. The HMM, equipped with the basic repertoire of 3D actions created just once, drives the action level behavior of a 3D game character.

Keywords: 3D SIFT, HMM, Motion Capture, SVM.

1 Introduction

New generations of 3D games are revolutionizing the gaming industry with particular emphasis on increased immersion in the game world for enhanced gaming experience. Avatar personalization as well as realistic performances by plot-essential non-player characters (NPCs), contribute significantly towards enhancing the gaming experience [1-3]. This is especially so when the game is based on films and the actors' styles are distinctive. Motion capture is one popular solution to this challenge. Motion capture has been used many a time to successfully reproduce the stylistic movements of an individual. There are many activities in which individual style is also present at a higher level in the sequence in which actions are performed. For reproducing this style level, new action sequences have to be generated. Captured motion sequences are difficult to modify to obtain new action sequence patterns. Capture of yet another new motion sequence is today the best available solution. Thus, given the difficulty in capturing a large number of different action sequences, most of the games and/or simulations today rely on repeating a few prerecorded and predefined action sequences.

Individual styles are present in human actions at various levels. At lower levels, it is seen in the manner in which low-level component motions are performed, such as lift of leg, turning of the head, bending of the torso, *etc*. Various dance or martial arts forms are among the activities in which style is seen in the sequence in which basic actions are performed. It is this level of human performance styles which forms the concern of this paper. The goal of this paper is to develop an innovative framework to learn aforementioned human style from videos and incorporate into a 3D game character's performance. A database consisting of motion capture data for basic actions is created. New action sequences are generated by stringing together the motion data of these basic actions in a pattern learned from the video. These generated sequences will have learned styles embedded in them. The proposed framework consists of three phases as follows. (1) The first phase is for feature word vocabulary creation. For this, we identify salient points on the silhouette, compute 3D SIFT local descriptors and then quantize them to obtain a feature word vocabulary. (2) The second phase is devoted to construction of the semantic representatives of basic actions and to train Support Vector Machine (SVM) classifiers for action detection. (3) The third phase is for action sequence identification from an individual performer's video, followed by unsupervised Hidden Markov Model (HMM) parameter learning from recognized action sequences, and finally 3D animation sequence generation. Since the video acquisition today is so much easier than full 3D motion capture, learning from video makes the whole process highly attractive for a wide range of applications. We demonstrate the proposed framework on kickboxing action sequences owing to the fact that this form of martial arts admits individual styles through improvisation in the sequence in which basic actions are performed. This framework could be easily extended to all other activities which also have a repertoire of well defined basic actions.

The rest of this paper is organized as follows: Section 2 provides a brief overview of the prior work in related areas. Section 3 describes the proposed framework in detail. Section 4 demonstrates the application of this framework to kickboxing activity, and finally section 5 contains conclusions and the future work.

2 Related Work

Extensive previous researches have been conducted towards the goal of creating realistic 3D game character animation. Capturing of low-level motion patterns such as the stylistic variation of motions are presented in [5], [6]. An intuitive animation interface to control the animation of avatars through measuring foot pressure exerted by performers is presented in [7]. An approach that uses video cameras and a small set of retro-reflective markers to capture performance animations is presented in [8]. All the afore-mentioned research attempts focus on capturing performers' low-level motion patterns, such as individual's walking styles, and on applying them to a 3D model. However, automatic extraction of high-level motion patterns, such as the sequence in which the actions are performed has been researched to a very much lesser extent. This is essential to provide character to NPCs and consequently to provide unique gaming experiences to players. A statistic model for motion synthesis is presented in [4]. However, the approach is best suited for motions consisting of highly repeating sequences such as in disco dance activity. Masaki *et. el.* [9] used real players to play the role of characters and logged discrete executed motion types (*e.g.* move, attack, defend). Based on the control log obtained, the parameters of SVM were learned, which was then used to control autonomous characters. A related framework presented in [10] includes both video motion analysis and motion pattern learning. Besides the strict requirements on the video data and rather complicated design of HMM models, the goal of their research was also different - to transfer the exact motion sequence from video data into 3D animation.

In comparison to all the above research in action sequence generation, the goal of our framework is to embed a performer's high-level style into the 3D game character to generate non-repetitive action sequences in that performer's style. To the best of our knowledge, ours is the first attempt to propose such a framework. However, on the individual components used in our framework there is considerable previous research and is discussed more below.

2.1 Action Categorization from Video Data

Recognizing actions from video data, though well researched is still a very difficult problem. A detailed survey on vision-based human action recognition is presented in [11], [12]. The difficulty stems from the fact that the low level features extracted from video frames bear little or no semantics of the action performed. Seeking to bridge the gap, Csurka *et. el.* [13] took the concept of "bag of words" from the field of document classification and introduced a "bag of key points" approach to visual categorization. It is also shown in [13] that Scale-Invariant Feature Transform (SIFT) [14], as a local descriptor, can achieve scale and translation invariance and is therefore more suitable for image classification, compared to its alternatives. Based on [13], [14], Scovanner *et. el.* [15] introduced 3D SIFT descriptors for video action recognition and extended the 'bag of words" paradigms from 2D to 3D. Klaser, in [16], presented a novel local descriptor which is achieved by computing 3D gradients for arbitrary scales and 3D orientation quantization based on regular polyhedrons. In our framework too, we use the "bag of words" paradigm and 3D SIFT as local descriptor which is proven to be stable and robust in video recognition. We do differ somewhat from other work in our choice of the key points.

Another important factor in action categorization is the selection of a classifier; we employ the SVM classifier for its simplicity and excellent reputation for classifying feature vectors of extremely high dimensions.

2.2 Learning Action Patterns

Using machine learning to capture the motion style from a large set of motion capture sequences has been the current trend for the control of characters in games. In [5-6] methods for the capture of motion patterns in terms of stylistic variations of motions are presented. As mentioned earlier, this is not the style level that is the concern in our framework. In [17], Marco Gilles introduced HMM which uses the transition probabilities of Finite State Machines to govern the character's behavior. Machine learning method and the motion graph driving component of our framework builds on the work presented in [17]. One major difference between the preceding approaches and the proposed framework lies in the fact that learning in our case is done through video data, an easy to acquire media type, as compared to motion capture data used in the above works.

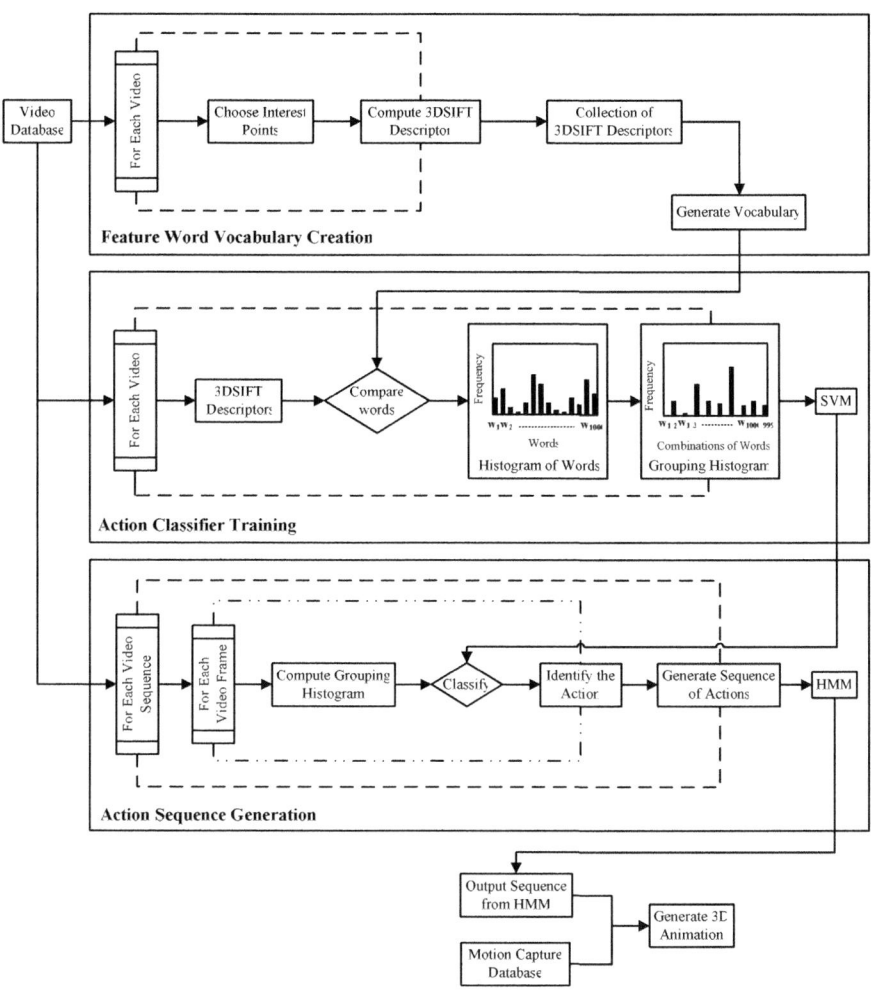

Fig. 1. Framework for the proposed approach depicting various phases involved

2.3 Motion Graphs

Motion graphs presented in [18] are used for creating realistic and controllable motion with smooth transitions. In [19] a similar approach is adapted for connecting a collection of motion sequences employing a directed graph. In addition, a framework allowing interactive synthesis of realistic looking motions is presented. A combination of physical model along with motion graphs, to create natural reaction of motions, is presented in [20]. Our focus in this paper is on action sequence generation for realistic animation and not on creating smooth transitions between actions. Of course smooth transitions are required and it will be easily possible to incorporate any of the above described approaches in our framework.

3 Proposed Framework

The overall flowchart of the proposed framework illustrating various phases involved is shown in Fig. 1.

3.1 Feature Word Vocabulary Creation

We start with a set of training videos of various actors performing the kickboxing activity chosen from the video database. The first critical step in phase I involves selection of interest points or key points. In the literature, there are various methods presented for the selection of these interest/salient points [21]. We choose to select salient points on the silhouette of the actor. From the perspective of an action, salient points are those with significant movement within that action. For this, we implement the optical flow algorithm on silhouettes extracted from videos. Silhouette is extracted by subtracting previous frame from the next frame. We choose two salient points per frame that have the highest optical flow values, and are sufficiently apart. 3D SIFT descriptors that are used as feature vectors for classification process are computed for these interest points. All the videos of the set are processed one at a time to generate a collection of 3D SIFT descriptors. These obtained 3D SIFT descriptors are quantized by clustering them into a pre-specified number of clusters. Hierarchical k-means, an unsupervised clustering method, is employed for clustering purpose. Cluster centers are referred to as "words", while the set of all these cluster centers is referred to as 'vocabulary'. We tried several vocabulary sizes and vocabulary size of 1000 yielded the best results. Therefore, in our experiments with kickboxing, we created 1000 (w_1, w_2, ..., w_{1000}) words or clusters from our training video input.

3.2 Action Classifier Training

This phase, similar to phase I, works using a distinct set of training videos from the video database. However, these videos are categorized based on the actions performed in them. Videos pertaining to the same actions are grouped together. Two salient points per frame are chosen as before and 3D SIFT descriptors are computed. These 3D SIFT descriptors are compared against "words" from the vocabulary generated in phase I, and the frequency of the words in each video is accumulated into a histogram. This word frequency histogram is just the initial representation of the video. A co-occurrence matrix is constructed to facilitate the process of finding the relationship amongst the words, which is used to build a more discriminative representation of the given action video. The test for finding word co-occurrences is carried out as follows. If the size of the vocabulary is N then the co-occurrence matrix will have dimensions N x N. When the frequency of occurrence of a word is above a certain threshold, the entire corresponding row is replaced with the number one, indicating the possibility of this word co-occurring with all the other words. Non-zero entries in the co-occurrence matrix contribute to construct a feature grouping histogram by adding corresponding frequency counts of the words from their initial histogram. This process of populating the co-occurrence matrix and constructing the feature grouping histogram is different from the one presented in [15]. The goal is also different as discussed next.

In [15] the co-occurrence matrix is populated with frequencies of "words". Each row of the co-occurrence matrix is then considered as contextual distribution of the word with other words. When a similar distribution is found, two words are considered correlated and their frequency counts are combined to build an entry in feature grouping histogram. Their process seemed to work for video clip level classification, in which the testing video signatures are in the relatively similar scale with the training videos. In contrast, our goal is to simultaneously segment and label the video frame sequence into actions. We achieve this by automatically labeling frames with actions. A continuous sequence of frames with the same action label automatically segments the video into actions. For labeling, we train one against all SVM classifier and one against one SVM classifier using the resulting feature grouping histograms as the discriminators for each action category. One against one SVM classifier is employed in cases when one against all SVM classifier generates multiple positive results. For example, when one against all SVM classifiers for kickboxing actions Jab and Hoop both have positive results for one particular frame, one against one SVM classifier for Jab and Hoop is used to further distinguish the action.

Note that the interest points could also be extracted from video content using other methods [21]. We chose 3D SIFT as local descriptor and adapted the implementation code available in [13]. It yields a local descriptor with 640 dimensions.

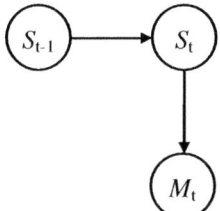

Fig. 2. An input-output HMM

3.3 Action Sequence Generation

Unlike phase I and phase II, videos in phase III are kickboxing videos of any individual actor. These do not necessarily have to be of an actor whose videos are already present in the video database used in previous phases. The idea is to learn the action sequence style of any actor performing that activity. For this, action classification is performed at frame level. In the earlier phases, we have decided to choose just two salient points per frame. This was computationally efficient and also sufficient because we have a large number of input frames belonging to any single action, and together they will yield to us a larger set of salient points which help recognize that action. In this phase, since we do frame by frame labeling, we increase the number of salient points to 25 per frame. This way, we increase the probability of selecting salient points with good discrimination. These points are processed in a manner similar to that in phase I and phase II. 3D SIFT descriptors are first computed. Computation of histogram of words, co-occurrence matrix and grouping histogram is next done. With grouping histograms as feature vectors, every frame is classified into one of the six predefined actions. Thus, complete processing of a video yields a labeled sequence indicating the actions performed in that video. Windowing operation is then performed on this labeled sequence with an average size window. Various window sizes were experimented with and window size of 6 yielded the best results. To some extent, this operation rectifies the noise in labeling as well and increases the accuracy of action recognition by relabeling and/or neglecting some of the incorrectly classified frames. All performance videos showing the individual style of an actor are processed this way, and a set of action sequences are recognized and outputted. These action sequences are then fed to HMM in order to help HMM learn the individual style of the actor. Once the HMM is trained, it can synthesize any number of new action sequences following the style of the actor.

HMM is a variant of probabilistic Finite State Machines (FSMs). An HMM usually includes an input alphabet and an output alphabet. The input alphabet contains input events that trigger transition in the states, while the output alphabet contains the set of responses of the model. In our HMM, we consider that the transitions are being made as the result of spontaneous choice of a character in a performance, with no external influence. As such, an input alphabet is not required in our case. The action sequences detected in videos are used to infer action to action transition probabilities of the HMM. The structure of FSM states and the learned probabilities of action to action transition embody the character's behavior pattern. An example of HMM is presented in Figure 2, depicting a transition from S_{t-1} to S_t and M_t is the motion data outputted from S_t.

For demonstration, we created a repertoire of 3D actions from motion capture data using Measurand Inc's Shapewrap system [22] (See Fig. 3). The repertoire consists of the basic actions captured for the actor under consideration. Using this repertoire, action sequences are generated by HMM based motion synthesizer. As mentioned earlier, smooth animation can be created using motion graphs for which sophisticated algorithms are available in the literature [18].

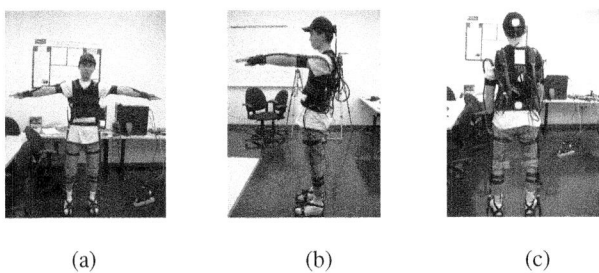

(a)   (b)   (c)

Fig. 3. Subject with Shape Wrap system in (a) front view, (b) side view, and (c) back view

4 Results

In this paper, we have presented a framework for embedding an individual actor's action sequence style captured through video into a 3D game character and demonstrated this for kickboxing activity. For evaluation of the proposed framework, we recorded a video database of kickboxing performed by various actors and also individual actions. We also created a repertoire of basic actions using our motion capture setup. Our video database contains 240 video sequences, consisting of six basic kickboxing actions - Jab, Hoop, Uppercut (UC), Defense (Def), Lower Kick (LK), and Roundhouse Kick (RK) recorded for 10 actors from 3 different views (front view and two side views). All the videos are taken with a stationary camera; however, backgrounds as well as actors' clothes/positions/orientations may differ.

The videos are down-sampled to spatial resolution of 160X120 pixels and a frame rate of 20fps. The primary reason for this choice of spatial resolution and frame rate is to reduce the computational cost and minimize training and learning time. The spatial resolution and the frame rate are proportional to the complexity of the action performed. These videos have an average length of 15 seconds. Fig. 4 shows sample frames from the kickboxing video database.

Fig. 4. Kickboxing video database (available on request): six predefined actions performed by several actors and captured from three different views

We implemented all major components of the framework which include silhouette detection, optical flow computation on the silhouettes, salient point identification, 3D SIFT computation, hierarchical k-means clustering for vocabulary creation, bag-of-words identification and feature grouping histogram, one against one and one against all SVM classifier for action recognition, action sequence generation, HMM learning, and finally, motion capture data based 3D game character action sequence generation. Accuracy of our SVM classifier to identify actions at video clip level is around 90%. Fig. 5 illustrates the video level action recognition along with the chosen salient points. Although our framework does not require action recognition at the video level, the accuracy achieved justifies the validity of using 3D SIFT as the descriptor for our framework. Moreover, it serves as a benchmark to evaluate the amount of recognition error introduced while trying to identify actions at the frame level.

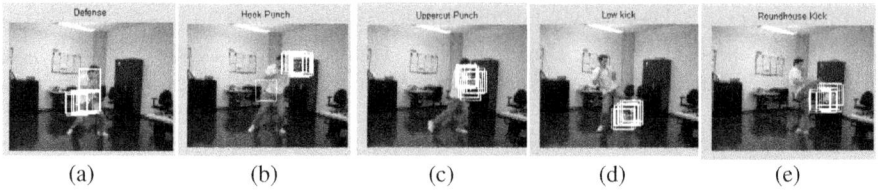

Fig. 5. Frame level action recognition results where (a), (b) , (c) , (d) , (e) are representative frames of original action videos

	Jab	UC	Hoop	Def	RK	LK
Jab	0.85	0	0.03	0	0.07	0.05
UC	0.17	0.56	0.15	0	0	0.12
Hoop	0	0.12	0.62	0	0.26	0
Def	0.06	0	0	0.66	0.06	0.22
RK	0.04	0.16	0	0.20	0.6	0
LK	0.04	0.21	0	0.20	0	0.55

(a)

	Jab	UC	Hoop	Def	RK	LK
Jab	0.90	0	0.02	0	0.04	0.04
UC	0.12	0.66	0.11	0	0	0.11
Hoop	0	0.12	0.70	0	0.18	0
Def	0.06	0	0	0.71	0.05	0.18
RK	0.04	0.14	0	0.20	0.62	0
LK	0.04	0.20	0	0.17	0	0.59

(b)

Fig. 6. Confusion matrices showcasing frame-wise action recognition accuracy for the overall framework (a) without windowing operation, and (b) with windowing operation

Accuracy of SVM classifier to identify actions performed in video sequences at frame level is found to be 64% and confusion matrix for the same is depicted in Fig. 6(a). A windowing operation is then performed with window size of 6 to remove noise and improve accuracy as average action length is found to be 6 frames. The window size selection depends on frame rate of the video and also complexity of the action being performed. This operation increases the overall accuracy to 70% and the confusion matrix obtained after windowing operation is illustrated in Fig. 6(b). This somewhat lower accuracy for frame level labeling does not affect the process of pattern learning with HMM negatively, as only the actions classified with certainty are fed to HMM. Fig. 7 depicts the synthesis output result of the proposed framework for a sample video sequence. A sample 3D game character's action sequence generated by HMM (shown as applied to a biped created in 3dsMax [23]) and video frame labeling results can be seen in the accompanying video.

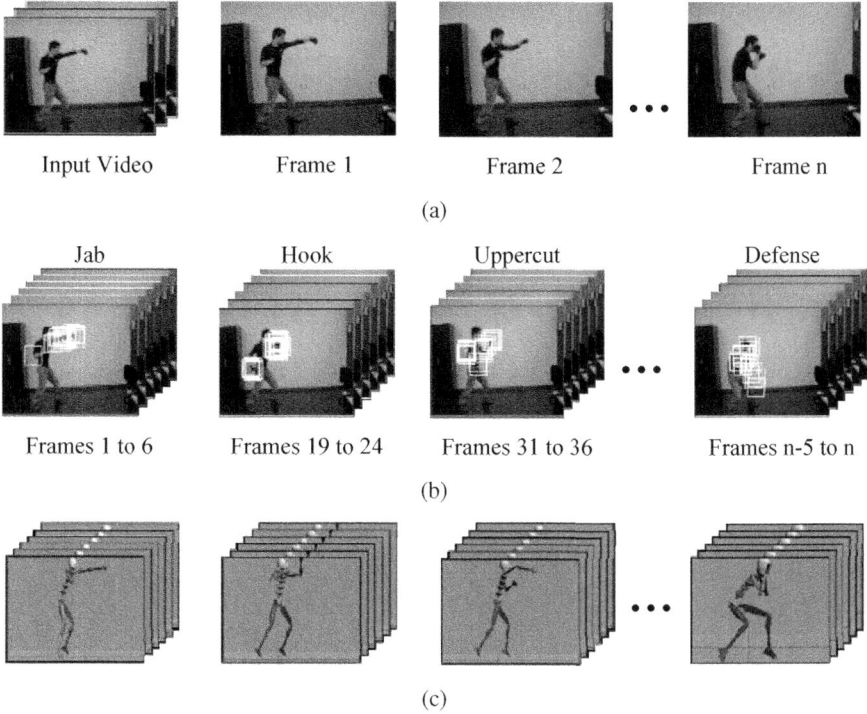

Fig. 7. Synthesis results of a sample video sequence, (a) input video, (b) actions recognized after windowing operation, and (c) generated 3D motion sequence

5 Discussion and Future Work

In this paper we have proposed an innovative framework to learn high-level human styles from videos of human actors performing the activity and to embed this style into 3D game characters. This has been achieved by recognizing the action sequence to suitably train an HMM for synthesizing the motion. Its demonstration on kickboxing action sequences confirms its usability. In the current implementation, our framework has a training and a learning phase that typically lasts for a few hours. Training and learning duration varies in proportion to the complexity of the actions. These are preprocessing tasks. A new video action sequence can then be synthesized in real time. Creating very large volumes of data required in the learning phases is a problem if that data has to be in 3D format. In our framework by using video data to learn, this problem is largely alleviated. Video is non-intrusive as well as so much easier to capture these days. Our choice of using salient points on the silhouette works considerably well for action recognition, as the SVM classifier for recognizing actions has yielded a satisfactory accuracy of about 90% at the video level and of about 70% at individual frame level. 70% accurate rate at the frame level means that the generated action sequences could have occasional mistakes. This may prevent our framework to be used in situations that require highly accurate reconstruction of action styles. As a future work of this framework we plan to explore the possibilities of using other descriptors. We also would like to deploy other known faster clustering techniques [24] to identify optimum vocabulary size for the data under consideration. Techniques such as relevance feedback have the potential to improve accuracy of action recognition and we would like to incorporate these into our framework in the future.

References

1. Hogue, A., Gill, S., Jenkin, M.: Automated Avatar Creation for 3D Games. In: Future Play, Toronto, Canada, pp. 174–180 (2007)
2. Hou, J., Wanga, X., Xua, F., Nguyena, V.D., Wua, L.: Humanoid personalized avatar through multiple natural language processing. World Academy of Science, Engineering and Technology 59, 230–235 (2009)
3. Sucontphunt, T., Deng, Z., Neumann, U.: Crafting personalized facial avatars using editable portrait & photograph example. In: IEEE Virtual Reality Conference, Lafayette, LA, USA, pp. 259–260 (2009)
4. Li, Y., Wang, T., Shum, H.: Motion texture- a two level statistical model for character motion synthesis. In: Proceedings of the 29th Annual Conference on Computer Graphics and Interactive Techniques, SIGGRAPH, San Antonio, Texas, USA, pp. 465–472 (2002)
5. Brand, M., Hertzmann, A.: Style machines. In: Proceedings of the 27th annual conference on Computer Graphics and Interactive Techniques, SIGGRAPH, New Orleans, Louisiana, USA, pp. 183–192 (2000)
6. Hsu, E., Pulli, K., Popović, J.: Style translation for human motion. In: Proceedings of the 32nd Annual Conference on Computer Graphics and Interactive Techniques, SIGGRAPH, Los Angeles, USA, pp. 1082–1089 (2005)
7. Yin, K., Pai, D.: Footsee- an interactive animation system. In: Proceedings of the 2003 ACM, SIGGRAPH/Eurographics Symposium on Computer Animation, San Diego, California, USA, pp. 329–338 (2003)
8. Chai, J., Hodgins, J.: Performance animation from low-dimensional control signals. ACM Transaction on Graphics (24), 686–696 (2005)
9. Oshita, M., Yoshiya, T.: Learning motion rules for autonomous characters from control logs using support vector machine. In: International Conference on Computer Animation and Social Agents, Saint-Malo, France (2010) (to appear)
10. Ofli, F., Erzin, E., Yemez, Y., Tekalp, A., Erdem, C.: Unsupervised dance figure analysis from video for dancing avatar animation. In: IEEE International Conference on Image Processing, San Diego, CA, USA, pp. 1484–1487 (2008)
11. Poppe, R.: A survey on vision-based human action recognition. Image and Vision Computing (28), 976–990 (2010)
12. Turaga, P., Chellapa, R., Subramhanian, V., Udrea, O.: Machine recognition of human activities- A Survey. IEEE Transactions on Circuits and Systems for Video Technology (18), 1473–1488 (2008)
13. Csurka, G., Dance, C., Fan, L., Willamowski, J., Bray, C.: Visual categorization with bags of keypoints. In: ECCV Workshop on Statistical Learning in Computer Vision, Prague, Czech Republic, pp. 59–74 (2004)
14. Lowe, D.: Distinctive Image Features from Scale-Invariant Keypoints. International Journal Computer Vision 2(60), 91–110 (2004)
15. Scovanner, P., Ali, S., Shah, M.: A 3-dimensional sift descriptor and its application to action recognition. In: ACM Multimedia, Augsburg, Germany, pp. 357–360 (2007)

16. Klaser, A., Marszałek, M., Schmid, C.: A spatio-temporal descriptor based on 3D-gradients. In: British Machine Vision Conference, Leeds, UK, pp. 995–1004 (2008)
17. Gillies, M.: Learning finite-state machine controllers from motion capture data. IEEE Transactions on Computational Intelligence and AI in Games 1(1), 63–72 (2009)
18. Kovar, L., Gleicher, M., Pighin, F.: Motion Graphs. ACM Transactions on Graphics 21(3), 473–482 (2002)
19. Arikan, O., Forsyth, D.: Interactive motion generation from examples. ACM Transaction on Graphics 21(3), 483–490 (2002)
20. Zordan, V., Majkowska, A., Chiu, B., Fast, M.: Dynamic response for motion capture animation. ACM Transaction on Graphics 24(3), 697–701 (2005)
21. Niebles, J., Wang, H., Li, F.: Unsupervised learning of human action categories using Spatial-Temporal words. In: British Machine Vision Conference, Edinburgh, UK (2006)
22. Shapewrap Motion Capture System, http://www.motion-capture-system.com/shapewrap.html (retrieved)
23. Autodesk 3ds Max, http://usa.autodesk.com/adsk/servlet/pc/index?siteID=123112&id=13567410 (retrieved)
24. Suryavanshi, B.S., Shiri, N., Mudur, S.P.: An Efficient Technique for Mining Usage Profiles Using Relational Fuzzy Subtractive Clustering. In: IEEE WIRI (Web Information Retrieval and Integration), Washington, DC, pp. 23–29 (2005)

Author Index

Allbeck, Jan M.   182
Allen, Brian F.   48
Amato, Nancy M.   82
André, Elisabeth   206

Bade, Abdullah   351
Bekris, Kostas E.   121
Bhattacharya, Prabir   422
Boulic, Ronan   59
Britton, Ontario   158

Cani, Marie-Paule   170
Carter, Chris   326
Charalambous, Panayiotis   35
Chen, Xianshun   375
Chen, XiaoLong   422
Chrysanthou, Yiorgos   35
Cig, Cagla   278
Cooper, Simon   326
Courty, Nicolas   290

Dell'Acqua, Pierangelo   339
Denny, Jory   82
Dutoit, Thierry   363

Egges, Arjan   11, 278
El Rhalibi, Abdennour   326
Erra, Ugo   194

Faloutsos, Petros   48, 386
Ferreira, José Antônio Salini   410
Fischer, Leonardo   410
Frola, Bernardino   194

Geijtenbeek, Thomas   11
Geng, Weidong   313
Geraerts, Roland   94
Gibet, Sylvie   290
Gross, Markus   325

Hahmann, Stefanie   170
Hall, Anthony   158
Harrison, Joseph F.   218
Ho, Choon Sing   375
Hoogeveen, Han   94

Hoyet, Ludovic   266, 313
Huang, Wenjia   36
Huang, Yazhou   242

Johansson, Anja   339
Jones, Michael   158

Kallmann, Marcelo   230, 242
Kang, Young-Min   301
Kapadia, Mubbasir   36
Kasap, Zerrin   278
Kistler, Felix   206
Komura, Taku   23, 266
Kravtsov, Denis   146
Krontiris, Athanasios   121
Kuffner, James   70

Lau, Manfred   70
Lecuyer, Anatole   266
Liang, Xiubo   313
Lien, Jyh-Ming   134, 218
Long, Jie   158
Louis, Sushil   121

Magnenat-Thalmann, Nadia   278
Maupu, Damien   59
Mendhurwar, Kaustubha   422
Merabti, Madjid   326
Mudur, Sudhir   422
Multon, Franck   266, 313

Nedel, Luciana   410
Neff, Michael   48
Nguyen, Quang Huy   375

Ong, Yew-Soon   375
Oskam, Thomas   325

Pasko, Alexander   146
Pasko, Galina   146
Peinado, Manuel   59
Poyart, Eduardo   386
Prestes, Edson   410
Prins, Corien   94

Author Index

Radhakrishnan, Thiruvengadam 422
Raunhardt, Daniel 59
Reimschussel, Cory 158
Rodriguez, Samuel 82
Rohmer, Damien 170

Scarano, Vittorio 194
Sharbini, Hamizan 351
Shiratori, Takaaki 23
Shum, Hubert P.H. 23
Silveira, Renato 410
Sok, Kwang Won 254
Sumner, Robert W. 325

Takagi, Shu 23
Tarapata, Zbigniew 398

Terzopoulos, Demetri 1, 36
Thuerey, Nils 325
Tilmanne, Joëlle 363

van Basten, Ben J.H. 11
van de Panne, Michiel 106
van den Akker, Marjan 94
van den Bogert, Antonie J. 11
Vo, Christopher 134, 218

Wißner, Michael 206
Wroclawski, Stefan 398

Yamane, Katsu 254

Zordan, Victor 109
Zourntos, Takis 82

GPSR Compliance

The European Union's (EU) General Product Safety Regulation (GPSR) is a set of rules that requires consumer products to be safe and our obligations to ensure this.

If you have any concerns about our products, you can contact us on ProductSafety@springernature.com

In case Publisher is established outside the EU, the EU authorized representative is:

Springer Nature Customer Service Center GmbH
Europaplatz 3
69115 Heidelberg, Germany

Batch number: 09478804

Printed by Printforce, the Netherlands